I0064389

Essentials of Genomics

Essentials of Genomics

Editor: Jamie Spooner

www.callistoreference.com

Callisto Reference,
118-35 Queens Blvd., Suite 400,
Forest Hills, NY 11375, USA

Visit us on the World Wide Web at:
www.callistoreference.com

ISBN: 978-1-64116-263-0 (Hardback)

Cataloging-in-Publication Data

Essentials of genomics / edited by Jamie Spooner.
p. cm.
Includes bibliographical references and index.
ISBN 978-1-64116-263-0
1. Genomics. 2. Genomes. 3. Molecular genetics. I. Spooner, Jamie.
QH447 .E87 2020
572.86--dc23

Table of Contents

Preface

This book has been a concerted effort by a group of academicians, researchers and scientists, who have contributed their research works for the realization of the book. This book has materialized in the wake of emerging advancements and innovations in this field. Therefore, the need of the hour was to compile all the required researches and disseminate the knowledge to a broad spectrum of people comprising of students, researchers and specialists of the field.

Genomics is an interdisciplinary field of biology that delves into the structure, mapping, evolution, function and editing of genomes. It aims at the collective quantification and characterization of genes. This is done through the analysis and sequencing of genomes with the aid of high throughput DNA sequencing and bioinformatics. The studies of intragenomic phenomena like pleiotropy, heterosis, epistasis and other interactions between alleles and loci within the genome are also within the purview of this field. Genomics plays a crucial role in the fields of anthropology, medicine, biotechnology and other sciences. Metagenomics, functional genomics, structural genomics, epigenomics and model systems are the primary areas of research in genomics. This book includes some of the vital pieces of work being conducted across the world, on various topics related to genomics. It strives to provide a fair idea about this discipline and to help develop a better understanding of the latest advances within this field. This book is appropriate for students seeking detailed information in this area as well as for experts.

At the end of the preface, I would like to thank the authors for their brilliant chapters and the publisher for guiding us all-through the making of the book till its final stage. Also, I would like to thank my family for providing the support and encouragement throughout my academic career and research projects.

Editor

1

Characterization of Transcription Termination-Associated RNAs: New Insights into their Biogenesis, Tailing, and Expression in Primary Tumors

Ilaria Laudadio,[1] Sara Formichetti,[1] Silvia Gioiosa,[2] Filippos Klironomos,[3] Nikolaus Rajewsky,[3] Giuseppe Macino,[1] Claudia Carissimi,[1] and Valerio Fulci[1]

[1]*Dipartimento di Biotecnologie Cellulari ed Ematologia, Sez Genetica Molecolare, Sapienza Università di Roma, Rome, Italy*
[2]*Istituto di Biomembrane e Bioenergetica (IBBE), CNR, Bari, Italy*
[3]*Laboratory for Systems Biology of Gene Regulatory Elements, Berlin Institute for Medical Systems Biology, Max-Delbrück Center for Molecular Medicine, Berlin, Germany*

Correspondence should be addressed to Claudia Carissimi; claudia.carissimi@uniroma1.it

Academic Editor: Marco Gerdol

Next-generation sequencing has uncovered novel classes of small RNAs (sRNAs) in eukaryotes, in addition to the well-known miRNAs, siRNAs, and piRNAs. In particular, sRNA species arise from transcription start sites (TSSs) and the transcription termination sites (TTSs) of genes. However, a detailed characterization of these new classes of sRNAs is still lacking. Here, we present a comprehensive study of sRNAs derived from TTSs of expressed genes (TTSa-RNAs) in human cell lines and primary tissues. Taking advantage of sRNA-sequencing, we show that TTSa-RNAs are present in the nuclei of human cells, are loaded onto both AGO1 and AGO2, and their biogenesis does not require DICER and AGO2 endonucleolytic activity. TTSa-RNAs display a strong bias against a G residue in the first position at 5′ end, a known feature of AGO-bound sRNAs, and a peculiar oligoA tail at 3′ end. AGO-bound TTSa-RNAs derive from genes involved in cell cycle progression regulation and DNA integrity checkpoints. Finally, we provide evidence that TTSa-RNAs can be detected by sRNA-Seq in primary human tissue, and their expression increases in tumor samples as compared to nontumor tissues, suggesting that in the future, TTSa-RNAs might be explored as biomarker for diagnosis or prognosis of human malignancies.

1. Introduction

In the last ten years, the technologies for RNA profiling strongly improved. These advances disclosed pervasive transcription of more than 70% of the human genome, with protein-coding genes accounting for less than 2% of the total RNA [1]. Thus, a dominant portion of the transcribed regions on the human genome originates from non-protein-coding genes (noncoding RNAs (ncRNAs)). Increasing evidences are being obtained, indicating that noncoding RNAs possess essential regulatory functions and could be one of the major contributors to the complex traits of the organisms [2].

Small RNAs (sRNAs) of~ 20–30 nucleotides (nt) constitute a large family of ncRNAs that regulate gene expression. Through base pairing, sRNAs recognize their complementary RNAs as their targets and mediate posttranscriptional and/or transcriptional silencing [3]. The best characterized sRNA classes are microRNAs (miRNAs), small interfering RNAs (siRNAs), and PIWI-interacting RNAs (piRNAs). All these classes of sRNAs are loaded onto a member of Argonaute family proteins. Argonaute proteins typically have a molecular weight of~ 100 kDa and can be divided into two subfamilies: AGO subfamily (e.g., AGO1, AGO2, AGO3, and AGO4 in mammals) that binds to miRNAs and siRNAs and PIWI subfamily (HIWI/PIWIL1, HILI/PIWIL2, HIWI2/PIWIL4, and HIWI3/PIWIL3) that binds to piRNAs [4].

sRNA profiling by next-generation sequencing (sRNA-Seq) is the method of choice for the identification of lowly abundant sRNA classes, in addition to the miRNA, siRNA, and piRNA families. Indeed, sRNAs with varying lengths of between 18 and 200 nucleotides have been reported to be derived from specific genomic regions in higher eukaryotes. At least three classes of sRNAs derived from regions mapping around the 5′ termini of genes have been described: transcription initiation RNAs (tiRNAs) [5], transcription start site-associated RNAs (TSSa-RNAs) [6–8], and promoter-associated small RNAs (PASRs) [9, 10]. The origin and function of these RNAs are uncertain, but preliminary evidence suggests that they are involved in epigenetic control of gene expression. Recently it has been described a new family of sRNAs, termed DNA damage-response RNAs (DDRNAs) [11] or double-strand break- (DSB-) induced RNAs (diRNAs) [12], that are generated at sites of DNA damage and control the DNA damage response.

Finally, sRNA mapping around 3′ termini of genes of higher eukaryotes have also been described, such as termini-associated sRNAs (TASRs) [10, 13], antisense TASRs (aTASRs) [14], and transcription termination site-associated RNAs (TTSa-RNAs) [15]. These sRNA classes can vary in length from 22 to 200 nt and their biological roles and mechanisms of biogenesis remain to be elucidated.

Over the years, sRNAs have become the focus of biomarker research, an approach that has been favorably used in the prediction and early detection of disease and in the investigation of response to treatment for several medical conditions. miRNAs are the most frequently assessed for their potential role as biomarkers, such as in cancer [16], liver [17], and cardiovascular disease [18] among many others. Moreover, dysregulation of other sRNA species such as piRNAs, small nucleolar RNAs (snoRNAs), and small nuclear RNAs (snRNAs) is being found to have relevance to tumorigenesis, neurological, cardiovascular, developmental, and other diseases [19]. Thus, they might be potentially assessed as biomarkers of disease.

Here, we present a comprehensive characterization of TTSa-RNAs [15] in human cells. We show that TTSa-RNAs are present in both AGO1 and AGO2 complexes immunopurified from human cell nuclei. Moreover, TTSa-RNAs can be detected in both soluble and chromatin-associated nuclear extract. Biogenesis of this class of sRNAs is independent of both DICER and AGO2 endonucleolytic activity. TTSa-RNAs show a very strong bias against a G residue in the first position at 5′ end, which is a specific feature of AGO-bound sRNAs and a peculiar oligoA tail at 3′ end. Finally, AGO-bound TTSa-RNAs derive from genes involved in cell cycle progression regulation and DNA integrity checkpoints. Interestingly, this class of sRNAs can be detected in primary human tissues, and their expression increases in tumor samples as compared to nontumor tissues.

2. Materials and Methods

2.1. Cell Culture and Transfection. HeLaS3 cells were grown in DMEM medium supplemented with 10% (*v*/*v*) fetal bovine serum, 2 mM L-glutamine and penicillin-streptomycin. HCT 116 WT and DICER^{Ex5} cells [20] were grown in McCoy's 5A medium supplemented with 10% (*v*/*v*) fetal bovine serum, 2 mM L-glutamine and penicillin-streptomycin.

HeLaS3 AGO2KO monoclonal cell line was obtained by using two specific Zinc Finger Nucleases (ZNF1 and ZNF2, CompoZr® Knockout Zinc Finger Nucleases, Sigma). HeLaS3 cells were plated in a 6-well plate and transfected using Lipofectamine® 2000 (Thermo Fisher Scientific) with 1.5 μg of each ZNF expression plasmid. After two sequential transfections with the AGO2-specific ZNF expression plasmids, individual clones were isolated by serial dilution and assayed by Western Blot and RT-qPCR for loss of AGO2 expression.

2.2. Nuclear Fractionation. Nuclear fractionation was performed similar to [21]. 12 x 10^6 adherent HeLaS3 cells were washed in 1xPBS and then recovered by scraping and centrifugation. Cell pellets were resuspended in 200 μl of Buffer A (10 mM HEPES pH 7.5, 10 mM KCl, 10% glycerol, 340 mM sucrose, 4 mM MgCl2, 1 mM DTT,1 x Protease Inhibitor Cocktail, Sigma), and then an equal volume of Buffer A with 0.2% (*v*/*v*) Triton X-100 was added and the mixture was incubated on ice for 12 minutes to lyse cells, followed by centrifugation (1200 ×g, 5 min, 4°C). The crude nuclear pellet was resuspended in 300 μl NRB (20 mM HEPES pH 7.5, 50% glycerol, 75 mM NaCl, 1 mM DTT, 1 x protease inhibitor cocktail), transferred to a microcentrifuge tube, and centrifuged (500 ×g, 5 min, 4°C) to wash. The pellet was resuspended in 100 μl NRB, and then an equal volume of NUN buffer (20 mM HEPES, 300 mM NaCl, 1 M urea, 1% NP-40 Substitute, 10 mM MgCl2, 1 mM DTT, 1 x Protease Inhibitor Cocktail) was added and incubated 5 minutes on ice and then centrifuged (1200 ×g, 5 min, 4°C). The soluble nuclear extract supernatant was transferred to another tube, and the depleted nuclear pellet was resuspended in 1 ml Buffer A to wash, transferred to another microcentrifuge tube, and centrifuged (1200 ×g, 5 min, 4°C). Resulting purified chromatin pellets were resuspended in 200 μl Buffer A. TriFast (0.6 ml) (EuroClone) was added to 140 μl of each fraction. As a spike-in control, 2 fmol of a synthetic nonhuman microRNA (ath-miR-159a) was added to each fraction.

2.3. RNA Isolation and RT-qPCR. RNA from HeLa S3, HeLa S3 AGO2KO, HCT 116 WT, and DICER^{Ex5} cells, as well as fractions from nuclear fractionation protocol, was isolated using Direct-zol™ RNA MiniPrep Kit (Zymo Research) according to the manufacturer's protocol.

For RT-qPCR of AGO2, ACTB, and MALAT-1, RNA was retrotranscribed using GoScript™ Reverse Transcriptase and random primers (Promega). Quantitative real-time PCR was performed using GoTaq® qPCR Master Mix (Promega). Primer sequences are as follows:

AGO2F ggaggtctgtaacattgtgg, AGO2R gcaatagctttatttc ctgccc; ACTB F cctggcacccagcacaat, ACTB R gggccggactcgt catact; and MALAT-1 F aaagcaaggtctccccacaag, MALAT-1 R ggtctgtgctagatcaaaaggca.

TTSa-RNA (Table 1) expression was checked by using Custom TaqMan® Small RNA Assay (Thermo Fisher Scientific). RNA was retrotranscribed using SuperScript II Reverse Transcriptase (Thermo Fisher Scientific), and qPCR was performed using iTaq Universal Probes Supermix (Bio-Rad). Quantification was normalized to small nucleolar RNA U44 (for total RNA) or to ath-miR-159a spike-in

TABLE 1: Sequences of TTSa-RNAs selected for RT-qPCR analysis.

TTSa-RNA	Sequence (5′-> 3′)	Rank of TTSa-RNAs associated to each gene based on their abundance in each library		
		HeLaS3 AGO2-IP	HeLaS3 AGO1-IP	HCT116 AGO2-IP
BUB3_TTSa-RNA	CTAATAAACGAGATGCAGAACCCT	4	4	3
WEE1_TTSa-RNA	CATATTAAAAGTCACTCTGAGCT	13	13	15
PSMB1_TTSa-RNA	TTTATTAAAAGAGAAACCTGAAG	26	39	8
CDKNA1_TTSa-RNA	CTCAATAAATGATTCTTAGTGACT	47	32	21
YWHAG_TTSa-RNA	CAGTGACGAGGAACTCCCGAGA	86	47	101

(for SNE and CPE fractions), amplified by TaqMan Small RNA Assay (Thermo Fisher Scientific).

2.4. Western Blot. For Western blot analyses, the following antibodies were used: anti-AGO2 (11A9, Millipore), anti-Histone H3 antibody (FL-136, Santa Cruz), anti-BAF155 (Abcam), antibeta Tubulin (Sigma).

2.5. Small RNA Sequencing. sRNA-Seq was performed from total RNA isolated from HeLaS3 and AGO2KO HeLaS3 cells by IGA TECH (Udine, Italy). Library was prepared using TruSeq Small RNA Library Preparation Kit (Illumina). Sequencing was performed on HiSeq 2500 in a 50 bp single-read mode 10 million reads/sample. It is worth pointing out that the protocol we applied to generate sRNA libraries only captures 5′ phosphorylated RNA molecules and does not allow cloning of sRNAs harboring a 5′Cap.

2.6. TTSa-RNA Data Analysis from Cell Lines. RNA Immunoprecipitation and next-generation sequencing were performed as previously published [6, 22]. NGS data quality was checked using FastQC. tRNA and rRNA genomic coordinates on hg38 genome were retrieved from UCSC RepeatMasker track [23] using Table Browser tool [24]. By using bedtools [25], tRNA and rRNA sequences in FASTA format were retrieved from hg38 genome sequence and used to build a bowtie1 [26] reference index for tRNA and rRNA. miRNA hairpin FASTA sequence was downloaded from miRbase (v21) [27], and human hairpins were extracted. Using bowtie-build, a bowtie1 index was obtained for miRNA hairpins.

Each FASTQ file was processed as follows:

(1) Adapter sequences were removed using cutadapt; only reads longer than 17 nt were retained.

(2) Adapter-cleaned reads were aligned to the abovementioned miRNA hairpin bowtie1 index using bowtie1 with -n 2 -l 18 options.

(3) Reads without any valid alignment in step 2 were aligned to the abovementioned tRNA and rRNA bowtie1 index with -n 2 -l 18 options.

(4) Reads without any valid alignment in step 3 were aligned to hg38 human reference with bowtie1 with -n 0 -l 18 -m 1 options. Only reads with a single valid alignment were investigated further.

TTSa-RNAs were defined as those reads aligned in step 4, with at least 1 nt overlapping a 5 nt-long window centered on genomic coordinates of polyA sites (GENCODE v24, [26]). Only polyA sites with a 1 nt length were taken into account in all the analyses (99.8% of the 48535 polyA sites reported by GENCODE v24 annotation).

Within each sample, RPM was defined as the number of reads divided by the total number of reads that gave at least one valid alignment in either step 2, 3, or 4 of the alignment procedure $* 10^6$.

3′ tails of TTSa-RNAs and miRNAs were determined using Tailor v1.1 [28].

2.7. PAR-CLIP Data Analysis. PAR-CLIP libraries were analyzed using the same pipeline outlined above.

2.8. Read Coverage Plots. Cumulative coverage of sRNAs giving rise to a single alignment in step 4 was computed using bedtools in a window of 1000 nt centered on GENCODE v24 polyA sites. Each sRNA contributed to the coverage of the closest polyA site only. Three polyA sites (# 605212, 627827, and 582464) lying within 500 nt of loci giving rise to very large numbers of small RNA molecules (not-overlapping any polyA site) were omitted to obtain a clearer picture.

For PAR-CLIP datasets, only reads harboring a single mismatch to the reference genome were used to compute coverage.

2.9. GO Term Analysis. Ensembl GENE IDs were retrieved from GENCODE v24 annotation. Genes giving rise to TTSa-RNAs were defined as any Ensembl GENE overlapping a single TTSa-RNA. GO term analysis was performed on DAVID [29] by submitting the list of Ensembl GENE IDs overlapped by TTSa-RNA in each sample. As background, the entire set of Ensembl GENE IDs was used. The reported GO terms were obtained by selecting GO_term_BP_4 in the DAVID output and downloading the full table.

Significantly enriched GO terms (FDR < 0.01) were sorted for decreasing "Fold Enrichment" and the Fold Enrichment was plotted for the top 10 GO Terms.

2.10. Correlation of mRNA Expression with TTSa-RNA Counts. mRNA expression in HeLaS3 was retrieved from ENA: EXR352930, EXR352926. Reads were pseudoaligned to the human transcriptome, and transcripts were quantified in transcript per million (TPM) using Kallisto v 0.43.1 [30]. Ensembl GENE IDs corresponding to the Ensembl

TRANSCRIPT IDs were retrieved using BioMart, and the TPM for all transcripts of each gene were summed, giving a summarized TPM value for each gene. The genes with TPM below the first tertile were filtered out.

mRNA expression in 44 adrenocortical carcinoma primary samples was retrieved from GEO repository (GSE49278) and belongs to the same samples whose sRNA-seq were used for TTSa-RNA quantification as described below. The Affymetrix Human Gene 2.0 ST Array was analyzed in R (R package version 1.58.1) using Bioconductor packages *oligo* [31] and *genefilter* (using RMA as a background correction/normalization/summarization method). Genes with RMA expression values below the first tertile in more than 10 samples were filtered out; the PROBEIDs were mapped to Ensembl GENE IDs and the Ensembl GENE IDs mapped more than once (<10%) were excluded.

mRNA expression in 10 kidney tumor samples and 10 matched controls was retrieved from SRA project SRP003901 and belongs to the same samples whose sRNA-seq were used for TTSa-RNAs quantification. The digital gene expression (DGE) sequencing reads were trimmed from Illumina adapters using cutadapt [32], resulting in 17 nt reads, which were mapped using bowtie1 with -p 8 -v 0 -m 2 options, after adding the four nucleotides of NlaIII restriction site to the 5′ of each read. The DGE tags belonging to each transcript were quantified using htseq-count [33] (with intersection-strict option) and the htseq-counts were normalized using *DESeq* package [34]. Genes with no counts in more than the 10% of samples and genes with normalized counts below the first tertile were filtered out.

The correlations were computed using custom R scripts and the R^2 was computed from a Spearman correlation coefficient. In all cases, the genes with TTSa-RNA RPM in the first three quartiles were filtered out, because this corresponded to less than 4 TTSa-RNA reads per sample.

2.11. TTSa-RNAs Data Analysis from Primary Samples. A search in SRA was performed using miRNA and cancer/tumor as keywords. We retrieved 17 datasets containing sRNA-seq data for tumor tissue and matched controls. However, only 5 datasets were suitable for our analysis. In fact, we discarded a dataset that was obtained in colorspace (it could not be analyzed using our computational pipeline), a dataset with a sudden drop in Phred quality (median Phred score = 2 for all positions > 16 in all samples), a dataset which contained only one single healthy tissue control sample, three datasets because all reads were shorter than 18 nt (multiple alignments on the genome did not allow specific identification of TTSa-RNAs), and six datasets in which abundant mRNA fragments with random size distribution were retrieved among sRNA-seq reads, suggesting that mRNA degradation does not allow proper identification of bona fide TTSa-RNAs.

The 5 datasets analyzed correspond to SRA projects: SRP028291, SRP014142, SRP045645, SRP003902, and SRP048750. Analysis of these datasets was performed as outlined above for AGO protein-bound sRNAs, except for the application of an 18 to 26 nt size filter as discussed in the results and discussion section. All p values were computed using Wilcoxon sum-rank test (a paired version of the test was used for datasets containing paired normal tissue and tumor tissue from each patient) using R function wilcox.test (https://www.r-project.org/).

2.12. Availability of Data and Materials. The following datasets analyzed during the current study are available in the ENA repository:

(1) http://www.ebi.ac.uk/ena/data/view/ERX352930
(2) http://www.ebi.ac.uk/ena/data/view/ERX352926
(3) http://www.ebi.ac.uk/ena/data/view/ERX344794
(4) http://www.ebi.ac.uk/ena/data/view/ERX344797
(5) http://www.ebi.ac.uk/ena/data/view/ERX350060
(6) http://www.ebi.ac.uk/ena/data/view/ERX338767
(7) http://www.ebi.ac.uk/ena/data/view/ERX338764

The following datasets analyzed during the current study are available in the SRA repository;

(1) https://trace.ncbi.nlm.nih.gov/Traces/sra/sra.cgi?study=SRP038925
(2) https://trace.ncbi.nlm.nih.gov/Traces/sra/?study=SRP028291
(3) https://trace.ncbi.nlm.nih.gov/Traces/sra/sra.cgi?study=SRP014142
(4) https://trace.ncbi.nlm.nih.gov/Traces/sra/sra.cgi?study=SRP045645
(5) https://trace.ncbi.nlm.nih.gov/Traces/sra/sra.cgi?study=SRP003902
(6) https://trace.ncbi.nlm.nih.gov/Traces/sra/sra.cgi?study=SRP048750
(7) https://trace.ncbi.nlm.nih.gov/Traces/sra/?run=SRR650321
(8) https://trace.ncbi.nlm.nih.gov/Traces/sra/?run=SRR650318

3. Results and Discussion

3.1. Nuclear AGO1 and AGO2 Bind a Class of sRNAs Arising from Human Transcription Termination Sites. To obtain an overview of nuclear sRNA classes bound to AGO proteins in human cells, we reanalyzed a set of sRNA-seq libraries generated by immunopurifying AGO1 and AGO2 from nuclear extracts of HeLaS3 cell line [6]. We focused on less characterized classes of AGO bound sRNAs, that is reads not aligning to known human miRNAs (miRbase 21) or human tRNAs and rRNAs. All the reads surviving to this preliminary filtering were aligned to human hg38 reference genome requiring unambiguous mapping to a unique genomic position. Reads with multiple mappings on the hg38 reference genome were not further investigated (see Materials and Methods for details).

We found that in nuclei of human cells, both AGO1 and AGO2 bind a class of sRNAs (21–24 nt long, Figure 1(a)) arising from TTSs of human mRNA transcripts and lying on the sense strand of the gene they arise from (Figure 1(b)). Notably, 3′ end of these sRNAs in most cases (60% to 66% depending on the dataset analysed) maps within 2 nt of GENCODE-annotated polyA sites. We assumed that these sRNAs belong to the class of TTSa-RNAs. Indeed, TTSa-RNAs have been described as AGO1/2-associated,

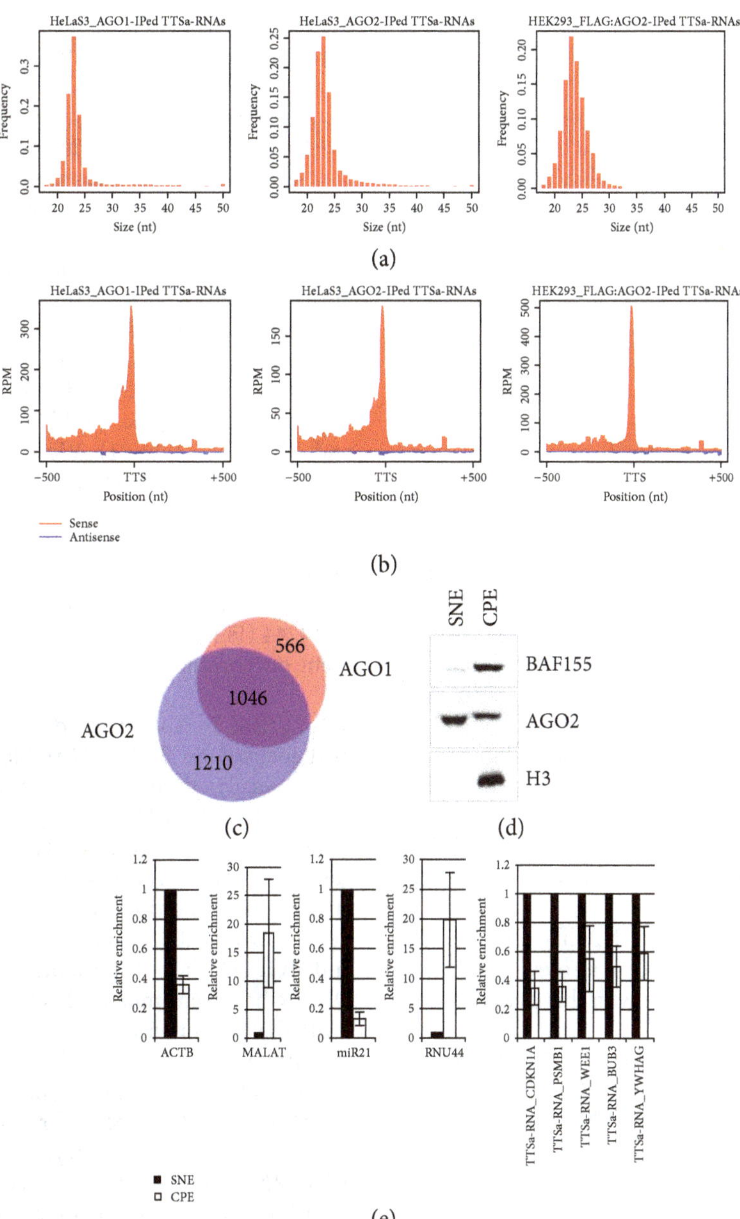

Figure 1: TTSa-RNAs are associated with nuclear AGO1 and AGO2. (b) Coverage of AGO1- and AGO2-immunoprecipitated TTSa-RNAs (HeLa S3) and FLAG:AGO2-immunoprecipitated TTSa-RNAs (HEK293) around GENCODE v25 annotated TTSs. Red and blue represent sRNAs in the sense and antisense orientation with respect to gene transcription, respectively. (c) Overlap of genes giving rise to AGO1- and AGO2-associated TTSa-RNAs. (d) Purified nuclei from HeLaS3 cells were extracted with a forcing urea/detergent buffer to yield a soluble nuclear extract (SNE) and chromatin pellet extract (CPE). SNE and CPE were analyzed by Western blot for the presence of chromatin-associated proteins (BAF-155 and H3) and AGO2. (e) RNA was isolated from HeLaS3 SNE and CPE and analyzed by qRT-PCR for the detection of ACTB mRNA (as a non-chromatin-associated RNA), MALAT-1 (as a chromatin-associated RNA), miR-21, RNU44, and 5 selected TTSa-RNAs ($n = 3$). Synthetic ath-miR159a was added to SNE and CPE fractions before RNA isolation and subsequently used for qRT-PCR normalization.

miRNA- (22–24 nt) sized sRNAs whose 3′ends aligned with the annotated polyadenylation sites and mapping on the sense strand [15]. Another class of sRNA mapping around 3′ termini of genes has been described in human cells and in plants, namely TASRs [10]. However, human TASRs are more heterogeneous in size as compared to TTSa-RNAs. Moreover, TASRs, whose 3′termini aligned with annotated polyadenylation sites, map antisense to mRNA transcription. In plants, TASRs are 23-24 nt in length and are bound by an AGO protein [13], similar to TTSa-RNAs. However, like human TASRs, plant TASRs seem to be a heterogeneous class mapping on the sense, on the antisense strand or on both strands of protein coding genes. Although we cannot exclude that TTSa-RNAs are a subclass of TASR, because of differences in size, orientation, and mapping with respect to polyadenylation sites, we believe that TTSa-RNAs are a distinct class of sRNAs.

In our experimental design [6], we attained a sequencing depth of 100 million reads per sample, which is substantially higher than the depth employed by Valen and colleagues [15]. This prompted us to investigate peculiar features of this class of sRNAs in order to return a detailed characterization.

The estimated abundance of TTSa-RNAs is about 100 reads per million (RPM) in libraries generated from the nuclear fraction of sRNAs in HeLaS3, reaching about 300 RPM and 150 RPM in AGO1 and AGO2-immunoprecipitated samples, respectively (Figure 1(b)). As shown in Figure 1(c), TTSa-RNAs arise from 2822 genes, and 37% of these genes give rise to TTSa-RNAs that are loaded on both AGO1 and AGO2. More interestingly, GO term analysis suggests that both genes giving rise to AGO1- and AGO2-bound TTSa-RNAs are significantly enriched for genes involved in cell cycle progression regulation and DNA integrity checkpoints (Supplementary Figure 1). A GO term analysis of genes giving rise to IgG-immunoprecipitated sRNAs lying on TTS did not highlight any significant enrichment for the same GO terms (data not shown).

A parallel analysis on a sRNA-seq library obtained by immunopurification of a FLAG-tagged AGO2 complexes from a HEK293 cell line expressing a FLAG-tagged AGO2 [22] yielded similar results, highlighting that TTSa-RNAs are found in different human cell lines (Figures 1(a) and 1(b)).

Not only TTSa-RNAs show a miRNA-like size that is a specific feature for AGO loading but we also evidenced that AGO2 and AGO1 physically interact with this class of sRNA (Supplementary Figure 1A and B) by taking advantage of PAR-CLIP-Seq datasets [35]. Indeed, PAR-CLIP shows that TTSa-RNAs are associated with AGO2 and AGO1. Interestingly, the abundance of TTSa-RNAs in PAR-CLIP libraries is higher compared to the one observed in noncrosslinked AGO IPs, suggesting that TTSa-RNAs are directly associated with AGO proteins. Notably, only reads containing one single mismatch with reference (mainly T->C transitions, Supplementary Figure 1C and D) were used to generate Supplementary Figure 1A and B, thus ensuring that only *bona fide* AGO crosslinked molecules are contributing to the signal observed.

Furthermore, we asked whether TTSa-RNA expression levels are correlated with the abundance of the corresponding mRNA. We looked at the correlation between gene expression levels (as assessed by RNA-seq) of each gene and the abundance of the TTSa-RNAs arising from TTSs of the same gene in HeLaS3 cells for both AGO1 (R^2<0.1) and AGO2 (R^2<0.1, $n = 2$). The R^2 values observed do not support any correlation between gene expression and TTSa-RNA abundance.

Differently from Valen and colleagues [15], we showed that TTSa-RNAs can be detected in libraries obtained from nuclear extracts of HeLaS3 cells. We also investigated whether TTSa-RNAs are recruited on chromatin. We extracted HeLaS3 nuclei to separate soluble and loosely bound material (soluble nuclear fraction (SNF)) from the chromatin pellet (chromatin pellet extract (CPE)), which retains tightly bound factors [21] (Figures 1(d) and 1(e)). As a control of a chromatin-associated RNAs, we measured MALAT [36]. Moreover, in order to provide evidence that this procedure also preserves chromatin localization of sRNAs, we verified that the small nucleolar RNA RNU44 was enriched in the CPE as compared to SNF. On the other hand, we included in the analysis a soluble nuclear-enriched mRNA, beta-actin (ACTB), and miR-21, which has been previously detected not only in the cytosol but also in the nucleus of human cells [37] (Figure 1(e)). We selected 5 abundant TTSa-RNAs arising from genes belonging to the enriched GO categories (*YWHAG*, *WEE1*, *PSMB1*, *CDKNA1*, and *BUB3*, see Table 1). We measured their abundance in different nuclear fractions by qRT-PCR using specific primers designed to selectively amplify sRNA molecules and not their precursors [38]. TTSa-RNAs display a profile of nucleoplasm/chromatin abundance very similar to the one of ACTB mRNA and miR-21 rather than MALAT and U44 (Figure 1(e)), arguing against their specific recruitment on chromatin.

3.2. DICER Is Not Involved in TTSa-RNA Biogenesis. We next focused on TTSa-RNAs biogenesis. Previously, it has been speculated that the miRNA biogenesis pathway was not involved in TTSa-RNA processing [15]. We aimed to experimentally verify this hypothesis and we first investigated whether TTSa-RNAs are processed by DICER. To answer this question, we reanalyzed sRNA-seq libraries derived from sRNAs bound to nuclear AGO2 in HCT116 and HCT116 Dicer^{EX5} (a subclone of HCT116 cells in which DICER has been impaired) cell lines [20] (Figures 2(a) and 2(b)) that we previously generated [6]. If DICER endonucleolytic activity is required for TTSa-RNA biogenesis, one would expect a dramatic global reduction of TTSa-RNA abundance as it happens for miRNA [20] and for TSSa-RNAs [6]. On the contrary, we found that AGO2-bound TTSa-RNA abundance is considerably increased in HCT116 Dicer^{EX5} cells compared to parental HCT116 cells (Figure 2(c)). To further validate these findings by qRT-PCR, we analyzed the expression of the same five TTSa-RNAs mentioned above by qRT-PCR in whole cell extracts in HCT116 Dicer^{EX5} cells as compared to parental cells. None of the five TTSa-RNAs tested by qRT-PCR displayed any decrease in HCT116 Dicer^{EX5} cells (Figure 2(d)). In agreement with NGS data, qPCR validation showed that for some of the tested TTSa-RNAs, abundance increases in HCT116 Dicer^{EX5} compared to parental cells. Therefore, in order to exclude that this increase was due to a higher expression of the corresponding gene in HCT116 Dicer^{EX5}, we took advantage of the GEO dataset GSE6427 which reports mRNA expression levels in HCT116 and HCT116 Dicer^{EX5} cells. As shown in Figure 2(e), the expression of the YWHAG, WEE1, PSMB1, CDKNA1, and BUB3 mRNAs is not affected by DICER ablation. We hypothesize that higher levels of TTSa-RNAs in DICER hypomorphic cells as compared to control, observed both in NGS data and in qPCR validations, might be explained taking into account

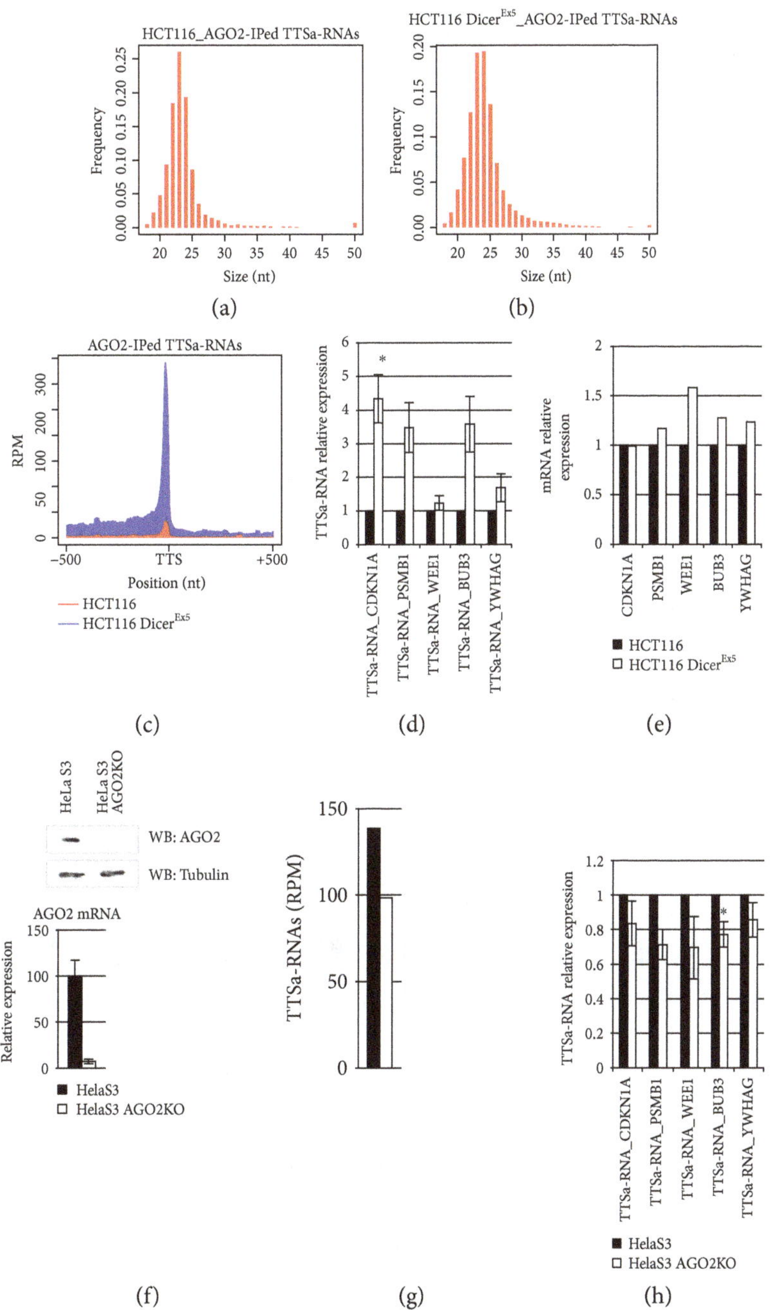

Figure 2: TTSa-RNA biogenesis is independent of both DICER and AGO2. (a) Size distribution of AGO2-immunoprecipitated TTSa-RNAs from HCT116 and HCT116 Dicer^{Ex5} (b) nuclear lysate. (c) Coverage of AGO2-bound TTSa-RNAs around GENCODE v25 annotated human TTSs in parental HCT116 (in red) and HCT116 Dicer^{Ex5} subclone (in blue). (d) qPCR analysis of selected TTSa-RNAs in parental HCT116 and HCT116 Dicer^{Ex5} subclone ($n = 4$). RNU44 was used for normalization. (e) Microarray analysis (GEO dataset GSE6427) of the CDKNA1, PSMB1, WEE1, BUB3, and YWHAG mRNA expression in parental HCT116 and HCT116 Dicer^{Ex5} subclone. (g) Abundance (RPM) of TTSa-RNAs in sRNA libraries generated from whole cell extract of parental HelaS3 and AGO2KO cells. (h) *AGO2* has been genetically knocked out (AGO2KO) in HeLaS3 cells by using specific Zinc Finger Nucleases. In these cells, AGO2 is undetectable at protein level (upper panel) and strongly reduced at mRNA level (lower panel). (f) qPCR analysis of selected TTSa-RNAs in parental HeLaS3 and HeLaS3 AGO2 KO subclone ($n = 5$). RNU44 was used for normalization. qPCR data are expressed as mean ± SEM. (*p value ≤ 0.05; paired *t test*).

that lack of mature miRNA may promote loading on AGO1 and AGO2 of TTSa-RNAs, perhaps increasing their stability. Overall, our data highlights that TTSa-RNA biogenesis is not dependent on DICER.

Since it has been reported that miR-451 is processed in a DICER-independent manner by AGO2 endonucleolytic activity [39], we wondered whether TTSa-RNAs might be processed by AGO2. In case AGO2 was required for

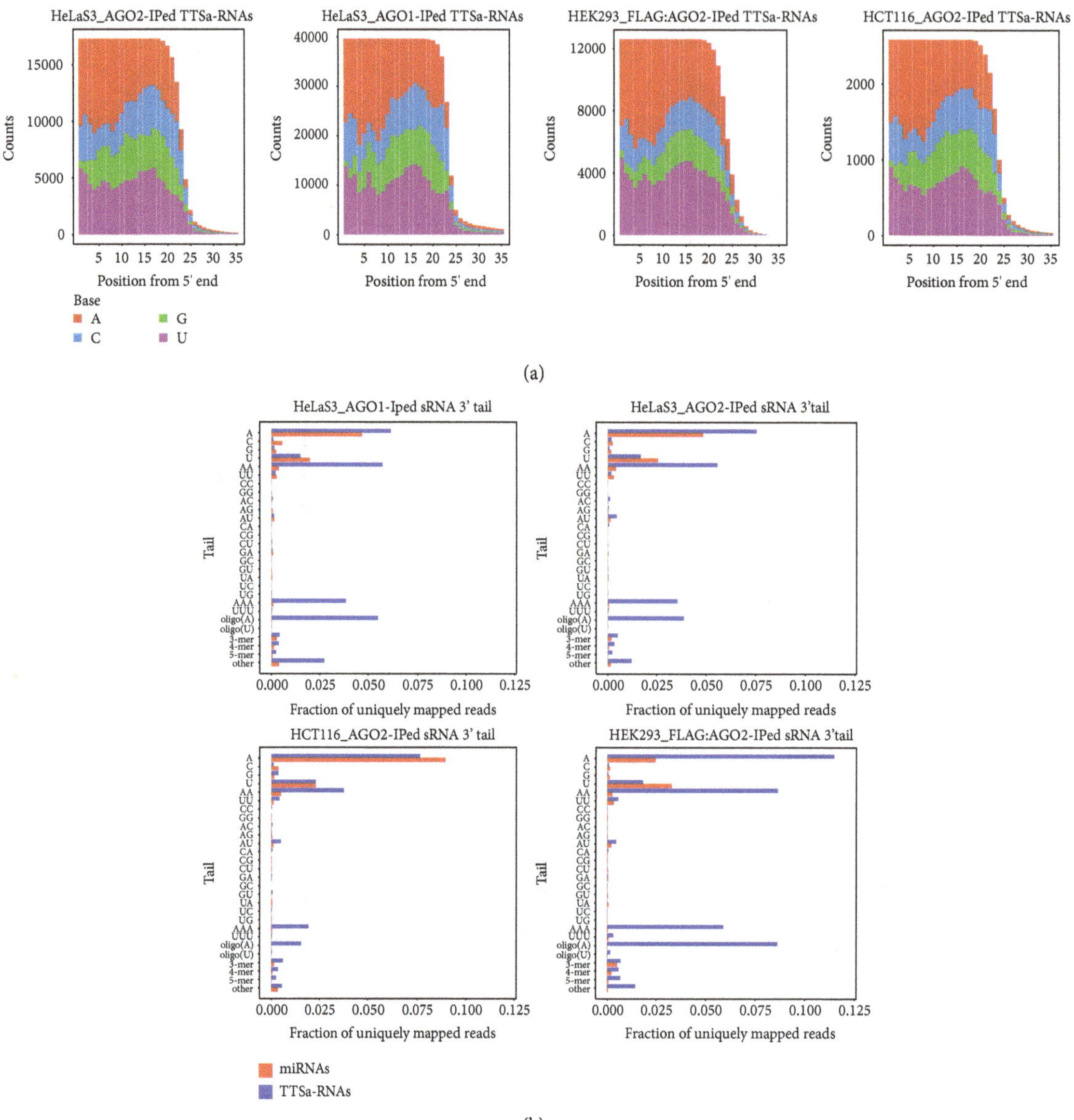

Figure 3: TTSa-RNAs display specific features in 5′ end nucleotide composition and in 3′ end tailing. (a) Nucleotide composition at each position of AGO-bound TTSa-RNAs in different cell lines. (b) 3′ end tail composition and frequency in AGO1- and AGO2-bound TTSa-RNAs (in blue) and miRNAs (in red) from HeLaS3 nuclear extracts, in AGO2-associated TTSa-RNAs and miRNAs in HCT116 nuclear lysate, and in FLAG:AGO2-bound TTSa-RNAs and miRNAs from HEK293T whole cell extracts.

TTSa-RNA biogenesis, AGO2 depletion should have resulted in a major decrease of TTSa-RNA abundance in whole cell extracts. We therefore profiled sRNAs in parental and AGO2KO HeLaS3 cells (obtained by genome editing mutation of *AGO2* gene, Figure 2(e)) by sRNA-Seq and we calculated TTSa-RNA abundance (Figure 2(g)). Moreover, we evaluated the abundance of the five selected TTSa-RNAs by qRT-PCR (Figure 2(h)). Both approaches showed a slight decrease in TTSa-RNA abundance when AGO2 is depleted. We concluded that even though AGO2 is not required for TTSa-RNA biogenesis, it might participate to TTSa-RNA stabilization, recapitulating what was previously seen for miRNAs [40].

The fact that both DICER and AGO2 are dispensable for TTSa-RNAs biogenesis is in agreement with our observation that the regions flanking TTSa-RNAs are not biased toward the formation of hairpin structures compared to randomly picked genomic regions (data not shown).

3.3. Sequence Characteristics of TTSa-RNAs. We set out to investigate whether any consensus nucleotide sequence could be found within TTSa-RNAs or in the surrounding genomic sequence. We could not identify a consensus sequence that might provide any hints on TTSa-RNA biogenesis. However, similarly to what was observed for other AGO-bound sRNAs [41, 42], TTSa-RNAs have a bias against a G residue in

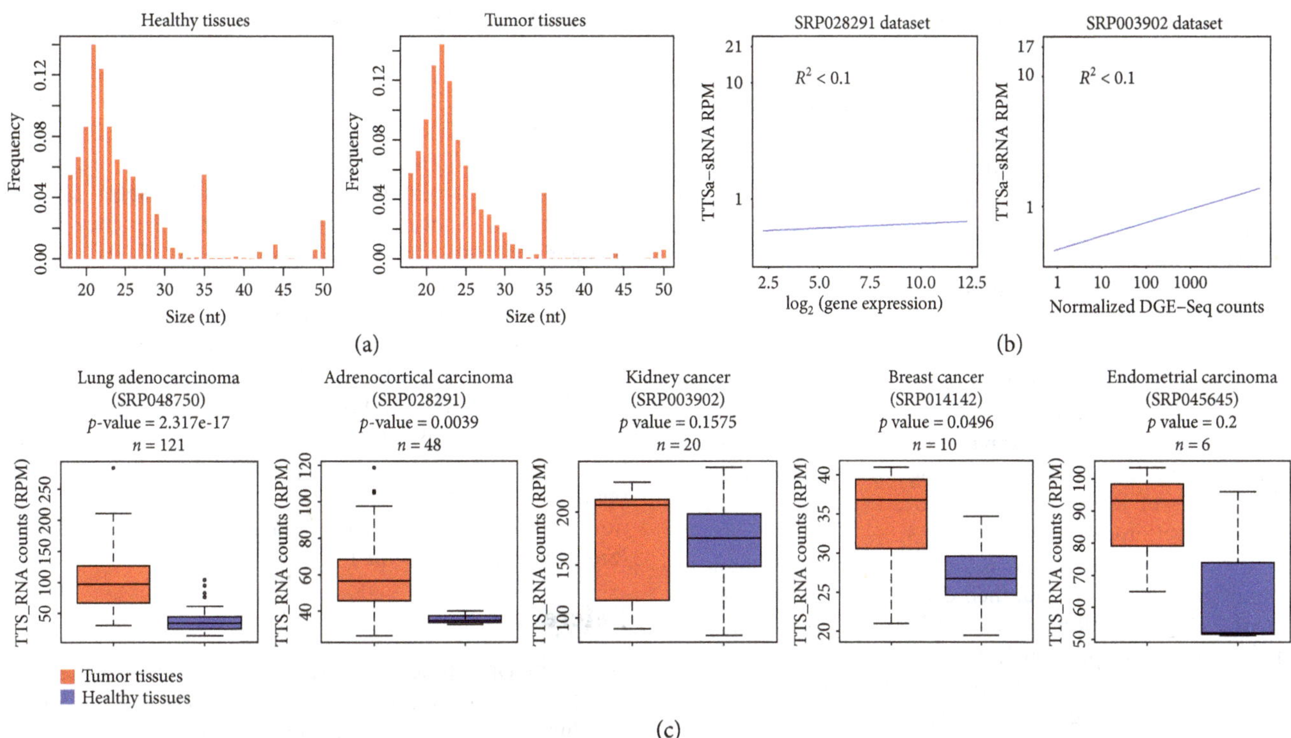

FIGURE 4: TTSa-RNAs are overexpressed in cancerous tissues compared to matched controls. (a) Size distribution of TTSa-RNAs from primary samples (healthy tissues, left panel; cancerous tissues, right panel). (b) Correlation between mRNA expression level and abundance of TTSa-RNAs in SRP028291 and SRP003902 datasets. Each dot represents a single gene in a single sample. (c) TTSa-RNA expression (RPM) in different tumor types compared to matched control tissues.

the first position at 5′ end if compared to other AGO2-bound sRNAs (Fisher Exact test p value < 2.2e-16) in all libraries generated from AGO-immunoprecipitated samples (Figure 3(a)). We noted that for all libraries, this bias is stronger in TTSa-RNAs (3.25%) as compared to other AGO-bound sRNAs (8.10%). This can be explained by the low GC content in the regions upstream of human TTSs.

We also checked for any tailing at 3′ end of TTSa-RNAs. We found that about 20% to 27% of TTSa-RNAs bound to endogenous AGO1 and AGO2 displayed a short 3′ tail (1 to 3 nt in most cases). Importantly, 3′ tailing of both AGO1- and AGO2-bound TTSa-RNAs displays a similar pattern. Furthermore, 41% of the TTSa-RNAs bound to FLAG-tagged AGO2 in HEK293T display a 3′ tail (Figure 3(b)). Since it has been reported that miRNAs are 3′ tailed [43, 44], we compared TTSa-RNAs and miRNA 3′ tailing in the same samples. In both sRNA classes, we found abundant monoadenylation and monouridylation. On the other hand, our data highlight a significant (p value < 2.2E-16) increase of oligoA tails (4 or more A) in TTSa-RNAs compared to miRNAs. This finding suggests that mature polyadenylated mRNAs might be the precursors of TTSa-RNAs. Although this observation might suggest that TTSa-RNAs are mere degradation by-products of mRNAs, the fact that they arise specifically from the region of mRNAs immediately upstream of polyA sites, the lack of any correlation with corresponding mRNA abundance and the specific loading on AGO proteins argue against this hypothesis. In particular, it is difficult to conceive a general or widespread mRNA degradation pathway leading to the specific accumulation only of degradation fragments flanking polyA sites. Therefore, we hypothesized that these sRNAs are not simply mRNA degradation products but are specific, biologically generated species.

3.4. TTSa-RNAs Are Overexpressed in Human Tumors. Up to now, the existence of sRNAs mapping around 3′ termini of mammalian genes was only reported for cultured cell lines [10, 14, 15]. Therefore, we extended our analysis of TTSa-RNAs to primary samples, and we set out to look for TTSa-RNAs in sRNA NGS libraries obtained from primary human samples. We searched the SRA repository for datasets in which microRNA profile was assessed by NGS. We retrieved five datasets suitable for TTSa-RNA analysis which contained publicly available sRNA expression profiles in tumor samples and matched control tissue. In all the five datasets analyzed, we could detect bona fide TTSa-RNAs displaying the expect size profiles (Figure 4(a)). A small number of reads mapping to TTS with a size of 35 nt were retrieved; therefore, we only counted molecules shorter than 28 nt as bona fide TTSa-sRNAs. Since AGO-bound TTSa-RNAs in cell lines do not correlate with the expression of the corresponding mRNA, we asked whether bona fide TTSa-RNAs that we identified in primary samples display any correlation with mRNA expression. To answer this question, we took advantage of the mRNA expression profiling available for two of the five datasets. As depicted in Figure 4(b), in both cases, we do not observe any significant correlation, highlighting that not only in cultured cell lines but also in human tissues, the amount of TTSa-RNAs is not correlated with the expression of the corresponding gene.

Since deregulation of other sRNA classes has been reported in human tumors [16, 19], we looked at TTSa-RNAs expression in tumor tissues and matched controls. In four out of five datasets analyzed, we found an increase of TTSa-RNAs in tumor samples compared to matched healthy tissues (Figure 4(c)). The lack of statistical significance of the result obtained in endometrial cancer dataset is likely due to the small number of samples in this dataset ($n = 6$). Interestingly, we found that the expression of most TTSa-RNAs is increased in cancer tissues, suggesting a general deregulation of this class of RNAs rather than overexpression of a few TTSa-RNAs (Supplementary Figure 3).

We concluded that TTSa-RNAs are not only detectable in human primary tissues, but their expression also increases in tumors. Accumulation of TTSa-RNAs in tumor samples might reflect the previously reported deregulation of the expression of RNAi components, such as DICER, AGO2, and AGO1 in cancer cells [45]. In particular, downregulation of DICER might impair miRNA processing, thus promoting TTSa-RNA loading on AGO proteins. On the other hand, AGO protein overexpression might stabilize TTSa-RNAs.

4. Conclusions

Here, we present a comprehensive study of sRNAs derived from TTS of expressed genes in human cell lines and primary tissues. Taking advantage of a sequencing depth of 100 million reads per sample, this characterization gets insights into previously unknown details of biogenesis, localization, sequence features, and expression of this new and poorly studied class. Our data demonstrate that TTSa-RNAs are a class of DICER-independent and AGO-bound sRNAs, which display an oligoA tailing at $3'$ end and which are expressed not only in cultured cell lines but also in human primary tissues. Even though the function of TTSa-RNAs still remains elusive, the fact that genes that give rise to TTSa-RNAs are involved in regulation of proliferation and DNA damage response and that TTSa-RNAs are overexpressed in human tumors suggest that TTSa-RNA expression is linked to tumorigenesis and they might be explored as biomarker for diagnosis or prognosis of human malignancies in the future.

Conflicts of Interest

The authors declare that there is no conflict of interest regarding the publication of this article.

Acknowledgments

This work was supported by an EPIGEN (MIUR-CNR) grant to Valerio Fulci and by a grant MIUR Progetti di Ateneo Sapienza Università di Roma to Claudia Carissimi.

Supplementary Materials

Figure 1: AGO proteins are physically associated with TTSa-RNAs. (a) Coverage of AGO2- and FLAG:AGO1-PAR-CLIPPed TTSa-RNAs (HEK293) around GENCODE v25 annotated TTSs. Red and blue represent sRNAs in the sense and antisense orientation with respect to gene transcription, respectively. (b) Mutation frequency of all possible base transitions for AGO2- and FLAG:AGO1-PAR-CLIPPed TTSa-RNAs. Figure 2: top-enriched GO term of genes giving rise to TTSa-RNAs. GO term analysis was performed for genes giving rise to HeLaS3 AGO1- and AGO2-IPed TTSa-RNAs, HCT116 AGO2-IPed TTSa-RNAs, and HEK293 FLAG:AGO2-IPed TTSa-RNAs. Figure 3: most TTSa-RNAs increase their expression in cancer samples. For each TTSa-RNA, the log2 ratio in cancer tissues relative to matched controls was computed. The distribution of the values obtained is depicted for each dataset. *(Supplementary Materials)*

References

[1] ENCODE Project Consortium, E. Birney, J. A. Stamatoyannopoulos et al., "Identification and analysis of functional elements in 1% of the human genome by the ENCODE pilot project," *Nature*, vol. 447, no. 7146, pp. 799–816, 2007.

[2] J. S. Mattick, "The genetic signatures of noncoding RNAs," *PLoS Genetics*, vol. 5, no. 4, article e1000459, 2009.

[3] S. A. Gorski, J. Vogel, and J. A. Doudna, "RNA-based recognition and targeting: sowing the seeds of specificity," *Nature Reviews Molecular Cell Biology*, vol. 18, no. 4, pp. 215–228, 2017.

[4] H. Kobayashi and Y. Tomari, "RISC assembly: coordination between small RNAs and Argonaute proteins," *Biochimica et Biophysica Acta (BBA) - Gene Regulatory Mechanisms*, vol. 1859, no. 1, pp. 71–81, 2016.

[5] R. J. Taft, E. A. Glazov, N. Cloonan et al., "Tiny RNAs associated with transcription start sites in animals," *Nature Genetics*, vol. 41, no. 5, pp. 572–578, 2009.

[6] C. Carissimi, I. Laudadio, E. Cipolletta et al., "ARGONAUTE2 cooperates with SWI/SNF complex to determine nucleosome occupancy at human Transcription Start Sites," *Nucleic Acids Research*, vol. 43, no. 3, pp. 1498–1512, 2015.

[7] A. C. Seila, J. M. Calabrese, S. S. Levine et al., "Divergent transcription from active promoters," *Science*, vol. 322, no. 5909, pp. 1849–1851, 2008.

[8] J. R. Zamudio, T. J. Kelly, and P. A. Sharp, "Argonaute-bound small RNAs from promoter-proximal RNA polymerase II," *Cell*, vol. 156, no. 5, pp. 920–934, 2014.

[9] K. Fejes-Toth, V. Sotirova, R. Sachidanandam et al., "Post-transcriptional processing generates a diversity of $5'$-modified long and short RNAs," *Nature*, vol. 457, no. 7232, pp. 1028–1032, 2009.

[10] P. Kapranov, J. Cheng, S. Dike et al., "RNA maps reveal new RNA classes and a possible function for pervasive transcription," *Science*, vol. 316, no. 5830, pp. 1484–1488, 2007.

[11] S. Francia, M. Cabrini, V. Matti, A. Oldani, and F. d'Adda di Fagagna, "DICER, DROSHA and DNA damage response RNAs are necessary for the secondary recruitment of DNA damage response factors," *Journal of Cell Science*, vol. 129, no. 7, pp. 1468–1476, 2016.

[12] W. Wei, Z. Ba, M. Gao et al., "A role for small RNAs in DNA double-strand break repair," *Cell*, vol. 149, no. 1, pp. 101–112, 2012.

[13] X. Ma, N. Han, C. Shao, and Y. Meng, "Transcriptome-wide discovery of PASRs (promoter-associated small RNAs) and TASRs (terminus-associated small RNAs) in *Arabidopsis thaliana*," *PLoS One*, vol. 12, no. 1, article e0169212, 2017.

[14] P. Kapranov, F. Ozsolak, S. W. Kim et al., "New class of gene-termini-associated human RNAs suggests a novel RNA copying mechanism," *Nature*, vol. 466, no. 7306, pp. 642–646, 2010.

[15] E. Valen, P. Preker, P. R. Andersen et al., "Biogenic mechanisms and utilization of small RNAs derived from human protein-coding genes," *Nature Structural & Molecular Biology*, vol. 18, no. 9, pp. 1075–1082, 2011.

[16] G. A. Calin and C. M. Croce, "MicroRNA signatures in human cancers".

[17] G. Szabo and S. Bala, "MicroRNAs in liver disease," *Nature Reviews Gastroenterology & Hepatology*, vol. 10, no. 9, pp. 542–552, 2013.

[18] A. Flemming, "Heart failure: targeting miRNA pathology in heart disease," *Nature Reviews Drug Discovery*, vol. 13, no. 5, p. 336, 2014.

[19] M. Esteller, "Non-coding RNAs in human disease," *Nature Reviews Genetics*, vol. 12, no. 12, pp. 861–874, 2011.

[20] J. M. Cummins, Y. He, R. J. Leary et al., "The colorectal microRNAome," *Proceedings of the National Academy of Sciences of the United States of America*, vol. 103, no. 10, pp. 3687–3692, 2006.

[21] M. S. Werner and A. J. Ruthenburg, "Nuclear fractionation reveals thousands of chromatin-tethered noncoding RNAs adjacent to active genes," *Cell Reports*, vol. 12, no. 7, pp. 1089–1098, 2015.

[22] A. Rybak-Wolf, M. Jens, Y. Murakawa, M. Herzog, M. Landthaler, and N. Rajewsky, "A variety of dicer substrates in human and *C. elegans*," *Cell*, vol. 159, no. 5, pp. 1153–1167, 2014.

[23] J. Jurka, "Repbase update: a database and an electronic journal of repetitive elements," *Trends in Genetics*, vol. 16, no. 9, pp. 418–420, 2000.

[24] D. Karolchik, A. S. Hinrichs, T. S. Furey et al., "The UCSC Table Browser data retrieval tool," *Nucleic Acids Research*, vol. 32, no. 90001, Supplement 1, pp. 493D–4496, 2004.

[25] A. R. Quinlan and I. M. Hall, "BEDTools: a flexible suite of utilities for comparing genomic features," *Bioinformatics*, vol. 26, no. 6, pp. 841-842, 2010.

[26] B. Langmead, C. Trapnell, M. Pop, and S. L. Salzberg, "Ultrafast and memory-efficient alignment of short DNA sequences to the human genome," *Genome Biology*, vol. 10, no. 3, article R25, 2009.

[27] A. Kozomara and S. Griffiths-Jones, "miRBase: annotating high confidence microRNAs using deep sequencing data," *Nucleic Acids Research*, vol. 42, D1, pp. D68–D73, 2014.

[28] M.-T. Chou, B. W. Han, C.-P. Hsiao, P. D. Zamore, Z. Weng, and J.-H. Hung, "Tailor: a computational framework for detecting non-templated tailing of small silencing RNAs," *Nucleic Acids Research*, vol. 43, no. 17, article e109, 2015.

[29] D. W. Huang, B. T. Sherman, and R. A. Lempicki, "Systematic and integrative analysis of large gene lists using DAVID bioinformatics resources," *Nature Protocols*, vol. 4, no. 1, pp. 44–57, 2009.

[30] N. L. Bray, H. Pimentel, P. Melsted, and L. Pachter, "Near-optimal probabilistic RNA-seq quantification," *Nature Biotechnology*, vol. 34, no. 5, pp. 525–527, 2016.

[31] B. S. Carvalho and R. A. Irizarry, "A framework for oligonucleotide microarray preprocessing," *Bioinformatics*, vol. 26, no. 19, pp. 2363–2367, 2010.

[32] M. Martin, "Cutadapt removes adapter sequences from high-throughput sequencing reads," *EMBnet.Journal*, vol. 17, no. 1, pp. 10–12, 2011.

[33] S. Anders, P. T. Pyl, and W. Huber, "HTSeq—a Python framework to work with high-throughput sequencing data," *Bioinformatics*, vol. 31, no. 2, pp. 166–169, 2015.

[34] S. Anders and W. Huber, "Differential expression analysis for sequence count data," *Genome Biology*, vol. 11, no. 10, article R106, 2010.

[35] S. Memczak, M. Jens, A. Elefsinioti et al., "Circular RNAs are a large class of animal RNAs with regulatory potency," *Nature*, vol. 495, no. 7441, pp. 333–338, 2013.

[36] J. A. West, C. P. Davis, H. Sunwoo et al., "The long noncoding RNAs NEAT1 and MALAT1 bind active chromatin sites," *Molecular Cell*, vol. 55, no. 5, pp. 791–802, 2014.

[37] G. Meister, M. Landthaler, A. Patkaniowska, Y. Dorsett, G. Teng, and T. Tuschl, "Human Argonaute2 mediates RNA cleavage targeted by miRNAs and siRNAs," *Molecular Cell*, vol. 15, no. 2, pp. 185–197, 2004.

[38] C. Chen, D. A. Ridzon, A. J. Broomer et al., "Real-time quantification of microRNAs by stem-loop RT-PCR," *Nucleic Acids Research*, vol. 33, no. 20, article e179, 2005.

[39] S. Cheloufi, C. O. Dos Santos, M. M. W. Chong, and G. J. Hannon, "A dicer-independent miRNA biogenesis pathway that requires Ago catalysis," *Nature*, vol. 465, no. 7298, pp. 584-589, 2010.

[40] J. Winter and S. Diederichs, "Argonaute proteins regulate microRNA stability: increased microRNA abundance by Argonaute proteins is due to microRNA stabilization," *RNA Biology*, vol. 8, no. 6, pp. 1149–1157, 2011.

[41] E. Elkayam, C.-D. Kuhn, A. Tocilj et al., "The structure of human argonaute-2 in complex with miR-20a," *Cell*, vol. 150, no. 1, pp. 100–110, 2012.

[42] F. Frank, N. Sonenberg, and B. Nagar, "Structural basis for 5′-nucleotide base-specific recognition of guide RNA by human AGO2," *Nature*, vol. 465, no. 7299, pp. 818–822, 2010.

[43] A. M. Burroughs, Y. Ando, M. J. L. de Hoon et al., "A comprehensive survey of 3′ animal miRNA modification events and a possible role for 3′ adenylation in modulating miRNA targeting effectiveness," *Genome Research*, vol. 20, no. 10, pp. 1398–1410, 2010.

[44] D. Koppers-Lalic, M. Hackenberg, I. V. Bijnsdorp et al., "Non-templated nucleotide additions distinguish the small RNA composition in cells from exosomes," *Cell Reports*, vol. 8, no. 6, pp. 1649–1658, 2014.

[45] S. Lin and R. I. Gregory, "MicroRNA biogenesis pathways in cancer," *Nature Reviews Cancer*, vol. 15, no. 6, pp. 321–333, 2015.

2

The Progress of Methylation Regulation in Gene Expression of Cervical Cancer

Chunyang Feng, Junxue Dong, Weiqin Chang, Manhua Cui, and Tianmin Xu

The Second Hospital of Jilin University, Jilin, Changchun 130041, China

Correspondence should be addressed to Tianmin Xu; xutianmin@126.com

Academic Editor: Yujing Li

Cervical cancer is one of the most common gynecological tumors in females, which is closely related to high-rate HPV infection. Methylation alteration is a type of epigenetic decoration that regulates the expression of genes without changing the DNA sequence, and it is essential for the progression of cervical cancer in pathogenesis while reflecting the prognosis and therapeutic sensitivity in clinical practice. Hydroxymethylation has been discovered in recent years, thus making 5-hmC, the more stable marker, attract more attention in the field of methylation research. As markers of methylation, 5-hmC and 5-mC together with 5-foC and 5-caC draw the outline of the reversible cycle, and 6-mA takes part in the methylation of RNA, especially mRNA. Furthermore, methylation modification participates in ncRNA regulation and histone decoration. In this review, we focus on recent advances in the understanding of methylation regulation in the process of cervical cancer, as well as HPV and CIN, to identify the significant impact on the prospect of overcoming cervical cancer.

1. Introduction

Cervical cancer, which is one of the three most common gynecological tumors, has been the fourth leading cause of cancer-associated death among women worldwide, as well as becoming the second most commonly diagnosed cancer in developing countries. According to statistics, newly diagnosed cases and cervical cancer-associated deaths are approximately 520,000 and 260,000, respectively, every year, which affected youth trends more clearly [1]. It is widely recognized that persistent infection of high-risk-HPV (hr-HPV) accounts for the process from cervical intraepithelial neoplasia (CIN) to neoplasms, and vaccines of HPV and application of screening methods contribute a lot towards cervical carcinoma prevention. However, for established infections, vaccines have limited function and full-type coverage has not been achieved yet [1]. Additionally, as the 5-year survival rate is about 15% among advanced patients, the prognosis still remains unoptimistic in the late stages [2, 3]. Hence, it cries out for investigating the underlying molecular mechanisms on different biological expression levels to understand the genesis and progression of cervical cancer.

While gene mutation theory is incapable of providing reasonable explanations for many biological changes in tumor development, epigenetic alteration is drawing more attention, which involves modifications such as methylations of DNA and RNA, acetylations of histone, and regulations of ncRNA and aberrant chromatin. Methylated modification is extensively studied these years. DNA methylation mainly occurs at CpG islands where the methyltransferase DNMT family mediates the transfer of a methyl group to cytosines, generating 5-methylcytosine (5-mC), which can be oxidized into 5-hydroxymethylcytosine (5-hmC), 5-foC, and 5-caC by TET proteins step by step, so that methylation is achieved reversibly [4, 5]. Methylation decoration in RNAs is as common as it is in DNAs. M6A is one of the markers in mRNA methylation, and modifications take place in nascent pre-mRNAs predominantly [6]. Additionally, miR-RNAs and lnc-RNAs take part in epigenetic modifications themselves, and their biological functions are affected by the methylation state at the same time.

In this article, we summarize several recent studies of methylation regulation in the field of cervical cancer and discuss the potential of these molecular mechanisms in the period of gene expression, to get some enlightenment in epigenetics to carry forward the prevention and treatment of cervical cancer.

2. Hydroxymethylation and Cervical Cancer

2.1. Hydroxymethylation and Its Regulations. In 1972, 5-hmC was initially found in bacteriophages and then in mammalian DNA. Currently, 5-hmC, a more stable epigenetic mark than 5-mC, plays an important role in epigenetics and works as an intermediate in demethylation [7]. It has been confirmed that the brain has the highest concentration of 5-hmC, while the rectum, liver, colon, and kidney are subordinate. In contrast, 5-hmC is at a low level in the lung, placenta, and breast [8]. The regulation of DNA hydroxymethylation is mediated by several factors, among which human ten-eleven translocation (TET) is identified as a dioxygenase for converting 5-mC to 5-hmC; meanwhile, αKG, Fe^{2+}, and ascorbate may activate the TET proteins as cofactors [9].

The TET protein family consists of TET1, TET2, and TET3, and their C-terminal catalytic domains come from a high degree of homology, which can be regulated by CXXC finger protein 1 (CFP1). Different CXXC domains have different functions; the CXXC5 domain of TET2 is able to downregulate TET2 with a 5-hmC decrease. But CXXC4 was found to be binding to the unmethylated DNA of TET1, TET2, and TET3, which then starts a caspase-dependent degradative process [10]. Some researchers found that in TET1-lacking cells, 5-hmC was reduced while 5-mC was increased. Moreover, TET1 can control 5-hmC by regulating hydroxylase activity to convert 5-mC to 5-hmC, which is HIF-1 dependent; at the same time, TET1 can also bind to CpG regions to stop some DNA methyltransferase activity [11]. It was demonstrated that TET3 is important for proper DNA repair, cell survival, and promotion of 5-hmC [12]. Besides, 5-hmC levels are also partly regulated by microRNAs. There are also some genes regulating 5-hmC, such as IDH1, IDH2, SDH, and FH [13]. Those factors are linked to the alteration of 5-hmC levels in cancer.

2.2. DNA Hydroxymethylation in Cervical Cancer and Other Cancers. To have cervical cancer treated and diagnosed precisely, many researches about 5-mC and other epigenetic modifications of cervical cancer aim to find treatment methods and diagnostic markers. But 5-hmC of cervical cancer is less researched, as there are only two articles about 5-hmC in cervical cancer.

Zhang et al. used immunohistochemistry to detect the expression of 5-hmC, 5-mC, and TET1/2/3 in 140 cervical squamous cell carcinoma (CSCC) tissues and 40 normal cervical tissues. They found that the expression of 5-hmC was an independent prognostic factor of squamous cell carcinoma, and compared with normal cervix tissues, the level of 5-mC was increased while 5-hmC was significantly decreased, which predicts poor prognosis of CSCC. Moreover, only the expression of TET2 was decreased in CSCC [14]. In contrast, Bhat et al. found that the 5-mC and 5-hmC levels were both significantly reduced in squamous cell carcinoma, but receiver operating characteristic curve analysis showed a significant difference in 5-mC and 5-hmC between normal and squamous cell carcinoma tissues. They also tested the promoter methylation of 33 genes; only PROX1, NNAT, ARHGAP6, HAND2, NKX2-2, PCDH10, DAPK1, RAB6C, and PITX2 could effectively tell the difference among the various stages of tumor with high sensitivity and specificity [15]. Expressions of 5-hmC and 5-mC in cervical cancer need further demonstrations, and these related results may serve as useful biomarkers for the early detection and accurate management of cervical cancer.

Although 5-hmC was studied little in cervical cancer, it is a noticeable part in other cancers; scientists have been making further studies for deeper mechanisms of 5-hmC as well.

It is demonstrated that TET1 and TET3 catalyze the conversion from 5-mC to 5-hmC by activating the TNFα-p38-MAPK signaling axis and inducing tumor malignancy and poor prognosis in breast cancer patients [16]. In prostate cancer, the androgen receptor decreases the expression of miR-29b which targets both TET2 and 5-hmC; 5-hmC represses FOXA1 activity, while its reduction activates the mTOR pathway and AR of prostate cancer [17]. In DLD1 cells, knockdown of TET1 will promote cancer cell growth, migration, invasion, and even epithelial-mesenchymal transition (EMT) which can also reduce UTX-1 but increase the EZH2 expression which can cause a loss of H3K27 methylation at the epithelial gene E-cadherin promoter [18]. In contrast, the levels of TETs are similar in colorectal tumor tissue and normal tissues. TET2 targets promoters marked by 5-hmC in normal tissue and turns it to colorectal cancer tissue [19].

3. DNA Methylation in Cervical Cancer and CIN

In cervical lesions, aberrant DNA methylation includes hypomethylation and hypermethylation. In cervical cancer and high-grade cervical intraepithelial neoplasia, most genes are hypermethylated; only three promoter regions are hypomethylated (Table 1).

3.1. Gene Hypomethylation in Cervical Cancer and CIN. Hypomethylation often occurs in the promoter region of genes, regardless if the gene is for a protein or RNA. The STK31 gene targets at oncogene E7 of HPV16. Its promoter/exon 1 is hypomethylated in HPV16/18-positive cervical cell lines, which induces an integration of HPV16E7/E6 [20]. The COL17A1 promoter is also hypomethylated in cervical cancer, and it precisely predicts both the increased invasive nature and patient outcome [21]. In CIN tissues, the rDNA promoter region reveals significant hypomethylation at cytosines in the context of CpG dinucleotides, which can result in an increase in rRNA synthesis in the development of human cervical cancer [22].

3.2. Gene Hypermethylation in Cervical Cancer and CIN

3.2.1. Genome-Wide Studies of Aberrant Gene Methylation. There are some genome-wide studies of aberrant gene expression and methylation profiles which reveal susceptibility genes and underlying mechanisms of cervical cancer. In one study, a total of 1357 DEGs as well as 666 cervical cancer-

Table 1: DNA methylation of CIN or cervical cancer in recent studies.

Name of gene	Methylation status	Methylation-variable position	Function/relevant pathway	Reference	Notes
STK31	Hypomethylation	Promoter/exon 1	HPV oncogene-E6/E7	[20]	CIN III and CCA
COL17A1	Hypomethylation	Promoter	Collagen XVII	[21]	CCA
Ribosomal DNA	Hypomethylation	Promoter	rRNA synthesis	[22]	CIN II-III, CCA
EDN3 and EDNRB	Hypermethylation	Promoter	MAPK signal pathway MITF-Wnt/β-catenin signal pathway	[20, 23]	
VIM	Hypermethylation	Promoter	Epithelial-mesenchymal transition and aggressiveness	[24]	Ib1 and IIa stages of CCA
AJAP1 and SOX17	Hypermethylation	Promoter	Wnt signal pathway	[25]	
SFRP1 and SFRP4	Hypermethylation	Promoter	Wnt/β-catenin signal pathway	[25]	
CDKN2A	Hypermethylation	Downstream region	p16(INK4A)/p14(ARF)	[26]	CIN and CCA
IFN-γ	Hypermethylation	Promoter	IFN-γ-cancer immunoediting	[27]	CIN II-III and CCA
SALL3	Hypermethylation	Promoter	hrHPV-induced immortalization and malignant transformation	[28]	HPV-infected
EPB41L3	Hypermethylation	Promoter	DAL-1 protein	[29]	CIN II-III
CADM1/MAL	Hypermethylation	Unmentioned	Lesion-specific	[30]	CIN II-III and CCA
PAX1	Hypermethylation	Promoter	Unclear yet	[32]	CIN and CCA
DAPK1	Hypermethylation	Promoter	Epithelial-mesenchymal transition	[34]	CIN III and CCA
Keap1	Hypermethylation	Promoter	NRF2	[35]	CCA
GPX3	Hypermethylation	Promoter	Repair oxidative damages and lymph node metastasis	[36]	CCA
LDOC1	Hypermethylation	Promoter	Nuclear transcription factor	[37]	CCA
RASSF	Hypermethylation	Promoter	Ras protein	[38, 42]	CCA or plasma of CCA
DOC2B	Hypermethylation	Promoter	AKT1 and ERK1/2 signal pathway	[40]	CIN and CCA
MEG3	Hypermethylation	Promoter	Proliferation and apoptosis	[41]	Plasma of CIN III and CCA

(CC-) related methylation sites were screened out and 26 DEGs with 35 CC-related methylation sites were identified; ACOX3, CYP39A1, and DPYS are potential risk markers in CC, which were significantly enriched in 25 subpathways of 6 major pathways. EDN3 and EDNRB might play important roles in the molecular mechanism of CC [23]. In another study, 32 genes that might be associated with prognosis in the stages between Ib1 and IIa cervical cancer are profiled, among which the VIM gene is frequently methylated in CSCC and VIM methylation might predict a favorable prognosis [24]. The 14 hypermethylated genes, including ADRA1D, AJAP1, COL6A2, EDN3, EPO, HS3ST2, MAGI2, POU4F3, PTGDR, SOX8, SOX17, ST6GAL2, SYT9, and ZNF614, are implicated in β-catenin signaling in cervical carcinogenesis [25].

3.2.2. Gene Hypermethylation Found in Cervical Cancer/CIN Tissue Cell Lines and Patients' Plasmas. Gene hypermethylation is found in CIN cervical cancer tissues, cervical cancer cells, and even cervical cancer patients' plasmas. The methylation rates of IFN-γ, FHIT, MGMT, CDKN2A, SALL3, and gene promoters were significantly higher in cervical cancer tissues than those in CIN and normal cervical tissues, which are related to the progression of cervical oncogenesis. CDKN2A methylation may lead to the development of malignant disease by increased p16(INK4A)/p14(ARF) expression [26–28]. LINE-1, HS3ST2, CCNA1, EPB41L3, EDNRB, LMX1, and DPYS were hypermethylated in cervical cancer tissues, CIN III and CIN II, versus normal tissues and CIN I, of which EPB41L3 seems to be the best marker. CADM1 is regulated by p53, and CADM1/MAL is hypermethylated in the HPV16/18-infected cell lines. The methylation status in cervical scrapes appears to represent the worst underlying lesion, particularly CIN III and cervical cancer. Results imply that hypermethylation of these genes may be highly associated with the development of cervical cancer [29–31]. Specific hypermethylated genes serve as the early prevention and prognostic prediction for cervical cancer. The different methylation statuses of all three genes PAX1, SOX1, and ZNF582 showed reasonable concordance in normal control samples as well as CIN I, CIN II, CIN III, and SCC samples [32, 33]. The promoter methylation statuses of DAPK1, MGMT, and RARB were positively correlated with the cervical disease grades, respectively. DAPK1 combined with the other two showed a significantly positive correlation with cervical disease grade as well [34]. The promoter hypermethylation of Keap1 significantly increased nuclear NRF2 expression in cervical cancer tissues, which is a marker of poor prognosis in patients with cervical cancer [35]. And the promoter of GPX3 is significantly downregulated due to its promoter hypermethylation in cervical cancer tissues; at the same time, GPX3 expression plays a role in the development of cervical squamous cell carcinoma and is significantly related to lymph node metastasis and prognosis in cervical cancer patients [36]. Promoter methylation and the loss of LDOC1 expression are frequent events in cervical cancer and could be potential molecular markers in cervical cancer [37]. Hypermethylation of RASSF2A and TSLC1 downregulating the expression of RASSF2A and TSLC1 was detected, which predicts a greater risk of progressing towards invasive cervical cancer [38, 39]. Hypermethylation of DOC2B promotes colony formation and cell proliferation, induces cell cycle arrest, and represses cell migration and invasion deeply; the promoter region of the DOC2B gene inhibiting AKT1 and ERK1/2 signaling is hypermethylated in premalignant and malignant cervical tissues and cervical cancer cell lines [40]. Those gene promoter methylations may be correlated with clinical stage and tumor grade and play a crucial role in cervical cancer progression.

The level of MEG3 methylation is significantly higher in cervical cancer tissues and patients' plasmas than in adjacent normal tissues and plasmas of healthy participants, respectively [41]. Promoter hypermethylation of some other genes like MYOD1, CALCA, hTERT, and RASSF1A can also be detected in serum samples of cervical cancer patients and are related to lymph node metastasis and FIGO stage [42, 43]. In conclusion, the present studies clearly showed that MEG3, MYOD1, CALCA, hTERT, and RASSF1A methylation in plasma can serve as diagnostic and prognostic biomarkers for cervical cancer patients, providing useful information for clinical management.

3.2.3. Gene Hypermethylation Found in Different Ethnicities. The hypermethylation status of genes in cervical cancer patients is associated with different countries. In the North Indian population, methylation of the p16 gene promoter which induces loss of tumor-suppressing activity and promotes the development of cervical cancer is observed significantly in FIGO stage III [44, 45]. Meanwhile, correlated with clinical parameters, promoter hypermethylation and expression loss of PARK-2, RARβ, and FHIT are significantly higher in cervical cancer than in CINs and normal tissues, resulting in a significant association with tumor stage and histological grade [46, 47]. In Uighur women, increased methylation was detected at 13 CpG sites, and a high methylation level was associated with the risk of CIN2$^+$; the strongest related site was 6650 [48]. The methylation level of the ERp57 gene promoter is higher in CSCC than in CIN, and normal tissues in Uighur women. Hypermethylation occurs only in certain CpG islands and sites, such as CpG1, CpG5, and CpG7, and it differs significantly in CSCC, CIN, or control groups [49]. In Uygur and Han, aberrant methylation of TFPI2 is present in a higher proportion of invasive cervical carcinoma (ICC) clinical samples [50]. Apart from that, hypermethylation is related to different age groups as well. Hypermethylation of the CDKN2A gene promoter is a frequent epigenetic change in younger patients with cervical carcinoma and implies a significant epigenetic role in tumor development in this age group [51].

3.3. The Relationship between HPV and Aberrant DNA Methylation in Cervical Cancer/CIN (Figure 1). On the one hand, HPV and aberrant host gene methylation contribute to CIN and CCA, respectively, methylation of HPV can prevent itself from cleaning to keep the persistent infection state, and the host methylation level can also reflect the level of HPV-associated CCA. On the other hand, the HPV genome and host act on each other by methylated regulation. HPV takes part in the methylation of host genomes such as FAM19A4 and LHX1; the methylation of HPV itself can also work with the methylation of PAX1 and SOX1 in the host to enhance transcription, both of which induce bad outcomes of the host cervix.

3.3.1. Methylation Status of HPV Genome in Cervical Cancer/CIN. HPV genome epigenetic alterations play an important role in cervical cancer progression. Among them, methylation of CpG sites in the L1, L2, and LCR regions in different types of HPV is studied most, and several deep relationships between the methylation of those regions and cervical cancer/CIN have been found out. HPV L1 gene methylation was the risk factor to cervical and elevated levels. HPV16 L1 methylation affects E6/E7 mRNA levels and can detect high-grade cervical lesions (CIN2$^+$) [52, 53]. It also prolongs the cleaning of HPV infection and increases the risk of HPV cleaning failure in premalignant cervical lesion patients [54]. Besides, a panel of 12 HPV16 CpG sites which are methylated in L1, L2, and E5 can work as an informative biomarker for the triage of women positive for HPV16 infection and is correlated with the severity of cervical neoplasia, even cervical cancer [55]. But some other evidence shows HPV16 L1/L2 DNA methylation weakly associated with cervical disease grade in young women, which means HPV DNA methylation as a biomarker must take into account women's age [56]. The L1 and L2 regions of other types of HPVs are methylated in cervical cancer/CIN. Aberrant methylation of CpG sites in the L1 and L2 regions of HPV18 and other high-risk HPV types including HPV31, HPV33, HPV45, HPV52, HPV51, and HPV58 relates with the progression from early-stage CINs and may be considered as a biomarker of the progression of cervical neoplasia [57, 58]. Another research shows that the methylation of L1 in HPV16, HPV18, and HPV52 does not only play an important role in cervical cancer alone. The methylation of most HPV types except HPV52 also works together with the methylation of host genes including PAX1 and SOX1, which leads to a more significant result of cervical cancer/CIN [59]. Combining HPV methylation with PAX1 methylation improves the clustering for CIN2$^+$ and methylated CpG sites in HPV31 LCR,

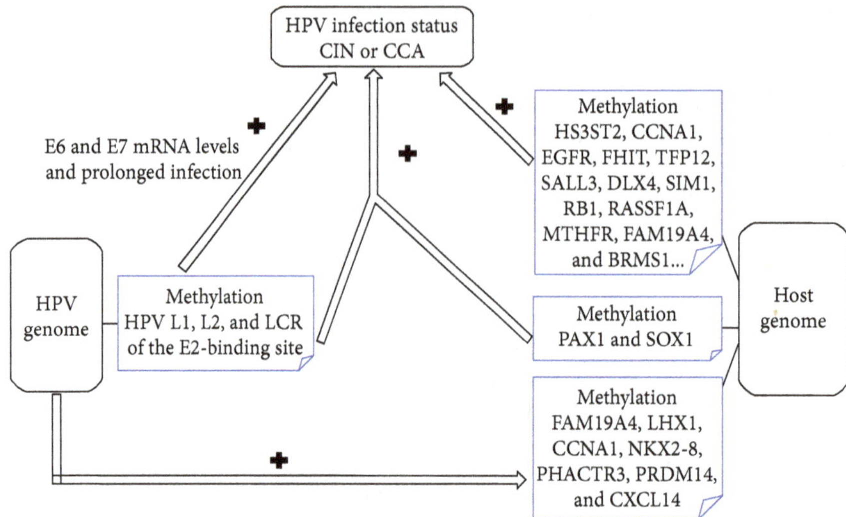

FIGURE 1: Methylation regulation between HPV and host genome in CIN or cervical cancer.

including position 7479 and/or 7485, which is the promoter distal E2-binding site, suggesting a potential regulatory mechanism for papillomavirus transcription [60].

3.3.2. The Interaction between HPV and Aberrant Methylation of Other Genes. In cervical cancer/CIN, methylated HPVs and other genes correlate with each other and serve as diagnostic and prognostic biomarkers for cervical cancer. Methylated HS3ST2, CCNA1, EGFR promoter, FHIT, TFPI2, CpG6, and CpG15 sites were associated with HPV16 infection in the progression of cervical cancer [61–63]. The results indicate that methylated genes may play important roles and be effective targets for the prevention and treatment of cervical cancer. HPV infection is also associated with hypermethylation of the promoter region of SALL3, DLX4, and SIM1 genes, which should be a significant progression marker for HPV infection in cervical cancer [64]. The methylation-mediated gene silencing of PRDM14, a regulator of NOXA and PUMA-mediated apoptosis, becomes an important factor in the development of hr-HPV-ICC (invasive cervical carcinoma) and offers a novel therapeutic target for HPV-induced cervical cancers [65]. In addition, that FAM19A4 promoter methylation even altered DNA methylation seems to be associated with HPV infection and high-risk types of HPV-induced carcinogenesis in the uterine cervix, CIN3$^+$, and may increase with disease progression [66]. Moreover, some of the methylated genes have been demonstrated as attractive markers for hr-HPV-positive women, with a high reassurance for the detection of cervical carcinoma and advanced CIN2/3 lesions, such as EPB41L3 and FAM19A4 [29, 67].

Not only can gene methylation affect HPV infection, but HPV also results in other genes' aberrant methylation. HPV can result in novel DNA methylation events, including FAM19A4, LHX1, NKX2–8, PHACTR3, and PRDM14 genes in cervical carcinogenesis [68]. Numerous pieces of evidence suggest that HPV16 E7 oncoprotein mediates DNA hypermethylation in the CCNA1 and CXCL14 promoter and suppresses gene expression. The data also shows that E7 induces CCNA1 methylation by forming a complex with Dnmt1 at the CCNA1 promoter [69, 70]. The potential carcinogenic mechanism of HPVs, including influencing the DNA methylation pathway to affect DNA methylation and mRNA expression levels of those genes, can be utilized not only as a biomarker for early detection, disease progression, diagnosis, and prognosis of cervical cancer but also to design effective therapeutic strategies.

3.3.3. Identification of Cervical Cancer by HPV and Gene Methylation Test. Currently, the HPV DNA test is one of the most vital tools to identify the risk of cervical cancer/CIN. Some studies show that detecting the methylation status of a few kinds of genes can also give evidence for diagnosing CIN2$^+$ or help the HPV test to improve the specificity and sensitivity in the detection of cervical cancer/CIN.

In an independent cohort test, the methylated PCDHA4 and PCDHA13 test is equally sensitive but more specific than the human papillomavirus (HPV) test in the diagnosis of CIN2$^+$ [71]. Combining the triage by MAL/miR-124-2 methylation analysis with threshold-80 and HPV16/18 genotyping can reach higher CIN3$^+$ sensitivity and identify women at the highest risk of cervical (pre)cancer [72, 73]. Combining parallel testing of PAX1, DAPK1, RARB, WIF1, and SLIT2 DNA methylation and HPV DNA increases specificity to identify cervical cancer and achieves better precision than single HPV DNA testing does [74–76]. Above all, methylation of some genes has a prospect to be an auxiliary biomarker for cervical cancer screening.

Now, cervical (pre)cancer is usually classified by histologic pathology, but cervical conization will lead to a high risk of premature delivery and abortion for patients. A quantitative measurement of HPV-type 16 L1/L2 DNA methylation has demonstrated its correlation with cervical disease grade. The best separation between normal and dyskaryotic samples is achieved by assessment of the L1/L2 CpGs at nucleotide positions 5600 and 5609 [77]. At the same time, CCNA1 promoter methylation serves as a potential marker for distinguishing between histologic LSIL (low-grade squamous intraepithelial lesion)/negative and HSIL (high-grade squamous intraepithelial lesion)/positive [78].

4. Methylation-Related Regulations on Other Levels in Cervical Cancer/CIN

According to the central dogma of molecular biology, epigenetic modifications also occur in the process of genetic information expression, such as the DNA level mentioned above, RNA level including mRNA and ncRNA (noncoding RNA, miR-RNA, and lnc-RNA are included), and protein level involving common protein or histone.

4.1. Pervasive Gene Expression Adjustment of Cervical Cancer/CIN at RNA Level

4.1.1. m6A Induces Methylated Regulation in mRNA. mRNAs carry genetic information by encoding polypeptides or proteins; that m6A methylates mRNA is widespread in eukaryotic cells. N^6-Methyladenosine (m6A), which is an abundant and conservative RNA modification, is involved in a series of biological processes such as differentiation, metabolism, immune tolerance, and neuronal signaling by impacting on mRNA splicing, export, localization, translation, and stability [79]. As the UV cross-linking immunoprecipitation and single-nucleotide resolution show, the distribution of m6A is not random in mature transcripts but concentrates around the $3'$ untranslated regions (UTRs), stop codons, and is within internal long exons [80]. The reversibility of m6A is accomplished by the orchestrated action of a battery of enzymes or proteins: as readers, proteins YTHDF and hnRNP recognize m6A-containing mRNA; as writers, METTL3, METTL14, and the WTAP complex support RNA methylation; and as erasers, FTO and ALKBH5 prop up RNA demethylation [79, 81].

As investigations about relationships between various tumors and m6A deepen, some crucial targets of tumor biological processes are found. Theories of m6A are elucidated increasingly in GSC, AML, HCC, BRC, and so on. Inhibition of FTO not only suppresses growth and self-renewal but also prolongs the lifespan of grafted mice and restrains tumor progression additionally compared with overexpression of METTL3 [82]. However, research of m6A about cervical cancer is rarely covered.

4.1.2. ncRNAs Play an Important Role in Methylation Regulation. As genomics analysis shows, there are numerous transcripts being generated in the human body; just 1–2% transcripts own the function of encoding polypeptides or proteins, and the remaining 98% noncoding products play vital roles in many biological process, including proliferation, differentiation, and apoptosis [83]. They are all hot topics in the field of apparent genetics.

With NGS and qRT-PCR applied, levels of various miR-RNAs in cervical cancer are evaluated: most miR-RNAs are downregulated and relevant downstream signal pathways or target genes and proteins are reported, such as SOX2 of miR-145, TCF of miR-212, Bcl-2 of miR-187, and NF-κb of miR-429, performing significant relationships with FIGO stage, lymph node metastasis and prognosis of patients in clinic, and colony formation, tumor size, proliferation, differentiation, apoptosis, and invasion on the lab research *in vivo* and *in vitro* [84–87]. There are still some miR-RNAs upregulated in CCA, such as miR-9 [88]. However, Zhang et al. reported that miR-9 is downregulated in cervical cancer on account of hypermethylation of miR-9 precursor promoters, which weakens the inhibiting effect on activity of the IL-6/Jak/STAT3 pathway [89]. These different outcomes may be induced by the potentially different methylation status in the objects.

Impacts of miR-RNA on the progress from HPV infection to cervical cancer are nonnegligible. Morel et al. reported that miR-375 could destabilize HPV16 early viral mRNA and contribute to the regulation of E6/E7 expression, which indicated the role of miR-RNA in high-risk HPV-associated carcinogenesis [90]. Yeung et al. revealed that HPV16 E6 takes part in epigenetic regulation of host gene-associated cervical cancer development; HPV16 E6 methylates the promoter region of the host gene of miR-23b, C9, or f3; and downregulated miR-23b enhances c-MET pathway-induced apoptosis of cervical cancer cells [91].

lnc-RNA interacting with miR-RNA regulates cervical cancer biological activity. lnc-RNA MEG3 is negatively relevant with FIGO stages, tumor size, lymphatic metastasis, and HR-HPV infection, and downexpressed MEG3 in cervical cancer reduces the inhibition effect on miR-21-5p expression, which leads to less apoptosis and more proliferation of cancer cells [92]. There are some cases about interactions between lnc-RNA, miR-RNA, and histone. For example, Zhang et al. explained the regulatory mechanism of lnc-RNA PVT1, which is overexpressed in cervical cancer: PVT1 binds with EZH2 directly to activate EZH2 to increase the histone H3K27 trimethylation level of the miR-200b promoter so that downexpressed miR-200b enhanced proliferation, cycle progression, and migration [93].

4.2. Methylation Research Related to Cervical Cancer Therapy Applications. Many mechanisms of methylation-associated regulations become the targets of therapy in the fields of chemotherapy and radiotherapy. It has been shown that cisplatin as well as 5-azacytidine touch off cytotoxic and growth inhibitory effects *in vitro* by demethylating the promoters of ESR1, BRCA1, RASSF1A, MLH1, MYOD1, hTERT, and DAPK1 to reexpress these tumor-associated genes [94]. Narayan et al. identified inactivation of decoy receptors TNFRSF10C and TNFRSF10D as major target genes at the 8p MDR region. On the one hand, the promoter hypermethylation of TNFRSF10C was an early event in cervical tumorigenesis; on the other hand, inactivation of decoy receptors induced extrinsic-apoptotic-pathway-dependent cell death in the cooperation of TRAIL and cisplatin in the presence of DNA-damaging drugs [95]. These covers above demonstrate that methylation-associated regulation offers an idea for developing new therapy targets.

Besides, methylation modifications impact on the sensitivity of chemotherapy and radiotherapy. Radiosensitization occurs when SiHa cells accept the therapeutic regimen combining DNA methylation inhibitor hydralazine with a histone deacetylase inhibitor valproic acid; unexpectedly, the efficacy of cisplatin chemoradiation was increased under the use of two epigenetic drugs [96]. Furthermore, epigenetic modifications also participate in therapeutic resistance. A univariate and hierarchical cluster analysis uncovered that standard chemoradiation resistance contacts closely with lower ESR1 transcript levels as well as unmethylated ESR1, unmethylated MYOD1, and methylated hTERT promoter [97]. In an article about the suppressor of cytokine signaling (SOCS) family and cervical cancer, Kim et al. found that DNA methylation contributed to SOCS1 downregulation, and histone deacetylation may be the mechanism of SOCS1 and SOCS3 regulation; in the meantime, ectopic expression of SOCS1 or SOCS3 could induce radioresistance of HeLa cells [98]. Similarly, a research about type-I ribosome-inactivating protein trichosanthin reported that Smac demethylation was subdued and Twist was upregulated in TCS-resistance cervical cells, which indicated that aberrant mitochondrial methylation may be partly the reason for drug resistance [99].

5. Conclusion

Cervical cancer is likely to be the first tumor which can receive idealized prevention and cure depending on the vital status of HPV in the pathological process. In spite of the astounding advances of screening plans and HPV vaccines, cervical cancer is still threatening the physical and psychological health of females with the absence of effective treatment, surveillance indexes, and fundamentally unclear molecular mechanisms. Over the past decades, methylation modification has been identified as having a significant role in the generation of cervical cancer. With the development of methylation-detecting techniques, there may be more convenient choices to explore it, not limited to cells and tissues, but techniques like liquid biopsy to advanced clinical transformation. We believe that it can not only enrich the markers for the early diagnosis and prognosis evaluation with other biomarkers to improve sensitivity and specificity in the clinic but also provide targets for exploiting new drugs as well as modifying the sensitivity in radiotherapy and chemotherapy for cervical cancer. However, there are still some items to be investigated deeply. Firstly, studies about the relationships between 5-hmC or 6mA and cervical cancer are rare, especially the aspect of HPV infection. Secondly, some researches find that methylation modification does not act itself but correlates with other epigenetic forms such as ncRNA regulation and histone decoration; therefore, the effective application of methylation relies on the simplification of key points. Additionally, from HPV infection via CIN to cervical cancer, relevant researches of the dynamic pathogenesis are inconsecutive. Moreover, it is recognized that methylation regulation is reversible, which is the unique advantage of therapy; only by enabling the reversibility controllable can we make full use of the characteristic. In conclusion, 5-hmC of hydroxymethylation, 5-mC of methylation, and 6-mA of RNA methylation are typical mechanisms of the methylation modification in gene expression; some ncRNA and histone regulations are involved in methylation in the meantime, and these investigations have profound instructive significance in the process of overcoming cervical cancer.

Disclosure

Chunyang Feng and Junxue Dong should be considered as co-first authors.

Conflicts of Interest

The authors declare that there is no conflict of interests regarding the publication of this paper.

Authors' Contributions

Chunyang Feng and Junxue Dong contributed to this work equally.

Acknowledgments

This work was supported by grants from the National Key R&D Program of China (2016YFC1302901), Jilin Province Science and Technology funds (20180201032YY, 20150204007YY, and 20140204022YY), Jilin Province Development and Reform Commission funds (2014G073 and 2016C046-2), and Education Department of Jilin Province funds (JJKH20170804KJ).

References

[1] L. A. Torre, F. Bray, R. L. Siegel, J. Ferlay, J. Lortet-Tieulent, and A. Jemal, "Global cancer statistics, 2012," *CA: a Cancer Journal for Clinicians*, vol. 65, no. 2, pp. 87–108, 2015.

[2] Y. B. Zhao, J. H. Wang, X. X. Chen, Y. Z. Wu, and Q. Wu, "Values of three different preoperative regimens in comprehensive treatment for young patients with stage Ib2 cervical cancer," *Asian Pacific Journal of Cancer Prevention*, vol. 13, no. 4, pp. 1487–1489, 2012.

[3] W. Yang, L. Hong, X. Xu, Q. Wang, J. Huang, and L. Jiang, "LncRNA GAS5 suppresses the tumorigenesis of cervical cancer by downregulating miR-196a and miR-205," *Tumor Biology*, vol. 39, no. 7, 2017.

[4] K. Williams, J. Christensen, and K. Helin, "DNA methylation: TET proteins—guardians of CpG islands?," *EMBO Reports*, vol. 13, no. 1, pp. 28–35, 2011.

[5] B. F. Yuan, "5-Methylcytosine and its derivatives," *Advances in Clinical Chemistry*, vol. 67, pp. 151–187, 2014.

[6] S. Ke, A. Pandya-Jones, Y. Saito et al., "m6A mRNA modifications are deposited in nascent pre-mRNA and are not required for splicing but do specify cytoplasmic turnover," *Genes & Development*, vol. 31, no. 10, pp. 990–1006, 2017.

[7] O. L. Kantidze and S. V. Razin, "5-Hydroxymethylcytosine in DNA repair: a new player or a red herring?," *Cell Cycle*, vol. 16, no. 16, pp. 1499–1501, 2017.

[8] W. Li and M. Liu, "Distribution of 5-hydroxymethylcytosine in different human tissues," *Journal of Nucleic Acids*, vol. 2011, Article ID 870726, 5 pages, 2011.

[9] S. R. M. Kinney and S. Pradhan, "Ten eleven translocation enzymes and 5-hydroxymethylation in mammalian

development and cancer," *Advances in Experimental Medicine and Biology*, vol. 754, pp. 57–79, 2013.

[10] J. An, A. Rao, and M. Ko, "TET family dioxygenases and DNA demethylation in stem cells and cancers," *Experimental & Molecular Medicine*, vol. 49, no. 4, p. e323, 2017.

[11] C. J. Mariani, A. Vasanthakumar, J. Madzo et al., "TET1-mediated hydroxymethylation facilitates hypoxic gene induction in neuroblastoma," *Cell Reports*, vol. 7, no. 5, pp. 1343–1352, 2014.

[12] G. Ficz and J. G. Gribben, "Loss of 5-hydroxymethylcytosine in cancer: cause or consequence?," *Genomics*, vol. 104, no. 5, pp. 352–357, 2014.

[13] L. I. Kroeze, B. A. van der Reijden, and J. H. Jansen, "5-Hydroxymethylcytosine: an epigenetic mark frequently deregulated in cancer," *Biochimica et Biophysica Acta (BBA) - Reviews on Cancer*, vol. 1855, no. 2, pp. 144–154, 2015.

[14] L. Y. Zhang, C. S. Han, P. L. Li, and X. C. Zhang, "5-Hydroxymethylcytosine expression is associated with poor survival in cervical squamous cell carcinoma," *Japanese Journal of Clinical Oncology*, vol. 46, no. 5, pp. 427–434, 2016.

[15] S. Bhat, S. P. Kabekkodu, V. K. Varghese et al., "Aberrant gene-specific DNA methylation signature analysis in cervical cancer," *Tumour Biology*, vol. 39, no. 3, pp. 1–16, 2017.

[16] M. Z. Wu, S. F. Chen, S. Nieh et al., "Hypoxia drives breast tumor malignancy through a TET–TNFα–p38–MAPK signaling axis," *Cancer Research*, vol. 75, no. 18, pp. 3912–3924, 2015.

[17] K. Takayama, A. Misawa, T. Suzuki et al., "Tet2 repression by androgen hormone regulates global hydroxymethylation status and prostate cancer progression," *Nature Communications*, vol. 6, no. 1, p. 8219, 2015.

[18] Z. Zhou, H.-S. Zhang, Y. Liu et al., "Loss of TET1 facilitates DLD1 colon cancer cell migration via H3K27me3-mediated down-regulation of E-cadherin," *Journal of Cellular Physiology*, vol. 233, no. 2, pp. 1359–1369, 2018.

[19] S. Uribe-Lewis, R. Stark, T. Carroll et al., "5-Hydroxymethylcytosine marks promoters in colon that resist DNA hypermethylation in cancer," *Genome Biology*, vol. 16, no. 1, p. 69, 2015.

[20] F. F. Yin, N. Wang, X. N. Bi et al., "Serine/threonine kinases 31(STK31) may be a novel cellular target gene for the HPV16 oncogene E7 with potential as a DNA hypomethylation biomarker in cervical cancer," *Virology Journal*, vol. 13, no. 1, p. 60, 2016.

[21] P. U. Thangavelu, T. Krenács, E. Dray, and P. H. G. Duijf, "In epithelial cancers, aberrant COL17A1 promoter methylation predicts its misexpression and increased invasion," *Clinical Epigenetics*, vol. 8, no. 1, p. 120, 2016.

[22] H. Zhou, Y. Wang, Q. Lv et al., "Overexpression of ribosomal RNA in the development of human cervical cancer is associated with rDNA promoter hypomethylation," *PLoS One*, vol. 11, no. 10, article e0163340, 2016.

[23] H. Lin, Y. Ma, Y. Wei, and H. Shang, "Genome-wide analysis of aberrant gene expression and methylation profiles reveals susceptibility genes and underlying mechanism of cervical cancer," *European Journal of Obstetrics, Gynecology, and Reproductive Biology*, vol. 207, pp. 147–152, 2016.

[24] M. K. Lee, E. M. Jeong, J. H. Kim, S. B. Rho, and E. J. Lee, "Aberrant methylation of the vim promoter in uterine cervical squamous cell carcinoma," *Oncology*, vol. 86, no. 5-6, pp. 359–368, 2014.

[25] Y. C. Chen, R. L. Huang, Y. K. Huang et al., "Methylomics analysis identifies epigenetically silenced genes and implies an activation of β-catenin signaling in cervical cancer," *International Journal of Cancer*, vol. 135, no. 1, pp. 117–127, 2014.

[26] N. A. Wijetunga, T. J. Belbin, R. D. Burk et al., "Novel epigenetic changes in CDKN2A are associated with progression of cervical intraepithelial neoplasia," *Gynecologic Oncology*, vol. 142, no. 3, pp. 566–573, 2016.

[27] D. Ma, C. Jiang, X. Hu et al., "Methylation patterns of the IFN-γ gene in cervical cancer tissues," *Scientific Reports*, vol. 4, no. 1, 2014.

[28] X. Wei, S. Zhang, D. Cao et al., "Aberrant hypermethylation of SALL3 with HPV involvement contributes to the carcinogenesis of cervical cancer," *PLoS One*, vol. 10, no. 12, article e0145700, 2015.

[29] N. Vasiljević, D. Scibior-Bentkowska, A. R. Brentnall, J. Cuzick, and A. T. Lorincz, "Credentialing of DNA methylation assays for human genes as diagnostic biomarkers of cervical intraepithelial neoplasia in high-risk HPV positive women," *Gynecologic Oncology*, vol. 132, no. 3, pp. 709–714, 2014.

[30] R. van Baars, J. van der Marel, P. J. F. Snijders et al., "CADM1 and MAL methylation status in cervical scrapes is representative of the most severe underlying lesion in women with multiple cervical biopsies," *International Journal of Cancer*, vol. 138, no. 2, pp. 463–471, 2016.

[31] H. J. Woo, S. J. Kim, K.-J. Song et al., "Hypermethylation of the tumor-suppressor cell adhesion molecule 1 in human papillomavirus-transformed cervical carcinoma cells," *International Journal of Oncology*, vol. 46, no. 6, pp. 2656–2662, 2015.

[32] J. Xu, L. Xu, B. Yang, L. Wang, X. Lin, and H. Tu, "Assessing methylation status of PAX1 in cervical scrapings, as a novel diagnostic and predictive biomarker, was closely related to screen cervical cancer," *International Journal of Clinical and Experimental Pathology*, vol. 8, no. 2, pp. 1674–1681, 2015.

[33] C. C. Chang, R. L. Huang, Y. P. Liao et al., "Concordance analysis of methylation biomarkers detection in self-collected and physician-collected samples in cervical neoplasm," *BMC Cancer*, vol. 15, no. 1, p. 418, 2015.

[34] Y. Sun, S. Li, K. Shen, S. Ye, D. Cao, and J. Yang, "DAPK1, MGMT and RARB promoter methylation as biomarkers for high-grade cervical lesions," *International Journal of Clinical and Experimental Pathology*, vol. 8, no. 11, pp. 14939–14945, 2015.

[35] J. Q. Ma, H. Tuersun, S. J. Jiao, J. H. Zheng, J. B. xiao, and A. Hasim, "Functional role of NRF2 in cervical carcinogenesis," *PLoS One*, vol. 10, no. 8, article e0133876, 2015.

[36] X. Zhang, Z. Zheng, S. Yingji et al., "Downregulation of glutathione peroxidase 3 is associated with lymph node metastasis and prognosis in cervical cancer," *Oncology Reports*, vol. 31, no. 6, pp. 2587–2592, 2014.

[37] M.-L. Buchholtz, J. Jückstock, E. Weber, I. Mylonas, D. Dian, and A. Brüning, "Loss of LDOC1 expression by promoter methylation in cervical cancer cells," *Cancer Investigation*, vol. 31, no. 9, pp. 571–577, 2013.

[38] Y. Lin, M. Cui, T. Xu, W. Yu, and L. Zhang, "Silencing of cyclooxygenase-2 inhibits the growth, invasion and migration of ovarian cancer cells," *Molecular Medicine Reports*, vol. 9, no. 6, pp. 2499–2504, 2014.

[39] X. Zhao, Y. Cui, Y. Li et al., "Significance of TSLC1 gene methylation and TSLC1 protein expression in the progression of cervical lesions," *Zhonghua Zhong Liu Za Zhi*, vol. 37, no. 5, pp. 356–360, 2015.

[40] S. P. Kabekkodu, S. Bhat, R. Radhakrishnan et al., "DNA promoter methylation-dependent transcription of the double C2-like domain β (DOC2B) gene regulates tumor growth in human cervical cancer," *The Journal of Biological Chemistry*, vol. 289, no. 15, pp. 10637–10649, 2014.

[41] J. Zhang, T. Yao, Z. Lin, and Y. Gao, "Aberrant methylation of MEG3 functions as a potential plasma-based biomarker for cervical cancer," *Scientific Reports*, vol. 7, no. 1, p. 6271, 2017.

[42] H. Wang, S. Y. Pan, Z. R. Pang et al., "Quantitative detection of APC/RASSF1A promoter methylation in the plasma of patients with cervical diseases," *Zhonghua Fu Chan Ke Za Zhi*, vol. 48, no. 12, pp. 929–934, 2013.

[43] A. K. Jha, V. Sharma, M. Nikbakht et al., "A comparative analysis of methylation status of tumor suppressor genes in paired biopsy and serum samples from cervical cancer patients among North Indian population," *Genetika*, vol. 52, no. 2, pp. 255–259, 2016.

[44] F. L. Wang, Y. Yang, Z. Y. Liu, Y. Qin, and T. Jin, "Correlation between methylation of the p16 promoter and cervical cancer incidence," *European Review for Medical and Pharmacological Sciences*, vol. 2017, no. 21, 6 pages, 2017.

[45] A. Gupta, M. K. Ahmad, A. A. Mahndi, R. Singh, and Y. Pradeep, "Promoter methylation and relative mRNA expression of the p16 gene in cervical cancer in North Indians," *Asian Pacific Journal of Cancer Prevention*, vol. 17, no. 8, pp. 4149–4154, 2016.

[46] R. Shu, J. He, C. Wu, and J. Gao, "The association between-RARβ and FHIT promoter methylation and the carcinogenesis of patients with cervical carcinoma: a meta-analysis," *Tumour Biology*, vol. 39, no. 6, 10 pages, 2017.

[47] A. Naseem, Z. I. Bhat, P. Kalaiarasan, B. Kumar, G. Gandhi, and M. M. A. Rizvi, "Genetic and epigenetic alterations affecting PARK-2 expression in cervical neoplasm among North Indian patients," *Tumour Biology*, vol. 39, no. 6, 2017.

[48] M. Niyazi, S. Sui, K. Zhu, L. Wang, Z. Jiao, and P. Lu, "Correlation between methylation of human papillomavirus-16 L1 gene and cervical carcinoma in Uyghur women," *Gynecologic and Obstetric Investigation*, vol. 82, no. 1, pp. 22–29, 2017.

[49] M. Abdula, G. Abudulajiang, R. Amiduo, A. Abudala, and A. Hasim, "Association of promoter methylation of ERp57 gene with the pathogenesis of cervical lesions in Uighur women," *Zhonghua Zhong Liu Za Zhi*, vol. 35, no. 8, pp. 600–603, 2013.

[50] Y. Dong, Q. Tan, L. Tao et al., "Hypermethylation of TFPI2 correlates with cervical cancer incidence in the Uygur and Han populations of Xinjiang, China," *International Journal of Clinical and Experimental Pathology*, vol. 8, no. 2, pp. 1844–1854, 2015.

[51] J. Y. Liau, S. L. Liao, C. H. Hsiao, M. C. Lin, H. C. Chang, and K. T. Kuo, "Hypermethylation of the CDKN2a gene promoter is a frequent epigenetic change in periocular sebaceous carcinoma and is associated with younger patient age," *Human Pathology*, vol. 45, no. 3, pp. 533–539, 2014.

[52] C. Qiu, Y. Zhi, Y. Shen, J. Gong, Y. Li, and X. Li, "High-resolution melting analysis of HPV-16L1 gene methylation: a promising method for prognosing cervical cancer," *Clinical Biochemistry*, vol. 48, no. 13-14, pp. 855–859, 2015.

[53] L. Mirabello, M. Schiffman, A. Ghosh et al., "Elevated methylation of HPV16 DNA is associated with the development of high grade cervical intraepithelial neoplasia," *International Journal of Cancer*, vol. 132, no. 6, pp. 1412–1422, 2013.

[54] F. Yang-Chun, C. Zhen-Zhen, H. Yan-Chun, and M. Xiu-Min, "Association between PD-L1 and HPV status and the prognostic value for HPV treatment in premalignant cervical lesion patients," *Medicine*, vol. 96, no. 25, article e7270, 2017.

[55] J. L. Brandsma, M. Harigopal, N. B. Kiviat et al., "Methylation of twelve CpGs in human papillomavirus type 16 (HPV16) as an informative biomarker for the triage of women positive for HPV16 infection," *Cancer Prevention Research*, vol. 7, no. 5, pp. 526–533, 2014.

[56] D. Bryant, S. Hibbitts, M. Almonte, A. Tristram, A. Fiander, and N. Powell, "Human papillomavirus type 16 L1/L2 DNA methylation shows weak association with cervical disease grade in young women," *Journal of Clinical Virology*, vol. 66, pp. 66–71, 2015.

[57] M. Kalantari, K. Osann, I. E. Calleja-Macias et al., "Methylation of human papillomavirus 16, 18, 31, and 45 L2 and L1 genes and the cellular DAPK gene: considerations for use as biomarkers of the progression of cervical neoplasia," *Virology*, vol. 448, pp. 314–321, 2014.

[58] V. Simanaviciene, V. Popendikyte, Z. Gudleviciene, and A. Zvirbliene, "Different DNA methylation pattern of HPV16, HPV18 and HPV51 genomes in asymptomatic HPV infection as compared to cervical neoplasia," *Virology*, vol. 484, pp. 227–233, 2015.

[59] Y. W. Hsu, R. L. Huang, P. H. Su et al., "Genotype-specific methylation of HPV in cervical intraepithelial neoplasia," *Journal of Gynecologic Oncology*, vol. 28, no. 4, p. e56, 2017.

[60] B. László, A. Ferenczi, L. Madar et al., "CpG methylation in human papillomavirus (HPV) type 31 long control region (LCR) in cervical infections associated with cytological abnormalities," *Virus Genes*, vol. 52, no. 4, pp. 552–555, 2016.

[61] Q. Zuo, W. Zheng, J. Zhang et al., "Methylation in the promoters of HS3ST2 and CCNA1 genes is associated with cervical cancer in Uygur women in Xinjiang," *The International Journal of Biological Markers*, vol. 29, no. 4, pp. e354–e362, 2014.

[62] W. Zhang, Y. Jiang, Q. Yu et al., "EGFR promoter methylation, EGFR mutation, and HPV infection in Chinese cervical squamous cell carcinoma," *Applied Immunohistochemistry & Molecular Morphology*, vol. 23, no. 9, pp. 661–666, 2015.

[63] L. X. Bai, J. T. Wang, L. Ding et al., "Folate deficiency and FHIT hypermethylation and HPV 16 infection promote cervical cancerization," *Asian Pacific Journal of Cancer Prevention*, vol. 15, no. 21, pp. 9313–9317, 2014.

[64] J. Sakane, K. Taniyama, K. Miyamoto et al., "Aberrant DNA methylation of DLX4 and SIM1 is a predictive marker for disease progression of uterine cervical low-grade squamous intraepithelial lesion," *Diagnostic Cytopathology*, vol. 43, no. 6, pp. 462–470, 2015.

[65] S. Snellenberg, S. A. G. M. Cillessen, W. Van Criekinge et al., "Methylation-mediated repression of PRDM14 contributes to apoptosis evasion in HPV-positive cancers," *Carcinogenesis*, vol. 35, no. 11, pp. 2611–2618, 2014.

[66] R. Luttmer, L. M. A. De Strooper, J. Berkhof et al., "Comparing the performance of FAM19A4 methylation analysis, cytology and HPV16/18 genotyping for the detection of cervical (pre)-cancer in high-risk HPV-positive women of a gynecologic

outpatient population (COMETH study)," *International Journal of Cancer*, vol. 138, no. 4, pp. 992–1002, 2016.

[67] L. M. A. De Strooper, C. J. L. M. Meijer, J. Berkhof et al., "Methylation analysis of the FAM19A4 gene in cervical scrapes is highly efficient in detecting cervical carcinomas and advanced CIN2/3 lesions," *Cancer Prevention Research*, vol. 7, no. 12, pp. 1251–1257, 2014.

[68] R. D. Steenbergen, M. Ongenaert, S. Snellenberg et al., "Methylation-specific digital karyotyping of HPV16E6E7-expressing human keratinocytes identifies novel methylation events in cervical carcinogenesis," *The Journal of Pathology*, vol. 231, no. 1, pp. 53–62, 2013.

[69] L. Cicchini, J. A. Westrich, T. Xu et al., "Suppression of antitumor immune responses by human papillomavirus through epigenetic downregulation of CXCL14," *MBio*, vol. 7, no. 3, pp. e00270–e00216, 2016.

[70] K. Chalertpet, W. Pakdeechaidan, V. Patel, A. Mutirangura, and P. Yanatatsaneejit, "Human papillomavirus type 16 E7 oncoprotein mediates CCNA1 promoter methylation," *Cancer Science*, vol. 106, no. 10, pp. 1333–1340, 2015.

[71] K. H. Wang, C. J. Lin, C. J. Liu et al., "Global methylation silencing of clustered proto-cadherin genes in cervical cancer: serving as diagnostic markers comparable to HPV," *Cancer Medicine*, vol. 4, no. 1, pp. 43–55, 2015.

[72] V. M. J. Verhoef, D. A. M. Heideman, F. J. van Kemenade et al., "Methylation marker analysis and hpv16/18 genotyping in high-risk HPV positive self-sampled specimens to identify women with high grade CIN or cervical cancer," *Gynecologic Oncology*, vol. 135, no. 1, pp. 58–63, 2014.

[73] V. M. J. Verhoef, R. P. Bosgraaf, F. J. van Kemenade et al., "Triage by methylation-marker testing versus cytology in women who test HPV-positive on self-collected cervicovaginal specimens (PROHTECT-3): a randomised controlled non-inferiority trial," *The Lancet Oncology*, vol. 15, no. 3, pp. 315–322, 2014.

[74] L.-Y. Kong, W. Du, L. Wang, Z. Yang, and H.-S. Zhang, "PAX1 methylation hallmarks promising accuracy for cervical cancer screening in Asians: results from a meta-analysis," *Clinical Laboratory*, vol. 61, no. 10, pp. 1471–1479, 2015.

[75] E. M. Siegel, B. M. Riggs, A. L. Delmas, A. Koch, A. Hakam, and K. D. Brown, "Quantitative DNA methylation analysis of candidate genes in cervical cancer," *PLoS One*, vol. 10, no. 3, article e0122495, 2015.

[76] Y. Chen, Z. Cui, Z. Xiao et al., "PAX1 and SOX1 methylation as an initial screening method for cervical cancer: a meta-analysis of individual studies in Asians," *Annals of Translational Medicine*, vol. 4, no. 19, pp. 365–365, 2016.

[77] D. Bryant, A. Tristram, T. Liloglou, S. Hibbitts, A. Fiander, and N. Powell, "Quantitative measurement of human papillomavirus type 16 L1/L2 DNA methylation correlates with cervical disease grade," *Journal of Clinical Virology*, vol. 59, no. 1, pp. 24–29, 2014.

[78] S. Chujan, N. Kitkumthorn, S. Siriangkul, and A. Mutirangura, "CCNA1 promoter methylation: a potential marker for grading Papanicolaou smear cervical squamous intraepithelial lesions," *Asian Pacific Journal of Cancer Prevention*, vol. 15, no. 18, pp. 7971–7975, 2014.

[79] A. Maity and B. Das, "N6-methyladenosine modification in mRNA: machinery, function and implications for health and diseases," *The FEBS Journal*, vol. 283, no. 9, pp. 1607–1630, 2016.

[80] S. Ke, E. A. Alemu, C. Mertens et al., "A majority of m6A residues are in the last exons, allowing the potential for 3′ UTR regulation," *Genes & Development*, vol. 29, no. 19, pp. 2037–2053, 2015.

[81] F. Li, S. Kennedy, T. Hajian et al., "A radioactivity-based assay for screening human m6A-RNA methyltransferase, METTL3-METTL14 complex, and demethylase ALKBH5," *Journal of Biomolecular Screening*, vol. 21, no. 3, pp. 290–297, 2016.

[82] Q. Cui, H. Shi, P. Ye et al., "m^6A RNA methylation regulates the self-renewal and tumorigenesis of glioblastoma stem cells," *Cell Reports*, vol. 18, no. 11, pp. 2622–2634, 2017.

[83] A. Granados López and J. López, "Multistep model of cervical cancer: participation of miRNAs and coding genes," *International Journal of Molecular Sciences*, vol. 15, no. 9, pp. 15700–15733, 2014.

[84] X. Zhou, Y. Yue, R. Wang, B. Gong, and Z. Duan, "MicroRNA-145 inhibits tumorigenesis and invasion of cervical cancer stem cells," *International Journal of Oncology*, vol. 50, no. 3, pp. 853–862, 2017.

[85] C. Zhou, D. M. Tan, L. Chen et al., "Effect of miR-212 targeting TCF7L2 on the proliferation and metastasis of cervical cancer," *European Review for Medical and Pharmacological Sciences*, vol. 21, no. 2, pp. 219–226, 2017.

[86] H. Li, Y. Sheng, Y. Zhang, N. Gao, X. Deng, and X. Sheng, "MicroRNA-138 is a potential biomarker and tumor suppressor in human cervical carcinoma by reversely correlated with TCF3 gene," *Gynecologic Oncology*, vol. 145, no. 3, pp. 569–576, 2017.

[87] W. Li, J. Liang, Z. Zhang et al., "MicroRNA-329-3p targets MAPK1 to suppress cell proliferation, migration and invasion in cervical cancer," *Oncology Reports*, vol. 37, no. 5, pp. 2743–2750, 2017.

[88] S. Azizmohammadi, A. Safari, S. Azizmohammadi et al., "Molecular identification of miR-145 and miR-9 expression level as prognostic biomarkers for early-stage cervical cancer detection," *QJM*, vol. 110, no. 1, pp. 11–15, 2017.

[89] J. Zhang, J. Jia, L. Zhao et al., "Down-regulation of microRNA-9 leads to activation of IL-6/Jak/STAT3 pathway through directly targeting IL-6 in HeLa cell," *Molecular Carcinogenesis*, vol. 55, no. 5, pp. 732–742, 2016.

[90] A. Morel, A. Baguet, J. Perrard et al., "5azadC treatment upregulates miR-375 level and represses HPV16 E6 expression," *Oncotarget*, vol. 8, no. 28, pp. 46163–46176, 2017.

[91] C. L. Yeung, T. Y. Tsang, P. L. Yau, and T. T. Kwok, "Human papillomavirus type 16 E6 suppresses microRNA-23b expression in human cervical cancer cells through DNA methylation of the host gene C9orf3," *Oncotarget*, vol. 8, no. 7, pp. 12158–12173, 2017.

[92] J. Zhang, T. Yao, Y. Wang, J. Yu, Y. Liu, and Z. Lin, "Long noncoding RNA MEG3 is downregulated in cervical cancer and affects cell proliferation and apoptosis by regulating miR-21," *Cancer Biology & Therapy*, vol. 17, no. 1, pp. 104–113, 2015.

[93] S. Zhang, G. Zhang, and J. Liu, "Long noncoding RNA PVT1 promotes cervical cancer progression through epigenetically silencing miR-200b," *APMIS*, vol. 124, no. 8, pp. 649–658, 2016.

[94] S. Sood and R. Srinivasan, "Alterations in gene promoter methylation and transcript expression induced by cisplatin in comparison to 5-azacytidine in HeLa and SiHa cervical cancer cell lines," *Molecular and Cellular Biochemistry*, vol. 404, no. 1-2, pp. 181–191, 2015.

[95] G. Narayan, D. Xie, G. Ishdorj et al., "Epigenetic inactivation of TRAIL decoy receptors at 8p12-21.3 commonly deleted region confers sensitivity to Apo2l/trail-cisplatin combination therapy in cervical cancer," *Genes, Chromosomes & Cancer*, vol. 55, no. 2, pp. 177–189, 2016.

[96] E. Mani, L. A. Medina, K. Isaac-Olivé, and A. Dueñas-González, "Radiosensitization of cervical cancer cells with epigenetic drugs hydralazine and valproate," *European Journal of Gynaecological Oncology*, vol. 35, no. 2, pp. 140–142, 2014.

[97] S. Sood, F. D. Patel, S. Ghosh, A. Arora, L. K. Dhaliwal, and R. Srinivasan, "Epigenetic alteration by DNA methylation of ESR1, MYOD1 and hTERT gene promoters is useful for prediction of response in patients of locally advanced invasive cervical carcinoma treated by chemoradiation," *Clinical Oncology*, vol. 27, no. 12, pp. 720–727, 2015.

[98] M. H. Kim, M. S. Kim, W. Kim et al., "Suppressor of cytokine signaling (SOCS) genes are silenced by DNA hypermethylation and histone deacetylation and regulate response to radiotherapy in cervical cancer cells," *PLoS One*, vol. 10, no. 4, article e0123133, 2015.

[99] L. Cui, J. Song, L. Wu et al., "Smac is another pathway in the anti-tumour activity of trichosanthin and reverses trichosanthin resistance in CaSki cervical cancer cells," *Biomedicine & Pharmacotherapy*, vol. 69, pp. 119–124, 2015.

3

Molecular Crosstalking among Noncoding RNAs: A New Network Layer of Genome Regulation in Cancer

Marco Ragusa,[1,2] Cristina Barbagallo,[1] Duilia Brex,[1] Angela Caponnetto,[1] Matilde Cirnigliaro,[1] Rosalia Battaglia,[1] Davide Barbagallo,[1] Cinzia Di Pietro,[1] and Michele Purrello[1]

[1]*BioMolecular, Genome and Complex Systems BioMedicine Unit (BMGS Unit), Section of Biology and Genetics G Sichel, Department of BioMedical Sciences and Biotechnology, University of Catania, Catania, Italy*
[2]*IRCCS Associazione Oasi Maria S.S., Institute for Research on Mental Retardation and Brain Aging, Troina, Enna, Italy*

Correspondence should be addressed to Michele Purrello; purrello@unict.it

Academic Editor: Brian Wigdahl

Over the past few years, noncoding RNAs (ncRNAs) have been extensively studied because of the significant biological roles that they play in regulation of cellular mechanisms. ncRNAs are associated to higher eukaryotes complexity; accordingly, their dysfunction results in pathological phenotypes, including cancer. To date, most research efforts have been mainly focused on how ncRNAs could modulate the expression of protein-coding genes in pathological phenotypes. However, recent evidence has shown the existence of an unexpected interplay among ncRNAs that strongly influences cancer development and progression. ncRNAs can interact with and regulate each other through various molecular mechanisms generating a complex network including different species of RNAs (e.g., mRNAs, miRNAs, lncRNAs, and circRNAs). Such a hidden network of RNA-RNA competitive interactions pervades and modulates the physiological functioning of canonical protein-coding pathways involved in proliferation, differentiation, and metastasis in cancer. Moreover, the pivotal role of ncRNAs as keystones of network structural integrity makes them very attractive and promising targets for innovative RNA-based therapeutics. In this review we will discuss: (1) the current knowledge on complex crosstalk among ncRNAs, with a special focus on cancer; and (2) the main issues and criticisms concerning ncRNAs targeting in therapeutics.

1. Introduction

When the Human Genome Project (HGP) began in the late 1990s, researchers hypothesized that our genome comprised about 100,000 protein-coding genes [1]. Over the years, this estimate has been continuously downsized. In 2001, the International Human Genome Sequencing Consortium (IHGSC) published the initial sequence of the human genome and proposed that the number of protein-coding genes was about 30,000 [2]. At the same time, Celera Genomics (a competitor group of IHGSC) estimated this number at 26,000 [3]. In 2004, when the final draft of the human genome was published, this number was further reduced to 24,500 [4], but in 2007 an additional analysis established that it was around 20,500 [5]. More recently, new studies updated the number of human protein-coding genes to 19,000 [6]. This estimate is particularly surprising, because it would suggest that less than 2% of the whole human genome encodes for proteins; accordingly, the keystone of *Homo sapiens* complexity could lie in the 98% of our DNA (the genome *dark matter*), which does not encode proteins but would be endowed with critical regulatory functions. In the last decade, two important scientific initiatives supported by the US National Institutes of Health (i.e., the projects ENCODE and Roadmap Epigenomics) reported seminal data on hundreds of thousands of functional regions in the human genome, whose function is to supervise gene expression [7, 8]. These data suggested that much more space in our genome is committed to regulatory than to structural functions. Moreover, these studies proposed that about 80% of the human genome is dynamically and pervasively transcribed, mostly as non-protein-coding RNAs (ncRNAs). The biological relevance of the noncoding transcriptome has become increasingly undeniable over the last few years. Studies of comparative genomics showed that the relative proportion of genome space, occupied by the proteome-encoding genome as opposed to the regulatory (non-protein-encoding) genome is very variable among evolutionarily distant species; for instance, the protein-coding genome represents almost the entire genome of the unicellular yeast *Saccharomyces cerevisiae*, whereas it constitutes only 2% of mammalian genomes [9]. Moreover and intriguingly, the noncoding transcriptome is frequently altered in major diseases, including cancer [10–12]. These observations strongly suggest that ncRNAs are closely related to the complexity of higher eukaryotes and that their dysfunction may result in pathological phenotypes. RNA is a structurally versatile molecule, able to perform several molecular functions. By simple base pairing with other nucleic acids, RNA can recognize and bind both DNA and RNA targets in a very specific manner and regulate their transcription, processing, editing, translation, or degradation. An intriguing field for future explorations is the tridimensional folding of RNA molecules, which confers them allosteric properties: this increases the range of potential molecular interactors (including proteins); additionally, dynamic conformational changes can be triggered by ligand binding. Moreover and different from proteins, RNA can be rapidly transcribed and degraded making it a very dynamic molecule that can be quite rapidly synthesized without additional time and energetic costs of translation [13]. For all these reasons, over the past few years, ncRNAs have been extensively studied because of the significant biological roles that they play in regulation of cellular mechanisms. Noncoding RNA genes can generally be divided into two major categories by their transcript sizes: (1) long noncoding RNAs (lncRNAs) are longer than 200 nucleotides; and (2) small noncoding RNAs have a length equal to or lower than 200 nucleotides [i.e., microRNAs (miRNAs), small interfering RNAs (siRNAs), small nuclear RNAs (snRNAs or U-RNAs), small nucleolar RNAs (snoRNAs), PIWI-interacting RNAs (piRNAs), and tRNAs] [14]. To date, most research efforts have been focused on how ncRNAs (in particular, miRNAs) modulate the expression of protein-coding genes and their roles in human pathophysiology. However, recent evidence has shown the existence of unexpected interplay among ncRNAs, which influences cell physiology and diseases. In addition to the canonical multilayered control of expression of protein-coding genes (briefly described below), ncRNAs can interact with and regulate each other through various molecular mechanisms generating a complex network including different species of RNAs. In such a regulatory network, ncRNAs also compete among each other for binding to mRNAs, thus acting as competing endogenous RNAs (ceRNAs). In this review, we will summarize the current knowledge on the complex crosstalk among ncRNAs (including miRNAs, lncRNAs, and circRNAs) and how they could reciprocally interact to regulate cancer progression and dissemination.

1.1. miRNAs. miRNAs are 18–25 nucleotides long, evolutionarily conserved, single-stranded RNAs, which negatively modulate the expression of their target mRNAs (more than 60% of protein-coding genes) by binding to the 3′-UTR of specific mRNA targets, leading either to their translational repression, cleavage, or decay [15–17]. This binding occurs through a specific miRNA region (named *seed region*), which is a contiguous string of at least 6 nucleotides beginning at position two of the 5′ of the molecule [18]. The block of translation is due to the inhibition of mRNA 5′-cap recognition and interference on the interaction between the mRNA and the 60S ribosomal subunit, while mRNA degradation is promoted by mechanisms of decapping and deadenylation [19]. These molecular mechanisms are mediated by an RNA-induced silencing complex (RISC) that includes proteins belonging to the Argonaute (AGO) family; specifically, RISC endonuclease activity depends exclusively on AGO2 protein [20]. A single miRNA can control the expression of several mRNAs, and a single mRNA may be targeted by more than one miRNA, thus creating a complex interplay of cooperative regulation [21]. To date, more than 2500 mature miRNAs have been included in the *miRbase* database [22].

Extensive studies have shown that miRNAs control pivotal cellular processes, (e.g., cell proliferation, differentiation, migration, cell death, and angiogenesis), thus contributing to the pathogenesis of diseases such as cancer. Indeed, several miRNAs have been identified as potential oncogenes or tumor suppressors in cancer development and progression [23]. In the last two decades, their mutations and altered expression were reported to be causally related to the neoplastic features of the cells, thus providing new perspectives for the understanding of the complex regulatory networks that rule tumor biology [24]. miRNA dysfunctions exert a pleiotropic effect on the expression of their mRNA targets impairing the functioning of biological networks. It has been convincingly demonstrated that different cancer histotypes display specific miRNA expression patterns: this phenomenon would be helpful to improve diagnosis of poorly differentiated tumors and predict prognosis in cancer [25, 26]. Moreover, multiple experimental evidence has shown that miRNAs can be also secreted by cancer cells into bodily fluids, sending oncogenic signals through circulation, which could advantageously mold the extracellular tumor environment [27]. These discoveries gave a new intriguing diagnostic and prognostic role to circulating miRNAs, paving the way for their potential use as noninvasive molecular RNA markers in clinical management of cancer patients [28–30].

1.2. lncRNAs. lncRNAs are the most heterogeneous class of non-protein-coding RNAs with lengths ranging from 200 nt to 100,000 nt. They include transcripts that may be classified as (a) intergenic lncRNAs, (b) intronic lncRNAs, (c) sense or antisense transcripts, (d) pseudogenes, and (e) retrotransposons [14]. Currently, LNCipedia 4.0 records more than 118,000 human lncRNAs, which are usually expressed in a developmental and tissue-specific manner [31]. lncRNAs regulate gene expression at different levels, including chromatin modification, alternative splicing, and protein localization and activity [32]. Such a wide range of mechanisms is due to their ability to bind to DNA, RNAs, and proteins. lncRNAs, thanks to their binding to promoter DNA, can prevent the access of transcription factors to their own promoter binding sites and impede the transcription of specific genes (e.g., DHFR) [33]. Some lncRNAs (e.g., HOTAIR) are associated with chromatin-modifying complexes (e.g., polycomb repressive complex 2) to regulate epigenetic silencing of target genes [34]. Much evidence has also shown that lncRNAs may work as molecular scaffolds to connect two or more proteins in functional complexes or can serve to localize protein complexes to appropriate cellular compartments [35]. Antisense lncRNAs can target, by direct sequence complementarity, their antisense mRNAs and, accordingly, modulate alternative splicing processes or protect 3′-UTR from miRNA binding, increasing the stability of mRNAs (e.g., ZEB2-AS1, BACE1-AS) [36, 37]. Several recent studies have shown that lncRNAs are critically involved in a wide range of biological processes, such as cell cycle regulation, pluripotency, differentiation, and cell death [38–41]. Dysregulation of lncRNA activity has been frequently reported in association to diseases, including several types of cancer. Specifically, upregulated lncRNAs in cancer seem to possess tumor-promoting abilities, whilst downregulated lncRNAs exhibit tumor-suppressive roles [42–47]. Although several lncRNAs have been reported to be dysregulated in neoplastic phenotypes, their mechanistic role in cancer biology has not been satisfactorily explained for most of them. However, scientific evidence strongly suggests a promising role for lncRNAs as cancer-related biomarkers and potential targets for innovative therapeutic approaches.

1.3. circRNAs. Circular RNAs (circRNAs) represent a recently discovered class of noncoding RNAs, composed of single-stranded, covalently closed, exonuclease-resistant circular transcripts [48]. Although the existence of circular RNAs has been known since the 70s [49], for a long time such molecules were considered only by-products of pre-mRNA processing and therefore interpreted as artifacts of aberrant RNA splicing [50]. However, recent advances in RNA sequencing technologies have revealed a ubiquitous, and in some cases abundant, expression of endogenous circRNAs in mammalian genomes [51, 52]. circRNAs are a circularized isoform of linear protein-coding genes generated through backsplicing, a molecular process that is different from the canonical splicing of linear RNAs. Circular RNA biogenesis can occur both from exons (exonic circRNAs or ecircRNAs), through different mechanisms of backsplicing and introns (intronic circRNAs or ciRNAs), when lariat introns escape typical debranching processes [53]. Currently, about 35,000 circRNAs are reported in the *circBase* database [54], but molecular functions and biological processes, in which they are involved, remain elusive for most of them. Recent emerging evidence convincingly suggests that circRNAs may play an important role in RNA-RNA interactions. In some instances, circRNAs exhibit multiple binding sites for the same miRNA and represent a potential *molecular sponge* for sequestering the most abundant miRNAs [55]. In other words, circRNAs may negatively regulate the function of miRNAs, and, thus, protect miRNA targets, by acting as competing endogenous RNAs. As some papers would suggest that ceRNA role of circRNAs could not be their main function in cell biology, other molecular functions have been proposed for circRNAs (a) to bind and sequester RNA binding proteins (RBPs) [56–58] and (b) to be translated into proteins when recognized by ribosomes in the presence of internal ribosome entry sites (IRESs) [59, 60]. As circRNAs are potentially able to control different layers of gene expression, it is not surprising that their dysregulation is associated with human pathologies, including cancer [61–63]. Most reports that connect circRNAs and tumors mainly concern comparative gene expression profiling studies between tumor and normal samples. These investigations have shown that circRNAs are frequently downregulated in several types of cancer (e.g., colorectal cancer, ovarian cancer, and gastric cancer) [64–66]. Just few of these studies attempted to functionally explain how abnormal expression of circRNAs could impair physiological cell homeostasis and thus promote cancer phenotypes [67–69].

2. Noncoding RNAs: Different Ways to Interplay among Each Other

Interplay between ncRNAs obviously occurs because of sequence complementarity; for instance, ncRNAs may share miRNA response elements (MREs) with mRNAs and thus be targeted in the same manner [70]. The effects of miRNAs binding to other ncRNAs (i.e., lncRNAs and circRNAs) could be twofold: on the one hand, miRNAs could be sequestered and prevented from acting on the protein-coding mRNAs; on the other hand, miRNA binding to lncRNAs and circRNAs could promote their decay, similarly to mRNAs. In the next paragraphs, we will discuss the different mechanisms of ncRNA interaction and their influence on cancer biology.

2.1. miRNAs Induce Degradation of lncRNAs. Several papers have reported that miRNAs can bind lncRNAs and promote their degradation contributing to cancer processes (Table 1). lncRNAs are structurally similar to mRNAs; indeed, they have 5′-caps and 3′-poly(A) tails [71]; accordingly, the proteins involved in the regulation of decapping, deadenylation, and degradation of mRNAs may also control the turnover of lncRNAs by binding of specific miRNAs.

UCA1 (urothelial cancer associated 1), an lncRNA upregulated in several tumors (i.e., bladder cancer, tongue squamous cell carcinoma, breast cancer, and ovarian cancer) [72–75], possesses two predicted binding sites for miR-1, a well-known tumor suppressor miRNA. The binding of miR-1 to UCA1 has been confirmed by luciferase reporter assay in bladder cancer and, accordingly, *in vitro* upregulation of miR-1 induced UCA1 downregulation and caused a decreased cell growth and migration and also an augmented apoptosis. Such functional effects were reverted after UCA1 overexpression and silencing of AGO2, suggesting that miR-1 was able to downregulate UCA1 expression in an AGO2-mediated manner [76].

MALAT1 (metastasis-associated lung adenocarcinoma transcript 1) is one of the most studied and abundant lncRNAs: its expression was initially associated with metastasis in non-small-cell lung carcinoma (NSCLC) [77], but then

TABLE 1: miRNAs inducing degradation of lncRNAs.

miRNA	lncRNA/circRNA target	Tumor	miRNA role	PMID
let-7b	lincRNA-p21	Cervical carcinoma	Tumor suppressor	22841487
let-7b, let-7i	HOTAIR	Cervical carcinoma	Tumor suppressor	24326307
miR-1	UCA1	Bladder cancer	Tumor suppressor	25015192
miR-9	MALAT1	Hodgkin lymphoma, glioblastoma	Tumor suppressor	23985560
miR-21	CASC2	Renal cell carcinoma	Oncogene	27222255
miR-21	CASC2	Glioblastoma	Oncogene	25446261
miR-21	GAS5	Breast cancer	Oncogene	23933812
miR-34a	HOTAIR	Prostate cancer	Tumor suppressor	23936419
miR-101	MALAT1	Esophageal squamous cell carcinoma	Tumor suppressor	25538231
miR-125b	HOTTIP	Hepatocellular carcinoma	Tumor suppressor	25424744
miR-125b	MALAT1	Bladder cancer	Tumor suppressor	24396870
miR-141	H19	Gastric cancer	Tumor suppressor	26160158
miR-141	HOTAIR	Renal carcinoma	Tumor suppressor	24616104
miR-217	MALAT1	Esophageal squamous cell carcinoma	Tumor suppressor	25538231
miR-671	CDR1AS	Glioblastoma	Oncogene	26683098

This table reports for each miRNA: (1) its lncRNAs/circRNA target; (2) tumor where such interaction was reported; (3) its function in cancer (oncogene or tumor suppressor); and (4) bibliographic reference reported as Pubmed ID (PMID).

its deregulation has been reported in several other neoplastic diseases [78–80]. The 3′ end of MALAT1 is cleaved by RNase P and RNase Z, producing a tRNA-like ncRNA, called mascRNA (MALAT1-associated small cytoplasmic RNA), which will be exported into the cytoplasm [81], while most of the MALAT1 molecules are localized to nuclear speckles where they regulate alternative splicing of specific pre-mRNAs [82]. Moreover, MALAT1 may bind CBX4 (chromobox 4), a component of polycomb repressive complex 1 (PRC1), and modulate its localization in interchromatin granules, leading to activation or inhibition of gene expression [83]. Through these molecular mechanisms, MALAT1 controls the expression of several genes related to cell cycle and metastatic processes, thus influencing cell proliferation, migration, and invasion. Recent publications reported that MALAT1 is a target of a number of tumor suppressor miRNAs, which could induce its degradation and suppress its oncogenic effects. Leucci et al. reported miRNA-mediated regulation of MALAT1 in the nucleus of Hodgkin lymphoma and glioblastoma cell lines through direct binding of miR-9 to two different MREs in an AGO2-dependent manner [84]. There is evidence of a posttranscriptional regulation of MALAT1 by miR-101 and miR-217 in esophageal squamous cell carcinoma (ESCC) cells [85]. MiR-101 and miR-217 are functionally involved in several cancers as tumor suppressors and exhibited a significant negative correlation with MALAT1 in ESCC tissue samples and adjacent normal tissues. Enforced expression of miR-101 and miR-217 significantly repressed MALAT1 expression, leading to inhibition of cell growth, invasion, and metastasis in ESCC cells [85]. In bladder cancer, MALAT1 is inversely expressed with miR-125b. This miRNA was partially complementary with MALAT1 and bound it in *in vitro* models. MiR-125b was downregulated in bladder cancer, and its overexpression decreased the expression of MALAT1, causing an inhibition of bladder cancer cell proliferation, motility, and activation of apoptosis [86].

Additionally, miR-125b was also identified as a posttranscriptional regulator of HOTTIP (HOXA distal transcript antisense RNA) in hepatocellular carcinoma (HCC) [87]. HOTTIP is one of the most upregulated lncRNAs in HCC, also in early stages of HCC onset, and maps in antisense position to the distal end of the HOXA gene cluster. HOTTIP promotes tumor growth and metastasis *in vitro* and *in vivo* through regulation of the expression of its neighboring HOXA genes (e.g., HOXA10, HOXA11, and HOXA13). MiR-125b has been reported to be frequently downregulated in HCC, and a negative correlation of expression between miR-125b and HOTTIP existed in such cancer. The interaction between miR-125b and HOTTIP was validated by luciferase reporter assay; this was confirmed by ectopic expression of miR-125b that induced downmodulation of HOTTIP [87].

HOTAIR (HOX antisense intergenic RNA) is one of the most intensively studied lncRNAs, as it is frequently associated with different neoplasias. HOTAIR exerts its oncogenic functions by working as a scaffold to assemble polycomb repressive complex 2 (PRC2) on the HOXD gene cluster and inducing the transcriptional silencing of multiple metastasis suppressor genes (e.g., the protocadherin gene family) [34, 88]. HOTAIR is posttranscriptionally destabilized by several tumor suppressor miRNAs in different cancers. Chiyomaru et al. reported a functional binding between miR-34a and HOTAIR in prostate cancer cell lines treated with genistein, an isoflavone with antitumor activity: miR-34a directly bound to two MREs within HOTAIR RNA and lowered its levels [89]. Yoon et al. reported that human antigen R (HuR), let-7b, let-7i, and AGO2 cooperatively bind HOTAIR and promote HOTAIR decay, thus inhibiting the processes of ubiquitination and proteolysis of Ataxin-1 and

Snurportin-1, promoted by HOTAIR [90]. Interestingly, HuR and let-7b/AGO2 complex also decreased the stability of lincRNA-p21, an oncogenic lncRNA that reduced translation of beta-catenin and JUNB (JunB proto-oncogene, subunit of transcription factor AP-1) mRNAs in human cervical carcinoma HeLa cells [91]; even if in other experiments HuR was not able to transfer let-7b to AGO2 [92]. In another paper by Chiyomaru et al., it was reported that HOTAIR expression is negatively correlated to that of miR-141 in renal carcinoma cells (RCC) [93]. MiR-141 belongs to the miRNA-200 family, which has been reported to inhibit epithelial-mesenchymal transition (EMT) by ZEB1 (zinc finger E-box-binding homeobox 1) repression and E-cadherin upregulation [94]. MiR-141 was able to target and cleave HOTAIR in an AGO2-dependent manner, and such molecular action downregulated the expression of ZEB2 (zinc finger E-box-binding homeobox 2) induced by HOTAIR [93].

Expression of miR-141 was also found to be negatively correlated to that of lncRNA H19 (H19, imprinted maternally expressed transcript) in gastric cancer [95]. H19, an oncofetal lncRNA, is highly expressed during embryogenesis [96] and is upregulated in several cancers, including gastric cancer [97]. H19 acts as the primary miRNA precursor of miR-675, which in turn targets and represses RB1 (RB transcriptional corepressor 1) mRNA [98]. Overexpression of H19 enhances tumor cell growth and induces EMT; additionally, H19 modulates miRNA processing through its interaction with proteins involved in this molecular process (i.e., Drosha, Dicer). MiR-141 was shown to bind H19 in gastric cancer, and suppress H19 expression and its tumor-promoting functions [95].

MiR-21 is the most commonly upregulated miRNA in cancer: its genetic locus is often amplified in solid tumors, and its expression is promoted by a variety of cancer-related *stimuli* [99]. MiR-21 enhances cell proliferation, migration, and invasion by targeting several tumor suppressor genes, such as CCL20, CDC25A, PDCD4, and PTEN [100–103]. Recent findings showed that some lncRNAs could be added to the *repertoire* of miR-21 targets. Zhang et al. reported that expression of miR-21 and lncRNA GAS5 (growth arrest-specific 5) is negatively correlated in breast cancer and that miR-21 binds a miR-21-binding site in exon 4 of GAS5, thus inducing AGO2-mediated suppression of GAS5 [104]. GAS5 is an lncRNA with tumor-suppressive properties: its overexpression sensitizes cancer cells to UV or doxorubicin and decreases tumor proliferation and cell invasion. Interestingly, GAS5 also negatively regulated miR-21 at the posttranscriptional level through the RISC complex, suggesting the existence of a reciprocal negative feedback loop between GAS5 and miR-21 [104]. In two different studies on renal cell carcinoma and glioblastoma, it has been shown that miR-21 targeted and suppressed the expression of the tumor suppressor lncRNA CASC2 (cancer susceptibility candidate 2) in an AGO2-dependent manner [105, 106]. Indeed, the overexpression of miR-21 abrogated the inhibition of proliferation, migration, and the induction of apoptosis promoted by CASC2. Notably, when CASC2 was upregulated, miR-21 expression decreased: this suggests reciprocal repression between miR-21 and CASC2 [106].

The first experimental evidence that lncRNAs may be targeted by miRNAs was reported for the antisense transcript of the cerebellar degeneration-related protein 1 (CDR1, also known as CiRS-7 or CDR1AS), which is a circular RNA produced by a backsplice event [107]. MiR-671, a nuclear-enriched miRNA, induced cleavage of CDR1AS in an AGO2-dependent manner. Repression of miR-671 promoted the upregulation of both CDR1AS and CDR1, suggesting that CDR1AS was able to stabilize the sense transcript CDR1. Currently, this represents the only report on circRNA targeted and degraded by a miRNA. The interaction between miR-671 and CDR1AS could affect the biopathological molecular asset of glioblastoma multiforme (GBM), the most prevalent and aggressive cancer originating in the central nervous system, mainly in the brain. Indeed, Barbagallo et al. demonstrated that miR-671-5p is significantly upregulated in GBM. Enforced expression of miR-671-5p increased migration and decreased proliferation rates of GBM cell lines, suggesting its potential role as a novel oncomiRNA in GBM [108]. Expression of miR-671 was inversely correlated to that of CDR1AS and CDR1 in GBM biopsies and the expression of CDR1AS and CDR1 decreased when the miR-671 mimic was used, suggesting that the interaction of these molecules could be functionally altered in a GBM model [108].

2.2. lncRNAs as Decoys of miRNAs. The most explored mechanism of functional interactions between lncRNAs and miRNAs is based on sharing the same miRNA target sequence in both lncRNAs and mRNAs. In this way, lncRNAs are able to sequester miRNAs away from mRNAs, functioning as "miRNA sponges" or "miRNA decoys." Through such a competitive endogenous mechanism of interaction, lncRNAs decrease the quantity of available miRNAs and increase, accordingly, translations of their mRNA targets. lncRNAs, working as *competitive endogenous* RNAs, have been extensively described in molecular circuits involved in tumors (Table 2).

EWSAT1 (Ewing sarcoma-associated transcript 1) is an lncRNA with oncogenic functions in Ewing's sarcoma and nasopharyngeal carcinoma (NPC). EWSAT1 has two MREs for the miR-326/330-5p cluster and promoted the development and progression of tumors functioning as a ceRNA for these miRNAs, which in turn induced the expression of Cyclin D1, target of miRNAs from the miR-326/330-5p cluster [109].

Xia et al. showed that both lncRNA FER1L4 (FER-1-like family member 4, pseudogene) and PTEN (phosphatase and tensin homolog) mRNA had binding sites for oncomiR miR-106a-5p and were downregulated in gastric cancer [110]. As FER1L4 behaved as a ceRNA for miR-106a-5p, FER1L4 downregulation released miR-106a-5p that targeted PTEN mRNA, reducing its expression. Dysregulation of FER1L4-miR-106a-5p-PTEN axis increased cell proliferation by promoting the G0/G1 to S phase transition [110].

FTH1P3 (ferritin heavy chain 1 pseudogene 3) has been shown to function as a molecular sponge for miR-224-5p in oral squamous cell carcinoma (OSCC) [111]. Overexpression of FTH1P3 promoted proliferation and colony formation in

Table 2: lncRNAs acting as decoy of miRNAs.

lncRNA	miRNA target	Tumor	lncRNA role	PMID
CCAT1	let-7	Hepatocellular carcinoma	Oncogene	25884472
EWSAT1	miR-326/−330-5p cluster	Nasopharyngeal carcinoma	Oncogene	27816050
FER1L4	miR-106a-5p	Gastric cancer	Tumor suppressor	26306906
FTH1P3	miR-224-5p	Squamous cell carcinoma	Oncogene	28093311
FTX	miR-374a	Hepatocellular carcinoma	Tumor suppressor	27065331
GAS5	miR-135b	Non-small cell lung cancer	Tumor suppressor	28117028
H19	let-7a, let-7b	Breast cancer	Oncogene	28102845
HOST2	let-7b	Epithelial ovarian cancer	Oncogene	25292198
HOTAIR	miR-1	Hepatocellular carcinoma	Oncogene	27895772
HOTAIR	miR-152	Gastric cancer	Oncogene	26187665
HULC	miR-372	Liver cancer	Oncogene	20423907
lincRNA-RoR	miR-145	Breast cancer	Oncogene	25253741
lincRNA-RoR	miR-145	Endometrial cancer	Oncogene	24589415
LOC100129148	miR-539-5p	Nasopharyngeal carcinoma	Oncogene	28328537
MALAT1	miR-1	Breast cancer	Oncogene	26676637
MALAT1	miR-145	Cervical cancer	Oncogene	26311052
NEAT1	miR-449-5p	Glioma	Oncogene	26242266
PVT1	miR-152	Gastric cancer	Oncogene	28258379
PVT1	miR-186	Gastric cancer	Oncogene	28122299
RMRP	miR-206	Gastric cancer	Oncogene	27192121
SPRY4-IT1	miR-101-3p	Bladder cancer	Oncogene	27998761
TUG1	miR-145	Bladder cancer	Oncogene	26318860
TUG1	miR-299	Glioblastoma	Oncogene	27345398
TUG1	miR-300	Gallbladder carcinoma	Oncogene	28178615
TUG1	miR-9-5p	Osteosarcoma	Oncogene	27658774
TUSC7	miR-10a	Hepatocellular carcinoma	Tumor suppressor	27002617
TUSC7	miR-211	Colon cancer	Tumor suppressor	23558749
TUSC7	miR-23b, miR-320d	Gastric cancer	Tumor suppressor	25765901
UCA1	miR-143	Breast cancer	Oncogene	26439035
UCA1	miR-16	Bladder cancer	Oncogene	26373319
UCA1	miR-204-5p	Colorectal cancer	Oncogene	27046651
UCA1	miR-216b	Hepatocellular carcinoma	Oncogene	25760077
UCA1	miR-485-5p	Epithelial ovarian cancer	Oncogene	26867765
UCA1	miR-507	Melanoma	Oncogene	27389544
XIST	miR-139-5p	Hepatocellular carcinoma	Oncogene	28231734
XIST	miR-181a	Hepatocellular carcinoma	Tumor suppressor	28388883
XIST	miR-34a-5p	Nasopharyngeal carcinoma	Oncogene	27461945
XIST	miR-92b	Hepatocellular carcinoma	Tumor suppressor	27100897

This table reports for each lncRNA: (1) miRNA sponged; (2) tumor where such interaction was reported; (3) its function in cancer (oncogene or tumor suppressor); and (4) bibliographic reference reported as Pubmed ID (PMID).

OSCC cells and the upregulation of FZD5 (frizzled class receptor 5), target of miR-224-5p and an oncogene involved in activation of Wnt/β-catenin signaling.

It has been demonstrated that lncRNA GAS5 acts as a tumor suppressor in NSCLC by targeting and suppressing miR-135b [112]. GAS5 is downregulated in NSCLC and its expression is inversely correlated to that of miR-135b. After exposure to irradiation, expression of GAS5 and miR-135b was altered, as GAS5 was overexpressed whereas miR-135b was downregulated. Ectopic overexpression of GAS5 led to miR-135b downregulation, repression of cell proliferation, invasion, and improved radiosensitivity [112].

High expression of lncRNA H19 in breast cancer stem cells (BCSCs) is functionally critical for stemness maintenance [113]. In these cells, H19 functions as a molecular sponge for let-7a/b, leading to upregulation of pluripotency factor LIN28, a let-7 target that is highly abundant in BCSCs. Intriguingly, H19 is reciprocally repressed by its targets let-7a/b, but this negative feedback loop can be interfered with by LIN28 because of its ability to inhibit let-7a/b expression [113]. Let-7b expression is also buffered by lncRNA HOST2 (human ovarian cancer-specific transcript 2) in ovarian cancer cells. By binding to let-7b, HOST2 negatively regulates its availability and induces the expression of its oncogenic targets that enhance cell growth and motility in ovarian cancer [114].

Let-7 decoy by lncRNAs was also reported by Deng et al. Upregulation of lncRNA CCAT1 (colon cancer associated transcript 1) in HCC tissues was associated with increased cell proliferation and migration [115]; these oncogenic activities were mediated by its molecular sponge function for let-7: inhibition of let-7 caused upregulated expression of let-7 targets: HMGA2 (high mobility group AT-hook 2) and MYC (MYC proto-oncogene, bHLH transcription factor). Interestingly, other studies reported that MYC, by binding to CCAT1 promoter, induces CCAT1 transcription in colon cancer and gastric carcinoma [116, 117], suggesting the existence of a positive feedback loop between CCAT1 and MYC mediated by let-7 decoy.

Recent works reported the inhibitory effect of HOTAIR on miRNAs functions in different neoplasias. Su et al. found that HOTAIR was highly expressed in HCC tissues and promoted HCC cell proliferation and progression of tumor xenografts [118]. These oncogenic effects were partially due to HOTAIR ability of repressing miR-1 expression. Moreover, also miR-1 was able to negatively regulate HOTAIR expression, thus generating a reciprocal repression feedback loop between these two ncRNAs [118]. Other experimental evidence showed that HOTAIR was capable of binding and downregulating miR-152 in gastric cancer [119]. HOTAIR overexpression in gastric cancer tissues led to decreased expression of miR-152 and to upregulation of its target, HLA-G (human leukocyte antigen G), which in turn facilitated tumor escape mechanisms [119]. Downregulation of miR-152 in gastric cancer could be also caused by PVT1 (plasmacytoma variant translocation 1), an oncogenic lncRNA that acts as a precursor of six miRNAs (i.e., miR-1204, miR-1205, miR-1206, miR-1207-5p, miR-1207-3p, and miR-1208) [120]. Indeed, PVT1 had three MREs for miR-152 and suppressed its expression inducing the upregulation of miR-152 targets (i.e., CD151, FGF2) [121]. Upregulation of PVT1 in gastric cancer was also associated with inhibition of miR-186 function. Indeed, PVT1 bound miR-186 and induced upregulation of HIF-1α (Hypoxia-inducible factor 1-alpha subunit), a target of miR-186 which was related to poor prognosis and invasiveness in gastric cancer [122].

Wang et al. studied in liver cancer the molecular sponge action of lncRNA HULC (highly upregulated in liver cancer). HULC was able to downregulate several miRNAs, including miR-372. Repression of miR-372 enhanced the translation of its target gene, PRKACB (protein kinase cAMP-activated catalytic subunit beta), which in turn promoted phosphorylation of protein CREB1 (cAMP responsive element-binding protein-1) and affected deacetylation and methylation of histones [123]. This process resulted in alterations of chromatin organization and increased expression of HULC, thus showing that HULC was involved in an autoregulatory loop that mantained its abundant expression in liver cancer [123].

Jin et al. reported an association between MALAT1 upregulation and tumor growth and metastasis in triple-negative breast cancer (TNBC) tissues [124]. These tumorigenic properties of MALAT1 were mediated by its ability to decoy miR-1 and, consequently, increase the expression of miR-1 target, SNAI2 (snail family transcriptional repressor 2), also named Slug, an oncogene involved in regulation of cancer cell invasion. Moreover, overexpression of miR-1 was able to reduce MALAT1 expression, demonstrating a reciprocal negative loop between lncRNA and miRNA [124]. The miRNA sponge function of MALAT1 was also reported for cervical cancer [125]. Indeed, MALAT1 levels were found to be more abundant in radio-resistant than in radio-sensitive cancers. Moreover, expression of MALAT1 and of its potential binding partner, miR-145, reverted in response to irradiation. The authors demonstrated that there was a reciprocal repression between MALAT1 and miR-145, which regulated the molecular mechanisms of radio-resistance of cervical cancer [125].

Notably, tumor suppressor miR-145 was frequently reported to be buffered by lncRNAs in cancer models. MiR-145 negatively regulated cell invasion in TNBC, and its downregulation was related to overexpression of lincRNA-RoR (long intergenic ncRNA Regulator of Reprogramming), which acted as competitive endogenous RNA for miR-145 [126]. LincRNA-RoR-mediated downregulation of miR-145 led to upregulation of ARF6 (ADP-ribosylation factor 6), which is strongly involved in metastatic processes; indeed, ARF6 affected E-cadherin localization and impaired cell-cell adhesion, promoting cell invasion in TNBCs [126]. Zhou et al. reported a further effective interaction between lincRNA-RoR and miR-145 in endometrial cancer. Linc-RoR functioned as a miR-145 sponge by repressing the miRNA-mediated degradation of core stem cell transcription factors (i.e., Nanog, Oct4, and Sox2), thereby maintaining the pluripotency of endometrial cancer stem cells [127].

Decoying of miR-145 was performed also by TUG1 (taurine upregulated 1), which is a well-known oncogenic lncRNA, frequently upregulated in cancer and functionally related to several aggressive features of tumors. In bladder cancer, TUG1 decreased the expression of miR-145 and caused upregulation of ZEB2, miR-145 target, promoting EMT, and increasing the metastatic proneness of bladder cancer cells [128]. The ceRNA role of TUG1 was also proved in other tumors. Overexpression of TUG1 was involved in glioblastoma angiogenesis by modulation of endothelial cell proliferation, migration, and tube formation. These cellular processes were mediated by TUG1 interaction with miR-299, which was downregulated in glioblastoma. In fact, knockdown of TUG1-induced upregulation of miR-299 and concomitant decrease of VEGFA (vascular endothelial growth factor A), target of miR-299. These molecular events resulted in a reduced tumor microvessel density in xenograft glioblastoma models [129]. Ma et al. showed that upregulation of TUG1 in gallbladder carcinoma (GBC) was related to GBC cell proliferation and metastasis, and such oncogenic activities were, at least partly, due to the sponge activity of

TUG1 that bound miR-300 and negatively regulated its expression [130]. In osteosarcoma, TUG1 acted as a ceRNA by sponging miR-9-5p, inducing the upregulation of transcription factor POU2F1 (POU class 2 homeobox 1) [131]. POU2F1 is frequently upregulated in osteosarcoma and is involved in cell proliferation, differentiation and immune and inflammatory processes. Because POU2F1 is a target of miR-9-5p, silencing of TUG1-inhibited cell proliferation and colony formation, while inducing G0/G1 cell cycle arrest and apoptosis. These cellular processes were mediated by upregulation of miR-9-5p and repression of POU2F1 expression [131].

The tumor suppressor TUSC7 (tumor suppressor candidate 7; also named LOC285194) is an lncRNA transcriptionally induced by TP53 (tumor protein 53); it was initially discovered as depleted in osteosarcoma, inducing abnormal proliferation of osteoblasts, and associated with poor survival of osteosarcoma patients. Competitive endogenous binding between TUSC7 and onco-miRNAs has been frequently reported as associated with cancer-related processes. Wang et al. studied the biopathological meaning of strong downregulation of TUSC7 in HCC [132]. They found that ectopic expression of TUSC7 inhibited cell metastasis, invasion, and EMT, by functioning as a competitive sponge for miR-10a. Moreover, this miRNA was able to promote the EMT process in HCC through directly binding and repressing EPHA4 (EPH tyrosine kinase receptor A4) [132]. Moreover, exon 4 of TUSC7 harbors two binding sites for miR-211 [133]. In colon cancer, miR-211 enhanced cell growth, but this effect was reverted by enforced expression of TUSC7, which buffered the activity of miR-211 [133]. The tumor suppressor role of TUSC7 was also demonstrated in gastric cancer. TUSC7, downregulated in gastric cancer, was an independent prognostic marker of disease-free survival in patients, and its ectopic expression suppressed cancer cell growth both in *in vitro* and *in vivo* models, in part by negatively regulating the expression of miR-23 [134].

Unquestionably, one of the most iconic lncRNA acting as miRNA sponge is UCA1, which was reported to bind and repress several miRNAs in multiple tumors. UCA1 binding to miR-143 was proved in breast cancer, where UCA1 was able to modulate cell growth and apoptosis by downregulating miR-143: this in turn led to upregulation of BCL2 (BCL2, apoptosis regulator) and ERBB3 (erb-b2 receptor tyrosine kinase 3) [135]. The role of UCA1 in bladder cancer was associated with ROS (reactive oxygen species) metabolism [136]. Silencing of UCA1 decreased ROS production and promoted mitochondrial glutaminolysis in bladder cancer cells. In these cells, UCA1 acted as a ceRNA by sponging and downregulating miR-16. This induced the upregulation of GLS2 (Glutaminase 2), one of the miR-16 targets, which enhanced glutamine uptake and the rate of glutaminolysis, which is known to increase in cancer cells. UCA1-induced GLS2 maintained the redox balance and protected cancer cells by reducing excessive ROS production [136]. Oncogenic activity of UCA1 in CRC was the result of its decoy function for miR-204-5p, a critical tumor-suppressive miRNA [137]. UCA1, upregulated in CRC, inhibited miR-204-5p activity, thus promoting the upregulation of miRNA targets CREB1, BCL2, and RAB22A (RAB22A, member RAS oncogene) and regulating cell proliferation and apoptosis [137]. UCA1 upregulation in HCC was associated to cell growth and metastasis; these processes were induced by UCA1 binding to miR-216b and resulted in miR-216b downregulation [138]. Decreased levels of miR-216b led to the derepression of its target FGFR1 (fibroblast growth factor receptor 1) and the activation of ERK pathway [138]. Association between UCA1 and metastatic process was also reported for epithelial ovarian cancer [139]. In fact, UCA1 promoted the expression of MMP14 (matrix metallopeptidase14), a key protein involved in cell invasion, by working as a molecular sponge of miR-485-5p, a miRNA targeting MMP14 [139]. FOXM1 (forkhead box protein M1) is a transcription factor critical for G2/M-phase transition and DNA damage response, and it is also a target of miR-507. UCA1-mediated regulation of FOXM1 was discovered, in melanoma cells, to be based on the ceRNA function of UCA1 for miR-507, resulting in an increased malignant ability of these cells [140].

Finally, a ceRNA role in cancer was also reported for XIST (X-inactivate specific transcript). XIST was the first lncRNA to be functionally characterized, and it is considered the major effector of the X inactivation process during development in female mammals [141]. Its dysregulation was found in several tumors (e.g., breast cancer, glioblastoma, and hepatocellular carcinoma), suggesting that XIST could have a potential diagnostic power in cancer [142–144]. *In vitro* downregulation or upregulation of XIST was associated with altered cell proliferation, metastasis, and apoptosis in several cancer models. Song et al. discovered that XIST overexpression was related to metastasis and poor prognosis of NPC patients [145]. XIST induced the upregulation of E2F3 (E2F transcription factor 3), which is a critical protein for tumor cell proliferation. The authors demonstrated that XIST-promoted activation of E2F3 was caused by the competitive sponge role of XIST for miR-34a-5p (a well-known tumor suppressor miRNA), which targets E2F3 [145]. On the other hand, Chang et al. showed that XIST acts as tumor suppressor and inhibits metastatization and progression in HCC by binding miR-181a and reducing its availability; XIST induces PTEN upregulation, thus decreasing cell proliferation, invasion, and migration [146].

2.3. circRNAs as miRNA Sponges. circRNAs are considered new potential players among ceRNAs: they may harbor shared MREs and compete for miRNA binding with mRNAs [69]. Indeed, circRNAs competitively suppress the activity of miRNAs by adsorbing and sequestering them. As miRNAs are strongly involved in nearly all aspects of cellular physiology and perform pivotal roles in initiation and progression of cancer, circRNAs could reasonably be considered as a new class of RNA molecules closely associated with regulation of proliferation, differentiation, and metastatic processes (Table 3).

Zheng et al. reported that circ-TTBK2 (tau tubulin Kinase 2) is significantly upregulated in glioma tissues and cell lines, differently from its linear counterpart [147]. Overexpression of circ-TTBK2 is associated with increased cell proliferation rate, invasion, and decreased apoptosis.

Table 3: circRNAs acting as miRNA sponges.

circRNA	miRNA target	tumor	circRNA role	PMID
circRNA_0005075	miR-23b-5p, miR-93-3p, miR-581, miR-23a-5p	Hepatocellular carcinoma	Oncogene	27258521
circRNA_001569	miR-145	Colorectal cancer	Oncogene	27058418
circRNA_100290	miR-29 family	Oral cancer	Oncogene	28368401
Cdr1as	miR-7	Hepatocellular carcinoma	Oncogene	27391479
cir-ITCH	miR-7, miR-20a	Colorectal cancer	Tumor suppressor	26110611
cir-ITCH	miR-7, miR-17, miR-214	Esophageal squamous cell carcinoma	Tumor suppressor	25749389
ciR-SRY	miR-138	Cholangiocarcinoma	Oncogene	27671698, 23446431
cir-TTBK2	miR-217	Glioma	Oncogene	28219405

This table reports for each circRNA: (1) miRNAs sponged; (2) tumor where such interaction was reported; (3) its function in cancer (oncogene or tumor suppressor); and (4) bibliographic reference reported as Pubmed ID (PMID).

Circ-TTBK2 harbors MREs for miR-217, which has a tumor-suppressive role in glioma cells. In fact, circ-TTBK2 and miR-217 interact with each other in an AGO2-dependent manner and upregulation of circ-TTBK2 induced the malignant behavior of glioma cells via downregulation of miR-217. Thus, HNF1β (HNF1 homeobox B), a direct target of miR-217, was derepressed and bound to the promoter of Derlin-1 increasing its expression. Finally, Derlin-1 was able to promote cell proliferation, migration, and invasion and inhibit apoptosis of glioma cells by activating PI3K/AKT and ERK pathways. Moreover, restoration of miR-217 expression reversed the circ-TTBK2-induced promotion of cancer progression, suggesting a reciprocal negative feedback between circ-TTBK2 and miR-217 [147].

MiR-145 is a well-known tumor suppressor miRNA in CRC targeting the oncogenes ERK5 (mitogen-activated protein kinase 7) and IRS1 (insulin receptor substrate 1); furthermore, its ability to predict survival of CRC patients was also shown. In a study by Xie et al., it was demonstrated that downregulation of miR-145 in CRC was mechanistically explained by the role of circ_001569 acting as a miRNA sponge to directly inhibit miR-145 action [148]. Circ_001569 was found to be upregulated in CRC tissues and correlated with progression and aggressiveness of the disease. Notably, circ_001569 did not directly affect miR-145 expression, but through a *sponge mechanism* it inhibited its posttranscriptional activity; accordingly, it upregulated its targets E2F5 (E2F transcription factor 5), BAG4 (BCL2-associated athanogene 4), and FMNL2 (formin-like 2), which were responsible for cell proliferation and invasion promotion by circ_001569 [148].

Further work on CRC, investigating the role of cir-ITCH on the biopathology of this cancer, found a potential interaction between cir-ITCH and either miR-7 or miR-20a [149]. Cir-ITCH was downregulated in CRC tissues and its ectopic expression led to decreased cell proliferation. This cellular effect was due to cir-ITCH sponge activity for miR-7 and miR-20a; both can bind the 3′-UTR of ITCH (Itchy E3 Ubiquitin Protein Ligase), which is the linear isoform of cir-ITCH. Cir-ITCH-induced upregulation of ITCH promoted the ubiquitination and degradation of phosphorylated DVL2 (dishevelled segment polarity protein 2) and, accordingly, inhibited the Wnt/β-catenin pathway, by repressing the expression of MYC and CCND1 (cyclin D1) [149]. Interestingly, other authors found very similar findings in ESCC: cir-ITCH worked as a miRNA sponge for miR-7, miR-17, and miR-214, increased ITCH expression, and promoted ubiquitin-mediated DVL2 degradation, thus inhibiting canonical Wnt signaling [150].

Besides the cir-ITCH-induced decoy function for miR-7 described above, sponging of miR-7 by CDR1AS was one of the earliest and the most studied ceRNA mechanisms in ncRNA biology, which is also related to cancer. Expression of CDR1AS was found to be elevated in HCC tissues and inversely correlated to miR-7 expression, which was poorly expressed in the same samples [151]. Despite the oncogenic role of miR-7 (previously reported for CRC and ESCC), this miRNA exhibited tumor-suppressive properties in HCC. CDR1AS has sixty-three MREs for miR-7 and strongly suppresses its activity. Knockdown of CDR1AS promoted the expression of miR-7 and suppressed its targets, CCNE1 (cyclin E1) and PIK3CD (phosphatidylinositol-4,5-bisphosphate 3-kinase catalytic subunit delta): this molecular cascade resulted in a reduction of cell proliferation and invasion in HCC [151].

By expression profiling in OSCC, Chen et al. identified the upregulation of a circRNA named circRNA_100290, which was functionally related to abnormal control of cell cycle and cellular proliferation in OSCC cells [152]. circRNA_100290 worked as a miRNA sponge for several members of the miR-29 family, decreasing the quantity of available miR-29s and, accordingly, promoting translation of one of their targets, CDK6 (cyclin-dependent kinase 6), which in turn could induce transition from G1 to S phase in cancer [152].

The first circular transcript identified was Sry circRNA: its encoding gene maps to the sex-determining region of human Y chromosome and was discovered as highly expressed in adult mouse testis [153]. Initially, Sry circRNA was considered an artifact of aberrant RNA splicing and no specific function was attributed to it. The role of Sry circRNA has recently begun to be investigated. Sry circRNA harbors sixteen putative target sites for miR-138 and its function as a miR-138 sponge was demonstrated by Hansen et al. [55]. Currently, no experimental evidence of Sry circRNA-miR-138 axis dysregulation has been reported in cancer; however, as reviewed by Zhao and Shen, miR-138 could target different cancer-related transcripts [154]. For instance,

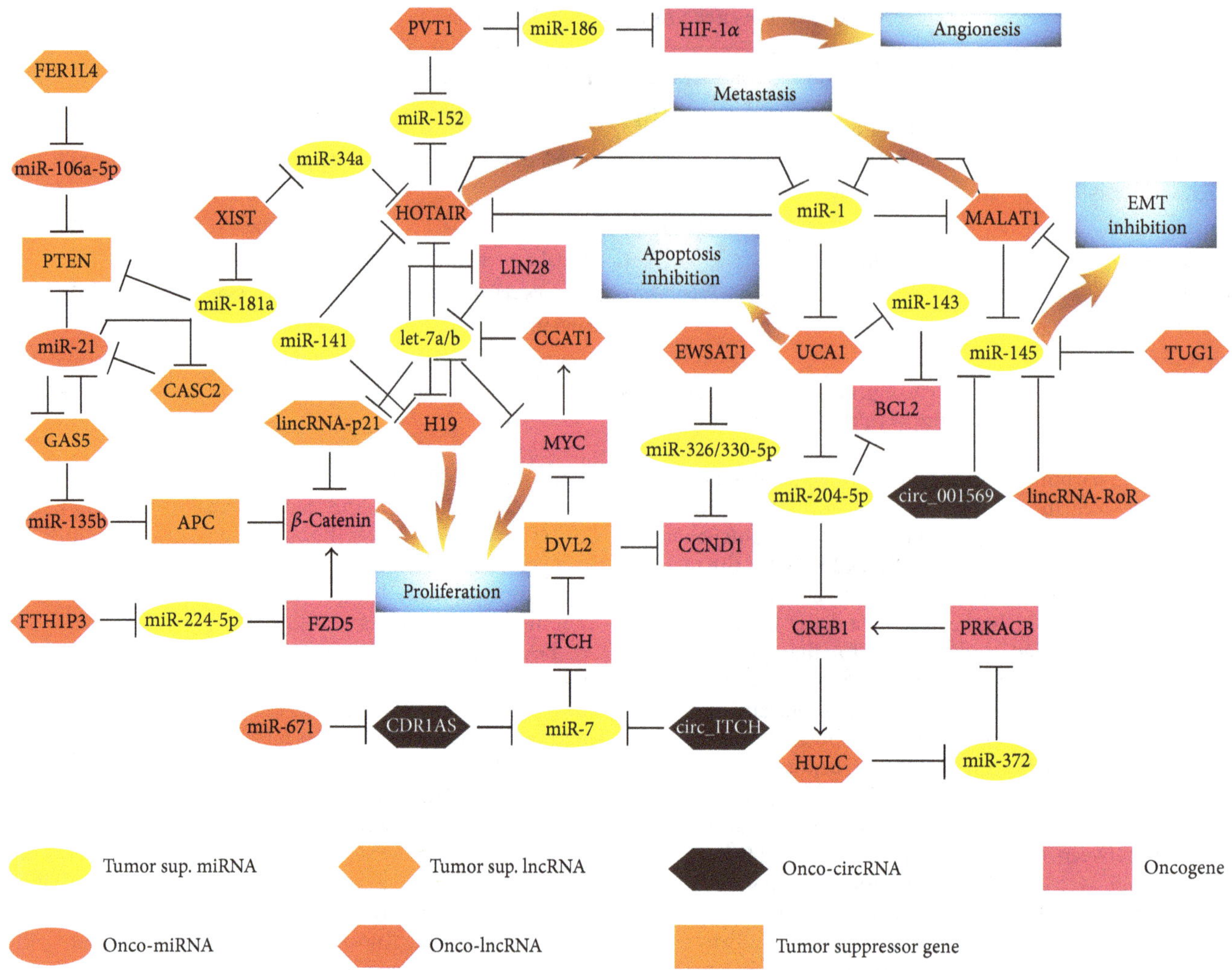

FIGURE 1: Network of noncoding crosstalking in cancer. Molecular interplay among ncRNAs (i.e., miRNAs, lncRNAs, and circRNAs) in cancer. RNA-RNA interactions were retrieved from papers cited in this review. Lines with arrowheads represent expression activation, those with bars represent expression inhibition.

downregulation of miR-138 promoted the malignant progression in cholangiocarcinoma by its target RhoC (ras homolog gene family, member C) [155]. These observations could suggest that the role of competitive endogenous binding between Sry circRNA and miR-138 would be worthy of in-depth analysis in cancer phenotypes.

3. Noncoding RNA Network: Future Perspectives for New Therapeutic Approaches

The existence of a complex RNA-based regulatory signaling, which controls cancer-related pathways, is evident from the experimental evidence collected to date. Such a partially hidden network of RNA-RNA interactions pervades and defines the correct functioning of canonical protein-coding pathways, classically involved in proliferation, differentiation, and invasion in cancer (Figure 1). The complexity of this *noncoding landscape* is dramatically expanded by the presence of several positive and negative regulatory loops: these make RNA signaling very robust and persistent, though complex and hard to functionally unveil. From a network biology point of view, it is possible to identify some *ncRNA hubs* that are a crossroad among different RNA-based circuits; accordingly, they represent a keystone of network structural integrity. For instance, the tumor suppressor miR-1 could repress and be sponged by the three most potent oncogenic lncRNAs, HOTAIR, UCA1, and MALAT1, which, in turn, could inhibit dozens of miRNAs with tumor-suppressive properties [76, 118, 124]. The signaling passing through let-7a/b appears extremely complex and pronged. Let-7a/b could be considered a crossroad of multiple interplays among cancer-related ncRNAs: let-7a/b and MYC are reciprocally negatively regulated through lncRNA CCAT1 [115], but MYC expression could be indirectly impaired by miR-7, which, in turn, is buffered by different circRNAs [156]. Moreover, let-7a/b could indirectly suppress the β-catenin pathway, which in a different way could be activated by lncRNA FTHIP3 [111], but also is regulated by molecular axis miR-21-GAS5-miR-135b [112]. This unexpected cross-talking between ncRNA signaling could shed a light on expression relationships among ncRNAs and mRNAs, which have been frequently reported in cancer literature, but to date have not been satisfactorily explained [157–159]. This *scenario* is made more complex by the tissue-specific expression pattern of all ncRNAs, which could effectively influence the occurrence of specific interactions among ncRNAs. In other words, specific and effective functional interplays among ncRNAs in a particular biological system could occur only if RNA molecules, binding each other, are present at appropriate concentrations. Effectiveness of Ago binding to miRNAs and their targets is dependent on the relative concentration of the miRNA and its target pool [160, 161]. Effective Ago binding occurs when the miRNA : target ratio is close to one but rises dramatically with increasing miRNA : target ratios [162]. Only the most abundant miRNAs show detectable activity, while poorly expressed miRNAs (<100 copies per cell) possess exiguous regulatory properties [163]. However, functional binding between a miRNA and its target can be perturbed by overexpression of other RNAs with multiple shared MREs (e.g., other mRNAs, lncRNAs, and circRNAs) [164]. Such competition among different RNA molecules occurs in a threshold-like manner [165]. Mathematical models predict ceRNA functional effects when miRNA and target levels are near equimolar [166]. However, when the target pool exceeds the threshold set by the buffering miRNA concentration plus the equilibrium dissociation constant (KD) of the miRNA : target interaction, smaller changes in target (i.e., ceRNA) concentration could result in remarkable changes in the concentration of free unrepressed targets [165, 167]. In fact, poorly expressed miRNAs appear to be more susceptible to ceRNA control than more abundant miRNAs. This phenomenon could explain why in *in vitro* experiments a specific miRNA, when ectopically overexpressed, degraded its lncRNA target, but at the same time the enforced upregulation of lncRNA suppressed miRNA activity (e.g., miR-1/MALAT1, miR-21/GAS5) [104, 124]. Taken together, these considerations strongly suggest that miRNA functionality and the switch to ceRNA-promoted repression of miRNAs would be based on the stoichiometric equilibrium among miRNAs and ceRNAs. Based on these observations, physiological ceRNA expression changes could not affect highly expressed miRNAs; however, the relationship between cellular abundance of RNAs and effectiveness of competitive endogenous interactions remains to be fully unveiled in pathological models, in which strong dysregulation of specific ceRNAs could be present [162, 166, 168].

In spite of unclear stoichiometric relationships among ncRNAs in cancer, multiple experimental evidence shows that *in vitro* and *in vivo* modulation of ncRNAs strongly impair aggressive properties of cancer cells. The emerging role of ncRNAs as key regulators of cancer-related signaling makes them very attractive and promising targets for novel, potentially groundbreaking therapeutic approaches. RNA-based therapeutics has several advantages compared to other strategies. RNAs are molecules more *druggable* than proteins, because their targeting is mainly based on nucleic acid complementarity; therefore, an RNA-based drug would be quite easy to design and inexpensive to synthesize (i.e., ASOs, ribozymes, and aptamers) [14]. It is worth stressing that the development of RNA therapeutic strategies has to challenge the redundancy and complexity of the multiple regulatory loops, present in the ncRNA network. It would be quite naive to hypothesize to slow down *in vivo* tumor progression by targeting a single ncRNA molecule: this would be very hard also for protein-based drugs. This axiom should lead researchers to develop multitargeted RNA therapies to improve their impact on oncogenic signaling. In theory, the β-catenin pathway, frequently hyperactivated in cancers, could be effectively attenuated by simultaneous silencing of miR-21, miR-135b, and FTH1P3 together with restoring physiologic levels of GAS5, CASC2, and miR-224-5p. Furthermore, simultaneous repression of HOTAIR, MALAT1, and UCA1 with reactivation of miR-1 would result in a pleiotropic favorable effect on different cancer-related processes, such as cancer growth, metastatic behavior, and cell death. Such a synergic approach based on simultaneous administration of miRNA mimics and siRNAs against ncRNAs in *in vitro* and *in vivo* models has already provided encouraging results. Ideally, such therapeutic approaches would be greatly improved by innovative knockout technologies (such as CRISPR/CAS9), which would avoid potential saturation of RISC complexes, typically occurring by using siRNAs or miRNA mimics [169]. The main issue related to ncRNA therapeutics is to develop efficient delivery systems, which should be able to maintain RNA stability in the circulation and guarantee an effective tissue-specific uptake, as well as minimize off-target side effects. Rapid progress in drug delivery technologies has provided promising chemical and nanotechnological resources well adaptable to RNA therapeutics: chemical modifications of antisense molecules (e.g., steroids and cholesterol) [170], adenoviral vectors [171], cationic liposomes [172], and polymer-based nanoparticles [173]. Recently, an exosomal-based miRNA delivery system has been developed. Such a system appears to be very promising because exosomes are less toxic and better tolerated by the organism and naturally protect their molecular cargo in the blood [174].

4. Conclusions

A better knowledge on the complex interplay among ncRNAs, together with the development of selective methods for RNA delivery to cancer cells, will provide great benefits for cancer treatment. Needless to say, researchers will have to overcome many technical challenges to develop effective RNA-based anticancer strategies realistically applicable to patients. Before ncRNA targeting is pervasively applied in clinical settings, it will be indispensable to organize large collaborative efforts between research institutes and industry to fully realize the clinical potential of this very promising approach.

Conflicts of Interest

The authors declare that there is no conflict of interests regarding the publication of this paper.

Acknowledgments

The authors wish to thank the Scientific Bureau of the University of Catania for language support.

References

[1] R. Nowak, "Mining treasures from 'junk DNA'," *Science*, vol. 263, no. 5147, pp. 608-610, 1994.

[2] E. S. Lander, L. M. Linton, B. Birren et al., "Initial sequencing and analysis of the human genome," *Nature*, vol. 409, no. 6822, pp. 860-921, 2001.

[3] J. C. Venter, M. D. Adams, E. W. Myers et al., "The sequence of the human genome," *Science*, vol. 291, no. 5507, pp. 1304-1351, 2001.

[4] International Human Genome Sequencing Consortium, "Finishing the euchromatic sequence of the human genome," *Nature*, vol. 431, no. 7011, pp. 931-945, 2004.

[5] M. Clamp, B. Fry, M. Kamal et al., "Distinguishing protein-coding and noncoding genes in the human genome," *Proceedings of the National Academy of Sciences of the United States of America*, vol. 104, no. 49, pp. 19428-19433, 2007.

[6] I. Ezkurdia, D. Juan, J. M. Rodriguez et al., "Multiple evidence strands suggest that there may be as few as 19,000 human protein-coding genes," *Human Molecular Genetics*, vol. 23, no. 22, pp. 5866-5878, 2014.

[7] ENCODE Project Consortium, E. Birney, J. A. Stamatoyannopoulos et al., "Identification and analysis of functional elements in 1% of the human genome by the ENCODE pilot project," *Nature*, vol. 447, no. 7146, pp. 799-816, 2007.

[8] B. E. Bernstein, J. A. Stamatoyannopoulos, J. F. Costello et al., "The NIH roadmap epigenomics mapping consortium," *Nature Biotechnology*, vol. 28, no. 10, pp. 1045-1048, 2010.

[9] J. Sana, P. Faltejskova, M. Svoboda, and O. Slaby, "Novel classes of non-coding RNAs and cancer," *Journal of Translational Medicine*, vol. 10, p. 103, 2012.

[10] T. Huang, A. Alvarez, B. Hu, and S. Y. Cheng, "Noncoding RNAs in cancer and cancer stem cells," *Chinese Journal of Cancer*, vol. 32, no. 11, pp. 582-593, 2013.

[11] R. Fatima, V. S. Akhade, D. Pal, and S. M. Rao, "Long noncoding RNAs in development and cancer: potential biomarkers and therapeutic targets," *Molecular and Cellular Therapies*, vol. 3, no. 1, p. 5, 2015.

[12] J. R. Prensner and A. M. Chinnaiyan, "The emergence of lncRNAs in cancer biology," *Cancer Discovery*, vol. 1, no. 5, pp. 391-407, 2011.

[13] S. Geisler and J. Coller, "RNA in unexpected places: long noncoding RNA functions in diverse cellular contexts," *Nature Reviews Molecular Cell Biology*, vol. 14, no. 11, pp. 699-712, 2013.

[14] M. Ragusa, C. Barbagallo, L. Statello et al., "Non-coding landscapes of colorectal cancer," *World Journal of Gastroenterology*, vol. 21, no. 41, pp. 11709-11739, 2015.

[15] R. C. Friedman, K. K. Farh, C. B. Burge, and D. P. Bartel, "Most mammalian mRNAs are conserved targets of microRNAs," *Genome Research*, vol. 19, no. 1, pp. 92-105, 2009.

[16] A. Wilczynska and M. Bushell, "The complexity of miRNA-mediated repression," *Cell Death and Differentiation*, vol. 22, no. 1, pp. 22-33, 2015.

[17] R. C. Lee, R. L. Feinbaum, and V. Ambros, "The C. Elegans heterochronic gene lin-4 encodes small RNAs with antisense complementarity to lin-14," *Cell*, vol. 75, no. 5, pp. 843-854, 1993.

[18] S. L. Ameres and P. D. Zamore, "Diversifying microRNA sequence and function," *Nature Reviews Molecular Cell Biology*, vol. 14, no. 8, pp. 475-488, 2013.

[19] A. Valinezhad Orang, R. Safaralizadeh, and M. Kazemzadeh-Bavili, "Mechanisms of miRNA-mediated gene regulation from common downregulation to mRNA-specific upregulation," *International Journal of Genomics*, vol. 2014, Article ID 970607, 15 pages, 2014.

[20] R. W. Carthew and E. J. Sontheimer, "Origins and mechanisms of miRNAs and siRNAs," *Cell*, vol. 136, no. 4, pp. 642-655, 2009.

[21] D. P. Bartel, "MicroRNAs: genomics, biogenesis, mechanism, and function," *Cell*, vol. 116, no. 2, pp. 281-297, 2004.

[22] A. Kozomara and S. Griffiths-Jones, "miRBase: annotating high confidence microRNAs using deep sequencing data," *Nucleic Acids Research*, vol. 42, Database issue, pp. D68-D73, 2014.

[23] L. A. Macfarlane and P. R. Murphy, "MicroRNA: biogenesis, function and role in cancer," *Current Genomics*, vol. 11, no. 7, pp. 537-561, 2010.

[24] C. M. Croce, "Causes and consequences of microRNA dysregulation in cancer," *Nature Reviews Genetics*, vol. 10, no. 10, pp. 704-714, 2009.

[25] S. Volinia, G. A. Calin, C. G. Liu et al., "A microRNA expression signature of human solid tumors defines cancer gene targets," *Proceedings of the National Academy of Sciences of the United States of America*, vol. 103, no. 7, pp. 2257-2261, 2006.

[26] C. Jay, J. Nemunaitis, P. Chen, P. Fulgham, and A. W. Tong, "miRNA profiling for diagnosis and prognosis of human cancer," *DNA and Cell Biology*, vol. 26, no. 5, pp. 293-300, 2007.

[27] M. Ragusa, L. Statello, M. Maugeri et al., "Highly skewed distribution of miRNAs and proteins between colorectal cancer cells and their exosomes following cetuximab treatment: biomolecular, genetic and translational implications," *Oncoscience*, vol. 1, no. 2, pp. 132-157, 2014.

[28] R. Ma, T. Jiang, and X. Kang, "Circulating microRNAs in cancer: origin, function and application," *Journal of Experimental & Clinical Cancer Research*, vol. 31, no. 1, p. 38, 2012.

[29] G. Cheng, "Circulating miRNAs: roles in cancer diagnosis, prognosis and therapy," *Advanced Drug Delivery Reviews*, vol. 81, pp. 75-93, 2015.

[30] M. Ragusa, C. Barbagallo, L. Statello et al., "miRNA profiling in vitreous humor, vitreal exosomes and serum from uveal melanoma patients: pathological and diagnostic implications," *Cancer Biology & Therapy*, vol. 16, no. 9, pp. 1387-1396, 2015.

[31] P. J. Volders, K. Helsens, X. Wang et al., "LNCipedia: a database for annotated human lncRNA transcript sequences and structures," *Nucleic Acids Research*, vol. 41, Database issue, pp. D246-D251, 2013.

[32] L. Nie, H. J. Wu, J. M. Hsu et al., "Long non-coding RNAs: versatile master regulators of gene expression and crucial players in cancer," *American Journal of Translational Research*, vol. 4, no. 2, pp. 127-150, 2012.

[33] I. Martianov, A. Ramadass, A. Serra Barros, N. Chow, and A. Akoulitchev, "Repression of the human dihydrofolate reductase gene by a non-coding interfering transcript," *Nature*, vol. 445, no. 7128, pp. 666-670, 2007.

[34] L. Li, B. Liu, O. L. Wapinski et al., "Targeted disruption of Hotair leads to homeotic transformation and gene derepression," *Cell Reports*, vol. 5, no. 1, pp. 3–12, 2013.

[35] J. M. Engreitz, N. Ollikainen, and M. Guttman, "Long noncoding RNAs: spatial amplifiers that control nuclear structure and gene expression," *Nature Reviews Molecular Cell Biology*, vol. 17, no. 12, pp. 756–770, 2016.

[36] M. Beltran, I. Puig, C. Pena et al., "A natural antisense transcript regulates Zeb2/Sip1 gene expression during Snail1-induced epithelial-mesenchymal transition," *Genes & Development*, vol. 22, no. 6, pp. 756–769, 2008.

[37] T. Liu, Y. Huang, J. Chen et al., "Attenuated ability of BACE1 to cleave the amyloid precursor protein via silencing long noncoding RNA BACE1AS expression," *Molecular Medicine Reports*, vol. 10, no. 3, pp. 1275–1281, 2014.

[38] A. Rosa and M. Ballarino, "Long noncoding RNA regulation of pluripotency," *Stem Cells International*, vol. 2016, Article ID 1797692, 9 pages, 2016.

[39] A. Fatica and I. Bozzoni, "Long non-coding RNAs: new players in cell differentiation and development," *Nature Reviews. Genetics*, vol. 15, no. 1, pp. 7–21, 2014.

[40] M. Kitagawa, K. Kitagawa, Y. Kotake, H. Niida, and T. Ohhata, "Cell cycle regulation by long non-coding RNAs," *Cellular and Molecular Life Sciences*, vol. 70, no. 24, pp. 4785–4794, 2013.

[41] Y. Su, H. Wu, A. Pavlosky et al., "Regulatory non-coding RNA: new instruments in the orchestration of cell death," *Cell Death & Disease*, vol. 7, no. 8, article e2333, 2016.

[42] P. Zhang, P. Cao, X. Zhu et al., "Upregulation of long noncoding RNA HOXA-AS2 promotes proliferation and induces epithelial-mesenchymal transition in gallbladder carcinoma," *Oncotarget*, vol. 8, no. 20, pp. 33137–33143, 2017.

[43] S. Deguchi, K. Katsushima, A. Hatanaka et al., "Oncogenic effects of evolutionarily conserved noncoding RNA ECONEXIN on gliomagenesis," *Oncogene*, vol. 36, no. 32, pp. 4629–4640, 2017.

[44] F. Yun-Bo, L. Xiao-Po, L. Xiao-Li, C. Guo-Long, Z. Pei, and T. Fa-Ming, "LncRNA TUG1 is upregulated and promotes cell proliferation in osteosarcoma," *Open Medicine*, vol. 11, no. 1, pp. 163–167, 2016.

[45] J. Tian, X. Hu, W. Gao et al., "Identification of the long noncoding RNA LET as a novel tumor suppressor in gastric cancer," *Molecular Medicine Reports*, vol. 15, no. 4, pp. 2229–2234, 2017.

[46] G. Luo, D. Liu, C. Huang et al., "LncRNA GAS5 inhibits cellular proliferation by targeting P27Kip1," *Molecular Cancer Research*, vol. 15, no. 7, pp. 789–799, 2017.

[47] Z. Li, C. Jin, S. Chen et al., "Long non-coding RNA MEG3 inhibits adipogenesis and promotes osteogenesis of human adipose-derived mesenchymal stem cells via miR-140-5p," *Molecular and Cellular Biochemistry*, vol. 433, no. 1-2, pp. 51–60, 2017.

[48] J. Liu, T. Liu, X. Wang, and A. He, "Circles reshaping the RNA world: from waste to treasure," *Molecular Cancer*, vol. 16, no. 1, p. 58, 2017.

[49] H. L. Sanger, G. Klotz, D. Riesner, H. J. Gross, and A. K. Kleinschmidt, "Viroids are single-stranded covalently closed circular RNA molecules existing as highly base-paired rod-like structures," *Proceedings of the National Academy of Sciences of the United States of America*, vol. 73, no. 11, pp. 3852–3856, 1976.

[50] C. Cocquerelle, B. Mascrez, D. Hetuin, and B. Bailleul, "Mis-splicing yields circular RNA molecules," *The FASEB Journal*, vol. 7, no. 1, pp. 155–160, 1993.

[51] J. Salzman, C. Gawad, P. L. Wang, N. Lacayo, and P. O. Brown, "Circular RNAs are the predominant transcript isoform from hundreds of human genes in diverse cell types," *PLoS One*, vol. 7, no. 2, article e30733, 2012.

[52] W. R. Jeck, J. A. Sorrentino, K. Wang et al., "Circular RNAs are abundant, conserved, and associated with ALU repeats," *RNA*, vol. 19, no. 2, pp. 141–157, 2013.

[53] I. Chen, C. Y. Chen, and T. J. Chuang, "Biogenesis, identification, and function of exonic circular RNAs," *Wiley Interdisciplinary Reviews RNA*, vol. 6, no. 5, pp. 563–579, 2015.

[54] P. Glazar, P. Papavasileiou, and N. Rajewsky, "circBase: a database for circular RNAs," *RNA*, vol. 20, no. 11, pp. 1666–1670, 2014.

[55] T. B. Hansen, T. I. Jensen, B. H. Clausen et al., "Natural RNA circles function as efficient microRNA sponges," *Nature*, vol. 495, no. 7441, pp. 384–388, 2013.

[56] E. Lasda and R. Parker, "Circular RNAs: diversity of form and function," *RNA*, vol. 20, no. 12, pp. 1829–1842, 2014.

[57] J. U. Guo, V. Agarwal, H. Guo, and D. P. Bartel, "Expanded identification and characterization of mammalian circular RNAs," *Genome Biology*, vol. 15, no. 7, p. 409, 2014.

[58] Y. Enuka, M. Lauriola, M. E. Feldman, A. Sas-Chen, I. Ulitsky, and Y. Yarden, "Circular RNAs are long-lived and display only minimal early alterations in response to a growth factor," *Nucleic Acids Research*, vol. 44, no. 3, pp. 1370–1383, 2016.

[59] C. Y. Chen and P. Sarnow, "Initiation of protein synthesis by the eukaryotic translational apparatus on circular RNAs," *Science*, vol. 268, no. 5209, pp. 415–417, 1995.

[60] N. R. Pamudurti, O. Bartok, M. Jens et al., "Translation of CircRNAs," *Molecular Cell*, vol. 66, no. 1, pp. 9–21.e7, 2017.

[61] M. Huang, Z. Zhong, M. Lv, J. Shu, Q. Tian, and J. Chen, "Comprehensive analysis of differentially expressed profiles of lncRNAs and circRNAs with associated co-expression and ceRNA networks in bladder carcinoma," *Oncotarget*, vol. 7, no. 30, pp. 47186–47200, 2016.

[62] W. W. Du, W. Yang, E. Liu, Z. Yang, P. Dhaliwal, and B. B. Yang, "Foxo3 circular RNA retards cell cycle progression via forming ternary complexes with p21 and CDK2," *Nucleic Acids Research*, vol. 44, no. 6, pp. 2846–2858, 2016.

[63] H. Li, X. Hao, H. Wang et al., "Circular RNA expression profile of pancreatic ductal adenocarcinoma revealed by microarray," *Cellular Physiology and Biochemistry*, vol. 40, no. 6, pp. 1334–1344, 2016.

[64] A. Bachmayr-Heyda, A. T. Reiner, K. Auer et al., "Correlation of circular RNA abundance with proliferation – exemplified with colorectal and ovarian cancer, idiopathic lung fibrosis, and normal human tissues," *Scientific Reports*, vol. 5, no. 1, 8057 pages, 2015.

[65] Y. Dou, D. J. Cha, J. L. Franklin et al., "Circular RNAs are down-regulated in KRAS mutant colon cancer cells and can be transferred to exosomes," *Scientific Reports*, vol. 6, no. 1, article 37982, 2016.

[66] W. Sui, Z. Shi, W. Xue et al., "Circular RNA and gene expression profiles in gastric cancer based on microarray chip technology," *Oncology Reports*, vol. 37, no. 3, pp. 1804–1814, 2017.

[67] L. Wan, L. Zhang, K. Fan, Z. X. Cheng, Q. C. Sun, and J. J. Wang, "Circular RNA-ITCH suppresses lung cancer proliferation via inhibiting the Wnt/β-catenin pathway," *BioMed Research International*, vol. 2016, Article ID 1579490, 11 pages, 2016.

[68] Z. Zhong, M. Lv, and J. Chen, "Screening differential circular RNA expression profiles reveals the regulatory role of circTCF25-miR-103a-3p/miR-107-CDK6 pathway in bladder carcinoma," *Scientific Reports*, vol. 6, article 30919, 2016.

[69] S. Memczak, M. Jens, A. Elefsinioti et al., "Circular RNAs are a large class of animal RNAs with regulatory potency," *Nature*, vol. 495, no. 7441, pp. 333–338, 2013.

[70] X. Su, J. Xing, Z. Wang, L. Chen, M. Cui, and B. Jiang, "microRNAs and ceRNAs: RNA networks in pathogenesis of cancer," *Chinese Journal of Cancer Research*, vol. 25, no. 2, pp. 235–239, 2013.

[71] J. J. Quinn and H. Y. Chang, "Unique features of long noncoding RNA biogenesis and function," *Nature Reviews Genetics*, vol. 17, no. 1, pp. 47–62, 2016.

[72] M. Xue, X. Li, W. Wu et al., "Upregulation of long noncoding RNA urothelial carcinoma associated 1 by CCAAT/enhancer binding protein alpha contributes to bladder cancer cell growth and reduced apoptosis," *Oncology Reports*, vol. 31, no. 5, pp. 1993–2000, 2014.

[73] Z. Fang, L. Wu, L. Wang, Y. Yang, Y. Meng, and H. Yang, "Increased expression of the long non-coding RNA UCA1 in tongue squamous cell carcinomas: a possible correlation with cancer metastasis," *Oral Surgery, Oral Medicine, Oral Pathology, Oral Radiology*, vol. 117, no. 1, pp. 89–95, 2014.

[74] J. Huang, N. Zhou, K. Watabe et al., "Long non-coding RNA UCA1 promotes breast tumor growth by suppression of p27 (Kip1)," *Cell Death & Disease*, vol. 5, article e1008, 2014.

[75] L. Zhang, X. Cao, L. Zhang, X. Zhang, H. Sheng, and K. Tao, "UCA1 Overexpression predicts clinical outcome of patients with ovarian cancer receiving adjuvant chemotherapy," *Cancer Chemotherapy and Pharmacology*, vol. 77, no. 3, pp. 629–634, 2016.

[76] T. Wang, J. Yuan, N. Feng et al., "Hsa-miR-1 downregulates long non-coding RNA urothelial cancer associated 1 in bladder cancer," *Tumour Biology*, vol. 35, no. 10, pp. 10075–10084, 2014.

[77] P. Ji, S. Diederichs, W. Wang et al., "MALAT-1, a novel noncoding RNA, and thymosin β4 predict metastasis and survival in early-stage non-small cell lung cancer," *Oncogene*, vol. 22, no. 39, pp. 8031–8041, 2003.

[78] Y. Huo, Q. Li, X. Wang et al., "MALAT1 predicts poor survival in osteosarcoma patients and promotes cell metastasis through associating with EZH2," *Oncotarget*, vol. 8, no. 29, pp. 46993–47006, 2017.

[79] Y. Li, Z. Wu, J. Yuan et al., "Long non-coding RNA MALAT1 promotes gastric cancer tumorigenicity and metastasis by regulating vasculogenic mimicry and angiogenesis," *Cancer Letters*, vol. 395, pp. 31–44, 2017.

[80] R. Lei, M. Xue, L. Zhang, and Z. Lin, "Long noncoding RNA MALAT1-regulated microRNA 506 modulates ovarian cancer growth by targeting iASPP," *OncoTargets and Therapy*, vol. 10, pp. 35–46, 2017.

[81] J. E. Wilusz, S. M. Freier, and D. L. Spector, "3' end processing of a long nuclear-retained noncoding RNA yields a tRNA-like cytoplasmic RNA," *Cell*, vol. 135, no. 5, pp. 919–932, 2008.

[82] V. Tripathi, J. D. Ellis, Z. Shen et al., "The nuclear-retained noncoding RNA MALAT1 regulates alternative splicing by modulating SR splicing factor phosphorylation," *Molecular Cell*, vol. 39, no. 6, pp. 925–938, 2010.

[83] J. A. West, C. P. Davis, H. Sunwoo et al., "The long noncoding RNAs NEAT1 and MALAT1 bind active chromatin sites," *Molecular Cell*, vol. 55, no. 5, pp. 791–802, 2014.

[84] E. Leucci, F. Patella, J. Waage et al., "microRNA-9 targets the long non-coding RNA MALAT1 for degradation in the nucleus," *Scientific Reports*, vol. 3, p. 2535, 2013.

[85] X. Wang, M. Li, Z. Wang et al., "Silencing of long noncoding RNA MALAT1 by miR-101 and miR-217 inhibits proliferation, migration, and invasion of esophageal squamous cell carcinoma cells," *The Journal of Biological Chemistry*, vol. 290, no. 7, pp. 3925–3935, 2015.

[86] Y. Han, Y. Liu, H. Zhang et al., "Hsa-miR-125b suppresses bladder cancer development by down-regulating oncogene SIRT7 and oncogenic long non-coding RNA MALAT1," *FEBS Letters*, vol. 587, no. 23, pp. 3875–3882, 2013.

[87] F. H. Tsang, S. L. Au, L. Wei et al., "Long non-coding RNA HOTTIP is frequently up-regulated in hepatocellular carcinoma and is targeted by tumour suppressive miR-125b," *Liver International*, vol. 35, no. 5, pp. 1597–1606, 2015.

[88] R. A. Gupta, N. Shah, K. C. Wang et al., "Long non-coding RNA HOTAIR reprograms chromatin state to promote cancer metastasis," *Nature*, vol. 464, no. 7291, pp. 1071–1076, 2010.

[89] T. Chiyomaru, S. Yamamura, S. Fukuhara et al., "Genistein inhibits prostate cancer cell growth by targeting miR-34a and oncogenic HOTAIR," *PLoS One*, vol. 8, no. 8, article e70372, 2013.

[90] J. H. Yoon, K. Abdelmohsen, J. Kim et al., "Scaffold function of long non-coding RNA HOTAIR in protein ubiquitination," *Nature Communications*, vol. 4, p. 2939, 2013.

[91] J. H. Yoon, K. Abdelmohsen, S. Srikantan et al., "LincRNA-p21 suppresses target mRNA translation," *Molecular Cell*, vol. 47, no. 4, pp. 648–655, 2012.

[92] J. H. Yoon, M. H. Jo, E. J. White et al., "AUF1 promotes let-7b loading on Argonaute 2," *Genes & Development*, vol. 29, no. 15, pp. 1599–1604, 2015.

[93] T. Chiyomaru, S. Fukuhara, S. Saini et al., "Long non-coding RNA HOTAIR is targeted and regulated by miR-141 in human cancer cells," *The Journal of Biological Chemistry*, vol. 289, no. 18, pp. 12550–12565, 2014.

[94] S. M. Park, A. B. Gaur, E. Lengyel, and M. E. Peter, "The miR-200 family determines the epithelial phenotype of cancer cells by targeting the E-cadherin repressors ZEB1 and ZEB2," *Genes & Development*, vol. 22, no. 7, pp. 894–907, 2008.

[95] X. Zhou, F. Ye, C. Yin, Y. Zhuang, G. Yue, and G. Zhang, "The interaction between MiR-141 and lncRNA-H19 in regulating cell proliferation and migration in gastric cancer," *Cellular Physiology and Biochemistry*, vol. 36, no. 4, pp. 1440–1452, 2015.

[96] O. Lustig, I. Ariel, J. Ilan, E. Lev-Lehman, N. De-Groot, and A. Hochberg, "Expression of the imprinted gene H19 in the human fetus," *Molecular Reproduction and Development*, vol. 38, no. 3, pp. 239–246, 1994.

[97] H. Li, B. Yu, J. Li et al., "Overexpression of lncRNA H19 enhances carcinogenesis and metastasis of gastric cancer," *Oncotarget*, vol. 5, no. 8, pp. 2318–2329, 2014.

[98] W. P. Tsang, E. K. Ng, S. S. Ng et al., "Oncofetal H19-derived miR-675 regulates tumor suppressor RB in human colorectal cancer," *Carcinogenesis*, vol. 31, no. 3, pp. 350–358, 2010.

[99] J. Ribas, X. Ni, M. Castanares et al., "A novel source for miR-21 expression through the alternative polyadenylation of VMP1 gene transcripts," *Nucleic Acids Research*, vol. 40, no. 14, pp. 6821–6833, 2012.

[100] B. Vicinus, C. Rubie, S. K. Faust et al., "miR-21 functionally interacts with the 3'UTR of chemokine CCL20 and down-regulates CCL20 expression in miR-21 transfected colorectal cancer cells," *Cancer Letters*, vol. 316, no. 1, pp. 105–112, 2012.

[101] P. Wang, F. Zou, X. Zhang et al., "microRNA-21 negatively regulates Cdc25A and cell cycle progression in colon cancer cells," *Cancer Research*, vol. 69, no. 20, pp. 8157–8165, 2009.

[102] I. A. Asangani, S. A. Rasheed, D. A. Nikolova et al., "MicroRNA-21 (miR-21) post-transcriptionally downregulates tumor suppressor Pdcd4 and stimulates invasion, intravasation and metastasis in colorectal cancer," *Oncogene*, vol. 27, no. 15, pp. 2128–2136, 2008.

[103] S. Roy, Y. Yu, S. B. Padhye, F. H. Sarkar, and A. P. Majumdar, "Difluorinated-curcumin (CDF) restores PTEN expression in colon cancer cells by down-regulating miR-21," *PLoS One*, vol. 8, no. 7, article e68543, 2013.

[104] Z. Zhang, Z. Zhu, K. Watabe et al., "Negative regulation of lncRNA GAS5 by miR-21," *Cell Death and Differentiation*, vol. 20, no. 11, pp. 1558–1568, 2013.

[105] Y. Cao, R. Xu, X. Xu, Y. Zhou, L. Cui, and X. He, "Down-regulation of lncRNA CASC2 by microRNA-21 increases the proliferation and migration of renal cell carcinoma cells," *Molecular Medicine Reports*, vol. 14, no. 1, pp. 1019–1025, 2016.

[106] P. Wang, Y. H. Liu, Y. L. Yao et al., "Long non-coding RNA CASC2 suppresses malignancy in human gliomas by miR-21," *Cellular Signalling*, vol. 27, no. 2, pp. 275–282, 2015.

[107] T. B. Hansen, E. D. Wiklund, J. B. Bramsen et al., "miRNA-dependent gene silencing involving Ago2-mediated cleavage of a circular antisense RNA," *The EMBO Journal*, vol. 30, no. 21, pp. 4414–4422, 2011.

[108] D. Barbagallo, A. Condorelli, M. Ragusa et al., "Dysregulated miR-671-5p / CDR1-AS / CDR1 / VSNL1 axis is involved in glioblastoma multiforme," *Oncotarget*, vol. 7, no. 4, pp. 4746–4759, 2016.

[109] P. Song and S. C. Yin, "Long non-coding RNA EWSAT1 promotes human nasopharyngeal carcinoma cell growth in vitro by targeting miR-326/-330-5p," *Aging*, vol. 8, no. 11, pp. 2948–2960, 2016.

[110] T. Xia, S. Chen, Z. Jiang et al., "Long noncoding RNA FER1L4 suppresses cancer cell growth by acting as a competing endogenous RNA and regulating PTEN expression," *Scientific Reports*, vol. 5, article 13445, 2015.

[111] C. Z. Zhang, "Long non-coding RNA FTH1P3 facilitates oral squamous cell carcinoma progression by acting as a molecular sponge of miR-224-5p to modulate fizzled 5 expression," *Gene*, vol. 607, pp. 47–55, 2017.

[112] Y. Xue, T. Ni, Y. Jiang, and Y. Li, "LncRNA GAS5 inhibits tumorigenesis and enhances radiosensitivity by suppressing miR-135b expression in non-small cell lung cancer," *Oncology Research*, vol. 25, no. 6, pp. 1027–1037, 2017.

[113] F. Peng, T. T. Li, K. L. Wang et al., "H19/Let-7/LIN28 reciprocal negative regulatory circuit promotes breast cancer stem cell maintenance," *Cell Death & Disease*, vol. 8, no. 1, article e2569, 2017.

[114] Y. Gao, H. Meng, S. Liu et al., "LncRNA-HOST2 regulates cell biological behaviors in epithelial ovarian cancer through a mechanism involving microRNA let-7b," *Human Molecular Genetics*, vol. 24, no. 3, pp. 841–852, 2015.

[115] L. Deng, S. B. Yang, F. F. Xu, and J. H. Zhang, "Long non-coding RNA CCAT1 promotes hepatocellular carcinoma progression by functioning as let-7 sponge," *Journal of Experimental & Clinical Cancer Research*, vol. 34, p. 18, 2015.

[116] X. He, X. Tan, X. Wang et al., "C-Myc-activated long noncoding RNA CCAT1 promotes colon cancer cell proliferation and invasion," *Tumour Biology*, vol. 35, no. 12, pp. 12181–12188, 2014.

[117] F. Yang, X. Xue, J. Bi et al., "Long noncoding RNA CCAT1, which could be activated by c-Myc, promotes the progression of gastric carcinoma," *Journal of Cancer Research and Clinical Oncology*, vol. 139, no. 3, pp. 437–445, 2013.

[118] D. N. Su, S. P. Wu, H. T. Chen, and J. H. He, "HOTAIR, a long non-coding RNA driver of malignancy whose expression is activated by FOXC1, negatively regulates miRNA-1 in hepatocellular carcinoma," *Oncology Letters*, vol. 12, no. 5, pp. 4061–4067, 2016.

[119] B. Song, Z. Guan, F. Liu, D. Sun, K. Wang, and H. Qu, "Long non-coding RNA HOTAIR promotes HLA-G expression via inhibiting miR-152 in gastric cancer cells," *Biochemical and Biophysical Research Communications*, vol. 464, no. 3, pp. 807–813, 2015.

[120] G. B. Beck-Engeser, A. M. Lum, K. Huppi, N. J. Caplen, B. B. Wang, and M. Wabl, "Pvt1-encoded microRNAs in oncogenesis," *Retrovirology*, vol. 5, no. 1, 4 pages, 2008.

[121] T. Li, X. L. Meng, and W. Q. Yang, "Long noncoding RNA PVT1 acts as a "sponge" to inhibit microRNA-152 in gastric cancer cells," *Digestive Diseases and Sciences*, 2017.

[122] T. Huang, H. W. Liu, J. Q. Chen et al., "The long noncoding RNA PVT1 functions as a competing endogenous RNA by sponging miR-186 in gastric cancer," *Biomedicine & Pharmacotherapy*, vol. 88, pp. 302–308, 2017.

[123] J. Wang, X. Liu, H. Wu et al., "CREB up-regulates long non-coding RNA, HULC expression through interaction with microRNA-372 in liver cancer," *Nucleic Acids Research*, vol. 38, no. 16, pp. 5366–5383, 2010.

[124] C. Jin, B. Yan, Q. Lu, Y. Lin, and L. Ma, "Reciprocal regulation of Hsa-miR-1 and long noncoding RNA MALAT1 promotes triple-negative breast cancer development," *Tumour Biology*, vol. 37, no. 6, pp. 7383–7394, 2016.

[125] H. Lu, Y. He, L. Lin et al., "Long non-coding RNA MALAT1 modulates radiosensitivity of HR-HPV+ cervical cancer via sponging miR-145," *Tumour Biology*, vol. 37, no. 2, pp. 1683–1691, 2016.

[126] G. Eades, B. Wolfson, Y. Zhang, Q. Li, Y. Yao, and Q. Zhou, "lincRNA-RoR and miR-145 regulate invasion in triple-negative breast cancer via targeting ARF6," *Molecular Cancer Research*, vol. 13, no. 2, pp. 330–338, 2015.

[127] X. Zhou, Q. Gao, J. Wang, X. Zhang, K. Liu, and Z. Duan, "Linc-RNA-RoR acts as a "sponge" against mediation of the differentiation of endometrial cancer stem cells by microRNA-145," *Gynecologic Oncology*, vol. 133, no. 2, pp. 333–339, 2014.

[128] J. Tan, K. Qiu, M. Li, and Y. Liang, "Double-negative feedback loop between long non-coding RNA TUG1 and miR-145 promotes epithelial to mesenchymal transition and radioresistance in human bladder cancer cells," *FEBS Letters*, vol. 589, no. 20, Part B, pp. 3175–3181, 2015.

[129] H. Cai, X. Liu, J. Zheng et al., "Long non-coding RNA taurine upregulated 1 enhances tumor-induced angiogenesis through inhibiting microRNA-299 in human glioblastoma," *Oncogene*, vol. 36, no. 3, pp. 318–331, 2017.

[130] F. Ma, S. H. Wang, Q. Cai et al., "Long non-coding RNA TUG1 promotes cell proliferation and metastasis by negatively regulating miR-300 in gallbladder carcinoma," *Biomedicine & Pharmacotherapy*, vol. 88, pp. 863–869, 2017.

[131] C. H. Xie, Y. M. Cao, Y. Huang et al., "Long non-coding RNA TUG1 contributes to tumorigenesis of human osteosarcoma by sponging miR-9-5p and regulating POU2F1 expression," *Tumour Biology*, vol. 37, no. 11, pp. 15031–15041, 2016.

[132] Y. Wang, Z. Liu, B. Yao et al., "Long non-coding RNA TUSC7 acts a molecular sponge for miR-10a and suppresses EMT in hepatocellular carcinoma," *Tumour Biology*, vol. 37, no. 8, pp. 11429–11441, 2016.

[133] Q. Liu, J. Huang, N. Zhou et al., "LncRNA loc285194 is a p53-regulated tumor suppressor," *Nucleic Acids Research*, vol. 41, no. 9, pp. 4976–4987, 2013.

[134] P. Qi, M. D. Xu, X. H. Shen et al., "Reciprocal repression between TUSC7 and miR-23b in gastric cancer," *International Journal of Cancer*, vol. 137, no. 6, pp. 1269–1278, 2015.

[135] Y. L. Tuo, X. M. Li, and J. Luo, "Long noncoding RNA UCA1 modulates breast cancer cell growth and apoptosis through decreasing tumor suppressive miR-143," *European Review for Medical and Pharmacological Sciences*, vol. 19, no. 18, pp. 3403–3411, 2015.

[136] H. J. Li, X. Li, H. Pang, J. J. Pan, X. J. Xie, and W. Chen, "Long non-coding RNA UCA1 promotes glutamine metabolism by targeting miR-16 in human bladder cancer," *Japanese Journal of Clinical Oncology*, vol. 45, no. 11, pp. 1055–1063, 2015.

[137] Z. Bian, L. Jin, J. Zhang et al., "LncRNA-UCA1 enhances cell proliferation and 5-fluorouracil resistance in colorectal cancer by inhibiting miR-204-5p," *Scientific Reports*, vol. 6, no. 1, article 23892, 2016.

[138] F. Wang, H. Q. Ying, B. S. He et al., "Upregulated lncRNA-UCA1 contributes to progression of hepatocellular carcinoma through inhibition of miR-216b and activation of FGFR1/ERK signaling pathway," *Oncotarget*, vol. 6, no. 10, pp. 7899–7917, 2015.

[139] Y. Yang, Y. Jiang, Y. Wan et al., "UCA1 functions as a competing endogenous RNA to suppress epithelial ovarian cancer metastasis," *Tumour Biology*, vol. 37, no. 8, pp. 10633–10641, 2016.

[140] Y. Wei, Q. Sun, L. Zhao et al., "LncRNA UCA1-miR-507-FOXM1 axis is involved in cell proliferation, invasion and G0/G1 cell cycle arrest in melanoma," *Medical Oncology*, vol. 33, no. 8, p. 88, 2016.

[141] C. J. Brown, B. D. Hendrich, J. L. Rupert et al., "The human XIST gene: analysis of a 17 kb inactive X-specific RNA that contains conserved repeats and is highly localized within the nucleus," *Cell*, vol. 71, no. 3, pp. 527–542, 1992.

[142] P. C. Schouten, M. A. Vollebergh, M. Opdam et al., "High XIST and low 53BP1 expression predict poor outcome after high-dose alkylating chemotherapy in patients with a BRCA1-like breast cancer," *Molecular Cancer Therapeutics*, vol. 15, no. 1, pp. 190–198, 2016.

[143] Y. Yao, J. Ma, Y. Xue et al., "Knockdown of long non-coding RNA XIST exerts tumor-suppressive functions in human glioblastoma stem cells by up-regulating miR-152," *Cancer Letters*, vol. 359, no. 1, pp. 75–86, 2015.

[144] L. K. Zhuang, Y. T. Yang, X. Ma et al., "MicroRNA-92b promotes hepatocellular carcinoma progression by targeting Smad7 and is mediated by long non-coding RNA XIST," *Cell Death & Disease*, vol. 7, article e2203, 2016.

[145] P. Song, L. F. Ye, C. Zhang, T. Peng, and X. H. Zhou, "Long non-coding RNA XIST exerts oncogenic functions in human nasopharyngeal carcinoma by targeting miR-34a-5p," *Gene*, vol. 592, no. 1, pp. 8–14, 2016.

[146] S. Chang, B. Chen, X. Wang, K. Wu, and Y. Sun, "Long non-coding RNA XIST regulates PTEN expression by sponging miR-181a and promotes hepatocellular carcinoma progression," *BMC Cancer*, vol. 17, no. 1, p. 248, 2017.

[147] J. Zheng, X. Liu, Y. Xue et al., "TTBK2 circular RNA promotes glioma malignancy by regulating miR-217/HNF1beta/Derlin-1 pathway," *Journal of Hematology & Oncology*, vol. 10, no. 1, 52 pages, 2017.

[148] H. Xie, X. Ren, S. Xin et al., "Emerging roles of circRNA_001569 targeting miR-145 in the proliferation and invasion of colorectal cancer," *Oncotarget*, vol. 7, no. 18, pp. 26680–26691, 2016.

[149] G. Huang, H. Zhu, Y. Shi, W. Wu, H. Cai, and X. Chen, "Cir-ITCH plays an inhibitory role in colorectal cancer by regulating the Wnt/beta-catenin pathway," *PLoS One*, vol. 10, no. 6, article e0131225, 2015.

[150] F. Li, L. Zhang, W. Li et al., "Circular RNA ITCH has inhibitory effect on ESCC by suppressing the Wnt/beta-catenin pathway," *Oncotarget*, vol. 6, no. 8, pp. 6001–6013, 2015.

[151] L. Yu, X. Gong, L. Sun, Q. Zhou, B. Lu, and L. Zhu, "The circular RNA Cdr1as act as an oncogene in hepatocellular carcinoma through targeting miR-7 expression," *PLoS One*, vol. 11, no. 7, article e0158347, 2016.

[152] L. Chen, S. Zhang, J. Wu et al., "circRNA_100290 plays a role in oral cancer by functioning as a sponge of the miR-29 family," *Oncogene*, vol. 36, no. 32, pp. 4551–4561, 2017.

[153] B. Capel, A. Swain, S. Nicolis et al., "Circular transcripts of the testis-determining gene *Sry* in adult mouse testis," *Cell*, vol. 73, no. 5, pp. 1019–1030, 1993.

[154] Z. J. Zhao and J. Shen, "Circular RNA participates in the carcinogenesis and the malignant behavior of cancer," *RNA Biology*, vol. 14, no. 5, pp. 514–521, 2017.

[155] Q. Wang, H. Tang, S. Yin, and C. Dong, "Downregulation of microRNA-138 enhances the proliferation, migration and invasion of cholangiocarcinoma cells through the upregulation of RhoC/p-ERK/MMP-2/MMP-9," *Oncology Reports*, vol. 29, no. 5, pp. 2046–2052, 2013.

[156] T. B. Hansen, J. Kjems, and C. K. Damgaard, "Circular RNA and miR-7 in cancer," *Cancer Research*, vol. 73, no. 18, pp. 5609–5612, 2013.

[157] X. Zhou, X. Xu, J. Wang, J. Lin, and W. Chen, "Identifying miRNA/mRNA negative regulation pairs in colorectal cancer," *Scientific Reports*, vol. 5, article 12995, 2015.

[158] S. Zadran, F. Remacle, and R. D. Levine, "miRNA and mRNA cancer signatures determined by analysis of expression levels in large cohorts of patients," *Proceedings of the National Academy of Sciences of the United States of America*, vol. 110, no. 47, pp. 19160–19165, 2013.

[159] J. Seo, D. Jin, C. H. Choi, and H. Lee, "Integration of microRNA, mRNA, and protein expression data for the identification of cancer-related microRNAs," *PLoS One*, vol. 12, no. 1, article e0168412, 2017.

[160] G. Mullokandov, A. Baccarini, A. Ruzo et al., "High-throughput assessment of microRNA activity and function using microRNA sensor and decoy libraries," *Nature Methods*, vol. 9, no. 8, pp. 840–846, 2012.

[161] A. Arvey, E. Larsson, C. Sander, C. S. Leslie, and D. S. Marks, "Target mRNA abundance dilutes microRNA and siRNA activity," *Molecular Systems Biology*, vol. 6, p. 363, 2010.

[162] A. D. Bosson, J. R. Zamudio, and P. A. Sharp, "Endogenous miRNA and target concentrations determine susceptibility to potential ceRNA competition," *Molecular Cell*, vol. 56, no. 3, pp. 347–359, 2014.

[163] B. D. Brown, B. Gentner, A. Cantore et al., "Endogenous microRNA can be broadly exploited to regulate transgene expression according to tissue, lineage and differentiation state," *Nature Biotechnology*, vol. 25, no. 12, pp. 1457–1467, 2007.

[164] M. S. Ebert, J. R. Neilson, and P. A. Sharp, "MicroRNA sponges: competitive inhibitors of small RNAs in mammalian cells," *Nature Methods*, vol. 4, no. 9, pp. 721–726, 2007.

[165] S. Mukherji, M. S. Ebert, G. X. Zheng, J. S. Tsang, P. A. Sharp, and A. van Oudenaarden, "MicroRNAs can generate thresholds in target gene expression," *Nature Genetics*, vol. 43, no. 9, pp. 854–859, 2011.

[166] U. Ala, F. A. Karreth, C. Bosia et al., "Integrated transcriptional and competitive endogenous RNA networks are cross-regulated in permissive molecular environments," *Proceedings of the National Academy of Sciences of the United States of America*, vol. 110, no. 18, pp. 7154–7159, 2013.

[167] C. Bosia, A. Pagnani, and R. Zecchina, "Modelling competing endogenous RNA networks," *PLoS One*, vol. 8, no. 6, article e66609, 2013.

[168] M. Jens and N. Rajewsky, "Competition between target sites of regulators shapes post-transcriptional gene regulation," *Nature Reviews Genetics*, vol. 16, no. 2, pp. 113–126, 2015.

[169] A. L. Jackson and P. S. Linsley, "Recognizing and avoiding siRNA off-target effects for target identification and therapeutic application," *Nature Reviews Drug Discovery*, vol. 9, no. 1, pp. 57–67, 2010.

[170] C. Lorenz, P. Hadwiger, M. John, H. P. Vornlocher, and C. Unverzagt, "Steroid and lipid conjugates of siRNAs to enhance cellular uptake and gene silencing in liver cells," *Bioorganic & Medicinal Chemistry Letters*, vol. 14, no. 19, pp. 4975–4977, 2004.

[171] M. B. Mowa, C. Crowther, and P. Arbuthnot, "Therapeutic potential of adenoviral vectors for delivery of expressed RNAi activators," *Expert Opinion on Drug Delivery*, vol. 7, no. 12, pp. 1373–1385, 2010.

[172] S. Mallick and J. S. Choi, "Liposomes: versatile and biocompatible nanovesicles for efficient biomolecules delivery," *Journal of Nanoscience and Nanotechnology*, vol. 14, no. 1, pp. 755–765, 2014.

[173] E. Miele, G. P. Spinelli, E. Miele et al., "Nanoparticle-based delivery of small interfering RNA: challenges for cancer therapy," *International Journal of Nanomedicine*, vol. 7, pp. 3637–3657, 2012.

[174] D. Ha, N. Yang, and V. Nadithe, "Exosomes as therapeutic drug carriers and delivery vehicles across biological membranes: current perspectives and future challenges," *Acta Pharmaceutica Sinica B*, vol. 6, no. 4, pp. 287–296, 2016.

4

Expression Characterization of Six Genes Possibly Involved in Gonad Development for Stellate Sturgeon Individuals (*Acipenser stellatus*, Pallas 1771)

Alexandru Burcea,[1] Gina-Oana Popa,[1] Iulia Elena Florescu (Gune),[1] Marilena Maereanu,[2] Andreea Dudu,[1] Sergiu Emil Georgescu,[1] and Marieta Costache[1]

[1]*Department of Biochemistry and Molecular Biology, Faculty of Biology, University of Bucharest, Bucharest 050095, Romania*
[2]*S.C. Danube Research-Consulting S.R.L., Isaccea 825200, Romania*

Correspondence should be addressed to Sergiu Emil Georgescu; georgescu_se@yahoo.com

Academic Editor: Lior David

Nowadays, in sturgeon's aquaculture, there is a necessity for sex identification at early stages in order to increase the efficiency of this commercial activity. The basis for a correct identification is studying the different factors that influence the gonad development. The research has been directed towards molecular methods that have been employed with various degrees of success in identifying genes with different expression patterns between male and female sturgeons during their development stages. For the purpose of understanding the sexual development of 4-year-old stellate sturgeon *(Acipenser stellatus)* individuals, we have selected six genes (*foxl2, cyp17a1, ar, dmrt1, sox9*, and *star*). We analysed the gene expression of the selected genes for gonads, anal fin, liver, body kidney, and white muscle. The *cyp17a1, ar, dmrt1,* and *sox9* genes have a significant higher expression in male gonads than in female gonads, while the data shows no significant differences in the expression of the investigated genes in the other organs. We investigate these genes to shed light on aquaculture sturgeon sexual development.

1. Introduction

Sturgeons and paddlefish form the order Acipenseriformes, which is a group of fish that have existed since at least the Lower Jurassic and nowadays are threatened in their entire range of distribution [1].

Caviar is harvested from sturgeons and it has been on high demand, even though they are threatened species, making them economically valuable in world trade [2]. The sevruga caviar is harvested from *A. stellatus* females and is one of the highest-priced caviar types. For this type of caviar to be harvested naturally, the females must reach maturity, which is at 8–10 years. In contrast, males reach maturity at 5-6 years [3, 4]. There is also a period of more than one year between subsequent spawning [5, 6].

Even though aquaculture can provide a supply of sturgeon-derived products, such as caviar and meat, and can also help in conservation efforts providing offspring for repopulation [7], the early sex identification of individuals could further help the aquaculture efforts through correctly separating the males from more economically profitable females. More affordable sturgeon products could lower the pressure of poaching, sustaining the *ex situ* conservation efforts. Also, aquaculture could benefit from the knowledge of genetic architecture and sex determination in sturgeons [8]. For distinguishing females from males, a good understanding of the processes of sexual development is needed. Therefore, studies of the possible sex-determining mechanisms are important for identifying a starting point.

In nature, there are different types of conserved patterns of sex-determining mechanisms which range from male heterogamety (XY) found in the majority of mammals and female heterogamety (ZW) found in most birds to environmental sex determination (ESD) expressed most commonly as temperature sex determination (TSD) found in reptiles and some fish species [9]. Fish species have complex patterns of sex determination which range from hermaphroditism, with individuals presenting both female and male gonads, and gonochorism, where each individual presents only female or male gonads, to TSD. There is also the possibility of natural gynogenesis, where the female ovule is penetrated by a sperm cell, but the sperm nucleus does not contribute to embryo formation which in turn inherits only the female chromosomes [10]. Genetic sex determination was described in *Oryzias latipes* which has an early type of XY genetic system [11]. Cytogenetic studies of fish chromosomes showed that the karyotype can be employed in a low number of fish species for identifying the sex of the individual [12]; furthermore, in the case of sturgeons, this method did not identify any sex-specific chromosomes [13].

The number of fish species that present male heterogamety compared to female heterogamety is higher [12]. One hypothesis is that the XY and ZW systems originated in ancestral species with environmental sex determination in which either the male or the female had the size advantage as an adult; thus, the XY system is present in species where fitness correlates with male size, and the ZW system is found in species where females have the size advantage [14]. It has also been observed that the system of sex determination may depend on reproduction advantage [15]. Wild *Acipenser transmontanus* individuals have a 1 : 1 sex ratio between males and females [16] which means that the ratio is not influenced by the environment. This creates the hypothesis that the system of sex determination in sturgeons is genetic and not environmental, which is backed up by gynogenesis experiments suggesting sex ratios in conformity with the ZW system where gynogenesis may produce ZZ males, WW superfemales, and ZW females. This is the case of *Acipenser baerii* [17], *Acipenser brevirostrum* [18], *Acipenser nudiventris* [19], *A. transmontanus* [20], Bester hybrid (*Huso* female × *Acipenser ruthenus* male) [21], and *Polyodon spathula* [22] even though there is no direct evidence from karyotype morphology [23].

Sturgeons do not present sexual dimorphism that could help in the early selection of females from less economically valuable males [13]; therefore, an extensive plethora of methods have been utilized with the objective of identifying the sex of sturgeon individuals, such as biopsy for *H. huso* [24, 25]; endoscopy for *Acipenser gueldenstaedtii* [26], *H. huso* [26], and *Acipenser oxyrinchus* [27]; and ultrasonography for *A. stellatus* [6]. Even though these methods have high accuracy, the lowest age at which the sex of an individual could be identified is between 1 and 3 years through invasive methods or ultrasonography. Also, noninvasive molecular methods, such as random amplified polymorphic DNA (RAPD) for *A. baerii*, *Acipenser naccarii*, *A. ruthenus* [28], *H. huso* [29], and *Acipenser fulvescens* [30]; amplified fragment length polymorphism (AFLP) for *A. baerii* and *A. gueldenstaedtii* [28]; and intersimple sequence repeats (ISSR) for *A. naccarii*, *A. baerii*, and *A. gueldenstaedtii* [28], have been utilized in order to see if any DNA fragments are specific to females or males, because of the high possibility that the sex-determining system is genetic. These methods did not provide any results that could discern between females and males. Because of this, the studies have shifted towards next-generation sequencing (NGS) studies. In the case of sturgeons, which have complex genomes, NGS allows the transcriptome analysis and gives information on possible differences between males and females in expression and presence of various genes (*dmrt1*, *tra-1*, *wt1*, *lhx1*, *cyp19A1*, *fhl3*, *fem1a*, *gsdf*, *foxl2*, *ar*, *emx2*, *cyp17a1*, etc.) that may be involved in the sexual development processes in different sturgeon species [7, 31–33]. With NGS laying the groundwork for gene discovery, the next step was the analysis of gene pathways using qPCR studies; therefore, the most-studied genes in vertebrates that are involved in male pathway were *sox9*, *amh*, *nr5a1*, *nrob1*, and *wt1* [34–36]. In the case of sturgeons, there have been a series of studies regarding genes involved in male or female sex development: Sertoli cell factors *dmrt1* and *sox9*, Leydig cell factors *star* and *cyp17a1*, and other genes such as *lh*, *ar*, *vtg*, *foxl2*, *fsh*, and *igf1* [35–41], showing that the genes are expressed differently in the various tissues investigated. Even if there are studies that investigate sturgeon sex development, the underlying mechanisms are still poorly understood, and no master sex gene has been identified as of yet.

In this study, we investigated the expression of six genes that are known from literature to be steroid-related Leydig cell factors (*cyp17a1*, *ar*, and *star*), male gonad genes that are Sertoli cell factors (*dmrt1* and *sox9*), and a female gene for ovarian differentiation (*foxl2*). These genes were chosen to investigate if they are involved in the sexual development of specific tissues in *A. stellatus*. Different organs were analysed alongside gonads in order to showcase if the studied genes are specific to gonad development. This study aims at observing the expression pattern between males and females and between the different organs (gonads, white muscle, body kidney, anal fin, and liver) in order to establish if these genes are involved in the sexual development of *A. stellatus* and if they are gonad specific. This is the first time that the expression of these genes was investigated through qPCR for this species.

2. Materials and Methods

2.1. Samples. Four female and four male *A. stellatus* 4-year-old individuals from aquaculture, kept in ponds with water recirculation and temperature of 22°C, were anesthetized with clove oil : ethanol (1 : 10) in water; their branchial arches were sectioned, and they died of blood loss. After dissection, the individuals were sexed by observing the presence of lamellae on the female gonads and the relatively smooth male gonads as in Flynn and Benfey [42]. At the age of 4 years, the individuals were at early maturation stage, undergoing gametogenesis. Females had previtellogenic oocytes, while males presented primary spermatocytes. The liver, white muscle, body kidney, gonads, and anal fin fragments were collected in RNAlater Stabilization Reagent (Qiagen).

2.2. RNA Isolation. The isolation of total RNA was performed from 10 mg of tissue using the PureLink RNA Mini Kit (Thermo Fisher) for the liver, body kidney, and gonads, while the RNeasy Fibrous Tissue Mini Kit (Qiagen) was used for the 20 mg of white muscle and anal fin following the manufacturer's protocol. After isolation, 10 μL of RNA solution was digested with DNase I (Qiagen) for removal of contaminant genomic DNA using 10 μL RDD buffer, 2.5 μL DNase I, and 77.5 μL RNase-free ultrapure water in a 100 μL final volume. The quantity and purity of total RNA samples were determined using the NanoDrop 8000 (Thermo Scientific). RIN values of the RNA samples were determined using Agilent RNA 6000 Nano Kit (Agilent) and Agilent 2100 Bioanalyzer using the manufacturer's protocol. Samples with RIN values smaller than 8 were not included in further analysis.

2.3. Reverse Transcription. For cDNA synthesis, 1000 ng of total RNA was reverse transcribed using the iScript Reverse Transcription Supermix for RT-qPCR (BioRad) following the manufacturer's protocol.

2.4. Primer Design. Because of the lack of data regarding genes involved in the sexual development of *A. stellatus*, GenBank sequences from other sturgeon species were used to design pairs of primers (Table 1). The pairs of primers were designed in silico using Primer BLAST NCBI, tested for annealing temperature using the Primer3 online software, and tested for specificity by using BLAST NCBI. The *foxl2*, *cyp17a1*, *ar*, *dmrt1*, *sox9*, *star*, *β-actin*, *gapdh*, and *28S rRNA* amplicons were sequenced using the primers in Table 1 on the ABI Prism 3130 Genetic Analyzer (Applied Biosystems), using the BigDye Terminator v3.1 Cycle Sequencing Kit (Applied Biosystems), and the cDNA sequences were deposited in GenBank with the following accession numbers: KX420678-KX420686 (Table 1).

2.5. qPCR. The qPCR was carried out on the iCycler iQ Real-Time PCR Detection System (BioRad) using the iQ SYBR Green Supermix Kit (BioRad) in 25 μL final volume with 400 nM of each primer from Table 1, 100 ng of cDNA, 12.5 μL of iQ SYBR Green Supermix 2X, and 10.5 μL of DNase-free ultrapure water. The incubation program was comprised of initial denaturation (95°C for 5 minutes), cycling stages (35 cycles of 95°C denaturation for 30 seconds, 58°C annealing for 30 seconds, and 72°C extension for 30 seconds with data collection after each extension), and a melt curve stage (from 55°C to 97.5°C with an increment of 0.5°C for 10 seconds and data collection at each increment). The raw data points were represented by the quantification cycle (C_q) as stated in "The MIQE Guidelines: Minimum Information for Publication of Quantitative Real-Time PCR Experiments" [43]. The analysis was performed in triplicate for each gene, and the validity of the qPCR was confirmed by analysing the melting curves. The amplification efficiency (E) of the investigated genes was situated between 94% and 105%, while the linear standard curve (r^2) was higher than 0.994 for all genes.

2.6. Data Analysis. The reference genes' (*gapdh*, *β-actin*, and *28S rRNA*) C_q results for each group were tested for stability using NormFinder 20 [44]. The best single result was observed for *β-actin* (0.088), while the best result for a combination of genes was for the arithmetic mean of *gapdh* and *β-actin* (0.061), the latter combination being used in the subsequent analysis because of the good score.

The relative expression value ($2^{-\Delta Cq}$) was obtained by normalization, subtracting the arithmetic mean of the *β-actin* and *gapdh* reference genes from each gene of interest [45]. The normal distribution of the dataset groups was tested using the Shapiro-Wilk test implemented in IBM SPSS 19.0 [46, 47]. We used the one-way ANOVA test ($p \le 0.05$) with Tukey correction for multiple comparisons implemented in GraphPad Prism 6.01 in order to investigate the difference in gene expression between 4-year-old *A. stellatus* males and females.

3. Results

The *β-actin* and *gapdh* arithmetic mean was chosen to normalize the data from the six genes of interest. The expression levels for the genes of interest are presented in Table 2 in the form of means and ±SD, along with one-way ANOVA results for the comparison between 4-year-old *A. stellatus* males and females.

A statistically significant difference ($p \le 0.01$) between females and males was observed only in gonads for the *cyp17a1*, *ar*, *dmrt1*, and *sox9* genes, for which the males presented higher levels of expression than the females (Figure 1(a)). In the case of body kidney, white muscle, and liver, the expression of *cyp17a1* was not observed. The expression of *cyp17a1* was observed only in gonads and anal fin (Figures 1(a) and 1(b)). The expression of the *star* gene was not detected for white muscle, but it was observed in the rest of the investigated organs (gonads, anal fin, body kidney, and liver) (Figure 1). The *foxl2* and *star* genes did not present any statistically significant difference in expression between the sexes for this organ. The difference in expression between females and males for the investigated genes was not statistically significant for the rest of the organs (anal fin, body kidney, liver, and white muscle) (Figures 1(b)–1(e)).

For males, a difference in the expression ($p \le 0.0001$) for the *cyp17a1* gene between the gonad and anal fin was observed, with higher levels in the gonad. For the *ar* gene, the expression in the male gonad was statistically different from that in the anal fin ($p \le 0.05$), in the body kidney ($p \le 0.01$), and in the white muscle ($p \le 0.01$), the highest levels being found in the gonad but not different from that in liver (Figure 2(a)). Higher levels of *dmrt1* gene were found in the testicle than in the anal fin, body kidney, liver, and white muscle ($p \le 0.01$). The *sox9* levels were statistically higher ($p \le 0.0001$) in the gonads than in the anal fin, body kidney, liver, and white muscle (Figure 2(a)). For the *foxl2* and *star* genes, no difference in expression between the testicle and the other organs was observed. For females, no statistically significant difference in expression between the gonads and the other organs was observed (Figure 2(b)).

4. Discussion

There could be a difference in expression between males and females for the reference genes that could be explained by the cease in expression that occurs during spermiogenesis

Table 1: Primers for qPCR analysis.

GenBank	Gene	PCR product length (bp)	Primer name	Primer sequence
KX420683	*ar—androgen receptor*	197	ar-F	5′-CKTGACTCCCCGAACAATCA-3′
			ar-R	5′-AAGGTAGCACGCTGGAACTC-3′
KX420684	*dmrt1—doublesex and mab-3 related transcription factor 1*	129	dmrt1-F	5′-CCACCCTGTTCCACTTCCAG-3′
			dmrt1-R	5′-GAAGWGGATGGTGCTGTGCT-3′
KX420685	*sox9—sex-determining region Y-box 9*	115	sox9-F	5′-AGGCCGATTCCYCTCACTCT-3′
			sox9-R	5′-TGCAYGTCTGTTTTGGGAGT-3′
KX420686	*foxl2—forkhead box L2*	120	foxl2-F	5′-GCCCACCTCGTACAATCCTT-3′
			foxl2-R	5′-CTTAGCTGCTGAGGGTGGTG-3′
KX420678	*cyp17a1—cytochrome P450 family 17 subfamily A polypeptide 1*	134	cyp17a1-F	5′-CCGTCGCTTACCTCCTACAC-3′
			cyp17a1-R	5′-CCGTATCGTTGCTTCCAGGT-3′
KX420679	*star—steroidogenic acute regulatory protein*	111	star-F	5′-AGTACCCTGACCGCCTGTAT-3′
			star-R	5′-TTGTGTCCTGCCCAATCCTC-3′
KX420681	*β-actin*	161	β-actin-F	5′-TGACCCTGAAGTAYCCMATC-3′
			β-actin-R	5′-CTTCTCTCTGTTRGCYTTGG-3′
KX420682	*gapdh—glyceraldehyde 3-phosphate dehydrogenase*	114	gapdh-F	5′-AGACACCCGCTCNTCHATCT-3′
			gapdh-R	5′-TCCACGACTCTGTTGCTGTA-3′
KX420680	*28S rRNA—28S ribosomal RNA*	160	28S-F	5′-TGTTTGTGAATGCAGCCCAA-3′
			28S-R	5′-GACCCCATCCGTTTACCTCT-3′

because of chromatin construction or transcription factors that inhibit the transcription of certain genes [38, 48, 49]. Therefore, the statistically significant difference in expression between males and females, regarding the *cyp17a1*, *ar*, *dmrt1*, and *sox9* genes could be due to a difference in expression of the reference genes. But because of prior observed differences reported on various sturgeon species using NGS techniques [7, 31–33] alongside our approach of using two reference genes to try and limit this possibility of different expression of reference genes, we consider this a good method for determining the true state of expression of the investigated genes.

It is very likely that the determination and consequent differentiation of sturgeon gonads take place during the first year of life, studies showing gonad surface differentiation at 4 months for *A. baerii*, at 4 months for *A. gueldenstaedtii*, at 180 days for *A. naccarii*, at 6 months for *A. brevirostrum*, at 8 months for *A. ruthenus*, and at 6 months for the Bester hybrid [50]. There are no studies regarding the onset of gonadal differentiation in *A. stellatus* individuals, but because the majority of sturgeons present first year differentiation of gonads, it could be feasible that *A. stellatus* also presents this characteristic. At 4 years old, the individuals have undergone the onset of differentiation and are in the primary spermatocyte stage, in the case of males (spermatogenesis), and previtellogenic oocytes, in the case of females (oogenesis).

There is no evidence for a master sex-determining gene present only for males or females, observed through NGS in the case of *A. fulvescens* studies, even though there is evidence of differential gene expression between males and females [7, 31]. On the other hand, *dmrt3*, *igf-1*, *lhx1*, and *sox11* genes were found to be specific to the testicle transcriptome, while *cyp19A1a*, *foxl2*, *gnrhr*, and *nanos3b* were only found in the ovary transcriptome of *Acipenser sinensis* [33]. For *A. naccarii*, five genes were found to be specific for male (*wt1*, *lhx1*, *cyp19A1*, *fhl3*, and *fem1a*) and two (*ar*, *emx2*) for female libraries [32]. Contrary to these studies, we have found an expression of *foxl2* and *ar* in all organs tested from both females and males which rules them out as master sex-determining genes in *A. stellatus* individuals. Because of the fact that all the genes we investigated were present both in males and females, we consider that they are not master sex-determining genes for *A. stellatus* individuals.

For the system of sex determination, the lack of *dmrt1* expression in females is consistent with an XY system which appears not to be the case in sturgeons [40], where both females and males present the expression of this gene, which is in pattern with the ZW system of sex determination, where in birds, *dmrt1* plays the role of master sex-determining gene [9]. It could be feasible that the *dmrt1* gene is expressed at higher levels in sturgeon males, independent of the number of copies that are present in the genome due to double dosage which appears for ZZ/ZW species. This has been observed in our study and in other research involving sturgeon species, such as *A. fulvescens* 13-14-year-old individuals (adolescent female and mature male) [31], *A. baerii* 3-year-old individuals

Table 2: The $2^{-\Delta Cq}$ of each gene. The arithmetic mean of *gapdh* and *β-actin* reference genes was used for normalization. The values are represented by arithmetic means ± SD. One-way ANOVA was used to compare gene expression levels of males versus females for the same organ. NS: not significant.

Organ	*foxl2*			*cyp17a1*			*ar*			*dmrt1*			*sox9*			*star*		
	Mean	±SD	p value	Mean	±SD	p value	Mean	±SD	p value	Mean	±SD	p value	Mean	±SD	p value	Mean	±SD	p value
White muscle ♂	4.75E-02	2.17E-02	NS				1.35E-02	6.46E-03		1.35E-02	7.89E-03	NS	1.53E-03	8.10E-04	NS			NS
White muscle ♀	6.94E-02	3.32E-02					3.58E-02	2.19E-02		3.58E-02	1.10E-02		3.92E-03	2.10E-03				
Liver ♂	3.35E-02	7.33E-03	NS				5.48E-02	2.68E-03		5.48E-02	1.25E-03	NS	8.67E-03	1.27E-03	NS	7.70E-04	1.62E-04	NS
Liver ♀	4.27E-02	2.83E-02					6.93E-02	5.22E-02		6.93E-02	2.39E-02		1.88E-02	7.82E-03		9.23E-04	2.09E-04	
Body kidney ♂	3.38E-03	8.74E-04	NS				8.90E-03	1.02E-03		8,90E-03	1.60E-04	NS	4.24E-03	9.62E-04	NS	2.18E-04	5.30E-05	NS
Body kidney ♀	2.10E-03	1.20E-03					7.64E-03	9.64E-04		7.64E-03	4.17E-04		3.64E-03	8.22E-04		7.80E-03	1.53E-02	
Testicle	6.59E-03	3.44E-03	NS	2.67E-01	7.10E-02	$p \leq 0.0001$	1.48E-01	9.26E-02	$p \leq 0.01$	1.48E-01	2.03E-01	$p \leq 0.01$	1.76E-01	8.62E-03	$p \leq 0.0001$	1.28E-02	4.17E-04	NS
Ovary	2.99E-02	2.05E-02		3.29E-04	2.62E-04		5.40E-03	2.25E-03		5.40E-03	7.03E-03		2.02E-02	1.35E-02		1.42E-03	8.71E-04	
Anal fin ♂	2.18E-02	8.82E-03	NS	2.38E-04	2.72E-05	NS	1.81E-02	1.18E-03	NS	1.81E-02	4.24E-03	NS	3.74E-02	3.06E-03	NS	2.39E-04	1.42E-04	NS
Anal fin ♀	2.77E-02	2.72E-02		4.69E-04	1.97E-04		1.72E-02	4.27E-03		1.72E-02	1.23E-03		3.86E-02	2.91E-03		8.61E-04	7.13E-04	

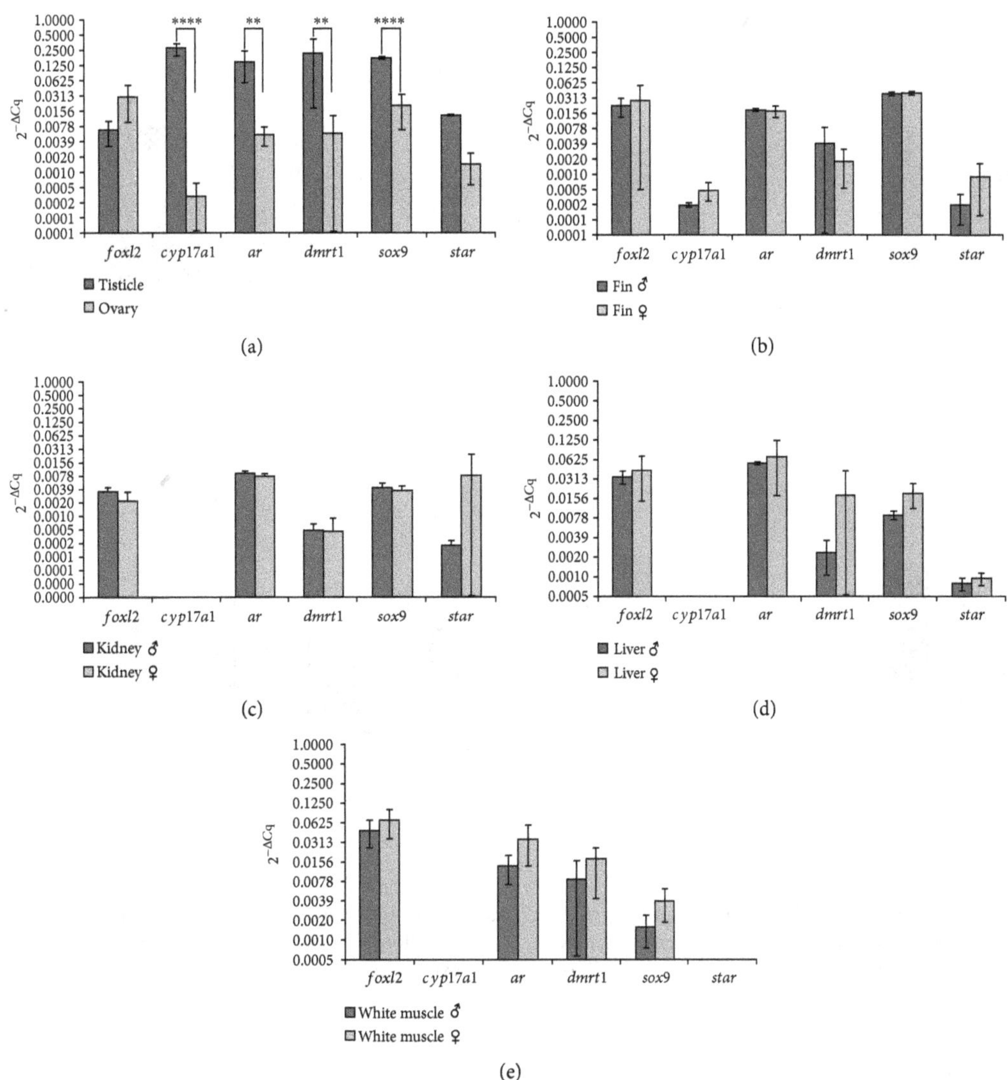

FIGURE 1: The $2^{-\Delta Cq}$ comparison of *foxl2, cyp17a1, ar, dmrt1, sox9*, and *star* genes between males and females in the gonads (a), anal fin (b), body kidney (c), liver (d), and white muscle (e). The arithmetic mean of *gapdh* and *β-actin* reference genes was used for normalization. The data points are represented by arithmetic means ± SD on a logarithmic scale in base two. The statistical significance of the difference in expression was tested with one-way ANOVA using Tukey correction ($^{**}p \leq 0.01$ and $^{****}p \leq 0.0001$).

with immature gonads [36, 39], *A. baerii* 16–18-month-old juveniles [35, 39], *A. gueldenstaedtii* 550-day-old and 1600-day-old individuals [40], and *A. sinensis* 3- and 4-year-old individuals. This could be due to specific targeting of *dmrt1* gene by an unknown factor that determines a higher expression in males. It could also be feasible that in females, there are factors that modulate the expression of *dmrt1* so that it is lower than in males. One such factor could be a microRNA molecule that specifically targets the *dmrt1* gene in females, silencing or reducing its expression as also observed in silkworm where a piRNA molecule can inhibit the *masc* gene leading to the formation of female individuals [51].

There is a lack of sexual dimorphism regarding the *dmrt1* gene in the case of *A. gueldenstaedtii* 9-month-old juveniles with undifferentiated gonads [7], *A. sinensis* 3-year-old gametogenetic juveniles [33], and *Scaphirhynchus platorynchus* above 2-year-old individuals with fully developed gonads [38], which could be due to the developmental stage at which the gonads were sampled. The lack of expression of the *dmrt1* gene could be due to incomplete coverage in NGS or the developmental stage at which the 6-month-old *A. naccarii* individuals were at the time of sampling [32]. The different developmental stages and species that were investigated show a nonspecific time of expression of the *dmrt1* gene in the case of sturgeons. Each species may have a specific moment in the development of the gonads at which the expression of *dmrt1* occurs in different mRNA levels between females and males. The *dmrt1* gene, a candidate for double expression in males, is mostly expressed in higher levels in males than in females, but there are developmental stages of sturgeons in which the *dmrt1* gene is not expressed in different levels [7, 33, 38]. Even though this pattern of expression could be species specific, it could be argued that the expression of *dmrt1* starts at the onset of differentiation and decreases to a stable level after the gonads are fully developed. This is why no difference in expression could be

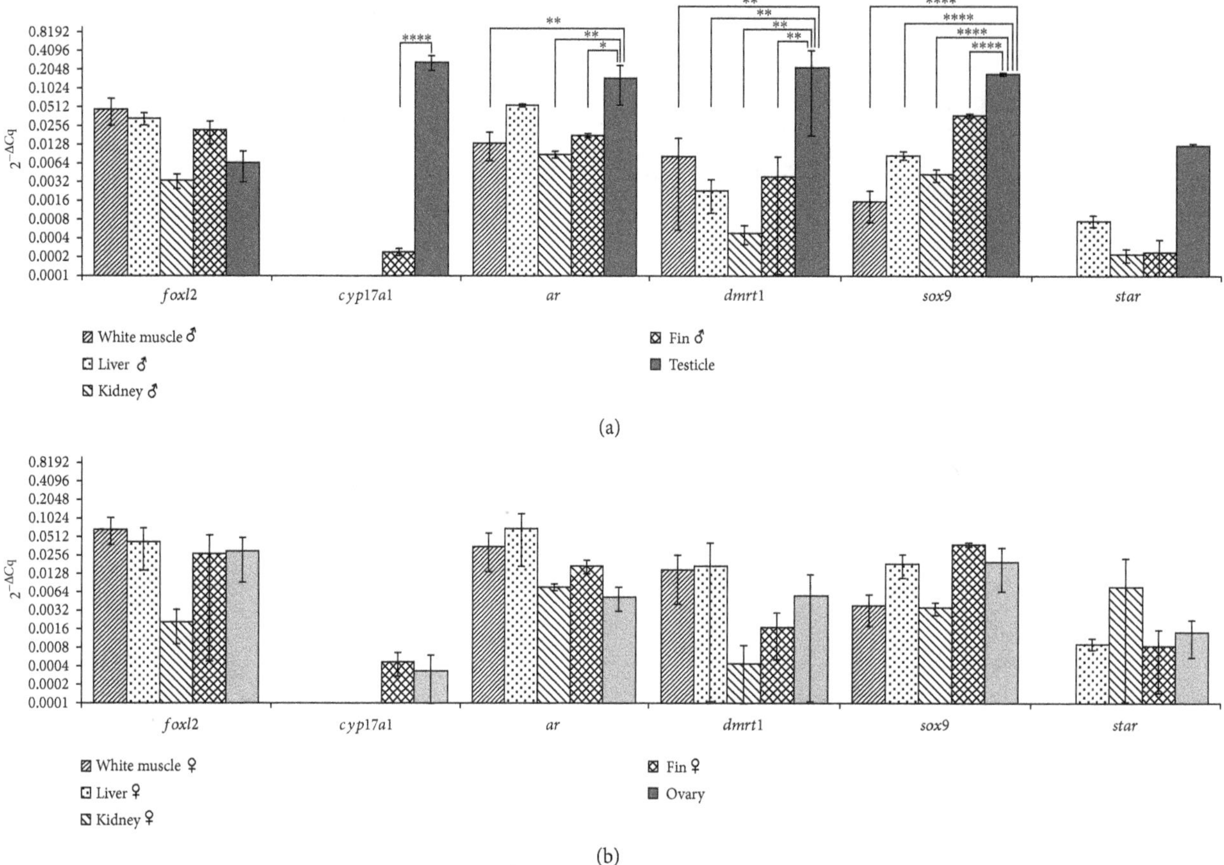

FIGURE 2: The $2^{-\Delta Cq}$ comparison of *foxl2*, *cyp17a1*, *ar*, *dmrt1*, *sox9*, and *star* genes between organs from males (a) and females (b). The arithmetic mean of *gapdh* and *β-actin* reference genes was used for normalization. The data points are represented by arithmetic means ± SD on a logarithmic scale in base two. The statistical significance of the expression comparison was tested with one-way ANOVA using Tukey correction ($^{*}p \leq 0.05$, $^{**}p \leq 0.01$, and $^{****}p \leq 0.0001$).

distinguished for *A. gueldenstaedtii* 9-month-old individuals with undifferentiated gonads [7] and for adult *S. platorynchus* individuals with fully developed gonads [38]. The fact that for *A. stellatus* male individuals the expression of *dmrt1* is higher than in females shows that this gene is involved in the development of the male gonads at the spermatogenesis stage.

The *sox9* gene has been found in *Acipenser sturio* and *A. fulvescens* in the genome of both males and females, through sequencing [31, 52], and is overexpressed in the testicles when compared to ovaries in immature 3-year-old *A. baerii* individuals [36, 39]. We observed a higher expression of *sox9* in male gonads that might indicate that the gene is involved in the development of testicles. In the case of 16-month-old juvenile *A. baerii* individuals with differentiated gonads, the expression of *sox9* in gonads is not different between females and males [35, 39], the same as in *A. gueldenstaedtii* 9-month-old individuals which indicates that the *sox9* gene is not involved in the onset of differentiation of sturgeon gonads. In *A. sinensis* 3-year-old individuals, the gene was present in both females and males through NGS [33], similar to 6-month-old *A. naccarii* individuals regarding the same gene [32]. For the *A. stellatus* individuals, the expression of *sox9* at 4 years old shows that this gene is involved in gametogenesis.

The *foxl2* gene has a higher expression in females for *S. platorynchus* individuals [38] and for *A. gueldenstaedtii* 9-month-old juvenile presumptive females [7]. This is in contrast to our observation that the difference in expression is not statistically significant between males and females. The *foxl2* gene is involved in the onset of female gonad differentiation [7], and in the case of *S. platorynchus* [38], it appears to be involved in female gonad differentiation because of the age of the sturgeons investigated. These results may be due to the different species studied; in our case, the individuals were well past starting differentiation and well into gonadal stages like spermatogenesis and oogenesis; therefore, we rule out the possibility of *foxl2* involvement in development at this stage for *A. stellatus* even though it is in contrast to what was previously described for *S. platorynchus*.

The *cyp17a1* gene is known to be a Leydig cell factor modulating androgen production, and a higher expression in male gonads than in female gonads is to be expected. We have observed this higher expression between males and females, and we also observed that it is not expressed in other tissues except for anal fin. The expression in the anal fin is lower than that of male gonads but not different from that of female gonads. In immature *A. baerii* gonads (16-month-old and 3-year-old individuals), the expression of *cyp17a1* was higher in males; however, different from our discovery is that, in this case, the expression was also detected in the muscle, liver, and kidney [39], which was not observed in our study. This pattern of expression, higher in male gonads than in female gonads, could be due to the need for steroidogenic hormones in males during gametogenesis for which the *cyp17a1* gene is responsible. The expression of the *cyp17a1* gene, which is involved in converting progestins into androgens in testicles, is correlated with the expression of the *ar* gene, which encodes for an androgen receptor. Both being in higher levels in males than in females in the case of *A. baerii* and in our own study, this coupled with the age at which this pattern is observed could indicate that androgens are the main mediators of vertebrate masculinization [39, 53, 54] and in this case sturgeon male gonad development, after testicular differentiation especially for *A. stellatus* of 4 years of age where the lamellae are clearly visible for female gonads and the surface differentiation already took place. The *cyp17a1*, *ar*, *dmrt1*, and *sox9* genes have different expression patterns in male gonads than in female gonads, which could indicate that these are probably involved in the development of male gonads. The *foxl2* gene is involved in female gonad differentiation in other species, but no significant difference in expression was detected so that we can conclude the same thing for the stage at which the *A. stellatus* individuals were when sampling took place. The *star* gene is normally involved in the male developmental pathway; however, the individuals we investigated may be over the age at which the difference in expression occurs.

Besides the *cyp17a1* and *ar* genes that are possibly involved in the male gonad development pathway, we also observed the expression of the *star* gene. The expression of the *star* gene in our study proved it was not significantly different between females and males, which is in contrast to the situation of *A. baerii* individuals [39]. The *star* gene is involved in steroid synthesis and should have a higher expression after gonadal differentiation. For our samples, this is correlated with reaching gonadal maturity at four years which is near the five-year mark for *A. stellatus* male maturity [3, 4].

5. Conclusions

We have observed that *cyp17a1*, *ar*, *dmrt1*, and *sox9* genes are involved in the gonad development of *A. stellatus* because of the significant difference in expression between males and females. The fact that the *cyp17a1*, *ar*, *dmrt1*, and *sox9* genes have a higher expression in male gonads than in the other organs tested could imply that these genes are involved in the male pathway for gonad development.

No significant difference in the expression of the investigated genes was detected in tissues that could be used for noninvasive sex identification in early stages (white muscle and anal fin).

Pinpointing the developmental stage at which these genes start presenting a difference in expression could help in the study of sexual development in sturgeons; the data presented here could be the groundwork for future studies that will focus on younger individuals of the same species. Our results are in agreement with other sturgeon research that has attempted to find a master sex-determining gene which would suggest that it has yet to be discovered.

Conflicts of Interest

The authors declare that they have no conflicts of interest.

Acknowledgments

This work was supported by the Executive Agency for Higher Education, Research, Development and Innovation Funding project 53PTE/2016 "Technology for selection and genetic improvement in order to increase profitability of sturgeon's aquaculture" and the European Cooperation in Science and Technology 2016-2020 (CA15219) "Developing new genetic tools for bioassessment of aquatic ecosystems in Europe."

References

[1] R. Billard and G. Lecointre, "Biology and conservation of sturgeon and paddlefish," *Reviews in Fish Biology and Fisheries*, vol. 10, no. 4, pp. 355–392, 2000.

[2] A. Ludwig, "A sturgeon view on conservation genetics," *European Journal of Wildlife Research*, vol. 52, no. 1, pp. 3–8, 2006.

[3] T. N. Shubina, A. A. Popova, and V. P. Vasil'ev, "Acipenser stellatus Pallas, 1771," in *The Freshwater Fishes of Europe VoL 1, Part 2. General Introduction to Fishes, Acipenseriformes*, J. Holčik, Ed., AULA Verlag, Wiesbaden, 1989.

[4] V. J. Birstein, "Sturgeons and paddlefishes: threatened fishes in need of conservation," *Conservation Biology*, vol. 7, no. 4, pp. 773–787, 1993.

[5] J. Holčik, *The freshwater Fishes of Europe. General Introduction to Fishes*, vol. 1-2, AULA Verlag, Wiesbaden, 1989.

[6] M. Moghim, A. R. Vajhi, A. Veshkini, and M. Masoudifard, "Determination of sex and maturity in *Acipenser stellatus* by using ultrasonography," *Journal of Applied Ichthyology*, vol. 18, no. 4–6, pp. 325–328, 2002.

[7] S. Hagihara, R. Yamashita, S. Yamamoto et al., "Identification of genes involved in gonadal sex differentiation and the dimorphic expression pattern in undifferentiated gonads of Russian sturgeon *Acipenser gueldenstaedtii* Brandt & Ratzeburg, 1833," *Journal of Applied Ichthyology*, vol. 30, no. 6, pp. 1557–1564, 2014.

[8] P. Martínez, A. M. Viñas, L. Sánchez, N. Díaz, L. Ribas, and F. Piferrer, "Genetic architecture of sex determination in fish: applications to sex ratio control in aquaculture," *Frontiers in Genetics*, vol. 5, 2014.

[9] A. E. Quinn, S. D. Sarre, T. Ezaz, J. A. Marshall Graves, and A. Georges, "Evolutionary transitions between mechanisms of sex determination in vertebrates," *Biology Letters*, vol. 7, no. 3, pp. 443–448, 2011.

[10] I. E. Samonte-Padilla, C. Eizaguirre, J. P. Scharsack, T. L. Lenz, and M. Milinski, "Induction of diploid gynogenesis in an evolutionary model organism, the three-spined stickleback (*Gasterosteus aculeatus*)," *BMC Developmental Biology*, vol. 11, no. 1, p. 55, 2011.

[11] M. Schartl, "A comparative view on sex determination in medaka," *Mechanisms of Development*, vol. 121, no. 7-8, pp. 639–645, 2004.

[12] R. H. Devlin and Y. Nagahama, "Sex determination and sex differentiation in fish: an overview of genetic, physiological, and environmental influences," *Aquaculture*, vol. 208, no. 3-4, pp. 191–364, 2002.

[13] S. Keyvanshokooh and A. Gharaei, "A review of sex determination and searches for sex-specific markers in sturgeon," *Aquaculture Research*, vol. 41, no. 9, pp. e1–e7, 2010.

[14] S. B. M. Kraak and E. M. A. De Looze, "A new hypothesis on the evolution of sex determination in vertebrates; big females ZW, big males XY," *Netherlands Journal of Zoology*, vol. 43, no. 3, pp. 260–273, 1992.

[15] N. F. Parnell and J. T. Streelman, "Genetic interactions controlling sex and color establish the potential for sexual conflict in Lake Malawi cichlid fishes," *Heredity*, vol. 110, no. 3, pp. 239–246, 2013.

[16] F. A. Chapman, J. P. Van Eenennaam, and S. I. Doroshov, "The reproductive condition of white sturgeon, Acipenser transmontanus, in San Francisco Bay, California," *Fishery Bulletin*, vol. 94, no. 4, pp. 628–634, 1996.

[17] D. Fopp-Bayat, "Meiotic gynogenesis revealed not homogametic female sex determination system in Siberian sturgeon (*Acipenser baeri Brandt*)," *Aquaculture*, vol. 305, no. 1–4, pp. 174–177, 2010.

[18] S. R. Flynn, M. Matsuoka, M. Reith, D. J. Martin-Robichaud, and T. J. Benfey, "Gynogenesis and sex determination in shortnose sturgeon, Acipenser brevirostrum Lesuere," *Aquaculture*, vol. 253, no. 1-4, pp. 721-727, 2006.

[19] M. H. Saber and A. Hallajian, "Study of sex determination system in ship sturgeon, *Acipenser nudiventris* using meiotic gynogenesis," *Aquaculture International*, vol. 22, no. 1, pp. 273–279, 2014.

[20] A. L. Van Eenennaam, J. P. Van Eenennaam, J. F. Medrano, and S. I. Doroshov, "Brief communication. Evidence of female heterogametic genetic sex determination in white sturgeon," *Journal of Heredity*, vol. 90, no. 1, pp. 231–233, 1999.

[21] N. Omoto, M. Maebayashi, S. Adachi, K. Arai, and K. Yamauchi, "Sex ratios of triploids and gynogenetic diploids induced in the hybrid sturgeon, the bester (*Huso huso female×Acipenser ruthenus male*)," *Aquaculture*, vol. 245, no. 1–4, pp. 39–47, 2005.

[22] W. L. Shelton and S. D. Mims, "Evidence for female heterogametic sex determination in paddlefish *Polyodon spathula* based on gynogenesis," *Aquaculture*, vol. 356-357, pp. 116–118, 2012.

[23] R. Arai, *Fish Karyotypes: a Check List*, Publisher Springer, Japan, 2011.

[24] M. Bahmani and R. Kazemi, "Histological study of gonad in young cultured sturgeon," *Shilat Science Magazine*, vol. 7, pp. 1–16, 1998.

[25] B. Falahatkar, S. R. Akhavan, A. Abbasalizadeh, and M. H. Tolouei, "Sex identification and sexual maturity stages in farmed great sturgeon, *Huso huso* L. through biopsy," *Iranian Journal of Veterinary Research*, vol. 14, no. 2, pp. 133–139, 2013.

[26] A. Hurvitz, K. Jackson, G. Degani, and B. Levavi-Sivan, "Use of endoscopy for gender and ovarian stage determinations in Russian sturgeon (*Acipenser gueldenstaedtii*) grown in aquaculture," *Aquaculture*, vol. 270, no. 1–4, pp. 158–166, 2007.

[27] S. J. Hernandez-Divers, R. S. Bakal, B. H. Hickson et al., "Endoscopic sex determination and gonadal manipulation in Gulf of Mexico sturgeon (*Acipenser oxyrinchus desotoi*)," *Journal of Zoo and Wildlife Medicine*, vol. 35, no. 4, pp. 459–470, 2004.

[28] S. Wuertz, S. Gaillard, F. Barbisan et al., "Extensive screening of sturgeon genomes by random screening techniques revealed no sex-specific marker," *Aquaculture*, vol. 258, no. 1–4, pp. 685–688, 2006.

[29] S. Keyvanshokooh, M. Pourkazemi, and M. R. Kalbassi, "The RAPD technique failed to identify sex-specific sequences in beluga (*Huso huso*)," *Journal of Applied Ichthyology*, vol. 23, no. 1, pp. 1-2, 2007.

[30] C. R. McCormick, D. H. Bos, and J. A. DeWoody, "Multiple molecular approaches yield no evidence for sex-determining genes in lake sturgeon (*Acipenser fulvescens*)," *Journal of Applied Ichthyology*, vol. 24, no. 6, pp. 643–645, 2008.

[31] M. C. Hale, J. R. Jackson, and J. A. DeWoody, "Discovery and evaluation of candidate sex-determining genes and xenobiotics in the gonads of lake sturgeon (*Acipenser fulvescens*)," *Genetica*, vol. 138, no. 7, pp. 745–756, 2010.

[32] M. Vidotto, A. Grapputo, E. Boscari et al., "Transcriptome sequencing and *de novo* annotation of the critically endangered Adriatic sturgeon," *BMC Genomics*, vol. 14, no. 1, p. 407, 2013.

[33] H. Yue, C. Li, H. Du, S. Zhang, and Q. Wei, "Sequencing and *de novo* assembly of the gonadal transcriptome of the endangered Chinese sturgeon (*Acipenser sinensis*)," *PLoS One*, vol. 10, no. 6, article e0127332, 2015.

[34] D. Baron, R. Houlgatte, A. Fostier, and Y. Guiguen, "Large-scale temporal gene expression profiling during gonadal differentiation and early gametogenesis in rainbow trout," *Biology of Reproduction*, vol. 73, no. 5, pp. 959–966, 2005.

[35] J. Berbejillo, A. Martinez-Bengochea, G. Bedó, and D. Vizziano-Cantonnet, "Molecular characterization of testis differentiation in the Siberian sturgeon, *Acipenser baerii*," *Indian Journal of Science and Technology*, vol. 4, no. S8, pp. 71-72, 2011.

[36] J. Berbejillo, A. Martinez-Bengochea, G. Bedó, and D. Vizziano-Cantonnet, "Expression of *dmrt1* and *sox9* during gonadal development in the Siberian sturgeon (*Acipenser baerii*)," *Fish Physiology and Biochemistry*, vol. 39, no. 1, pp. 91–94, 2013.

[37] A. Hurvitz, G. Degani, D. Goldberg, S. Y. Din, K. Jackson, and B. Levavi-Sivan, "Cloning of FSHβ, LHβ, and glycoprotein α subunits from the Russian sturgeon (*Acipenser gueldenstaedtii*), β-subunit mRNA expression, gonad development, and steroid levels in immature fish," *General and Comparative Endocrinology*, vol. 140, no. 1, pp. 61–73, 2005.

[38] J. J. Amberg, R. Goforth, T. Stefanavage, and M. S. Sepúlveda, "Sexually dimorphic gene expression in the gonad and liver of shovelnose sturgeon (*Scaphirhynchus platorynchus*)," *Fish Physiology and Biochemistry*, vol. 36, no. 4, pp. 923–932, 2010.

[39] J. Berbejillo, A. Martinez-Bengochea, G. Bedo, F. Brunet, J. N. Volff, and D. Vizziano-Cantonnet, "Expression and phylogeny of candidate genes for sex differentiation in a primitive fish species, the Siberian sturgeon, *Acipenser baerii*," *Molecular Reproduction and Development*, vol. 79, no. 8, pp. 504–516, 2012.

[40] M. Fajkowska, M. Rzepkowska, D. Adamek, T. Ostaszewska, and M. Szczepkowski, "Expression of *dmrt1* and *vtg* genes during gonad formation, differentiation and early maturation in cultured Russian sturgeon *Acipenser gueldenstaedtii*," *Journal of Fish Biology*, vol. 89, no. 2, pp. 1441–1449, 2016.

[41] X. Q. Leng, H. J. Du, C. J. Li, and H. Cao, "Molecular characterization and expression pattern of *dmrt1* in the immature

Chinese sturgeon *Acipenser sinensis*," *Journal of Fish Biology*, vol. 88, no. 2, pp. 567–579, 2016.

[42] S. R. Flynn and T. J. Benfey, "Sex differentiation and aspects of gametogenesis in shortnose sturgeon *Acipenser brevirostrum* Lesueur," *Journal of Fish Biology*, vol. 70, no. 4, pp. 1027–1044, 2007.

[43] S. A. Bustin, V. Benes, J. A. Garson et al., "The MIQE guidelines: *m*inimum *i*nformation for publication of *q*uantitative real-time PCR *e*xperiments," *Clinical Chemistry*, vol. 55, no. 4, pp. 611–622, 2009.

[44] C. L. Andersen, J. L. Jensen, and T. F. Ørntoft, "Normalization of real-time quantitative reverse transcription-PCR data: a model-based variance estimation approach to identify genes suited for normalization, applied to bladder and colon cancer data sets," *Cancer Research*, vol. 64, no. 15, pp. 5245–5250, 2004.

[45] B. Levavi-Sivan, H. Safarian, H. Rosenfeld, A. Elizur, and A. Avitan, "Regulation of gonadotropin-releasing hormone (GnRH)-receptor gene expression in tilapia: effect of GnRH and dopamine," *Biology of Reproduction*, vol. 70, no. 6, pp. 1545–1551, 2004.

[46] S. S. Shapiro and M. B. Wilk, "An analysis of variance test for normality (complete samples)," *Biometrika*, vol. 52, no. 3-4, pp. 591–611, 1965.

[47] N. M. Razali and Y. B. Wah, "Power comparisons of Shapiro-Wilk, Kolmogorov-Smirnov, Lilliefors and Anderson-Darling tests," *Journal of Statistical Modeling and Analytics*, vol. 2, no. 1, pp. 21–33, 2011.

[48] K. Steger, "Haploid spermatids exhibit translationally repressed mRNAs," *Anatomy and Embryology*, vol. 203, no. 5, pp. 323–334, 2001.

[49] H. Tanaka and T. Baba, "Gene expression in spermiogenesis," *CMLS Cellular and Molecular Life Sciences*, vol. 62, no. 3, pp. 344–354, 2005.

[50] M. Rzepkowska and T. Ostaszewska, "Proliferating cell nuclear antigen and Vasa protein expression during gonadal development and sexual differentiation in cultured Siberian (*Acipenser baerii* Brandt, 1869) and Russian (*Acipenser gueldenstaedtii* Brandt & Ratzeburg, 1833) sturgeon," *Reviews in Aquaculture*, vol. 6, no. 2, pp. 75–88, 2014.

[51] T. Kiuchi, H. Koga, M. Kawamoto et al., "A single female-specific piRNA is the primary determiner of sex in the silkworm," *Nature*, vol. 509, no. 7502, pp. 633–636, 2014.

[52] A. K. Hett, C. Pitra, I. Jenneckens, and A. Ludwig, "Characterization of sox9 in European Atlantic sturgeon (*Acipenser sturio*)," *Journal of Heredity*, vol. 96, no. 2, pp. 150–154, 2005.

[53] B. Borg, "Androgens in teleost fishes," *Comparative Biochemistry and Physiology Part C: Pharmacology, Toxicology and Endocrinology*, vol. 109, no. 3, pp. 219–245, 1994.

[54] J. Brennan and B. Capel, "One tissue, two fates: molecular genetic events that underlie testis versus ovary development," *Nature Reviews. Genetics*, vol. 5, no. 7, pp. 509–521, 2004.

5

Transcriptomic Profiling of Fruit Development in Black Raspberry *Rubus coreanus*

Qing Chen,[1] Xunju Liu,[1] Yueyang Hu,[1] Bo Sun,[1] Yaodong Hu,[2] Xiaorong Wang,[3] Haoru Tang,[1] and Yan Wang[3]

[1]*College of Horticulture, Sichuan Agricultural University, Chengdu, Sichuan 611130, China*
[2]*Science and Technology Management Division, Sichuan Agricultural University, Chengdu, Sichuan 611130, China*
[3]*Institute of Pomology and Olericulture, Sichuan Agricultural University, Chengdu, Sichuan 611130, China*

Correspondence should be addressed to Yan Wang; wangyanwxy@163.com

Academic Editor: Marco Gerdol

The wild *Rubus* species *R. coreanus*, which is widely distributed in southwest China, shows great promise as a genetic resource for breeding. One of its outstanding properties is adaptation to high temperature and humidity. To facilitate its use in selection and breeding programs, we assembled de novo 179,738,287 *R. coreanus* reads (125 bp in length) generated by RNA sequencing from fruits at three representative developmental stages. We also used the recently released draft genome of *R. occidentalis* to perform reference-guided assembly. We inferred a final 95,845-transcript reference for *R. coreanus*. Of these genetic resources, 66,597 (69.5%) were annotated. Based on these results, we carried out a comprehensive analysis of differentially expressed genes. Flavonoid biosynthesis, phenylpropanoid biosynthesis, plant hormone signal transduction, and cutin, suberin, and wax biosynthesis pathways were significantly enriched throughout the ripening process. We identified 23 transcripts involved in the flavonoid biosynthesis pathway whose expression perfectly paralleled changes in the metabolites. Additionally, we identified 119 nucleotide-binding site leucine-rich repeat (NBS-LRR) protein-coding genes, involved in pathogen resistance, of which 74 were in the completely conserved domain. These results provide, for the first time, genome-wide genetic information for understanding developmental regulation of *R. coreanus* fruits. They have the potential for use in breeding through functional genetic approaches in the near future.

1. Introduction

The genus *Rubus* L. comprises 900–1000 species and has a worldwide distribution (excluding Antarctica) with various climatic adaptations [1]. Plants used in fruit production are mainly from two subgenera, *Rubus* and *Idaeobatus*. Blackberries and raspberries are the most commonly cultivated fruits in these two subgenera. They are deemed functional fruits, mainly due to being rich sources of health-promoting antioxidant or "nutraceutical" compounds (i.e., anthocyanins, phenolics, and ellagic acid) in fresh fruits [2] and anticancer properties in fruit extracts [3]. Historically, they have been used in traditional Chinese medicine and are mentioned in the Compendium of Materia Medica (Bencao Gangmu) compiled by Li Shizhen (1518–1593) during the Ming Dynasty. Chinese or Korean black raspberry *R. coreanus* Miq., in the subgenus *Idaeobatus*, is named for the dark red (or black) color of its fruits when mature. Earlier investigations found that black raspberry fruits contained higher concentrations of the nutritional ingredients mentioned above than either red raspberry or blackberry [4]. However, Chinese black raspberry is not as popular as the other two species as much less effort is given to its cultivation and there is only a limited choice of cultivars available. After a thorough investigation of the biochemical components in fruit [5], researchers from South Korea provided the first transcriptome analysis of what they believed to be *R. coreanus* [6]. Unfortunately, contrary to what is reported in their paper [7], the species they studied was the commercially grown North American black raspberry

R. occidentalis, often confused with *R. coreanus*. In 2016, the first draft genome for *R. occidentalis*, 243 Mb in size, became publicly available [8]. It is the most useful *Rubus* genetic resource to date.

R. coreanus has been used as a valuable genetic donor in several *Rubus*-breeding programs [9, 10] because of its outstanding disease resistance and high productivity. *R. coreanus* cultivars are also commercialized in South Korea [11]. The lack of a genetic reference for *R. coreanus* has become a barrier to application of modern breeding techniques, such as marker-assisted selection and transgenic strategies. Within the past ten years, we have made a comprehensive study of *Rubus* species in China, mostly those endemic to the country, focusing especially on wild species distributed in the south [12]. The relatively high temperature and humidity of the area may have led to the development of unique disease resistance characteristics in these *Rubus* species. Given that viral and fungal diseases have become one of the main threats hindering the development of commercial cultivation of *Rubus* [13], exploring underlying mechanisms for disease resistance and finding new candidate genetic resources could facilitate selection and breeding.

Fruit maturation is a complex process involving genetic regulation closely linked with environmental signaling of palatability. Pigment deposition and the resulting color change, and sugar accumulation, usually coupled with depletion of titratable acid leading to formation of specific sensory traits, are common signs of fruit ripeness [14]. Dissection of the intrinsic genetic changes of fruit maturation could bring new insights into understanding the consequences of application of a particular agronomic practice at a particular time. For example, *Fragaria × ananassa* has traditionally been classified as nonclimacteric because its ripening process is not governed by ethylene. However, global analysis of transcriptome data and the *ethylene response factor* (ERF) gene family has identified involvement of ethylene in ripening of the receptacle at specific tissue/developmental stages [15]. Moreover, unveiling the gene expression atlas of fruit maturation could enable greater understanding of biosynthesis of bioactive compounds, a necessary step toward breeding new varieties for health benefits. Considerable effort has been made in this regard for fruits such as grapes [16], blueberries [17], and tomatoes [18]; in comparison, very little effort has been made for *Rubus* species.

Therefore, as a first step towards understanding gene expression during fruit ripening in *R. coreanus*, this study presents results of a comprehensive analysis of transcriptome data from fruits at three representative developmental stages. Both de novo and reference-guided assembly were carried out to maximize the possibility of finding potential transcripts. We also investigated the genes for long noncoding RNAs. Differentially expressed genes, specifically, (1) genes leading to flavonoid biosynthesis and (2) plant nucleotide-binding site leucine-rich repeat (NBS-LRR) genes, which contribute to biotic and abiotic stress responses, were analyzed. It is hoped that exploration of the genetics of *R. coreanus* may prove to be a profitable endeavor by providing valuable information for *Rubus* breeding.

2. Results and Discussion

2.1. Transcriptome Sequencing and Sequence Assembly. Although *Rubus* species are some of the most popular functional fruits in the world, it is only recently that genomic resources have become available for the genus [8]. In Hyun et al.'s study of *R. occidentalis* (which they mistook for *R. coreanus* [6]) from the perspective of fruit morphology and phenological traits [7], transcriptome analysis involved mRNA isolated only from fruits sampled 20 days after anthesis at an intermediate stage of ripening [6]. This may have underestimated the genetic information for the species. In the present study, 179.74 million 125 bp paired-end raw reads were generated from fruit libraries of three developmental stages. After trimming adapter-related reads and filtering low-quality reads, 65.27 million bases were subjected to error correction. Finally, 174.79 million reads comprising 43.7 gigabases were used to assemble the reference. In total, 78.80 million bases were assembled into 95,845 transcripts, with an N50 length of 1242 bp (Table 1). The generated 125 bp paired-end reads are available at NCBI Sequence Read Archive SRR6001072 to SRR6001077 associated with BioProject PRJNA401210.

To assess the quality of the assembly further, Bowtie (v2.2.9) [19] was employed to align all reads back to the reference. Of the reads, 83.66% could be aligned, with 64.17% aligning concordantly and uniquely to the final version of the reference. In contrast, only 52.97% of the total reads could be aligned to the genome-guided assembly, indicating high divergence between *R. coreanus* and *R. occidentalis*. This result may partly explain the previously observed phenomenon that although these two species are easy to cross, the F1 progenies are completely sterile [20]. Therefore, it is reasonable that a proportion of our transcripts could not be mapped to the reference. In addition, when evaluated against the complete 1440 plant-specific orthologs in the Benchmarking Universal Single-Copy Orthologs (BUSCO) database [21], the largest proportion of the assembly (95.3%) was complete, with only 27 (1.9%) fragmented and 41 (2.8%) missing orthologs. These results indicate high completeness of our assembly.

Taking expression values into consideration, we plotted the transcripts per million (TPM) distribution patterns of all transcripts (Figure 1(a)). Predominant portions of the transcripts were at low abundance. If using three TPM as a threshold, each fraction of 39,039 transcripts could be viewed as from one genuine gene. This number is within the range of gene numbers from the *R. occidentalis* genome project [8]. Taking this read coverage information before abundance filtering, the N50 value for the top Ex% transcripts was calculated (Figure 1(b)). The maximum N50 value (2142 bp) was reached when taking 96% of the upper expressed gene products.

2.2. Functional Annotations for the R. coreanus Transcriptome. Annocript pipeline [22] was employed to annotate transcripts and coding peptides. Searches for homologous counterparts in the manually annotated, nonredundant protein sequence database Swiss-Prot (SP) and the subset UniProt

Table 1: Overview of the assembly.

	Rubus occidentalis genome guided	De novo	Final reference
Total number of transcripts	47,239	296,591	95,845
Total nucleotides	80,446,066	214,031,901	78,800,996
Average length (bp)	1703	813	822
Minimum length (bp)	102	201	102
Maximum length (bp)	21,369	14,054	17,356
N50 (bp)	2496	1603	1242

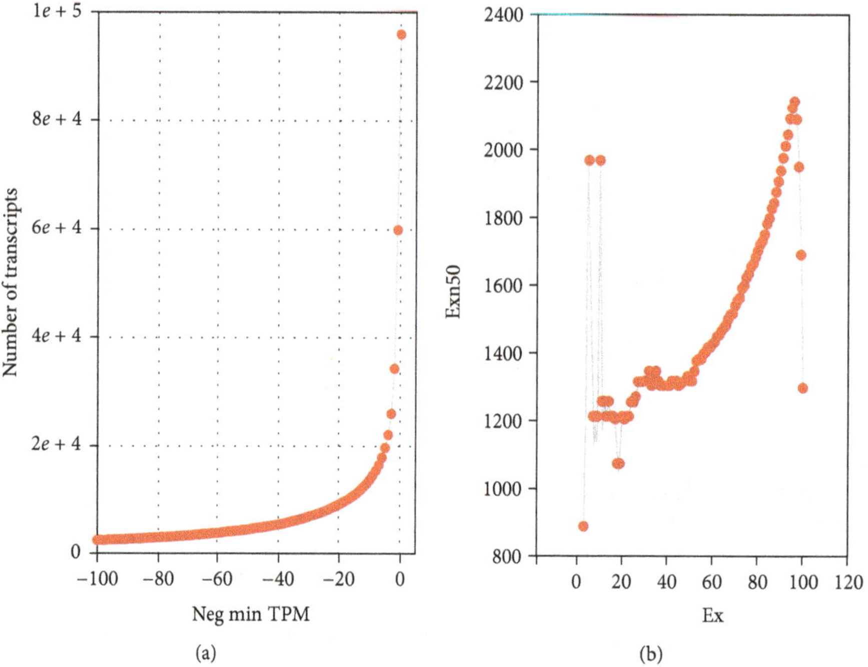

Figure 1: Expression statistics for all transcripts (a) and ExN50 distribution of the assembly (b). Neg min TPM in (a) indicates the negative value of a given minimum expression level as transcripts per million (TPM) reads. Ex indicates that x% of the assembled transcript nucleotides can be found in contigs that are at least of ExN50 length.

Reference Clusters Uniref90 (UF90) database by the blastx algorithm resulted in 47,090 (49.99%) of the raw transcripts generating hits in SP and 66,520 (69.40%) of the transcripts with homologs in UF90. More specifically, 15.60% of the SP hits and 25.43% of the UF90 hits were covered over 80% by the enquiry sequences. In the case of gene ontology (GO) assignment, 51,520 (53.758%) transcripts could be classified into Biological Process (28,547), Cellular Component (31,728), or Molecular Function (41,971) categories. Searches in the Kyoto Encyclopedia of Genes and Genomes (KEGG) Orthology (KO) database against related plants resulted in KO identifiers of 42,769 of the transcripts being assigned to the corresponding pathways. Through PORTRAIT noncoding potential evaluation using a new support vector machine-based algorithm [23], 2178 long noncoding RNA- (lncRNA-) coding genes were also discovered. Taken together, these results suggest that the de novo assembled reference covered a wide range of *Rubus* genetic information, which provides a valuable resource for facilitating the discovery of novel genes involved in specific physiological and developmental processes.

2.3. Analysis of Differential Gene Expression across the Three Developmental Stages of R. coreanus Fruit. We mapped all reads from each fruit developmental stage and estimated transcript counts against the reference using the RSEM method [24]. Transcripts with less than three TPM across the three stages were filtered in the subsequent differentially expressed gene (DEG) assay based on the above analysis. Three different expression analysis packages were used for DEG detection: (1) DESeq2 [25], which uses a Wald test; (2) edgeR [26], which uses a likelihood ratio test; and (3) limma-voom [27], which uses a moderated t-test, to compare expression differences between fruit stages. In the consensus results, 211 transcripts were expressed differentially in red fruits compared to green fruits. Among these genes, 49 were downregulated and 162 were upregulated. Between black (mature) and red fruit stages, 1141 genes were upregulated and 1423 downregulated. Variation in expression was observed in 2363 genes between black and green fruits. Although the strict criteria we used in this analysis may overlook other gene products, they can be viewed as generating the most reliable DEGs.

GO and KO enrichment analyses were carried out to consider more closely these differentially expressing genes. When testing for GO terms detected from differentially expressed genes in green versus red, red versus black, and green versus black fruits, no significantly enriched genes were found by GOEAST (http://omicslab.genetics.ac.cn/GOEAST/tools.php). However, several biological pathways were found to be significantly perturbed. Sixteen pathways were enriched across the whole fruit developmental process, including those of genes involved in flavonoid biosynthesis, phenylpropanoid biosynthesis, plant hormone signal transduction, and cutin, suberin, and wax biosynthesis, among others (Figures 2–4). In addition to these commonly impacted pathways, alterations in "degradation of aromatic compounds" and "MAPK signaling pathway - plant" were detected specifically in the early stages (change from small green to red fruit). In contrast, "bisphenol degradation" and "polycyclic aromatic hydrocarbon degradation" pathways were affected in the later stages (change from red to fully ripe black fruit). Fruit ripening is a process of highly coordinated and genetically programed physiological, biochemical, and organoleptic changes in the reproductive organs. In *Rubus* fruits, predominant changes in ripening include (1) depolymerization of carbohydrates, specifically, degradation of starches into sucrose and then into glucose and fructose; (2) decrease in organic acids, including amino acids; (3) production of volatile compounds, such as alcohols and aldehydes; and (4) accumulation of anthocyanins but depletion of cinnamic, ferulic, protocatechuic, and vanillic acids and epicatechin [5]. These changes may be evident in the metabolic pathway profiling in our study. The starch and sucrose metabolism pathway, significantly enriched in the two early stages of fruit ripening, adding to the degradation of aromatic compounds, can lead to the formation of special flavor and aroma of ripening *Rubus* fruits. A mixture of compounds, including ketones, alcohols, esters, and mainly terpenoids, constitutes the volatile flavor of most, if not all, fruits [28]. Some *Rubus* species have a special aroma to their fruit, but some do not [29]. Degradation of aromatic compounds could have a partial impact on these aroma volatiles. Another obvious sign of maturation of soft fruits is the decrease in firmness, which is the result of degradation of cell wall components and/or loss of integrity of the cell cuticular/wax layer [30, 31]. In strawberries, cell wall disassembly is characterized by solubilization of pectins, slight depolymerization of covalently bound pectins, and loss of galactose and arabinose, as well as a reduction in hemicellulose content [32]. Pectin content of mature fruit reduced dramatically in two raspberry cultivars, "Glen Clova" and "Glen Prosen" [33]. Further examination of *R. idaeus* cell wall fraction indicated that fruit ripening was associated with increased solubilization of pectin first and then depolymerization at the last stage [34]. In support of this, DEGs for cutin, suberin, and wax biosynthesis were found to be significantly enriched across the three fruit-ripening stages in our study. Only two DEGs (omega-hydroxypalmitate O-feruloyl transferase and peroxygenase) were common to all three stages. Progressive modulation of these particular genes may be the molecular basis of programing of the fruit-softening process.

2.4. Flavonoid Biosynthesis Genes and Their Expression. Anthocyanin, the most important metabolite in flavonoid production, is an essential nutritional component in raspberry fruits and their products [35, 36]. In *R. coreanus*, cyanidin-3-glucoside, cyanidin-3-rutinoside, and pelargonidin-3-glucoside have been recognized as the major anthocyanins [35]. Anthocyanins are first detected in green-red fruit but increase at the greatest rate to the highest amount in the last developmental stage [5]. The same trend has been observed for flavonols such as quercetin-glucuronide and quercetin-3-O-rutinoside. In contrast, flavanols and proanthocyanidins are accumulated at the very beginning of fruit set [5]. All these compounds are final products typical of the flavonoid biosynthesis, anthocyanin, and flavonol synthesis pathways. In our study, both flavonoid and anthocyanin pathways were significantly enriched during fruit development. This is in accordance with findings for other fruits such as grapes [16] and bayberries [37]. Confirming our prediction from the KEGG pathway enrichment above, we were able to manually identify 23 transcripts involved in the flavonoid biosynthesis pathway leading to flavonols, anthocyanins, or proanthocyanidins. The expression of most of these transcripts perfectly parallels the changes in the metabolites (Figure 5). Among these genes, five have alternative transcripts/multigene members including two *phenylalanine ammonia-lyases* (PAL), six *4-coumarate:CoA ligases* (4CL), three *chalcone synthases* (CHS), two *flavonol synthases* (FLS), and two *dihydroflavonol 4-reductases* (DFR). Major players among the transcripts from the same gene/multigene could be identified from expression patterns. For example, among the six *4CL* transcripts, transcript_52752 may be the key actor, whose abundance increased highly in parallel with fruit maturation. In comparison, although their roles could not be ruled out, most other transcripts of *4CL* exhibited very low levels of expression throughout the three fruit developmental stages. Functional diversification could be deduced from the results if multigene copies existed. Examples include *PAL* (transcript_21400 and transcript_22918) and *DFR* (transcript_1703 and transcript_61515). One member of *PAL* or *DFR* had a completely opposite expression mode compared to the other (Figure 5). In strawberries, the two copies of *DFR* have different substrate affinities, exerted at different stages for producing different types of anthocyanin [38]. Therefore, the function of the *DFR*s isolated in *R. coreanus* needs further investigation. Also noteworthy is the absence of the *F3′5′H* gene in the transcriptome, which implies that the synthesis of delphinidin-derived anthocyanins is blocked in *R. coreanus* fruits.

2.5. Identification and Abundance Estimation of NBS-LRR-Encoding Genes. Fungal and viral diseases are two worldwide threats to commercial cultivation of *Rubus*. Given the requirement for reducing pesticide use, cultivars with robust

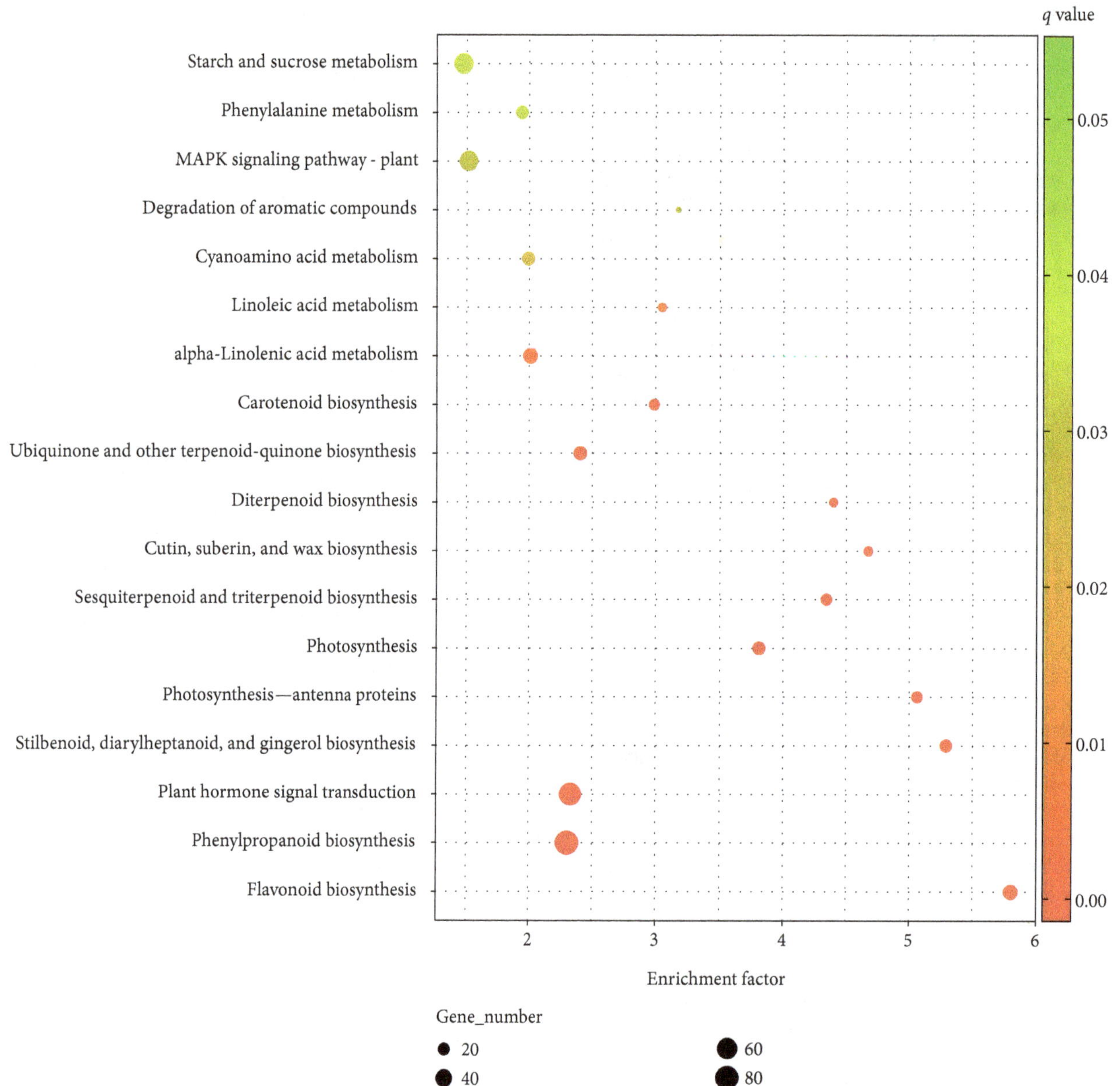

FIGURE 2: Pathway enrichment of differentially expressed genes between green and red *Rubus coreanus* fruits.

disease resistance become increasingly important. Fungal pathogens attack every part of *Rubus*, including the roots, canes, leaves, and fruits. Several fungal diseases can cause pre- or postharvest fruit rot in raspberries, leading to a short shelf life and limited sales of fresh fruit to distant markets. Gray mold (*Botrytis cinerea* Pers.:Fr.) is the most serious pathogen of fruit. Variation in susceptibility to it has been observed in fruits from different raspberry cultivars [39]. It is well documented that *R. coreanus* derivatives have strong resistance to cane diseases caused by *Elsinoë veneta*, *Didymella applanata*, and *B. cinerea* [40]. *R. coreanus* has also been recommended as a resource for promoting fruit rot resistance [10, 41, 42]. Proteins that contain a nucleotide-binding site and leucine-rich repeat (NBS-LRR) domains consist of the largest class of known plant resistance (R) gene products, conferring resistance to a diverse spectrum of pathogens [43, 44]. Recent advances have revealed that NBS-LRR R proteins are able to inhibit *B. cinerea* development [45]. In the family Rosaceae, NBS-LRR-coding genes form a large proportion of the genome, from 1.05% in strawberries to 1.52% in peaches [46]. However, except of a few studies, the availability of R gene resources in *Rubus* species is limited. Samuelian et al. [47] characterized 75 LRR genes from *R. idaeus* using degenerate primers designed from other Rosaceae species. Afanador-Kafuri and colleagues [48] obtained 47 LRR proteins using a similar strategy from Colombian *Rubus* genotypes. In a further exploration of our transcriptome data, initial screening via hmmsearch of the new reference uncovered 411 candidate NBS-encoding transcripts. Thereafter, through domain hunting, 119 NBS-LRR-domain-coding transcripts were identified, among which 74 had hits in the completely conserved NBS domain

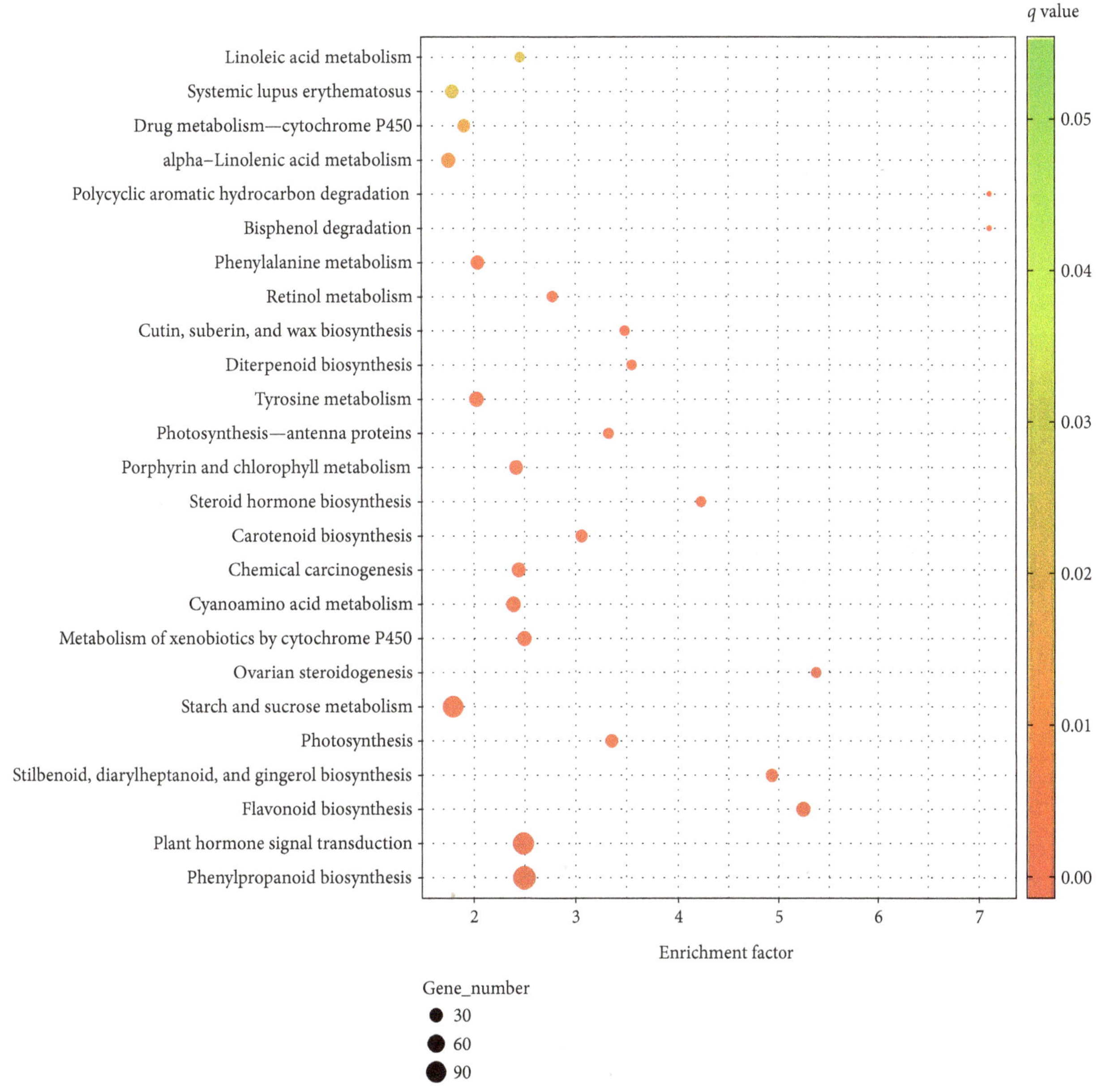

FIGURE 3: Pathway enrichment of differentially expressed genes between red and black *Rubus coreanus* fruits.

(Supplementary Table 1). We believe our resources greatly enrich the genetic information for *Rubus* breeding. Most of these plant resistance protein-coding genes have low abundance (less than five) estimated as trimmed mean of *M* values (TMM). This appears reasonable because a very high expression of R proteins could bring lethal effects to plant cells [49]. Twenty-four NBS transcripts are presented in Figure 6. Two of the transcripts show relatively high expression values (transcript_24284 and transcript_72010) in fruits at almost all three stages. Transcript_47133, with the highest abundance, functions mainly in the last stage, when fruits are fully ripe and are more vulnerable to pathogen attack. Closer examination of this resistance gene found that it is in the class of NBS-LRR (NL) proteins lacking additional N-terminal domains. Its closest ortholog in *R. occidentalis* is the gene *Bras_G19818*, which shares 62.58% sequence identity. Some of the *NBS-LRR* genes have tissue-specific expression properties [50] and can even confer different resistance reactions with different alleles from the same gene [51]. These three highly expressed or stage-specific factors could be interesting candidates for more detailed investigation in the future.

2.6. Cloning and Real-Time Quantitative PCR (RT-qPCR) Validation of Representative Genetic Information. Seven representative genes (*ANR*, *CER*, *CHI*, *CYP86B1*, *DFR*, *GPAT*, and *MYB44*), which encode key enzymes or regulators involved in anthocyanin/proanthocyanidin biosynthesis, amino acid metabolism, or plant cell wall wax formation, were successfully cloned and validated by sequencing. All these gene sequences corresponding to the full length of coding sequence with various lengths of 5′ or 3′ untranslated

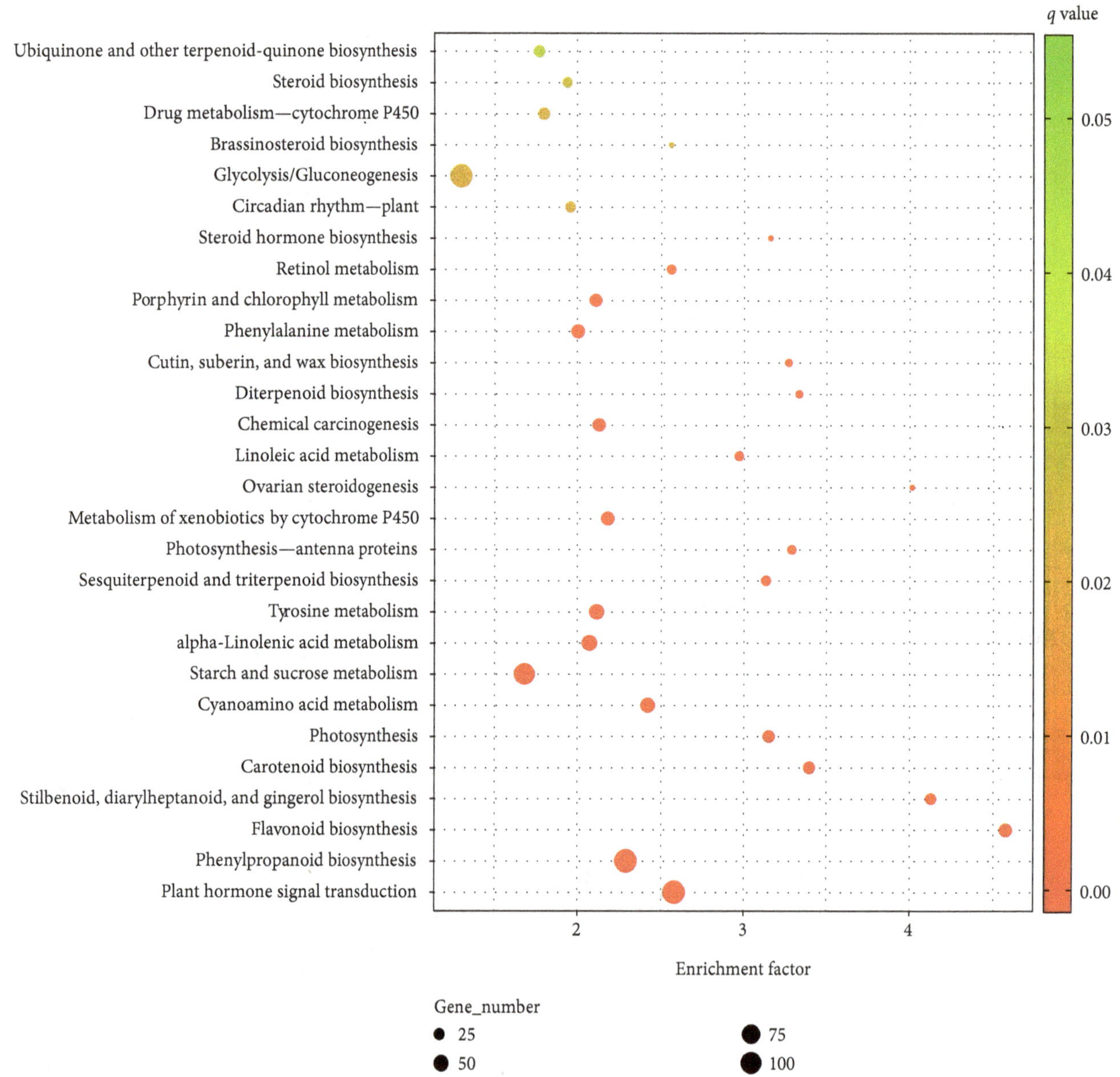

Figure 4: Pathway enrichment of differentially expressed genes between green and black *Rubus coreanus* fruits.

region (UTR) were identical to those deduced from the RNA-seq results. Similar expression patterns between RNA-seq and RT-qPCR were also observed (Figure 7), thus further validating the RNA-seq expression data.

3. Conclusions

This is the first transcriptomic profile, through RNA-seq investigation of sequence and transcript abundance, for *R. coreanus*. The transcriptomic analysis provides, for the first time, a 95,845-transcript reference for the species. Of these genetic resources, 69.5% were annotated. Differentially expressed genes in fruit developmental stages were mainly involved in flavonoid biosynthesis, plant cell wall formation, and aroma compound degradation. We identified 23 transcripts involved in the flavonoid biosynthesis pathway whose expression perfectly paralleled changes in the metabolites. Additionally, we identified 119 nucleotide-binding site leucine-rich repeat (NBS-LRR) protein-coding genes, involved in pathogen resistance, of which 74 were in the completely conserved domain. We believe that our study provides useful genetic information for *Rubus* breeding.

4. Materials and Methods

4.1. Sample Collection and RNA Preparation. Fruits of *R. coreanus* ($2n = 2x = 14$) [12] were collected from the wild at Ya'an city, Sichuan province (29°58′24.5″N, 103°00′18.7″E). Fruit set occurs in April and fruits mature in mid-June in this area. Fruits of three representative stages of ripening (green, red, and mature black) were harvested in 2015. For each stage, a total of about 30 fruits from no more than three canes

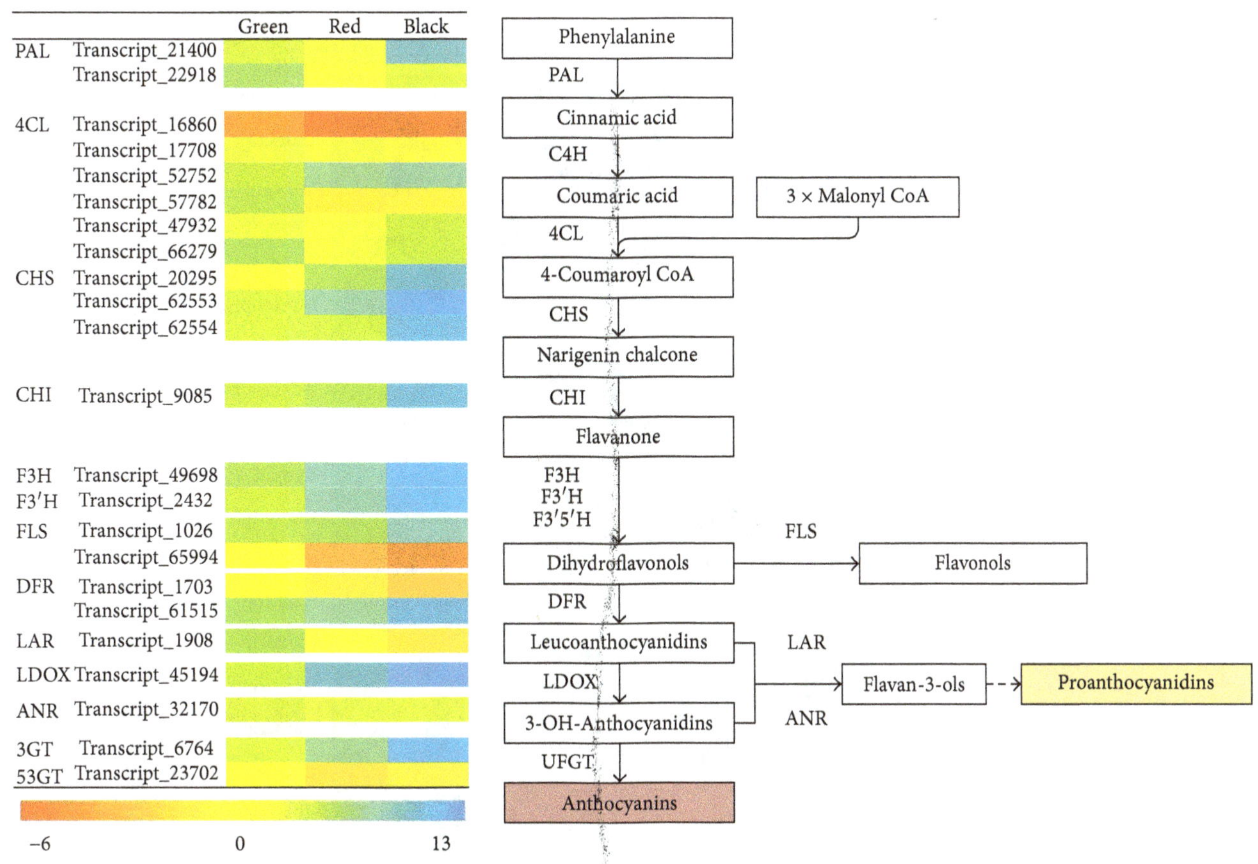

FIGURE 5: Flavonoid-synthesis-associated transcripts and their expression patterns during fruit ripening. Expression values are presented as log2-transformed trimmed mean of M value (TMM) derived from edgeR analysis. PAL: phenylalanine ammonia-lyase; 4CL: 4-coumarate:CoA ligase; CHS: chalcone synthase; CHI: chalcone flavanone isomerase; F3H: flavanone 3-hydroxylase; F3′H: flavonoid 3′-hydroxylase; F3′5′H: flavonoid 3′,5′-hydroxylase; FLS: flavonol synthase; DFR: dihydroflavonol 4-reductase; LDOX: leucoanthocyanidin dioxygenase; LAR: leucoanthocyanidin reductase; ANR: anthocyanidin reductase; 3GT: anthocyanin 3-O-glucosyltransferase; 53GT: anthocyanin 3,5-O-glucosyltransferase.

were collected in order to decrease background variation. They were frozen immediately in liquid nitrogen on collection and stored at −80°C until use. Two biological replicated samples were collected for each stage due to the limited yield of fruits.

Total RNA was isolated by using a modified cetyltrimethylammonium bromide method [52]. Genomic DNA was eliminated by using RNase-free DNase I (TaKaRa, Dalian, China). After monitoring RNA integrity and purity on 1% agarose gels and NanoPhotometer spectrophotometer (Implen, CA, USA), the Agilent 2100 Bioanalyzer system (Agilent Technologies, CA, USA), supplemented with RNA 6000 Nano Kit, was used to confirm the results. RNA concentration was measured using Qubit RNA Assay Kit in Qubit 2.0 Fluorometer (Life Technologies, CA, USA).

4.2. cDNA Synthesis, Library Construction, and Sequencing. As input material, 3 μg of RNA per sample was used. Sequencing libraries were generated using NEBNext Ultra Directional RNA Library Prep Kit for Illumina (NEB, USA) according to the manufacturer's instructions. Briefly, mRNA was purified from total RNA using poly(T) oligo-attached magnetic beads. Fragmentation was carried out using divalent cations under elevated temperature in NEBNext First Strand Synthesis reaction buffer (5x). First-strand cDNA was synthesized with random hexamer primer and M-MuLV Reverse Transcriptase (RNase H). Second-strand cDNA was synthesized by DNA polymerase I and RNase H. Remaining overhangs were blunted via exonuclease/polymerase activities. After adenylation of 3′ ends of DNA fragments, NEBNext Adaptors with a hairpin loop structure were ligated. AMPure XP system (Beckman Coulter, Beverly, USA) was used to purify cDNA fragments selectively at the correct size. Then, 3 μl of USER Enzyme (NEB, USA) was used with size-selected, adaptor-ligated cDNA at 37°C for 15 min followed by 5 min at 95°C. PCR was then performed with Phusion High-Fidelity DNA Polymerase, universal PCR primers, and index primers. Finally, the products were purified (AMPure XP system) and library quality was assessed on the Agilent 2100 Bioanalyzer system. Clustering and sequencing were carried out with the prepared libraries by Novogene (Beijing, China) using the Illumina HiSeq 2500 platform.

4.3. Transcriptome Assembly, Annotation, and Differential Expression and Enrichment Analysis. The raw reads were

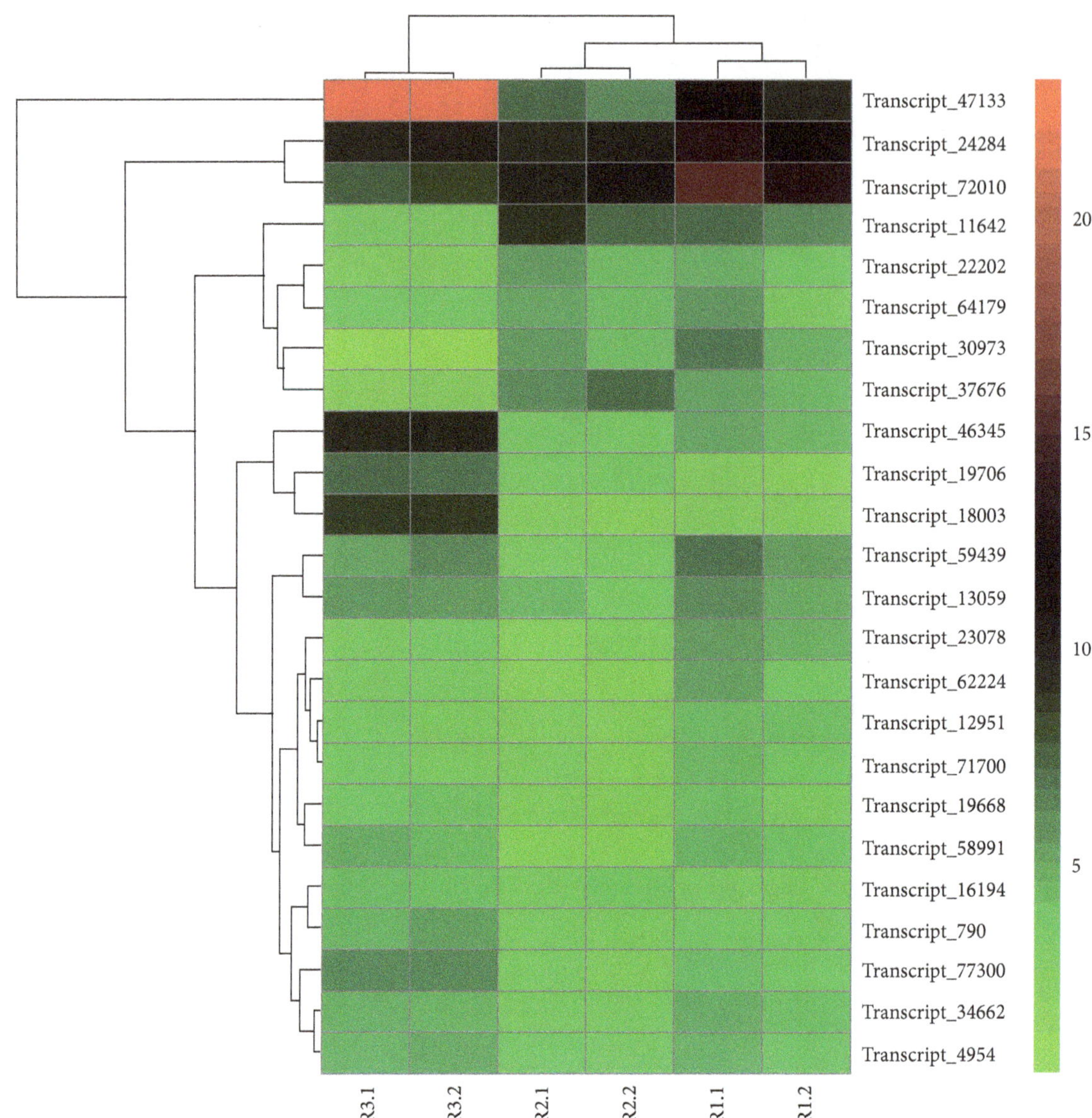

Figure 6: Heat map of the top 24 NBS-LRR genes expressed in *Rubus coreanus*. Normalized expression values are presented as trimmed mean of *M* value (TMM) derived from the edgeR package.

cleaned by removing adapter sequences and ambiguous reads (with "*N*" > 10%). Low-quality bases were trimmed, and reads that were too short were filtered through Trimmomatic (LEADING:3 TRAILING:3 SLIDINGWINDOW:4:15 MINLEN:50) [53], as were the corresponding read pairs. After trimming/filtering low-quality reads, SEECER (v0.1.3) [54] was used for error correction. All downstream analyses were based on high-quality clean data.

To facilitate the use of the recently published North American black raspberry genome information, we adopted the strategies of genome-guided transcript expression analysis by using the protocol of Trapnell et al. [55]. All reads were first mapped to the *R. occidentalis* genome (v1.0) with TopHat2 (allowing two bases of mispairing and multiple hits ≤ 20) and then assembled by using the Cufflinks suite with default parameters.

To evaluate divergence between *R. coreanus* and *R. occidentalis*, we also carried out de novo assembly of transcripts. Trinity (v2.2.0) [56] was used with default parameters except that the minimum contig length was set to 200 bp, reads were first normalized with maximum coverage 50 before putting in the assembly pipeline, and k-mer coverage was set to a minimum level of two. Redundancy in the de novo transcriptome was minimized with CD-HIT-EST (v4.6.4) [57] using an identity cutoff at 0.99. EvidentialGene tr2aacds pipeline [58] was used to combine both genome-guided and de novo assemblies. Nonredundant transcripts were also obtained. To evaluate the quality of the reference, all assemblies were searched against BUSCO for plants [21].

All reads in each sample were mapped back to the transcriptome using Bowtie 2 [19] (default parameters used, but end to end, allowing two bases of mispairing and multiple hits ≤ 20) and then used to estimate expression values for each transcript by RSEM [24]. Given that many of the very lowly expressed transcripts could be questionable due to

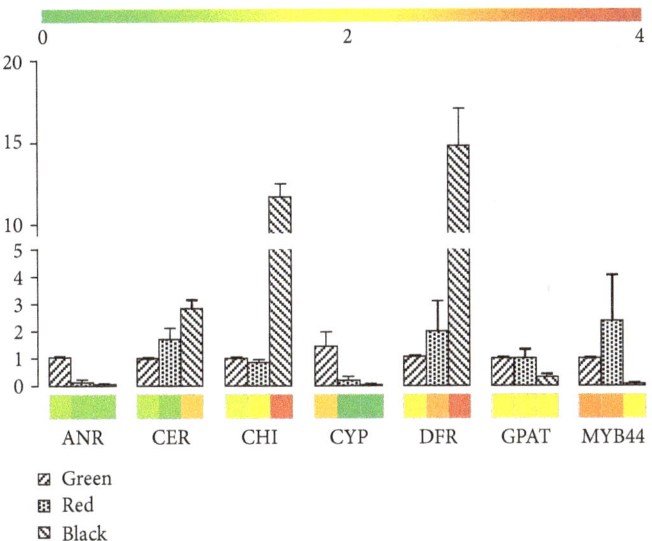

FIGURE 7: Similarities in expression patterns of seven genes between RNA-seq (heat map) and RT-qPCR (bar plot). ANR: anthocyanidin reductase; CER: ECERIFERUM; CHI: chalcone flavanone isomerase; CYP: cytochrome P450; DFR: dihydroflavonol 4-reductase, (transcript_61515 was chosen for DFR); GPAT: glycerol-3-phosphate acyltransferases. MYB44 is the transcript most resembling AtMYB44 in *Arabidopsis thaliana*.

our very high deep-sequencing coverage (exceeding 100x), we filtered the transcripts by setting transcripts per million (TPM) lower than three before conducting differential gene expression analysis.

Gene annotations were carried out according to the Annocript (v1.1.3) pipeline [21] using all assembled transcripts. We performed similarity searches through blastx against UniRef90 and Swiss-Prot (v201706, word_size = 4; e-value = 0.00001), rpsblast against the Conserved Domain Database (CDD) profiles (ftp://ftp.ncbi.nih.gov/pub/mmdb/cdd/little_endian/Cdd_LE.tar.gz, e-value = 0.00001, num_descriptions = 20, and num_alignments = 20), and blastn against Rfam and rRNAs (e-value = 0.00001, num_descriptions = 1, num_alignments = 1, and num_threads = 4). For each sequence, the best hit, if any, was chosen. For gene ontology (GO) functional classification, Enzyme Commission IDs were associated to the corresponding matches. KEGG Orthology (KO) assistant pathway assignment was implemented via KOBAS 3.0 [59] using the default parameters. The dna2pep tool implemented in the Annocript package [21] was used to identify the longest open reading frame (ORF) of each transcript. PORTRAIT software [23], which was developed for detecting noncoding RNA from poorly characterized species, was used to identify the noncoding potential of each query sequence by using a new support vector machine-based algorithm.

To investigate differential expression (DE) of transcripts, we used DESeq2 [25], which uses a Wald test; edgeR [26], which uses a likelihood ratio test; and limma-voom [27], which uses a moderated t-test to conduct pairwise comparison of the three fruit-ripening stages. Each of the comparisons was based on different statistical models. Differentially expressed genes were selected using $\log2FC \geq 1$ or $\log FC \leq -1$ and FDR (false discovery rate) < 0.01 in the three methods. Consensus DE results were obtained by comparing the outcomes of the three methods, which were used to present the most reliable differentially expressed transcripts. These transcripts associated with their corresponding GO or KO annotations were tested against the whole transcriptome as background gene sets for enrichment analysis. GO categories were checked using GOEAST (http://omicslab.genetics.ac.cn/GOEAST/tools.php) with an FDR (Benjamini–Yekutieli method) value of ≤0.05 as the cutoff to identify enriched terms by the hypergeometric test. Transcripts with a KO number were also tested by hypergeometric statistics to find significantly overperturbed pathways through a Perl in-house script.

4.4. Expression Patterns of Genes Involved in Flavonoid Synthesis. From the gene differential expression analysis, flavonoid biosynthesis pathway genes appeared to be extremely perturbed in both green versus red fruits and red versus black fruits. We identified all genes involved in the pathway from the assembled reference and manually curated by blasting against the nonredundant protein database at the National Center for Biotechnology Information site, coupling the annotation from Annocript described above. Afterwards, expression patterns of these genes were presented as heat maps after log2 transformation of the among-sample normalized count values by using edgeR.

4.5. NBS-Encoding Gene Retrieval and Expression Analysis. Based on the Annocript-deduced peptide collection, we identified potential NBS-encoding genes using the procedures described by Arya et al. [60]. Specifically, the hidden Markov model (HMM) profile for the NBS domain (http://pfam.xfam.org/family/PF00931) was used to search against the complete set of the predicted *R. coreanus* proteins using hmmsearch in HMMER (v3.1b) [61] with e-value < 0.00001. All the protein sequences identified were further subjected to CDD detection (https://www.ncbi.nlm.nih.gov/Structure/bwrpsb/bwrpsb.cgi) using 0.01 as a cutoff value to confirm the presence of NBS domains. Expression pattern analysis was carried out the same way as for flavonoid genes.

4.6. Cloning and RT-qPCR for Validation of Gene Expression Patterns. To verify the validity of the genetic information obtained, we selected seven representative genes (*ANR, CER1, CHI, CYP, DFR, GPAT*, and *MYB44*), which encode key enzymes involved in anthocyanin/proanthocyanidin biosynthesis, amino acid metabolism, plant cell wall wax formation, or stress response regulators. The deduced full coding sequences were cloned experimentally, and their expression values were determined using RT-qPCR. All the candidate sequences were amplified in a 20 μl reaction mixture, containing 10 ng first-strand cDNA, 10 pmol each primer (Supplementary Table 2), and 10 μl 2x PrimeSTAR HS premix (TaKaRa, Dalian, China). Following one cycle of 20 seconds at 98°C, 35 PCR cycles of 10 s at 98°C, 10 s at 60°C, and 90 s at 72°C were performed in the thermal cycler PTC-200 (Bio-Rad, Hercules, CA). Amplified products were purified using E.Z.N.A. Gel Extraction Kit (Omega, GA, USA). The enriched PCR product was cloned into pEASY-Blunt vectors (TransGen, Beijing, China) and transformed into JM109 competent *Escherichia coli* cells. Finally, positive clones were sequenced using the BigDye Terminator Cycle Sequencing Kit on an ABI PRISM 3730 automated DNA sequencer (Applied Biosystems, Foster City, CA, USA). For quantitative PCR, 10 μl reaction mixture is composed of 5 μl 2x SYBR Green mixture (TaKaRa, Dalian, China), 1 μl diluted cDNA, and 1 μl specific forward and reverse primer for each gene (Supplementary Table 1). The reaction was conducted on a CFX96 Real-Time PCR Detection System (Bio-Rad, US). Expression values were expressed as $2^{-\Delta\Delta CT}$ using *beta-actin* [52] as an internal control.

Conflicts of Interest

The authors declare that the research was conducted in the absence of any commercial or financial relationships that could be construed as a potential conflict of interest.

Authors' Contributions

Qing Chen and Yan Wang designed and supervised the experiment. Qing Chen and Xunju Liu conducted the experiment. Yueyang Hu assisted in RNA isolation. Xunju Liu and Yueyang Hu conducted the RT-qPCR analysis. Bo Sun and Xiaorong Wang helped in field plant identification and collection. Yaodong Hu and Haoru Tang provide guidance in bioinformatic analysis pipeline. Qing Chen and Yan Wang drafted and all authors revised the manuscript. Qing Chen and Xunju Liu contributed equally to this work.

Acknowledgments

The work was financially supported by the Basic Research Programs of Sichuan Province (2015JY0020) and the National Natural Science Foundation of China (31600232). The authors also would like to thank the staff members at the computational division of the Institute of Pomology and Olericulture of Sichuan Agricultural University for computation assistance.

References

[1] M. M. Thompson, "Survey of chromosome numbers in Rubus (Rosaceae: Rosoideae)," *Annals of the Missouri Botanical Garden*, vol. 84, no. 1, pp. 128–164, 1997.

[2] R. Bobinaitė, P. Viškelis, and P. R. Venskutonis, "Variation of total phenolics, anthocyanins, ellagic acid and radical scavenging capacity in various raspberry (*Rubus* spp.) cultivars," *Food Chemistry*, vol. 132, no. 3, pp. 1495–1501, 2012.

[3] N. P. Seeram, L. S. Adams, Y. Zhang et al., "Blackberry, black raspberry, blueberry, cranberry, red raspberry, and strawberry extracts inhibit growth and stimulate apoptosis of human cancer cells in vitro," *Journal of Agricultural and Food Chemistry*, vol. 54, no. 25, pp. 9329–9339, 2006.

[4] L. C. Torre and B. H. Barritt, "Quantitative evaluation of rubus fruit anthocyanin pigments," *Journal of Food Science*, vol. 42, no. 2, pp. 488–490, 1977.

[5] H.-S. Kim, S. J. Park, S.-H. Hyun et al., "Biochemical monitoring of black raspberry (*Rubus coreanus* Miquel) fruits according to maturation stage by ^{1}H NMR using multiple solvent systems," *Food Research International*, vol. 44, no. 7, pp. 1977–1987, 2011.

[6] T. K. Hyun, S. Lee, Y. Rim et al., "De-novo RNA sequencing and metabolite profiling to identify genes involved in anthocyanin biosynthesis in Korean black raspberry (*Rubus coreanus* Miquel)," *PLoS One*, vol. 9, no. 2, article e88292, 2014.

[7] J. Lee, M. Dossett, and C. E. Finn, "Mistaken identity: clarification of *Rubus coreanus* Miquel (Bokbunja)," *Molecules*, vol. 19, no. 7, pp. 10524–10533, 2014.

[8] R. VanBuren, D. Bryant, J. M. Bushakra et al., "The genome of black raspberry (*Rubus occidentalis*)," *The Plant Journal*, vol. 87, no. 6, pp. 535–547, 2016.

[9] V. H. Knight, D. L. Jennings, and R. J. McNicol, "Progress in the UK raspberry breeding programme," *Acta Horticulturae*, vol. 262, no. 262, pp. 93–104, 1989.

[10] C. Finn, J. R. Ballington, C. Kempler, H. Swartz, and P. P. Moore, "Use of 58 *Rubus* species in five north American breeding programmes—breeders notes," *Acta Horticulturae*, no. 585, pp. 113–119, 2002.

[11] S. H. Kim, H. G. Chung, and J. Han, "Breeding of Korean black raspberry (*Rubus coreanus* Miq.) for high productivity in Korea," *Acta Horticulturae*, no. 777, pp. 141–146, 2008.

[12] X. R. Wang, Y. Liu, B. F. Zhong et al., "Cytological and RAPD data revealed genetic relationships among nine selected populations of the wild bramble species, *Rubus parvifolius* and *R. coreanus* (Rosaceae)," *Genetic Resources and Crop Evolution*, vol. 57, no. 3, pp. 431–441, 2010.

[13] C. E. Finn, C. Kempler, and P. P. Moore, "Raspberry cultivars: what's new? What's succeeding? Where are breeding programs headed?," *Acta Horticulturae*, no. 777, pp. 33–40, 2008.

[14] G. B. Seymour, L. Østergaard, N. H. Chapman, S. Knapp, and C. Martin, "Fruit development and ripening," *Annual Review of Plant Biology*, vol. 64, no. 1, pp. 219–241, 2013.

[15] J. F. Sánchez-Sevilla, J. G. Vallarino, S. Osorio et al., "Gene expression atlas of fruit ripening and transcriptome assembly from RNA-seq data in octoploid strawberry (*Fragaria* × *ananassa*)," *Scientific Reports*, vol. 7, no. 1, article 13737, 2017.

[16] S. Zenoni, A. Ferrarini, E. Giacomelli et al., "Characterization of transcriptional complexity during berry development in *Vitis vinifera* using RNA-seq," *Plant Physiology*, vol. 152, no. 4, pp. 1787–1795, 2010.

[17] V. Gupta, A. D. Estrada, I. Blakley et al., "RNA-Seq analysis and annotation of a draft blueberry genome assembly identifies candidate genes involved in fruit ripening, biosynthesis of bioactive compounds, and stage-specific alternative splicing," *GigaScience*, vol. 4, no. 1, pp. 1–22, 2015.

[18] N. Fernandez-Pozo, Y. Zheng, S. I. Snyder et al., "The tomato expression atlas," *Bioinformatics*, vol. 33, no. 15, pp. 2397-2398, 2017.

[19] B. Langmead and S. L. Salzberg, "Fast gapped-read alignment with Bowtie 2," *Nature Methods*, vol. 9, no. 4, pp. 357–359, 2012.

[20] C. F. Williams, "Influence of parentage in species hybridization of raspberries," *Proceedings, American Society for Horticultural Science*, vol. 56, pp. 149–156, 1950.

[21] F. A. Simão, R. M. Waterhouse, P. Ioannidis, E. V. Kriventseva, and E. M. Zdobnov, "BUSCO: assessing genome assembly and annotation completeness with single-copy orthologs," *Bioinformatics*, vol. 31, no. 19, pp. 3210–3212, 2015.

[22] F. Musacchia, S. Basu, G. Petrosino, M. Salvemini, and R. Sanges, "Annocript: a flexible pipeline for the annotation of transcriptomes able to identify putative long noncoding RNAs," *Bioinformatics*, vol. 31, no. 13, pp. 2199–2201, 2015.

[23] R. T. Arrial, R. C. Togawa, and M. d. M. Brigido, "Screening non-coding RNAs in transcriptomes from neglected species using PORTRAIT: case study of the pathogenic fungus *Paracoccidioides brasiliensis*," *BMC Bioinformatics*, vol. 10, no. 1, p. 239, 2009.

[24] B. Li and C. N. Dewey, "RSEM: accurate transcript quantification from RNA-Seq data with or without a reference genome," *BMC Bioinformatics*, vol. 12, no. 1, p. 323, 2011.

[25] M. I. Love, W. Huber, and S. Anders, "Moderated estimation of fold change and dispersion for RNA-seq data with DESeq2," *Genome Biology*, vol. 15, no. 12, p. 550, 2014.

[26] M. D. Robinson, D. J. McCarthy, and G. K. Smyth, "edgeR: a bioconductor package for differential expression analysis of digital gene expression data," *Bioinformatics*, vol. 26, no. 1, pp. 139-140, 2010.

[27] M. E. Ritchie, B. Phipson, D. Wu et al., "*limma* powers differential expression analyses for RNA-sequencing and microarray studies," *Nucleic Acids Research*, vol. 43, no. 7, article e47, 2015.

[28] Q. Yu, A. Plotto, E. A. Baldwin et al., "Proteomic and metabolomic analyses provide insight into production of volatile and non-volatile flavor components in mandarin hybrid fruit," *BMC Plant Biology*, vol. 15, no. 1, p. 76, 2015.

[29] K. Klesk and M. Qian, "Preliminary aroma comparison of Marion (*Rubus* spp. *hyb*) and Evergreen (*R. laciniatus* L.) blackberries by dynamic headspace/OSME technique," *Journal of Food Science*, vol. 68, no. 2, pp. 697–700, 2003.

[30] M. Saladie, A. J. Matas, T. Isaacson et al., "A reevaluation of the key factors that influence tomato fruit softening and integrity," *Plant Physiology*, vol. 144, no. 2, pp. 1012–1028, 2007.

[31] A. Payasi, N. N. Mishra, A. L. S. Chaves, and R. Singh, "Biochemistry of fruit softening: an overview," *Physiology and Molecular Biology of Plants*, vol. 15, no. 2, pp. 103–113, 2009.

[32] S. Posé, J. A. García-Gago, N. Santiago-Doménech, F. Pliego-Alfaro, M. A. Quesada, and J. A. Mercado, "Strawberry fruit softening: role of cell wall disassembly and its manipulation in transgenic plants," *Genes, Genomes and Genomics*, vol. 5, no. 1, pp. 40–48, 2011.

[33] D. Stewart, P. P. M. Iannetta, and H. V. Davies, "Ripening-related changes in raspberry cell wall composition and structure," *Phytochemistry*, vol. 56, no. 5, pp. 423–428, 2001.

[34] A. R. Vicente, C. Ortugno, A. L. T. Powell, L. Carl Greve, and J. M. Labavitch, "Temporal sequence of cell wall disassembly events in developing fruits. 1. Analysis of raspberry (*Rubus idaeus*)," *Journal of Agricultural and Food Chemistry*, vol. 55, no. 10, pp. 4119–4124, 2007.

[35] S. Heo, D. Y. Lee, H. K. Choi et al., "Metabolite fingerprinting of *bokbunja* (*Rubus coreanus* Miquel) by UPLC-qTOF-MS," *Food Science and Biotechnology*, vol. 20, no. 2, pp. 567–570, 2011.

[36] D. Y. Lee, S. Heo, S. G. Kim et al., "Metabolomic characterization of the region- and maturity-specificity of *Rubus coreanus* Miquel (*Bokbunja*)," *Food Research International*, vol. 54, no. 1, pp. 508–515, 2013.

[37] C. Feng, M. Chen, C. J. Xu et al., "Transcriptomic analysis of Chinese bayberry (Myrica rubra) fruit development and ripening using RNA-Seq," *BMC Genomics*, vol. 13, no. 1, p. 19, 2012.

[38] S. Miosic, J. Thill, M. Milosevic et al., "Dihydroflavonol 4-reductase genes encode enzymes with contrasting substrate specificity and show divergent gene expression profiles in *Fragaria* species," *PloS One*, vol. 9, no. 11, article e112707, 2014.

[39] H. A. Daubeny and H. S. Pepin, "Variations among red raspberry cultivars and selections in susceptibility to the fruit rot causal organisms *Botrytis cinerea* and Rhizopus spp," *Canadian Journal of Plant Science*, vol. 54, no. 3, pp. 511–516, 1974.

[40] E. Keep, V. H. Knight, and J. H. Parker, "*Rubus coreanus* as donor of resistance to cane diseases and mildew in red raspberry breeding," *Euphytica*, vol. 26, no. 2, pp. 505–510, 1977.

[41] H. K. Hall, K. E. Hummer, A. R. Jamieson, S. N. Jennings, and C. A. Weber, "Raspberry breeding and genetics," in *Plant breeding reviews*, J. Janick, Ed., vol. 32, pp. 148–169, John Wiley & Sons, Inc., Hoboken, NJ, USA, 2009.

[42] H. K. Hall and C. Kempler, "Raspberry breeding," *Fruit, Vegetable and Cereal Science and Biotechnology*, vol. 5, Supplement 1, pp. 44–62, 2011.

[43] K. E. Hammond-Kosack and J. D. G. Jones, "Plant disease resistance genes," *Annual Review of Plant Physiology and Plant Molecular Biology*, vol. 48, no. 1, pp. 575–607, 1997.

[44] Z.-Q. Shao, J.-Y. Xue, P. Wu et al., "Large-scale analyses of angiosperm nucleotide-binding site-leucine-rich repeat genes reveal three anciently diverged classes with distinct evolutionary patterns," *Plant Physiology*, vol. 170, no. 4, pp. 2095–2109, 2016.

[45] J. S. Moreira, R. G. Almeida, L. S. Tavares et al., "Identification of botryticidal proteins with similarity to NBS-LRR proteins in rosemary pepper (Lippia sidoides Cham.) flowers," *Protein Journal*, vol. 30, no. 1, pp. 32–38, 2011.

[46] Y. Jia, Y. Yuan, Y. Zhang, S. Yang, and X. Zhang, "Extreme expansion of NBS-encoding genes in Rosaceae," *BMC Genetics*, vol. 16, no. 1, p. 48, 2015.

[47] S. K. Samuelian, A. M. Baldo, J. A. Pattison, and C. A. Weber, "Isolation and linkage mapping of NBS-LRR resistance gene analogs in red raspberry (*Rubus idaeus* L.) and classification among 270 Rosaceae NBS-LRR genes," *Tree Genetics & Genomes*, vol. 4, no. 4, pp. 881–896, 2008.

[48] L. Afanador-Kafuri, J. F. Mejía, A. González, and E. Álvarez, "Identifying and analyzing the diversity of resistance gene analogs in Colombian *Rubus* genotypes," *Plant Disease*, vol. 99, no. 7, pp. 994–1001, 2015.

[49] Y. Zhang, R. Xia, H. Kuang, and B. C. Meyers, "The diversification of plant *NBS-LRR* defense genes directs the evolution of microRNAs that target them," *Molecular Biology and Evolution*, vol. 33, no. 10, pp. 2692–2705, 2016.

[50] B. C. Meyers, M. Morgante, and R. W. Michelmore, "TIR-X and TIR-NBS proteins: two new families related to disease resistance TIR-NBS-LRR proteins encoded in *Arabidopsis* and other plant genomes," *The Plant Journal*, vol. 32, no. 1, pp. 77–92, 2002.

[51] H. Takahashi, J. Miller, Y. Nozaki et al., "*RCY1*, an *Arabidopsis thaliana RPP8/HRT* family resistance gene, conferring resistance to cucumber mosaic virus requires salicylic acid, ethylene and a novel signal transduction mechanism," *The Plant Journal*, vol. 32, no. 5, pp. 655–667, 2002.

[52] Q. Chen, H. Yu, H. Tang, and X. Wang, "Identification and expression analysis of genes involved in anthocyanin and proanthocyanidin biosynthesis in the fruit of blackberry," *Scientia Horticulturae*, vol. 141, pp. 61–68, 2012.

[53] A. M. Bolger, M. Lohse, and B. Usadel, "Trimmomatic: a flexible trimmer for Illumina sequence data," *Bioinformatics*, vol. 30, no. 15, pp. 2114–2120, 2014.

[54] H. S. Le, M. H. Schulz, B. M. McCauley, V. F. Hinman, and Z. Bar-Joseph, "Probabilistic error correction for RNA sequencing," *Nucleic Acids Research*, vol. 41, no. 10, article e109, 2013.

[55] C. Trapnell, A. Roberts, L. Goff et al., "Differential gene and transcript expression analysis of RNA-seq experiments with TopHat and Cufflinks," *Nature Protocols*, vol. 7, no. 3, pp. 562–578, 2012.

[56] B. J. Haas, A. Papanicolaou, M. Yassour et al., "De novo transcript sequence reconstruction from RNA-seq using the Trinity platform for reference generation and analysis," *Nature Protocols*, vol. 8, no. 8, pp. 1494–1512, 2013.

[57] W. Li and A. Godzik, "CD-HIT: a fast program for clustering and comparing large sets of protein or nucleotide sequences," *Bioinformatics*, vol. 22, no. 13, pp. 1658-1659, 2006.

[58] D. Gilbert, "Accurate & complete gene construction with EvidentialGene," *F1000Research*, vol. 5, p. 1567, 2016.

[59] C. Xie, X. Mao, J. Huang et al., "KOBAS 2.0: a web server for annotation and identification of enriched pathways and diseases," *Nucleic Acids Research*, vol. 39, Supplement 2, pp. W316–W322, 2011.

[60] P. Arya, G. Kumar, V. Acharya, and A. K. Singh, "Genome-wide identification and expression analysis of NBS-encoding genes in *Malus* × *domestica* and expansion of NBS genes family in Rosaceae," *PLOS One*, vol. 9, no. 9, article e107987, 2014.

[61] R. D. Finn, J. Clements, and S. R. Eddy, "HMMER web server: interactive sequence similarity searching," *Nucleic Acids Research*, vol. 39, Supplement_2, pp. W29–W37, 2011.

6

Endometriosis Malignant Transformation: Epigenetics as a Probable Mechanism in Ovarian Tumorigenesis

Jiaxing He, Weiqin Chang, Chunyang Feng, Manhua Cui, and Tianmin Xu

The Second Hospital of Jilin University, Jilin, Changchun 130041, China

Correspondence should be addressed to Tianmin Xu; xutianmin@126.com

Academic Editor: Changwon Park

Endometriosis, defined as the presence of ectopic endometrial glands and stroma outside the uterine cavity, is a chronic, hormone-dependent gynecologic disease affecting millions of women across the world, with symptoms including chronic pelvic pain, dysmenorrhea, dyspareunia, dysuria, and subfertility. In addition, there is well-established evidence that, although endometriosis is considered benign, it is associated with an increased risk of malignant transformation, with the involvement of various mechanisms of development. More and more evidence reveals an important contribution of epigenetic modification not only in endometriosis but also in mechanisms of endometriosis malignant transformation, including DNA methylation and demethylation, histone modifications, and miRNA aberrant expressions. In this present review, we mainly summarize the research progress about the current knowledge regarding the epigenetic modifications of the relations between endometriosis malignant transformation and ovarian cancer in an effort to identify some risk factors probably associated with ectopic endometrium transformation.

1. Introduction

Endometriosis is a chronic and hormone-dependent disease, defined as the presence of ectopic endometrial glands and stroma outside the uterine cavity [1]. The prevalence of endometriosis is widely different for various ethnic groups [2–4], which is likely to be about 5–15% of reproductive-age women and 3 to 5% of postmenopausal women [3]. To date, it is well established that endometriosis is a chronic inflammatory disease and the chronic inflammation is associated with pain and infertility [1]. As a common disease in reproductive woman, ectopic endometrium is predominantly detected in the pelvic compartment like the utero-sacral ligament, Douglas cavity, and ovary; moreover, the ectopic endometrial tissue can attach to other tissues including the bladder and ureter as well as the lung. Though several hypotheses are reported in order to explain the pathogenesis of endometriosis, mainly including coelomic metaplasia, retrograde menstruation, and lymphatic and vascular dissemination, none of them can explain all the different types of endometriosis.

Despite the fact that endometriosis is considered a benign condition because of its normal histology, the cellular, histologic, and molecular data strongly demonstrate that endometriosis has neoplastic characteristics [5, 6]. There is strong evidence that endometriosis shares striking features with malignancy [5]. Similar to cancer, ectopic endometrial tissue can result in normal tissue dissemination, invasion, and organ damage, as well as neoangiogenesis. It is reported that endometriosis is associated with ovarian cancer in all aspects of research fields including epigenetics; the link between endometriosis and ovarian cancer was reported for the first time as early as 1925 [7]. In the last nine decades, epidemiological investigation has been accumulated that endometriosis may contribute to the development and progression of ovarian cancer. In a cohort study by Melin et al., where 63,630 eligible women diagnosed with endometriosis entered, the risk of ovarian cancer (SIR 1.37) was moderately increased as compared with that of the general population [8]. As technology develops, multiple mechanisms about the occurrence of ovarian cancers associated with malignant transformation of endometriosis have been studied for a long time, but they still remain elusive. Currently, it is well demonstrated that epigenetic modifications contribute to ovarian tumorigenesis. Epigenetics is described as a heritable modification in gene expression without alteration of DNA sequence compared with gene mutation [9]. The epigenetic modifications so far involve DNA methylation, histone modifications, and non-coding microRNAs (miRNAs) [10–12]. In the present review, we mainly summarize the research progress regarding the epigenetic modifications of the relations between endometriosis malignant transformation and ovarian cancer in an effort to identify some risk factors probably associated with ectopic endometrium transformation.

2. DNA Methylation

DNA methylation, the most frequently studied epigenetic alteration, occurs at the carbon-5 position of cytosine residues, exclusively in CpG dinucleotide sequences, and inhibits gene transcription [13]. DNA methylation, referring to the addition of the methyl groups into the cytosines from *S*-adenosyl L-methionine, is mediated by a family of enzymes known as the DNA methyltransferases (DNMTs) including DNMT1, DNMT3a, and DNMT3b. DNA methylation is a heritable epigenetic occurrence that significantly regulates gene expression without changing DNA sequence [14]. Most CpG sites in the human genome are methylated. However, local CpG islands, the CpG-rich regions, founded in the promoter regions of widely expressed genes are in unmethylated conditions [15, 16]. It is well evidenced that hypermethylation of genes can result in inhibition of gene expression, whilst hypomethylation may give rise to increased transcription and protein activation. Furthermore, a considerable number of evidence have proved the positive relation between DNA methylation and tumor occurrence and progression. On the other hand, DNA hypomethylation also contributes to oncogenesis when previously inactivated oncogenes are transcriptionally activated [17].

2.1. Genes Involved in Endometriosis Malignant Transformation. A number of genes, which are silenced or activated by DNA methylation, have been investigated in malignant transformation of endometriosis. Moreover, some researches that were published demonstrate the common epigenetic alteration between endometriosis and ovarian cancers. It is testified that some major genes are actually involved in the malignant transformation of ovarian endometriosis; among these contributing genes, epigenetic inactivation of Runt-related transcription factor 3 (RUNX3) [18], human mutL homolog 1 (hMLH1) [19], E-cadherin (CDH1) [20], Ras-association domain family of gene 2 (RASSF2) [21], and P16 and phosphatase and tensin homolog deleted on chromosome 10 (PTEN) [22] by promoter hypermethylation was well observed; however, long interspersed nuclear element-1 (LINE-1) [23] and syncytin-1 [24] were hypomethylated and activated. An example of this is the study carried out by Guo et al. [18] in which RUNX3 promoter hypermethylation, which results in RUNX3 inactivation and decreased RUNX3 protein expression, has been identified in the 18 of 30 (60%) patients with endometriosis-associated ovarian carcinoma (EAOC). Besides, the degree of RUNX3 hypermethylation and decreased RUNX3 protein expression in the eutopic endometrium from the EAOC group was significantly higher than that in the endometriosis (EM) and control endometrium (CE) groups. It is probable that the tissue histology in the eutopic endometrium may appear normal and intrinsic molecular abnormalities have occurred. Furthermore, it is evidenced that patients with surgical stage IC EAOC have a higher degree of RUNX3 hypermethylation than those with stages IA and IB. This phenomenon suggests that RUNX3 is implicated in the progression of malignant transformation of ovarian EM. Therefore, RUNX3 gene hypermethylation is reputed to be an early event in the pathogenesis of EAOC. Another similar study by Ren et al. [19] exploring the relationship between hMLH1 hypermethylation and malignant transformation of ovarian endometriosis is consistent with the results of the aforementioned RUNX3 promoter hypermethylation research. hMLH1 is a member of the DNA mismatch repair (MMR) system which corrects errors in DNA replication during proliferation. Ren and his colleagues illustrated this point clearly that absence of hMLH1 protein expression that resulted from aberrant promoter methylation is associated with malignant evolution of ovarian endometrium.

Other gene methylation may play an equivalent crucial role in the malignant transformation of endometriosis similar to that mentioned above, although few researchers have been able to draw on systematic researches into DNA methylation conditions of those relevant genes. It is well demonstrated that some crucial gene hypermethylation are implicated in the pathogenesis of EAOC. A recent research by Ren et al. [21] screened differentially aberrant methylated candidate genes associated with the malignant transformation of ovarian endometriosis by MCA-RDA, and nine differentially methylated candidate genes emerged in the study of malignant transformation of ovarian endometriosis. Among these nine candidate genes, RASSF2, SPOCK2, and RUNX3 were proved in other researches; therefore, the remaining six candidate genes were further studied including GSTZ1, CYP2A, GBGT1, NDUFS1, and ADAM22, as well as TRIM36. On the basis of those gene functions, they may take part in the malignant evolution of ovarian endometriosis. For example, ARID1A, identified as a tumor suppressor gene, encodes BAF250a, a key component of the SWI-SNF chromatin remodeling complex. A large number of researches demonstrate that the loss of ARID1A expression has been noted in approximately 40% of endometriotic lesions [25]. This study identified mutations in the ARID1A gene in ovarian clear cell and endometrioid carcinomas; these results represent that mutations in ARID1A are an early event in the malignant transformation of endometriosis. In a similar study, Lakshminarasimhan and his colleagues discovered that the downregulation of ARID1A expression in an endometriosis cell line enhances colony formation capacity, cell adhesiveness, and invasiveness, suggesting that low ARID1A expression might be an early event in the malignant transformation of endometriosis to ovarian clear cell carcinoma (OCCC) [26]. Although the link of ARID1A expression and OCCC transformation is well established, whether DNA methylation of ARID1A significantly matters remains elusive. In addition, a study about hypermethylation of ARID1A in breast cancer by Zhang et al. [27] demonstrated that the promoter hypermethylation in the ARID1A gene is strongly associated with ARID1A gene low mRNA expression.

2.2. Are Hormones Useful for Endometriosis Transformation? It is known that endometriosis is an estrogen-sensitive and progesterone-resistant disease [28]. Estrogens have a paramount influence on various physiological processes including cell growth, reproduction, and differentiation and, in the meantime, also on pathological processes such as cancer, metabolic disease, and inflammation. The association between estrogen and various cancers is well reviewed [29]. Lots of clinical studies show that estradiol (E2) plays a key role in endometriosis. The role of E2 is regulated via the estrogen receptors (ERs) including estrogen receptors α (ER-α) and β (ER-β), which are, respectively, encoded by estrogen receptor gene 1 (ESR1) and estrogen receptor gene 2 (ESR2). Several studies investigated the expression of the ERs in the normal and ectopic endometrium of patients with endometriosis; a study reported by Cavallini et al. [30] confirmed the downregulation of ER-α and upregulation of ER-β in ovarian endometriotic tissue compared with eutopic tissue. Whether epigenetics such as DNA methylation is responsible for the expression of ERs in endometriotic cells needs to be further studied. Xue et al. [31] confirmed that the ESR2 gene promoter is hypomethylated in stromal endometriotic cells, which could be related to the upregulation of ER-β. However, low expression of the ER-β gene via promoter hypermethylation in tumors was observed [32]. Meyer et al. found [33] that ESR1 promoters (both ESR1A and ESR1B) are methylated, but the study reported by Toderow et al. [34] indicated that ER-α is not regulated by methylation of the promotor region in endometriosis. Without a doubt, estrogen-relevant genes and ER signal pathways are involved in the development of ovarian cancer [35–37]. Yamaguchi et al. [38], using MS-PCR to identify clear cell carcinoma-specific gene methylation, showed that 64 specific genes were involved in ER-associated pathways among the 276 hypermethylated genes. Representative ER pathway genes including ESR1, BMP4, DKK1, SOX11, SNCG, and MOSC1 are downregulated by promoter methylation, which are in accordance with decreased expression of ER-α. In addition, WT1, as one of the representative genes regulated by the ER-α signaling pathway, is downregulated in patients with endometriosis, consistent with the loss of WT1 expression in ovarian clear cell carcinoma [39, 40]. Furthermore, Akahane et al. [41] showed that decreased expression of ER-α occurred with progression from endometriosis to OCCC and that disappearance of hormone dependency might be associated with malignant transformation to OCCC. In conclusion, estrogen-relevant genes and pathways actually contribute to the malignant transformation of endometriosis, and the inconsistency of the ER-β gene expression between endometriosis and ovarian cancer and associated molecular mechanisms needs to be further investigated.

In the normal endometrium, progesterone strongly interacts with the activation of inflammatory pathways, recruits an influx of various immune cells, and mediates local inflammation [42]. Through binding to the nuclear receptors progesterone receptor isoform A (PRA) and progesterone receptor isoform B (PRB), which are members of the superfamily of ligand-activated transcription factors, the progesterone responses are regulated by directly binding to DNA and regulating the expression of target genes [43]. Several researches supported lower levels of protein expression of PRA and PRB in the eutopic endometrium and the ectopic lesions of patients compared with the normal endometrium of the control group [44]. In addition, another study, that by Fazleabas [45] using experimental animals as disease models, showed that the progesterone receptor (PR) and relevant signaling regulators exerted their effects in the early stages of endometriosis; however, with disease progression, PR expression and some targets of PR lost contact in the eutopic endometrium and the ectopic lesions of endometriosis. In short, it is well established that PR resistance plays an essential role in the occurrence, development, and progression of endometriosis, but it remains unclear whether epigenetic modifications such as DNA methylation contribute to alteration of PR-involved components. In a literature reported by Nie et al. [46], they investigated epigenetic modifications of hormones in endometriosis, and the results revealed that promoter of PRB was hypermethylated; additionally, treatment with both trichostatin A (TSA) and 5-aza-2'-deoxycytidine (ADC) increased PRB gene and protein expression in ectopic endometrial stromal cells but reduced cell viability of ectopic endometrial stromal cells. Another important study by Li et al. [47] investigating the consequences of inhibition of DNA methylation further revealed that lesion growth was ameliorated and PR and PR-target gene expression were restored; the results indicated a potential association between epigenetic regulation and PR-target signal pathways in the pathogenesis of endometriosis. In addition, one of the PR-involved components, Gata2, has been previously evidenced by Böhm et al. [48] as balancing the transcriptional activity of nuclear receptors including PGR; moreover, the expression of Gata2 gene also is consistent with that of PR in the epithelium and stroma of the uterus [49].

As indicated above, PR is dynamically associated with the occurrence of endometriosis and the progression of advanced endometriosis. Whether ovarian cancer involved in endometriosis malignant transformation also is controlled and regulated by epigenetic modification of progesterone related signaling pathways until now remains ambiguous. A previous research using immunohistochemical methods to compare the different expressions of tissues of endometriosis and EAOC indicated that the expression of PR and ER in EAOC was statistical significantly lower than that in endometriosis [50]. In addition, estrogen and progesterone regulated normal endometrial cells in proliferation and differentiation via mediating Wnt/β-catenin signaling whose main components include WNT7a, DKK-1, β-catenin, and GSK-3β [51]. However, abnormal activation the WNT/β-catenin signaling pathway via multiple regulators is reputed to be associated with ovarian cancer with epigenetic modification [52–56]. For instance, significant downregulation of the

Wnt antagonist SFRP5 is observed through the promoter hypermethylation associated with overall survival in ovarian cancer; moreover, epigenetic silence of SFRP5 expression leads to activation of the Wnt pathway and promotes ovarian cancer progression [56]. As indicated above, progesterone, together with other factors, probably plays a key role in indirectly regulating the progression and development of ovarian cancer. Though elaborate mechanisms about whether hormones actually contribute to endometriosis malignant transformation, to date, still remain mysterious, we will be capable of discovering and explaining the correlation between them in the future.

2.3. Oxidative Stress. A large body of literature has investigated that reactive oxygen species- (ROS-) mediated oxidative stress enacts a significant role in the pathophysiology of endometriosis [57–59]. ROS are a group of oxygen including chemically reactive molecules containing superoxide (O_2^-), hydroxyl (OH^-), hydrogen peroxide (H_2O_2), nitrogen oxide (NO), and nitrogen dioxide (NO_2). ROS are intermediaries produced by normal oxygen metabolism, whilst during excess of ROS release, the balance between ROS and antioxidants is broken and induces cellular damage through a variety of mechanisms, finally leading to harmful effects [58]. Many theories associated with ROS-induced endometriosis progression have been elaborated so far as possibly an important factor involved in the progression of endometriosis and malignant transformation. Oxidative stress can eliminate or induce specific DNA and histone methylation by regulating corresponding enzymes such as DNMTs and ten-eleven translocations (TETs); elaborate details are reviewed by Ito et al. in a recent literature [60]. For example, a latest study carried out by Xie et al. [61] which investigated the mechanism of the correlation between oxidative stress and ARID1A gene expression illustrated that ROS decreased the expression level of ARID1A gene via regulating the methylation of its promoter. In addition, another epigenetic enzyme, the TET family of hydroxylases of encoding genes (TET1, TET2, and TET3), significantly downregulated in endometriosis [62]. TET-mediated DNA demethylation may act as a protection against oxidative stress, which also can prove the link between oxidative stress and DNA methylation of endometriosis. Moreover, oxidative stress was extensively studied and reported in the amount of mechanisms of cancer including ovarian cancer [63–67]. Oxidative stress plays an effective role in carcinogenesis through epigenetic alterations. ROS lead to tumorigenesis by inhibiting or silencing of tumor suppressor genes through promoter hypermethylation. The intestine-specific transcription factor caudal type homeobox-1 (CDX1) is downregulated with the treatment of hydrogen peroxide (H_2O_2) because CDX1 promoter is hypermethylated and treatment with 5-aza-dC reversed this effect [68]. In ovarian cancer, a study reported by Hou et al. [69] showed that H_2O_2 downregulated miR-29b by directly targeting its mRNA 3′-UTR in ovarian cancer cells. Additionally, there is a new research proved by Mahalingaiah et al. [70] which demonstrated that a low level of chronic oxidative stress results in the malignant transformation of human renal tubular epithelial cells, and the potential role is the aberrant expression of epigenetic regulatory genes involved in DNA methylation (DNMTs) as well as histone modifications (HDAC1, HAT1) in human renal tubular epithelial cells malignantly transformed by chronic oxidative stress. Given all that, oxidative stress-mediated ovarian cancer malignantly transformed by endometriosis seems to hold great promise.

3. Histone Modifications

Histone modification exerts an equivalent effect on epigenetic regulation the same as DNA methylation. Histones are proteins that make up nucleosomes, which are the fundamental unit of chromatin. Epigenetic modifications such as acetylation, methylation, phosphorylation, and ubiquitylation regulate chromatin structure and gene expression. Histone acetylation is mediated by a class of enzymes called histone acetyl transferases (HATs), which allow chromatin to be in a more unstable state to accomplish gene expression. In contrast, histone deacetylation is regulated by the histone deacetylases (HDACs) and converts chromatin to a more condensed or transcriptionally repressive state for inhibiting gene expression [71]. For example, acetylation at lysine 9 (K9) of H3 is implicated in the transcriptionally active condition of chromatin [16, 72]. Unlike histone acetylation, histone methylation seems to be more elusive. Histone methylation can be either stimulatory or inhibitory to the condensed state of chromatin depending on the particular lysine residue modified. Furthermore, the extent of the methylation status (mono-, di-, and trimethylation) also remains implicated. For example, trimethylation of H3K4 is involved in the transcriptionally active condition of chromatin [16, 73]; however, inverse results present in the mono-, di-, and trimethylation of H3K9, which take part in repressing gene expression [74, 75]. Besides, other types of histone modifications (phosphorylation or ubiquitination) are associated with chromatin condensation status and regulating gene expression, forming a network of sophisticated crosstalk.

3.1. Aberrant Enzyme Expression. There is evidence to support the theory that aberrant HDAC pathways promote cancer growth and metastasis including ovarian cancer [76–78]. HDACs play a crucial role in regulating important cell processes such as cell growth, differentiation, and apoptosis. For instance, sirtuin 1 (SIRT1) is a promising family member and a class III HDAC, which regulates histone acetylation levels as well as the DNA repair [79]. There is a study that demonstrated that SIRT1 expression was significantly increased in epithelial ovarian carcinomas (EOCs) compared to benign tumors [80]. However, another study by Xiaomeng et al. revealed that SIRT1 expression level decreased in eutopic endometrium [81]. Moreover, this is supported by a recent study indicating that the expression of nuclear HDAC1, HDAC2, and HDAC3 proteins was increased in carcinomas compared with benign tumors [77]. A study investigating HDAC expression of endometriosis showed that levels of gene and protein expressions of HDAC1 and HDAC2 were higher in ectopic endometrium than in normal endometrium [82]. Beside, another crucial histone methylation is mediated by histone methyltransferases and histone demethylase. Abnormal expressions of corresponding enzymes are probable to promote oncogene expression and progression of malignant tumors. For example, H3-K27 methylation is regulated by the enhancer of zester homolog 2 (EZH2), which is a key histone methyltransferase that belongs to a subunit of polycomb repressive complex 2. The results published by Guo et al. [83] found that EZH2 expression was significantly higher in ovarian carcinoma than in benign and normal tissues. Another study proved by Kuang and coworkers [84] revealed that EZH2 expression was positively correlated to KDM2B, which controls gene expression by the demethylation of dimethyl histone H3 lysine 36 (H3K36me2) and trimethyl histone H3 lysine 4 (H3K4me3). Both of them play an important role in the development and progression of ovarian cancer.

3.2. Change of Relevant Genes and Signal Pathways. In a literature, it was indicated that global histone H4 acetylation and histone H3K4 methylation level decreased significantly in both eutopic and ectopic endometrium compared with controls. However, there was no difference in H3 acetylation between endometriosis patients and controls [81]. In addition, a study using a dominant-negative histone overexpression approach demonstrated that the tumor suppressor gene RASSF1 is directly downregulated by the methylation of H3-K27. Furthermore, the results suggest that targeted epigenetic therapies of H3-K27 methylation hold great promise [85]. Furthermore, it is well recognized that signal pathways are associated with various aspects of cancer progression. A study by Hurst et al. [86] investigated the molecular mechanism of G-protein-coupled receptor (GPCR) pathways. They discovered that in the regulator of G-protein signaling 2 (RGS2), as an inhibitor of GPCRs, its protein expression is downregulated in ovarian cancer progression. A relevant study by Cacan [87] showed that loss of histone acetylation at RGS2 promoter genes results in the loss of RGS2 expression and indicated that the downregulation of the RGS2 gene is partly due to accumulation of HDACs at the promoter region of RGS2 in chemoresistant ovarian cancer cells.

As mentioned above, we talked about the association between endometriosis and ovarian cancer, such as HDAC gene expression and the increase of proteins of both HDAC1 and HDAC2 in endometriosis and ovarian cancer. Therefore, HDAC expression may be responsible for the malignant transformation of endometriosis although the association between them is still ambiguous. On the other hand, more marked differences were observed in the results of two kinds of researches, such as SIRT1 upregulation in ovarian cancer and downregulation in endometriosis. In spite of those opposite results, the mechanisms of endometriosis malignant evolution will be completed in the future.

3.3. Inflammation and Immune Disorders Need to Be Further Investigated in Epigenetic Modification. Without a doubt, endometriosis is a complex, chronic inflammatory disease with variable symptoms in women. Inflammation and immune disorders absolutely play a key role in the pathobiology of endometriosis; therefore, inflammatory cells and inflammatory cytokines are regulated as target components in endometriosis patients; also, the immune system of women with endometriosis is also dysfunctional. The immune system contains a variety of immune cells, including macrophages, dendritic cells, natural killer cells, T helper cells, and B cells, which have been proved to be disordered in patients with endometriosis [88, 89]. Disorders of inflammatory cell populations in ectopic endometrium and their secretory products exert a harmful influence on normal microenvironment, inducing the development of the disease. Moreover, the immune system and endometrial cells secrete several cytokines and growth factors that promote invasion and growth of ectopic endometrium [90]; those cytokines and growth factors are representative inflammatory mediators, leading to inflammatory response and finally aggravating endometriosis, even causing ovarian cancer.

To date, the most robust pathogenic hypothesis involved in inflammation response is based on the so-called retrograde menstruation phenomenon. Through retrograde flow, viable endometrial fragments reach and implant onto the peritoneum and abdominal organs, leading to chronic inflammation with formation of adhesions and severe infertility. Chronic inflammation, in turn, also promotes proliferation and growth of ectopic endometrial tissue [28]. The presence of ectopic tissue is associated with secretion disorder of inflammatory cells and factors. Macrophages and associated signaling cascade are alternatively activated in patients with endometriosis, which were observed in the study reported by Mahdian et al. [91]. In addition, various inflammatory factors play different roles in infertility in patients with endometriosis [92, 93]. A study by Yang et al. [94] investigated the relations between exposure to pelvic microenvironments with overproduced inflammatory factors and structural or functional tissue abnormalities; their data showed that telocytes (TCs; interstitial Cajal-like cells (ICLCs)) were significantly decreased and interstitial fibrosis was observed, accompanied with an increased level of inducible nitric oxide synthase (iNOS), cyclooxygenase-2 (COX-2), lipid peroxide (LPO), and estradiol, which suggested that inflammation induced TC damage and fibrosis and dysmotility of the oviduct finally leading to subfertility or infertility. Yoshida et al. [95] indicated that interleukins 1 and 6 directly affect sperm mobility. In addition, Hosseini et al. indicated that epigenetic changes of CYP19A1 (aromatase) gene promoters may lead to poor oocyte and embryo condition by impairing follicular steroidogenesis in patients with endometriosis [96]. Tao et al., who studied the pathogenesis of endometriosis-associated infertility, confirmed the tight correlation between monocyte chemotactic protein- (MCP-) 1 and peritoneal leptin levels and infertility in the early stage of endometriosis [97]. Rathore et al. indicated that ghrelin and leptin might contribute to the pathophysiology of infertility, and leptin is associated with inflammatory factors such as IL-6 in patients with endometriosis [98] (Figure 1).

In recent years, cytokines caught the intense attention of researchers due to their involvement in the pathogenesis of endometriosis and cancer. Endometriosis is often accompanied by marked changes of inflammatory cytokines,

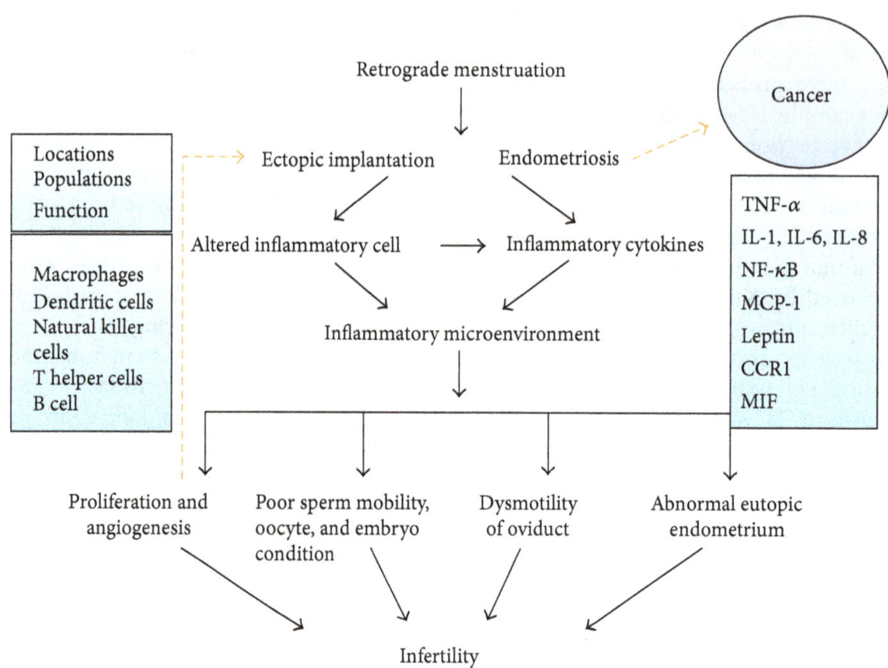

Figure 1: The potential inflammatory mechanisms between endometriosis and infertility. The figure indicates that various factors may result in infertility in patients with endometriosis. Inflammatory cytokines are secreted by inflammatory cells including TNF-α, IL-1, IL-6, IL-8, NF-κB, MCP-1, leptin, CCR1, MIF, and COX-2. Inflammatory responses depend on locations, populations, and functions of inflammatory cells, which include macrophages, dendritic cells, natural killer cells, T helper cells, and B cells. On the one hand, inflammatory responses alter microenvironment and influence various aspects of fertility; on the other hand, chronic exposure to microenvironments with overproduced inflammatory factors leads to ectopic implant proliferations and angiogenesis, which in turn promote growth and invasion of ectopic endometrium even in the development of cancer.

including epithelial cell-derived neutrophil-activating peptide-78 (ENA-78), macrophage migration inhibitory factor (MIF), high-sensitivity C-reactive protein (hs-CRP), tumor necrosis factors (TNF-α), interleukin-1β (IL-1β), IL-6, IL-8, interferon-induced protein 10 (IP-10), and chemokine receptor 1 (CCR1) [99, 100]. There is great expectations for one of them, IL-8, in the development and progression of endometriosis. The literature strongly suggests that IL-8 might play an important role in adhesion and growth of the endometrial implants [101, 102]. A literature by Ulukus et al. [103] showed higher epithelial IL-8 expression in eutopic endometrium of patients with endometriosis, as compared to normal women. In addition, the literature suggested that increased IL-8 expression levels in women with endometriosis might contribute to the development of endometriosis and finally progression of chronic inflammation, even probably malignant transformation [104, 105]. In ovarian cancer, epigenetic modifications are regarded as regulations and mediations of ovarian cancer development, and epigenetic therapies as inhibition of ovarian cancer cells. For instance, the study investigated the specific involvement of HDACs and HATs in the epigenetic regulation of IL-8 expression in ovarian cancer cells; the results indicated that the IL-8 expression in OC cells is regulated by CBP and might enhance effectiveness of HDAC inhibitors in OC treatment [106]. They previously showed that inhibition of histone deacetylase (HDAC) activity increased IL-8 expression in OC cells, resulting in their increased survival and proliferation [107].

The role of chemokine epigenetic regulations in endometriosis malignant transformation has so far remained ambiguous, but those researches suggest that chemokines might appear as a valuable tool to influence the correlation between endometriosis and ovarian cancers and perform early diagnosis of ovarian tumors.

4. MicroRNA Alteration

Over the past 20 years, another epigenetic regulation of gene expression has been discovered and well established, which participates in posttranscriptional gene downregulation mediated by small, non-protein-coding RNA molecules named microRNAs (miRNAs) [108, 109]. Since their initial discovery in 1993 [110], small noncoding RNAs or microRNAs have been intensely investigated across almost all biomedical fields including tumors. MicroRNAs, a class of single strand, are endogenously expressed, and noncoding RNAs, approximately 22 nucleotides in length, were found to mediate a series of essential biological processes including cell cycle, differentiation, development, and apoptosis as well as metabolism [111–114]. Plenty of work has been absorbed in investigating the biogenesis of miRNAs.

miRNA host gene is transcribed by RNA polymerase II, and transcription products are so-called pri-miRNA [115]. The pri-miRNA has to undergo two processing steps in order to become a mature miRNA. First, pri-miRNA is precisely recognized and cleaved by the enzyme Drosha which interacted with an RNA binding protein DGCR8, forming the so-called "microprocessor." The following processing step occurs in cytoplasm: the predominant enzyme called Dicer, also associated with an RNA binding protein named TRBP [116], cleaves the pri-miRNA to a short RNA duplex approximately 21–25 nucleotides in length, depending on the type of Dicer and miRNA [117]. The two strands of the duplex have their own independent influence: one strand of duplex is incorporated into RNA-induced silencing complex (RISC) as a mature miRNA exerting its function, and the remaining strand is usually degraded. However, both strands of some miRNAs are likely to be selected into RISC. The RISC or microRNAs are able to prevent mRNA translation and induce mRNA degradation by matching 3′ untranslated regions of target mRNAs; this phenomenon is known as RNA interference. MicroRNAs are strictly controlled in normal cells. Once microRNAs become deregulated, those aberrant productions can lead to the occurrence and progression of diseases. Researches showed that microRNAs may be associated with the pathogenesis of various human cancers.

4.1. Contributing MicroRNAs in Ovarian Cancer and Endometriosis. In ovarian cancer, the role of miRNAs is present in different biological processes including cell cycle, apoptosis, proliferation, invasion, and metastasis, and even chemoresistance. miRNAs are mediated by Drosha and Dicer; several reports showed that Dicer and Drosha mRNA expression levels and the corresponding proteins were decreased in the majority of ovarian cancers compared with normal tissues [118, 119]. Moreover, ovarian cancers were found to significantly upregulate four members of miR-200 family of miRNAs containing miR-200a, miR-141, miR-200c, and miR-200b, whereas miR-199a, miR-140, miR-145, and miR-125b1 were downregulated among most miRNAs [120]. On the other hand, alterations in the expression levels of different members of miR-200 family are differently associated with the distinct histotypes of ovarian carcinomas. miR-200a and miR-200c overexpressions occur in all the three histotypes including serous and endometrioid as well as clear cell, whereas miR-200b and miR-141 upmodulation exists in endometrioid and serous histotypes [120]. Additionally, another family of miRNAs, the let-7 (lethal-7) family, as tumor suppressor miRNAs, also gets widespread attention in multiple human tumors [121]. Remarkably reduced expressions of the let-7i were observed in tumors of cancer patients with poor survival [122, 123]. In the ovarian cancer, let-7i significantly reduced expression in chemotherapy-resistant patients as reported in a study; moreover, reduced let-7i expression significantly increased the resistance of ovarian cancer cells to the chemotherapy drug *cis*-platinum [123]. Several studies also unravel other miRNA expressions associated with ovarian cancers; for instance, miR-16, miR-20a, miR-21, miR-23a, miR-23b, miR-27a, miR-93, miR-30c and miR-30d, and miR-30e-3p were found to be overregulated, whereas miR-10b, miR-26a, miR-29a, miR-99a, miR-100, miR-125a, miR-125b, miR-143, miR-145, miR-199a, miR-214, miR-22, and miR-519a have opposite expressions [124–126]. Consequently, different miRNAs exert their influence on ovarian cancer development and progression.

Only few researches about malignant transformation were published. Several recent studies investigated the mechanisms of malignant transformation of endometriosis. Tissue inhibitor of metalloproteinases 3 (TIMP3), a proapoptotic protein [127], is proved by Qin and coworkers as a direct target of miR-191 [128]. The data of the study revealed that miR-191 expression was significantly higher in both endometriosis and EAOC and that TIMP3 expression was negatively correlated with miR-191 expression [129]. Additionally, another study, discussing whether cancer-associated miRNA single nucleotide polymorphisms (miRSNPs) accelerate endometriosis development and progression, demonstrated that MIR196A2 and MIR100 influenced endometriosis development and related clinical phenotypes [130].

4.2. EMT Affects Early Stage of the Oncogenic Transformation. Epithelial-mesenchymal transition (EMT) is a highly conserved cellular process that converts immotile and polarized epithelial cells to motile mesenchymal cells, which occurs in embryonic development, fibrosis, and wound healing; meantime, cancer development and progression that resemble embryonic development are regulated and controlled by EMT [131]. As usual, whether EMT occurs or not is detected by corresponding protein expression marks; E-cadherin and cytokeratins are the most common markers for the epithelial phenotypes and N-cadherin and vimentin for the mesenchymal [131]. Multiple noncoding RNAs are reputed to govern EMT; two major regulatory networks are demonstrated to be considered as the core regulatory machinery—the miR-34-SNAI1 and miR-200-ZEB1 axes—which are also controlled by various mediators.

EMT processes enhance migration and invasion of cells, which are prerequisites for the implantation of endometriotic lesions. A previous study reported by Bartley et al. [132] showed that the expression levels of N-cadherin, Twist and Snail, were significantly higher in endometriosis than in endometrium. However, in endometriosis, the expression of E-cadherin was inversely decreased in comparison with that in endometrium. Another study also proved that EMT-related processes might be involved in the pathogenesis of pelvic endometriosis [133]. Filigheddu et al. [134] showed the downregulation of miR-200b expression in the ectopic endometrium compared with the eutopic endometrium of endometriosis patients, together with enhancing of EMT. Eggers et al. [135], who investigated whether miR-200b expression contributes to EMT and invasive growth in endometriosis, indicated that upregulation of miR-200b reverts EMT and inhibits migration and invasion of cells of the endometriotic cell.

The significance of EMT during cancer progression has been commonly recognized, and EMT processes are thought to take part in many cancer cell metastases and progression. For instance, EMT occurs at the invasive front and single mesenchymal-like cells are detected to lose E-cadherin expression in colon carcinoma [136]. Knockdown of Linc-ROR in ovarian cancer cell lines prevents EMT processes through the repression of Wnt/β-catenin signaling; the results suggest that EMT can be an important phenomenon in the invasion and metastasis of ovarian cancer [137]. Another study showed that inhibition of miR-23a reduced the TGF-β1-induced EMT, invasion, and metastasis in breast cancer cells though directly targeting CDH1 that activated induced Wnt/β-catenin signaling [138]. This phenomenon is also discovered in other tumors, which reveals an EMT expression profile and shows increased vimentin and loss of E-cadherin.

As mentioned above, EMT is reputed to be an invasive behavior and enables normal cells to be more aggressive, further having a potential malignant tendency. EMT may exert its momentous transitional effects on endometriosis malignant transformation; furthermore, the occurrence of EMT is probable in an early stage event of endometriosis malignant ovarian cancer in the future.

5. Conclusion

The overall aim of this review was to summarize the epigenetic modifications of the relations between endometriosis malignant transformation and ovarian cancer; moreover, due to the constraints of research progress and attention, we talked more about the potential correlations between them by relevant literatures. Beyond the abovementioned epigenetic modifications, other epigenetics such as long noncoding RNA (lncRNA) and posttranslational modifications (PTMs) are associated with the pathogenesis of endometriosis and ovarian cancer [139–143]. As complex gynecologic disease is closely related to the cancer, enigmatic etiology of endometriosis and mechanism of endometriosis malignant transformation are absolute worth correctly uncovering and elucidating in the future. In recent years, researchers have made significant strides in understanding the disease-specific molecular pathways governing the development of endometriosis in ectopic locations by studying the blood, peritoneal fluid, and eutopic endometrium of women with the disease [144–147]. Vicente-Muñoz et al. [147] identified the plasma metabolites of endometriosis patients and found higher concentration of valine, fucose, choline-containing metabolites, lysine/arginine, and lipoproteins and lower concentration of creatinine than in healthy women, which can help to get a better understanding of the molecular mechanisms of endometriosis. Studying epigenetic modifications of endometriosis, as well as investigating the correlation between endometriosis and ovarian cancer, will propel our understanding of the pathogenesis of endometriosis malignant transformation, with the potential for early diagnostic interventions and new effective therapies.

Conflicts of Interest

The authors declare that there is no conflict of interest regarding the publication of this paper.

Authors' Contributions

Jiaxing He and Weiqin Chang contributed to this work equally and should be considered as co-first authors.

Acknowledgments

This study was supported by grants from the National Natural Science Foundation of China (81302242), Department of Science and Technology of Jilin Province (20150204007YY and 20140204022YY), Development and Reform Commission of Jilin Province (2014G073 and 2016C046-2), and Department of Education of Jilin Province (JJKH20170804KJ).

References

[1] P. Vercellini, P. Viganò, E. Somigliana, and L. Fedele, "Endometriosis: pathogenesis and treatment," *Nature Reviews Endocrinology*, vol. 10, no. 5, pp. 261–275, 2014.

[2] P. Viganò, F. Parazzini, E. Somigliana, and P. Vercellini, "Endometriosis: epidemiology and aetiological factors," *Best Practice & Research Clinical Obstetrics & Gynaecology*, vol. 18, no. 2, pp. 177–200, 2004.

[3] H. S. Kim, T. H. Kim, H. H. Chung, and Y. S. Song, "Risk and prognosis of ovarian cancer in women with endometriosis: a meta-analysis," *British Journal of Cancer*, vol. 110, no. 7, pp. 1878–1890, 2014.

[4] V. H. Eisenberg, C. Weil, G. Chodick, and V. Shalev, "Epidemiology of endometriosis: a large population-based database study from a healthcare provider with 2 million members," *BJOG: An International Journal of Obstetrics & Gynaecology*, vol. 125, no. 1, pp. 55–62, 2018.

[5] J. Siufi Neto, R. M. Kho, D. F. dos Santos Siufi, E. C. Baracat, K. S. Anderson, and M. S. Abrão, "Cellular, histologic, and molecular changes associated with endometriosis and ovarian cancer," *The Journal of Minimally Invasive Gynecology*, vol. 21, no. 1, pp. 55–63, 2014.

[6] P. S. Munksgaard and J. Blaakaer, "The association between endometriosis and ovarian cancer: a review of histological, genetic and molecular alterations," *Gynecologic Oncology*, vol. 124, no. 1, pp. 164–169, 2012.

[7] J. A. Sampson, "Endometrial carcinoma of the ovary, arising in endometrial tissue in that organ," *Archives of Surgery*, vol. 10, no. 1, pp. 1–72, 1925.

[8] A. Melin, P. Sparen, and A. Bergqvist, "The risk of cancer and the role of parity among women with endometriosis," *Human Reproduction*, vol. 22, no. 11, pp. 3021–3026, 2007.

[9] C. A. Barton, N. F. Hacker, S. J. Clark, and P. M. O'Brien, "DNA methylation changes in ovarian cancer: implications for early diagnosis, prognosis and treatment," *Gynecologic Oncology*, vol. 109, no. 1, pp. 129–139, 2008.

[10] F. W. Grimstad and A. Decherney, "A review of the epigenetic contributions to endometriosis," *Clinical Obstetrics and Gynecology*, vol. 60, no. 3, pp. 467–476, 2017.

[11] B. Borghese, K. T. Zondervan, M. S. Abrao, C. Chapron, and D. Vaiman, "Recent insights on the genetics and epigenetics of endometriosis," *Clinical Genetics*, vol. 91, no. 2, pp. 254–264, 2017.

[12] S. Gordts, P. Koninckx, and I. Brosens, "Pathogenesis of deep endometriosis," *Fertility and Sterility*, vol. 108, no. 6, pp. 872–885.e1, 2017.

[13] H. J. Smith, J. M. Straughn, D. J. Buchsbaum, and R. C. Arend, "Epigenetic therapy for the treatment of epithelial ovarian cancer: a clinical review," *Gynecologic Oncology Reports*, vol. 20, pp. 81–86, 2017.

[14] O. Koukoura, S. Sifakis, and D. A. Spandidos, "DNA methylation in endometriosis (review)," *Molecular Medicine Reports*, vol. 13, no. 4, pp. 2939–2948, 2016.

[15] J. Sandoval and M. Esteller, "Cancer epigenomics: beyond genomics," *Current Opinion in Genetics & Development*, vol. 22, no. 1, pp. 50–55, 2012.

[16] L. Maldonado and M. O. Hoque, "Epigenomics and ovarian carcinoma," *Biomarkers in Medicine*, vol. 4, no. 4, pp. 543–570, 2010.

[17] H. T. Nguyen, G. Tian, and M. M. Murph, "Molecular epigenetics in the management of ovarian cancer: are we investigating a rational clinical promise?," *Frontiers in Oncology*, vol. 4, p. 71, 2014.

[18] C. Guo, F. Ren, D. Wang et al., "RUNX3 is inactivated by promoter hypermethylation in malignant transformation of ovarian endometriosis," *Oncology Reports*, vol. 32, no. 6, pp. 2580–2588, 2014.

[19] F. Ren, D. Wang, Y. Jiang, and F. Ren, "Epigenetic inactivation of hMLH1 in the malignant transformation of ovarian endometriosis," *Archives of Gynecology and Obstetrics*, vol. 285, no. 1, pp. 215–221, 2012.

[20] Y. Li, D. An, Y. X. Guan, and S. Kang, "Aberrant methylation of the E-cadherin gene promoter region in endometrium and ovarian endometriotic cysts of patients with ovarian endometriosis," *Gynecologic and Obstetric Investigation*, vol. 82, no. 1, pp. 78–85, 2017.

[21] F. Ren, D. B. Wang, T. Li, Y. H. Chen, and Y. Li, "Identification of differentially methylated genes in the malignant transformation of ovarian endometriosis," *Journal of Ovarian Research*, vol. 7, no. 1, p. 73, 2014.

[22] M. Martini, M. Ciccarone, G. Garganese et al., "Possible involvement of *hMLH1*, $p16^{INK4a}$ and *PTEN* in the malignant transformation of endometriosis," *International Journal of Cancer*, vol. 102, no. 4, pp. 398–406, 2002.

[23] A. Senthong, N. Kitkumthorn, P. Rattanatanyong, N. Khemapech, S. Triratanachart, and A. Mutirangura, "Differences in LINE-1 methylation between endometriotic ovarian cyst and endometriosis-associated ovarian cancer," *International Journal of Gynecological Cancer*, vol. 24, no. 1, pp. 36–42, 2014.

[24] H. Zhou, J. Li, K. Podratz et al., "Hypomethylation and activation of syncytin-1 gene in endometriotic tissue," *Current Pharmaceutical Design*, vol. 20, no. 11, pp. 1786–1795, 2014.

[25] K. C. Wiegand, S. P. Shah, O. M. al-Agha et al., "ARID1A mutations in endometriosis-associated ovarian carcinomas," *The New England Journal of Medicine*, vol. 363, no. 16, pp. 1532–1543, 2010.

[26] R. Lakshminarasimhan, C. Andreu-Vieyra, K. Lawrenson et al., "Down-regulation of ARID1A is sufficient to initiate neoplastic transformation along with epigenetic reprogramming in non-tumorigenic endometriotic cells," *Cancer Letters*, vol. 401, pp. 11–19, 2017.

[27] X. Zhang, Q. Sun, M. Shan et al., "Promoter hypermethylation of ARID1A gene is responsible for its low mRNA expression in many invasive breast cancers," *PLoS One*, vol. 8, no. 1, article e53931, 2013.

[28] L. C. Giudice and L. C. Kao, "Endometriosis," *The Lancet*, vol. 364, no. 9447, pp. 1789–1799, 2004.

[29] M. Jia, K. Dahlman-Wright, and J.-Å. Gustafsson, "Estrogen receptor alpha and beta in health and disease," *Best Practice & Research Clinical Endocrinology & Metabolism*, vol. 29, no. 4, pp. 557–568, 2015.

[30] A. Cavallini, L. Resta, A. M. Caringella, E. Dinaro, C. Lippolis, and G. Loverro, "Involvement of estrogen receptor-related receptors in human ovarian endometriosis," *Fertility and Sterility*, vol. 96, no. 1, pp. 102–106, 2011.

[31] Q. Xue, Z. Lin, Y. H. Cheng et al., "Promoter methylation regulates estrogen receptor 2 in human endometrium and endometriosis," *Biology of Reproduction*, vol. 77, no. 4, pp. 681–687, 2007.

[32] I. Kyriakidis and P. Papaioannidou, "Estrogen receptor beta and ovarian cancer: a key to pathogenesis and response to therapy," *Archives of Gynecology and Obstetrics*, vol. 293, no. 6, pp. 1161–1168, 2016.

[33] J. L. Meyer, D. Zimbardi, S. Podgaec, R. L. Amorim, M. S. Abrão, and C. A. Rainho, "DNA methylation patterns of steroid receptor genes ESR1, ESR2 and PGR in deep endometriosis compromising the rectum," *International Journal of Molecular Medicine*, vol. 33, no. 4, pp. 897–904, 2014.

[34] V. Toderow, M. Rahmeh, S. Hofmann et al., "Promotor analysis of ESR1 in endometrial cancer cell lines, endometrial and endometriotic tissue," *Archives of Gynecology and Obstetrics*, vol. 296, no. 2, pp. 269–276, 2017.

[35] A. Bardin, N. Boulle, G. Lazennec, F. Vignon, and P. Pujol, "Loss of ERβ expression as a common step in estrogen-dependent tumor progression," *Endocrine-Related Cancer*, vol. 11, no. 3, pp. 537–551, 2004.

[36] S. Y. Jeon, K. A. Hwang, and K. C. Choi, "Effect of steroid hormones, estrogen and progesterone, on epithelial mesenchymal transition in ovarian cancer development," *The Journal of Steroid Biochemistry and Molecular Biology*, vol. 158, pp. 1–8, 2016.

[37] A. Ciucci, G. F. Zannoni, M. Buttarelli et al., "Multiple direct and indirect mechanisms drive estrogen-induced tumor growth in high grade serous ovarian cancers," *Oncotarget*, vol. 7, no. 7, pp. 8155–8171, 2016.

[38] K. Yamaguchi, Z. Huang, N. Matsumura et al., "Epigenetic determinants of ovarian clear cell carcinoma biology," *International Journal of Cancer*, vol. 135, no. 3, pp. 585–597, 2014.

[39] M. J. Worley Jr., S. Liu, Y. Hua et al., "Molecular changes in endometriosis-associated ovarian clear cell carcinoma," *European Journal of Cancer*, vol. 51, no. 13, pp. 1831–1842, 2015.

[40] S. Matsuzaki, M. Canis, C. Darcha, P. J. Déchelotte, J. L. Pouly, and G. Mage, "Expression of WT1 is down-regulated in eutopic endometrium obtained during the midsecretory phase from patients with endometriosis," *Fertility and Sterility*, vol. 86, no. 3, pp. 554–558, 2006.

[41] T. Akahane, A. Sekizawa, T. Okuda, M. Kushima, H. Saito, and T. Okai, "Disappearance of steroid hormone dependency during malignant transformation of ovarian clear cell cancer," *International Journal of Gynecological Pathology*, vol. 24, no. 4, pp. 369–376, 2005.

[42] J. J. Brosens, M. S. C. Wilson, and E. W.-F. Lam, "FOXO transcription factors: from cell fate decisions to regulation of human female reproduction," *Advances in Experimental Medicine and Biology*, vol. 665, pp. 227–241, 2009.

[43] M. Al-Sabbagh, E. W.-F. Lam, and J. J. Brosens, "Mechanisms of endometrial progesterone resistance," *Molecular and Cellular Endocrinology*, vol. 358, no. 2, pp. 208–215, 2012.

[44] A. Hayashi, A. Tanabe, S. Kawabe et al., "Dienogest increases the progesterone receptor isoform B/A ratio in patients with ovarian endometriosis," *Journal of Ovarian Research*, vol. 5, no. 1, p. 31, 2012.

[45] A. Fazleabas, "Progesterone resistance in a baboon model of endometriosis," *Seminars in Reproductive Medicine*, vol. 28, no. 1, pp. 075–080, 2010.

[46] J. Nie, X. Liu, and S.-W. Guo, "Promoter hypermethylation of progesterone receptor isoform B (PR-B) in adenomyosis and its rectification by a histone deacetylase inhibitor and a demethylation agent," *Reproductive Sciences*, vol. 17, no. 11, pp. 995–1005, 2010.

[47] Y. Li, M. K. Adur, A. Kannan et al., "Progesterone alleviates endometriosis via inhibition of uterine cell proliferation, inflammation and angiogenesis in an immunocompetent mouse model," *PLoS One*, vol. 11, no. 10, article e0165347, 2016.

[48] M. Böhm, W. J. Locke, R. L. Sutherland, J. G. Kench, and S. M. Henshall, "A role for GATA-2 in transition to an aggressive phenotype in prostate cancer through modulation of key androgen-regulated genes," *Oncogene*, vol. 28, no. 43, pp. 3847–3856, 2009.

[49] C. A. Rubel, H. L. Franco, J. W. Jeong, J. P. Lydon, and F. J. DeMayo, "GATA2 is expressed at critical times in the mouse uterus during pregnancy," *Gene Expression Patterns*, vol. 12, no. 5-6, pp. 196–203, 2012.

[50] M. G. del Carmen, A. E. Smith Sehdev, A. N. Fader et al., "Endometriosis-associated ovarian carcinoma: differential expression of vascular endothelial growth factor and estrogen/progesterone receptors," *Cancer*, vol. 98, no. 8, pp. 1658–1663, 2003.

[51] A. Pazhohan, F. Amidi, F. Akbari-Asbagh et al., "The Wnt/β-catenin signaling in endometriosis, the expression of total and active forms of β-catenin, total and inactive forms of glycogen synthase kinase-3β, WNT7a and DICKKOPF-1," *European Journal of Obstetrics & Gynecology and Reproductive Biology*, vol. 220, pp. 1–5, 2018.

[52] J. A. MacLean, M. L. King, H. Okuda, and K. Hayashi, "WNT7A regulation by miR-15b in ovarian cancer," *PLoS One*, vol. 11, no. 5, article e0156109, 2016.

[53] S. Yoshioka, M. L. King, S. Ran et al., "WNT7A regulates tumor growth and progression in ovarian cancer through the WNT/β-catenin pathway," *Molecular Cancer Research*, vol. 10, no. 3, pp. 469–482, 2012.

[54] M. L. King, M. E. Lindberg, G. R. Stodden et al., "WNT7A/β-catenin signaling induces FGF1 and influences sensitivity to niclosamide in ovarian cancer," *Oncogene*, vol. 34, no. 26, pp. 3452–3462, 2015.

[55] Z. Deng, L. Wang, H. Hou, J. Zhou, and X. Li, "Epigenetic regulation of IQGAP2 promotes ovarian cancer progression via activating Wnt/β-catenin signaling," *International Journal of Oncology*, vol. 48, no. 1, pp. 153–160, 2016.

[56] H. Y. Su, H. C. Lai, Y. W. Lin et al., "Epigenetic silencing of SFRP5 is related to malignant phenotype and chemoresistance of ovarian cancer through Wnt signaling pathway," *International Journal of Cancer*, vol. 127, no. 3, pp. 555–567, 2010.

[57] M. Alizadeh, S. Mahjoub, S. Esmaelzadeh, K. Hajian, Z. Basirat, and M. Ghasemi, "Evaluation of oxidative stress in endometriosis: a case-control study," *Caspian Journal of Internal Medicine*, vol. 6, no. 1, pp. 25–29, 2015.

[58] J. Donnez, M. M. Binda, O. Donnez, and M. M. Dolmans, "Oxidative stress in the pelvic cavity and its role in the pathogenesis of endometriosis," *Fertility and Sterility*, vol. 106, no. 5, pp. 1011–1017, 2016.

[59] A. Van Langendonckt, F. Casanas-Roux, and J. Donnez, "Oxidative stress and peritoneal endometriosis," *Fertility and Sterility*, vol. 77, no. 5, pp. 861–870, 2002.

[60] F. Ito, Y. Yamada, A. Shigemitsu et al., "Role of oxidative stress in epigenetic modification in endometriosis," *Reproductive Sciences*, vol. 24, no. 11, pp. 1493–1502, 2017.

[61] H. Xie, P. Chen, H. W. Huang, L. P. Liu, and F. Zhao, "Reactive oxygen species downregulate ARID1A expression via its promoter methylation during the pathogenesis of endometriosis," *European Review for Medical and Pharmacological Sciences*, vol. 21, no. 20, pp. 4509–4515, 2017.

[62] F. J. Roca, H. A. Loomans, A. T. Wittman, C. J. Creighton, and S. M. Hawkins, "Ten-eleven translocation genes are downregulated in endometriosis," *Current Molecular Medicine*, vol. 16, no. 3, pp. 288–298, 2016.

[63] A. V. Bhat, S. Hora, A. Pal, S. Jha, and R. Taneja, "Stressing the (epi)genome: dealing with reactive oxygen species in cancer," *Antioxidants & Redox Signaling*, 2017.

[64] J. Zhong, L. Ji, H. Chen et al., "Acetylation of hMOF modulates H4K16ac to regulate DNA repair genes in response to oxidative stress," *International Journal of Biological Sciences*, vol. 13, no. 7, pp. 923–934, 2017.

[65] X. Gao, X. Gào, Y. Zhang, L. P. Breitling, B. Schöttker, and H. Brenner, "Associations of self-reported smoking, cotinine levels and epigenetic smoking indicators with oxidative stress among older adults: a population-based study," *European Journal of Epidemiology*, vol. 32, no. 5, pp. 443–456, 2017.

[66] T. Maser, M. Rich, D. Hayes et al., "Tolcapone induces oxidative stress leading to apoptosis and inhibition of tumor growth in neuroblastoma," *Cancer Medicine*, vol. 6, no. 6, pp. 1341–1352, 2017.

[67] W. Li, Y. Guo, C. Zhang et al., "Dietary phytochemicals and cancer chemoprevention: a perspective on oxidative stress, inflammation, and epigenetics," *Chemical Research in Toxicology*, vol. 29, no. 12, pp. 2071–2095, 2016.

[68] R. Zhang, K. A. Kang, K. C. Kim et al., "Oxidative stress causes epigenetic alteration of CDX1 expression in colorectal cancer cells," *Gene*, vol. 524, no. 2, pp. 214–219, 2013.

[69] M. Hou, X. Zuo, C. Li, Y. Zhang, and Y. Teng, "miR-29b regulates oxidative stress by targeting SIRT1 in ovarian cancer cells," *Cellular Physiology and Biochemistry*, vol. 43, no. 5, pp. 1767–1776, 2017.

[70] P. K. Mahalingaiah, L. Ponnusamy, and K. P. Singh, "Oxidative stress-induced epigenetic changes associated with malignant transformation of human kidney epithelial cells," *Oncotarget*, vol. 8, no. 7, pp. 11127–11143, 2017.

[71] M. A. Dawson and T. Kouzarides, "Cancer epigenetics: from mechanism to therapy," *Cell*, vol. 150, no. 1, pp. 12–27, 2012.

[72] M. Esteller, "Cancer epigenomics: DNA methylomes and histone-modification maps," *Nature Reviews Genetics*, vol. 8, no. 4, pp. 286–298, 2007.

[73] D. Schübeler, D. M. MacAlpine, D. Scalzo et al., "The histone modification pattern of active genes revealed through genome-wide chromatin analysis of a higher eukaryote," *Genes & Development*, vol. 18, no. 11, pp. 1263–1271, 2004.

[74] C. Martin and Y. Zhang, "The diverse functions of histone lysine methylation," *Nature Reviews Molecular Cell Biology*, vol. 6, no. 11, pp. 838–849, 2005.

[75] H. A. LaVoie, "Epigenetic control of ovarian function: the emerging role of histone modifications," *Molecular and Cellular Endocrinology*, vol. 243, no. 1-2, pp. 12–18, 2005.

[76] D. J. Marsh, J. S. Shah, and A. J. Cole, "Histones and their modifications in ovarian cancer—drivers of disease and therapeutic targets," *Frontiers in Oncology*, vol. 4, p. 144, 2014.

[77] A. Hayashi, A. Horiuchi, N. Kikuchi et al., "Type-specific roles of histone deacetylase (HDAC) overexpression in ovarian carcinoma: HDAC1 enhances cell proliferation and HDAC3 stimulates cell migration with downregulation of e-cadherin," *International Journal of Cancer*, vol. 127, no. 6, pp. 1332–1346, 2010.

[78] W. Weichert, "HDAC expression and clinical prognosis in human malignancies," *Cancer Letters*, vol. 280, no. 2, pp. 168–176, 2009.

[79] S. Ropero and M. Esteller, "The role of histone deacetylases (HDACs) in human cancer," *Molecular Oncology*, vol. 1, no. 1, pp. 19–25, 2007.

[80] K. Y. Jang, K. S. Kim, S. H. Hwang et al., "Expression and prognostic significance of SIRT1 in ovarian epithelial tumours," *Pathology*, vol. 41, no. 4, pp. 366–371, 2009.

[81] X. Xiaomeng, Z. Ming, M. Jiezhi, and F. Xiaoling, "Aberrant histone acetylation and methylation levels in woman with endometriosis," *Archives of Gynecology and Obstetrics*, vol. 287, no. 3, pp. 487–494, 2013.

[82] M. Colón-Díaz, P. Báez-Vega, M. García et al., "HDAC1 and HDAC2 are differentially expressed in endometriosis," *Reproductive Sciences*, vol. 19, no. 5, pp. 483–492, 2012.

[83] J. Guo, J. Cai, L. Yu, H. Tang, C. Chen, and Z. Wang, "EZH2 regulates expression of p57 and contributes to progression of ovarian cancer *in vitro* and *in vivo*," *Cancer Science*, vol. 102, no. 3, pp. 530–539, 2011.

[84] Y. Kuang, F. Lu, J. Guo et al., "Histone demethylase KDM2B upregulates histone methyltransferase EZH2 expression and contributes to the progression of ovarian cancer in vitro and in vivo," *OncoTargets and Therapy*, vol. 10, pp. 3131–3144, 2017.

[85] P. H. Abbosh, J. S. Montgomery, J. A. Starkey et al., "Dominant-negative histone H3 lysine 27 mutant derepresses silenced tumor suppressor genes and reverses the drug-resistant phenotype in cancer cells," *Cancer Research*, vol. 66, no. 11, pp. 5582–5591, 2006.

[86] J. H. Hurst, N. Mendpara, and S. B. Hooks, "Regulator of G-protein signalling expression and function in ovarian cancer cell lines," *Cellular and Molecular Biology Letters*, vol. 14, no. 1, pp. 153–174, 2009.

[87] E. Cacan, "Epigenetic regulation of RGS2 (regulator of G-protein signaling 2) in chemoresistant ovarian cancer cells," *Journal of Chemotherapy*, vol. 29, no. 3, pp. 173–178, 2017.

[88] L. Schulke, M. Berbic, F. Manconi, N. Tokushige, R. Markham, and I. S. Fraser, "Dendritic cell populations in the eutopic and ectopic endometrium of women with endometriosis," *Human Reproduction*, vol. 24, no. 7, pp. 1695–1703, 2009.

[89] I. Jeung, K. Cheon, and M. R. Kim, "Decreased cytotoxicity of peripheral and peritoneal natural killer cell in endometriosis," *BioMed Research International*, vol. 2016, Article ID 2916070, 6 pages, 2016.

[90] E. Oral, D. L. Olive, and A. Arici, "The peritoneal environment in endometriosis," *Human Reproduction Update*, vol. 2, no. 5, pp. 385–398, 1996.

[91] S. Mahdian, R. Aflatoonian, R. S. Yazdi et al., "Macrophage migration inhibitory factor as a potential biomarker of endometriosis," *Fertility and Sterility*, vol. 103, no. 1, pp. 153–159.e3, 2015.

[92] C. M. Kyama, L. Overbergh, A. Mihalyi et al., "Endometrial and peritoneal expression of aromatase, cytokines, and adhesion factors in women with endometriosis," *Fertility and Sterility*, vol. 89, no. 2, pp. 301–310, 2008.

[93] D. Hornung, I. P. Ryan, V. A. Chao, J. L. Vigne, E. D. Schriock, and R. N. Taylor, "Immunolocalization and regulation of the chemokine RANTES in human endometrial and endometriosis tissues and cells," *The Journal of Clinical Endocrinology & Metabolism*, vol. 82, no. 5, pp. 1621–1628, 1997.

[94] X. J. Yang, J. Yang, Z. Liu, G. Yang, and Z. J. Shen, "Telocytes damage in endometriosis-affected rat oviduct and potential impact on fertility," *Journal of Cellular and Molecular Medicine*, vol. 19, no. 2, pp. 452–462, 2015.

[95] S. Yoshida, T. Harada, T. Iwabe et al., "A combination of interleukin-6 and its soluble receptor impairs sperm motility: implications in infertility associated with endometriosis," *Human Reproduction*, vol. 19, no. 8, pp. 1821–1825, 2004.

[96] E. Hosseini, F. Mehraein, M. Shahhoseini et al., "Epigenetic alterations of *CYP19A1* gene in *Cumulus* cells and its relevance to infertility in endometriosis," *Journal of Assisted Reproduction and Genetics*, vol. 33, no. 8, pp. 1105–1113, 2016.

[97] Y. Tao, Q. Zhang, W. Huang, H. L. Zhu, D. Zhang, and W. Luo, "The peritoneal leptin, MCP-1 and TNF-α in the pathogenesis of endometriosis-associated infertility," *American Journal of Reproductive Immunology*, vol. 65, no. 4, pp. 403–406, 2011.

[98] N. Rathore, A. Kriplani, R. K. Yadav, U. Jaiswal, and R. Netam, "Distinct peritoneal fluid ghrelin and leptin in infertile women with endometriosis and their correlation with interleukin-6 and vascular endothelial growth factor," *Gynecological Endocrinology*, vol. 30, no. 9, pp. 671–675, 2014.

[99] H. Kobayashi, Y. Higashiura, H. Shigetomi, and H. Kajihara, "Pathogenesis of endometriosis: the role of initial infection and subsequent sterile inflammation (review)," *Molecular Medicine Reports*, vol. 9, no. 1, pp. 9–15, 2014.

[100] B. D. McKinnon, D. Bertschi, N. A. Bersinger, and M. D. Mueller, "Inflammation and nerve fiber interaction in endometriotic pain," *Trends in Endocrinology & Metabolism*, vol. 26, no. 1, pp. 1–10, 2015.

[101] J. Sikora, M. Smycz-Kubańska, A. Mielczarek-Palacz, and Z. Kondera-Anasz, "Abnormal peritoneal regulation of chemokine activation—the role of IL-8 in pathogenesis of endometriosis," *American Journal of Reproductive Immunology*, vol. 77, no. 4, 2017.

[102] N. Malhotra, D. Karmakar, V. Tripathi, K. Luthra, and S. Kumar, "Correlation of angiogenic cytokines-leptin and IL-8 in stage, type and presentation of endometriosis," *Gynecological Endocrinology*, vol. 28, no. 3, pp. 224–227, 2012.

[103] M. Ulukus, E. C. Ulukus, E. N. Tavmergen Goker, E. Tavmergen, W. Zheng, and A. Arici, "Expression of interleukin-8 and monocyte chemotactic protein 1 in women with endometriosis," *Fertility and Sterility*, vol. 91, no. 3, pp. 687–693, 2009.

[104] J. L. Herington, K. L. Bruner-Tran, J. A. Lucas, and K. G. Osteen, "Immune interactions in endometriosis," *Expert Review of Clinical Immunology*, vol. 7, no. 5, pp. 611–626, 2011.

[105] H. Cakmak, O. Guzeloglu-Kayisli, U. A. Kayisli, and A. Arici, "Immune-endocrine interactions in endometriosis," *Frontiers in Bioscience (Elite Edition)*, vol. 1, pp. 429–443, 2009.

[106] H. R. Gatla, Y. Zou, M. M. Uddin, and I. Vancurova, "Epigenetic regulation of interleukin-8 expression by class I HDAC and CBP in ovarian cancer cells," *Oncotarget*, vol. 8, no. 41, pp. 70798–70810, 2017.

[107] H. R. Gatla, Y. Zou, M. M. Uddin et al., "Histone deacetylase (HDAC) inhibition induces IκB kinase (IKK)-dependent interleukin-8/CXCL8 expression in ovarian cancer cells," *Journal of Biological Chemistry*, vol. 292, no. 12, pp. 5043–5054, 2017.

[108] R. Schickel, B. Boyerinas, S. M. Park, and M. E. Peter, "MicroRNAs: key players in the immune system, differentiation, tumorigenesis and cell death," *Oncogene*, vol. 27, no. 45, pp. 5959–5974, 2008.

[109] J. D. Kuhlmann, J. Rasch, P. Wimberger, and S. Kasimir-Bauer, "MicroRNA and the pathogenesis of ovarian cancer—a new horizon for molecular diagnostics and treatment?," *Clinical Chemistry and Laboratory Medicine*, vol. 50, no. 4, pp. 601–615, 2012.

[110] R. C. Lee, R. L. Feinbaum, and V. Ambros, "The *C. elegans* heterochronic gene *lin*-4 encodes small RNAs with antisense complementarity to *lin*-14," *Cell*, vol. 75, no. 5, pp. 843–854, 1993.

[111] Y. Wang and R. Blelloch, "Cell cycle regulation by microRNAs in stem cells," *Results and Problems in Cell Differentiation*, vol. 53, pp. 459–472, 2011.

[112] M. Carleton, M. A. Cleary, and P. S. Linsley, "MicroRNAs and cell cycle regulation," *Cell Cycle*, vol. 6, no. 17, pp. 2127–2132, 2007.

[113] S. Oliveto, M. Mancino, N. Manfrini, and S. Biffo, "Role of microRNAs in translation regulation and cancer," *World Journal of Biological Chemistry*, vol. 8, no. 1, pp. 45–56, 2017.

[114] N. Li, B. Long, W. Han, S. Yuan, and K. Wang, "MicroRNAs: important regulators of stem cells," *Stem Cell Research & Therapy*, vol. 8, no. 1, p. 110, 2017.

[115] Y. Lee, M. Kim, J. Han et al., "MicroRNA genes are transcribed by RNA polymerase II," *The EMBO Journal*, vol. 23, no. 20, pp. 4051–4060, 2004.

[116] S. M. Hammond, "An overview of microRNAs," *Advanced Drug Delivery Reviews*, vol. 87, pp. 3–14, 2015.

[117] M. Ha and V. N. Kim, "Regulation of microRNA biogenesis," *Nature Reviews Molecular Cell Biology*, vol. 15, no. 8, pp. 509–524, 2014.

[118] G. Pampalakis, E. P. Diamandis, D. Katsaros, and G. Sotiropoulou, "Down-regulation of dicer expression in ovarian cancer tissues," *Clinical Biochemistry*, vol. 43, no. 3, pp. 324–327, 2010.

[119] W. M. Merritt, Y. G. Lin, L. Y. Han et al., "Dicer, Drosha, and outcomes in patients with ovarian cancer," *The New England Journal of Medicine*, vol. 359, no. 25, pp. 2641–2650, 2008.

[120] M. V. Iorio, R. Visone, G. di Leva et al., "MicroRNA signatures in human ovarian cancer," *Cancer Research*, vol. 67, no. 18, pp. 8699–8707, 2007.

[121] A. E. Pasquinelli, B. J. Reinhart, F. Slack et al., "Conservation of the sequence and temporal expression of *let*-7 heterochronic regulatory RNA," *Nature*, vol. 408, no. 6808, pp. 86–89, 2000.

[122] S. Shell, S. M. Park, A. R. Radjabi et al., "Let-7 expression defines two differentiation stages of cancer," *Proceedings of the National Academy of Sciences of the United States of America*, vol. 104, no. 27, pp. 11400–11405, 2007.

[123] N. Yang, S. Kaur, S. Volinia et al., "MicroRNA microarray identifies *Let-7i* as a novel biomarker and therapeutic target in human epithelial ovarian cancer," *Cancer Research*, vol. 68, no. 24, pp. 10307–10314, 2008.

[124] H. Lee, C. Park, G. Deftereos et al., "MicroRNA expression in ovarian carcinoma and its correlation with clinicopathological features," *World Journal of Surgical Oncology*, vol. 10, no. 1, p. 174, 2012.

[125] T. H. Kim, Y. K. Kim, Y. Kwon et al., "Deregulation of miR-519a, 153, and 485-5p and its clinicopathological relevance in ovarian epithelial tumours," *Histopathology*, vol. 57, no. 5, pp. 734–743, 2010.

[126] A. Mahdian-shakib, R. Dorostkar, M. Tat, M. S. Hashemzadeh, and N. Saidi, "Differential role of microRNAs in prognosis, diagnosis, and therapy of ovarian cancer," *Biomedicine & Pharmacotherapy*, vol. 84, pp. 592–600, 2016.

[127] B. Shen, Y. Jiang, Y.-R. Chen et al., "Expression and inhibitory role of TIMP-3 in hepatocellular carcinoma," *Oncology Reports*, vol. 36, no. 1, pp. 494–502, 2016.

[128] S. Qin, Y. Zhu, F. Ai et al., "MicroRNA-191 correlates with poor prognosis of colorectal carcinoma and plays multiple roles by targeting tissue inhibitor of metalloprotease 3," *Neoplasma*, vol. 61, no. 1, pp. 27–34, 2014.

[129] M. Dong, P. Yang, and F. Hua, "miR-191 modulates malignant transformation of endometriosis through regulating TIMP3," *Medical Science Monitor*, vol. 21, pp. 915–920, 2015.

[130] C. Y. Chang, M. T. Lai, Y. Chen et al., "Up-regulation of ribosome biogenesis by *MIR196A2* genetic variation promotes endometriosis development and progression," *Oncotarget*, vol. 7, no. 47, pp. 76713–76725, 2016.

[131] M. A. Nieto, R. Y.-J. Huang, R. A. Jackson, and J. P. Thiery, "EMT: 2016," *Cell*, vol. 166, no. 1, pp. 21–45, 2016.

[132] J. Bartley, A. Jülicher, B. Hotz, S. Mechsner, and H. Hotz, "Epithelial to mesenchymal transition (EMT) seems to be regulated differently in endometriosis and the endometrium," *Archives of Gynecology and Obstetrics*, vol. 289, no. 4, pp. 871–881, 2014.

[133] S. Matsuzaki and C. Darcha, "Epithelial to mesenchymal transition-like and mesenchymal to epithelial transition-like processes might be involved in the pathogenesis of pelvic endometriosis," *Human Reproduction*, vol. 27, no. 3, pp. 712–721, 2012.

[134] N. Filigheddu, I. Gregnanin, P. E. Porporato et al., "Differential expression of microRNAs between eutopic and ectopic endometrium in ovarian endometriosis," *Journal of Biomedicine and Biotechnology*, vol. 2010, Article ID 369549, 29 pages, 2010.

[135] J. C. Eggers, V. Martino, R. Reinbold et al., "MicroRNA miR-200b affects proliferation, invasiveness and stemness of endometriotic cells by targeting ZEB1, ZEB2 and KLF4," *Reproductive Biomedicine Online*, vol. 32, no. 4, pp. 434–445, 2016.

[136] T. Brabletz, A. Jung, S. Reu et al., "Variable β-catenin expression in colorectal cancers indicates tumor progression driven

by the tumor environment," *Proceedings of the National Academy of Sciences of the United States of America*, vol. 98, no. 18, pp. 10356–10361, 2001.

[137] Y. Lou, H. Jiang, Z. Cui, L. Wang, X. Wang, and T. Tian, "Linc-ROR induces epithelial-to-mesenchymal transition in ovarian cancer by increasing Wnt/β-catenin signaling," *Oncotarget*, vol. 8, no. 41, pp. 69983–69994, 2017.

[138] F. Ma, W. Li, C. Liu et al., "miR-23a promotes TGF-β1-induced EMT and tumor metastasis in breast cancer cells by directly targeting CDH1 and activating Wnt/β-catenin signaling," *Oncotarget*, vol. 8, no. 41, pp. 69538–69550, 2017.

[139] J.-J. Qiu, L.-C. Ye, J.-X. Ding et al., "Expression and clinical significance of estrogen-regulated long non-coding RNAs in estrogen receptor α-positive ovarian cancer progression," *Oncology Reports*, vol. 31, no. 4, pp. 1613–1622, 2014.

[140] Y. Wang, Y. Chen, and J. Fang, "Post-transcriptional and post-translational regulation of central carbon metabolic enzymes in cancer," *Anti-Cancer Agents in Medicinal Chemistry*, vol. 17, no. 11, pp. 1456–1465, 2017.

[141] R. Asadollahi, C. A. C. Hyde, and X. Y. Zhong, "Epigenetics of ovarian cancer: from the lab to the clinic," *Gynecologic Oncology*, vol. 118, no. 1, pp. 81–87, 2010.

[142] P.-R. Sun, S.-Z. Jia, H. Lin, J.-H. Leng, and J.-H. Lang, "Genome-wide profiling of long noncoding ribonucleic acid expression patterns in ovarian endometriosis by microarray," *Fertility and Sterility*, vol. 101, no. 4, pp. 1038–1046.e7, 2014.

[143] C. Zhou, T. Zhang, F. Liu et al., "The differential expression of mRNAs and long noncoding RNAs between ectopic and eutopic endometria provides new insights into adenomyosis," *Molecular BioSystems*, vol. 12, no. 2, pp. 362–370, 2016.

[144] H. Liu and J. H. Lang, "Is abnormal eutopic endometrium the cause of endometriosis? The role of eutopic endometrium in pathogenesis of endometriosis," *Medical Science Monitor*, vol. 17, no. 4, pp. RA92–RA99, 2011.

[145] M. Zhou, J. Fu, L. Xiao et al., "miR-196a overexpression activates the MEK/ERK signal and represses the progesterone receptor and decidualization in eutopic endometrium from women with endometriosis," *Human Reproduction*, vol. 31, no. 11, pp. 2598–2608, 2016.

[146] H. Jørgensen, A. S. Hill, M. T. Beste et al., "Peritoneal fluid cytokines related to endometriosis in patients evaluated for infertility," *Fertility and Sterility*, vol. 107, no. 5, pp. 1191–1199.e2, 2017.

[147] S. Vicente-Muñoz, I. Morcillo, L. Puchades-Carrasco, V. Payá, A. Pellicer, and A. Pineda-Lucena, "Pathophysiologic processes have an impact on the plasma metabolomic signature of endometriosis patients," *Fertility and Sterility*, vol. 106, no. 7, pp. 1733–1741.e1, 2016.

7

The Maize *Corngrass1* miRNA-Regulated Developmental Alterations are Restored by a Bacterial ADP-Glucose Pyrophosphorylase in Transgenic Tobacco

Ayalew Ligaba-Osena,[1] Kay DiMarco,[2] Tom L. Richard,[3] and Bertrand Hankoua[1]

[1]*College of Agriculture and Related Sciences, Delaware State University, 1200 N DuPont Highway, Dover, DE 19901, USA*
[2]*2217 Earth and Engineering Sciences, Pennsylvania State University, University Park, PA 16802, USA*
[3]*Agricultural and Biological Engineering, Pennsylvania State University, 132 Land and Water Research Building, PA 16802, USA*

Correspondence should be addressed to Ayalew Ligaba-Osena; alosena@uncg.edu and Bertrand Hankoua; bhankoua@desu.edu

Academic Editor: Antonio Ferrante

Crop-based bioethanol has raised concerns about competition with food and feed supplies, and technologies for second- and third-generation biofuels are still under development. Alternative feedstocks could fill this gap if they can be converted to biofuels using current sugar- or starch-to-ethanol technologies. The aim of this study was to enhance carbohydrate accumulation in transgenic *Nicotiana benthamiana* by simultaneously expressing the maize *Corngrass1* miRNA (*Cg1*) and *E. coli* ADP-glucose pyrophosphorylase (*glgC*), both of which have been reported to enhance carbohydrate accumulation *in planta*. Our findings revealed that expression of *Cg1* alone increased shoot branching, delayed flowering, reduced flower organ size, and induced loss of fertility. These changes were fully restored by coexpressing *Escherichia coli glgC*. The transcript level of miRNA156 target *SQUAMOSA promoter binding-like* (*SPL*) transcription factors was suppressed severely in *Cg1*-expressing lines as compared to the wild type. Expression of *glgC* alone or in combination with *Cg1* enhanced biomass yield and total sugar content per plant, suggesting the potential of these genes in improving economically important biofuel feedstocks. A possible mechanism of the *Cg1* phenotype is discussed. However, a more detailed study including genome-wide transcriptome and metabolic analysis is needed to determine the underlying genetic elements and pathways regulating the observed developmental and metabolic changes.

1. Introduction

Global energy demand is predicted to grow by 37 percent by the year 2040 [1]. During the same period, the distribution of energy demand will change dramatically, triggered by faster-growing economies and rising consumption in Asia, Africa, the Middle East, and Latin America. To meet this demand, the consumption of petroleum and other liquid fuels is projected to increase from 3.78 billion gallons per day in 2012 to 5.08 billion by 2040 [2]. However, increasing risks of environmental pollution and climate change due to production and use of fossil fuels necessitate the quest for alternative energy sources [3].

Production of biofuels and other chemicals from lignocellulosic biomass has been impeded by biomass recalcitrance (the resistance of plant cell walls to enzymatic deconstruction) largely due to the presence of highly heterogenic polymer lignin, which is a major barrier to cost-effective conversion of biomass to biofuels and useful chemicals [4]. Lignin consists of three major phenylpropanoid units, syringyl, guaiacyl, and hydroxyphenyl units, and can interlock with cellulose and hemicelluloses, limiting the accessibility of these polysaccharides to cellulase and hemicellulase enzymes, respectively [5–7]. Over a period of decades, several pretreatment technologies have been developed to break down lignin in the biomass and increase conversion efficiency [8]. However, these technologies have various limitations and are not being commercialized at the pace needed to address the short-term demand for biofuels. In this context, alternative feedstocks with enhanced carbohydrate yield that are easily converted to fuels using current technology have great potential.

Advances in genetic engineering have greatly contributed to the improvement of desirable traits including enhanced biomass yields, polysaccharide content, and modification of the cell wall composition to reduce pretreatment costs [9]. For example, an increase in starch content has been achieved in transgenic potato [10] and cassava [11] tubers overexpressing the *Escherichia coli* ADP-glucose pyrophosphorylase (AGPase or *glgC, EC* 2.7.7.27), which catalyzes the first dedicated and rate-limiting step in starch biosynthesis. The *glgC* gene encodes a major enzyme controlling starch biosynthesis, catalyzing the conversion of glucose 1-phosphate and ATP to ADP-glucose (ADPGlc) and inorganic pyrophosphate, with the ADPGlc subsequently used by starch synthases to incorporate glucosyl units into starch [12, 13]. A mutant form of the enzyme *Glg*CG336D (the amino acid glycine at position 336 is mutated to aspartic acid), which has less sensitivity to inhibitors and activators [10, 11], was shown to enhance tuber yield in cassava by 260 percent as compared to the nontransformed wild-type plants [11]. This is likely achieved by increasing the *Glg*C-mediated sink strength for carbohydrates, increasing overall photosynthetic rate, and reducing feedback inhibition of carbohydrate assimilation [10, 11, 14]. Many starch-metabolizing enzymes are redox-regulated [15–17]. Thioredoxins (*EC* 1.8.1.9) are oxidoreductases that mediate the thiol-disulfide exchange of Cys residues and act as a reductant of the redox-regulated enzymes involved in carbohydrate metabolism [9]. Overexpression of plastidal *Trxf* gene in transgenic tobacco has been shown to increase carbohydrate biosynthesis (starch and soluble sugars) in leaves [9, 18]. Similarly, a photorespiratory bypass via posttranslational targeting of the *E. coli* glycolate catabolic pathway (consisting glycolate dehydrogenase (*EC* 1.1.99.14) subunits E, F, and G) expressed in potato [19] has been shown to increase biomass, rate of photosynthesis, and sugars (glucose, fructose, and sucrose) and transitory starch, suggesting reduced photorespiration [20–22].

In addition to these and other coding genes that produce enzymes, small noncoding RNAs (microRNAs or miRNAs) of approximately 19–24 nucleotides in length can serve as gene regulatory factors and have the potential to improve complex traits including biomass traits [22–24]. Transgenic expression of the maize tandem miRNA *Corngrass1*, which belongs to miR156, in switchgrass (*Panicum virgatum*) has been reported to completely inhibit flowering, increase perenniality, increase starch content, and improve biomass digestibility with or without pretreatment due to reduced lignin content [25, 26].

miR156 are known to target the *SPL* transcription factors, which are involved in various physiological processes including promotion of juvenile to adult phase change (heteroblasty), reproductive transition, control of male fertility, and stress responses [27, 28]. The *Arabidopsis* genome contains 16 *SPL* genes, the majority of which are targeted by miRNA156 [27, 29]. The *SPLs* control plant development by directly regulating downstream genes [30]. Based on their conserved DNA-binding domain, the *SPLs* are grouped into five clades: *SPL3/SPL4/SPL5*, *SPL9/SPL15*, *SPL2/SPL8/SPL10/SPL11*, *SPL6*, and *SPL13A/B* [23, 29, 31]. Moreover, gene expression analysis and gain-of-function and loss-of-function studies have revealed several functionally distinct groups [28] including *SPL* genes regulating control of juvenile-to-adult vegetative transition and the vegetative-to-reproductive transition (*SPL2, SPL9, SPL10, SPL11, SPL13*, and *SPL15*) and those that have been reported to play a role in promoting floral meristem identity transition (*SPL3, SPL4*, and *SPL5*). One of these genes, *SPL8*, has been reported to regulate male fertility/seed set, petal trichome production, and root growth [29, 32–34].

In this study, maize *Cg1* was expressed in *N. benthamiana* with or without the *E. coli glgC* with the purpose of enhancing carbohydrate content in transgenic biomass. Our findings revealed that overexpression of *Cg1* alone significantly modulated plant growth and development including delayed flowering, reduced floral organs, and complete loss of fertility. These phenotypes were restored by coexpressing the *E. coli glgC*. Possible mechanisms of phenotypic alterations in tobacco by *Cg1* and the observed antagonistic effect of *glgC* are discussed.

2. Material and Methods

2.1. Gene Cloning and Generation of Expression Constructs. The sequence of the maize *Corngrass1* (*Cg1*) which encodes two tandem miRNAs [25] (GenBank Acc. number EF541486.1) was synthesized as a gBlocks gene fragment at Integrated DNA Technologies (https://www.idtdna.com) flanked by *Eco*RI and *Kpn*I for subsequent cloning into the pSAT1 entry vector [35] under the control of the enhanced CaMV 35S promoter. The coding region of *E. coli* ADPGlc pyrophosphorylase (AGPase or *glgc*; GenBank Acc. number S58224) was amplified from the pO12 plasmid obtained from Dr. Tony Romeo (University of Florida) using sense (5′-aagg aaaggaCTCGAGatggcttctatgatatcctcttccgctgtgacaac-3′) and antisense primers, (aaggaCCCGGGgtggtgatgatgatgatgtcgctcc tgtttatgccctaac) containing *Xho*I and *Sma*I, respectively. A 57-amino acid pea chloroplast transit peptide was fused to the N-terminus of the sequence to target protein expression to the amyloplast, a nonpigmented organelle responsible for starch synthesis and storage. Since single-amino acid substitution (Gly336Asp) has been shown to reduce sensitivity of the enzyme to inhibitors and activators [10], the mutation was introduced by site-directed mutagenesis using overlap extension PCR as previously described [36, 37]. *glgc* was inserted into the pSAT4 entry vector [35] also under the control of an enhanced CaMV 35S promoter. Expression cassettes of *Corngrass1* (*Cg1*) and *glgc* were assembled into the binary vector pPZP-RCS2 [38] using *Asc*I and a homing endonuclease *I-Sce*I, respectively, singly or together for coexpression. The resulting binary vectors pPZP-*NPTII-Cg1*, pPZP-*NPTII-glgc*, and pPZP-*NPTII-Cg1-glgc* (Figure 1) were introduced into *Agrobacterium* strain LBA4404 for subsequent transformation of tobacco (*N. benthamiana*).

2.2. Tobacco Transformation and Generation of Transgenic Lines. Leaf explants (~0.5 mm^2) of 4–6-week-old tobacco were infiltrated with *Agrobacterium* harboring the expression

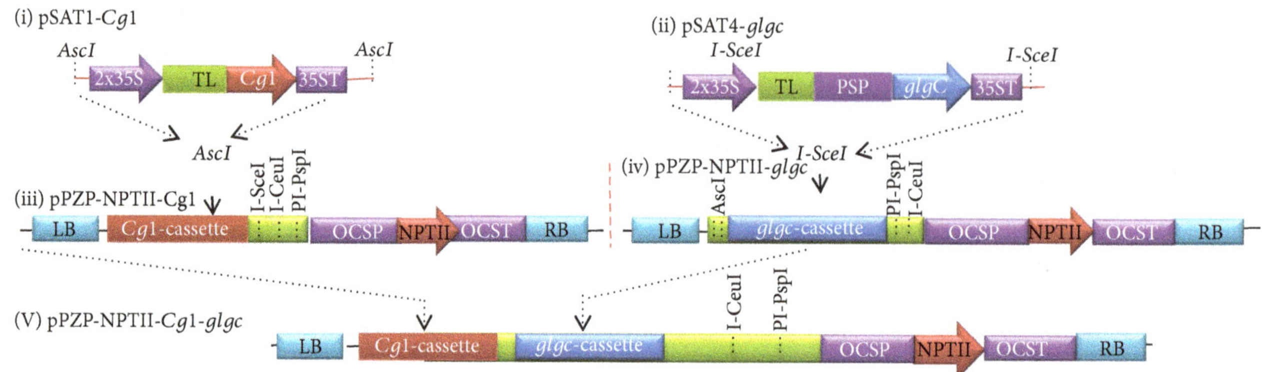

FIGURE 1: Constructs of maize *Cg1* (pSAT1-*Cg1*) (I) and *E. coli glgC* (pSAT4-*glgC*) (II) were generated in the pSAT shuttle vector under the control of enhanced 35S CaMV promoter (35S) and tobacco etch virus leader sequence (TL). The expression of *glgC* was targeted to amyloplast using pea transit peptide (PSP). Binary vectors with single and double expression cassettes pPZP-NPTII-*Cg1*, pPZP-NPTII-*glgC*, and pPZP-NPTII-*Cg1-glgC* were generated for subsequent transformation of tobacco.

vectors pPZP-*NPTII-Cg1*, pPZP-*NPTII-glgC*, and pPZP-*NPTII-Cg1-glgC* or the empty vector pPZP-NPTII for five minutes in the presence of 200 μM acetosyringone. Handling of transformed tissues, selection and regeneration, and maintenance of transgenic lines were performed based on Ligaba-Osena et al. [39].

2.3. Validation of Transgene Insertion. To verify the insertion of transgenes, genomic DNA was isolated from 100 mg of fresh leaves of wild-type or transgenic lines using the GeneJET Plant Genomic DNA Purification Mini Kit (Thermo Scientific). The DNA was used as a template in PCR reactions to amplify the selectable marker gene (neomycin phosphotransferase, *nptII*) and *Cg1* and *glgC* genes using sense and antisense primers in Supplementary Table S1. The PCR products were analyzed by agarose gel electrophoresis.

2.4. RNA Extraction. Total RNA was extracted from wild-type and independent transgenic lines (*Cg1*, *Cg1-glgC*, or *glgC*). Fully expanded leaves of two-month-old plants were collected and immediately frozen in liquid N_2 and ground to a fine powder using a mortar and pestle. Total RNA was isolated using Spectrum Plant Total RNA Kit (Sigma, St. Louis, USA). The RNA solution was stored at −80°C until it was used for first-strand complementary DNA synthesis.

2.5. Identification of Cg1 Target Genes. Given that miRNA156 has been implicated in the regulation of flowering via its downstream targets known as *SPL* transcription factors [22], we studied the expression of putative *Cg1* target genes in tobacco. By searching genome database and publications, we identified four *SPL* contiguous sequences (TC20466), EH36899 (GenBank Acc. number EH368993), and TC9706 and TC7909 [40]. EH36899 showed high homology with *SPL1*, while TC20466 is annotated as *SPL*12 (https://solgenomics.net). TC9706 (*SPL15*) and TC7909 (*SPL9*) were reported in Tang et al. [40]. To analyze the expression of target genes, primers were designed based on the contiguous sequences (Table S1). Potential *Cg1*-binding sites in the sequences of *SPL* genes were determined using the targetfinder.pl software previously developed at the Carrington lab [41].

2.6. Quantitative Real-Time RT-PCR. Expression of transgenes (*Cg1* and *glgC*) and putative *Cg1*-targets was studied by quantitative real-time RT-PCR (qPCR) as described previously [39]. Primers used for gene expression analysis are listed in Table S1; 18S RNA was used as an internal control. Relative expression level was calculated using the $\Delta\Delta C_T$ method available on SDS software (Applied Biosystems).

2.7. Determination of Starch Content. The shoot biomass was ground to 1 mm particle size, and 500 mg was used for the starch assay. The biomass was preextracted to remove free sugars by incubation at 40°C water bath and filtration using Whatman 41 filter paper. The starch content in the biomass was determined according to the Dairy One procedures (Dairy One Forage Laboratory, Ithaca, NY).

2.8. Biomass Saccharification. To see whether coexpression of the *Cg1* and *glgC* improves saccharification efficiency, the biomass was harvested at maturity, dried in an oven at 60°C for two days, and ground to powder. One gram of ground biomass was weighed into 15 mL Falcon tubes containing 9 mL of 50 mM sodium acetate pH 5.5, and the samples were vortexed for 2 min and then centrifuged at 9000 *g* for 10 min. The supernatant was recovered to determine initial sugar content, and the pellet was washed twice with 50 mM sodium acetate buffer and centrifuged again. The final pellet was suspended in 9 mL of the 50 mM sodium acetate buffer to which 50 μL of each *Trichoderma reesei* cellulase, *Aspergillus niger* glucosidase, and *Bacillus licheniformis* a-amylase (Sigma) [42] was added, with 100 μL of sodium azide (from 2% stock) added to suppress microbial growth [43]. The reactions were then incubated at 45°C for three days while shaking at 250 rpm. After three days, the samples were centrifuged at 10,000 *g* for 10 min and the resulting supernatant hydrolysate was filtered (0.22 μm). This hydrolysate was subsequently analyzed to determine the total reducing sugar yield or the specific sugar species released from the biomass.

2.9. Sugar Quantification. Sugars in the hydrolysate obtained after biomass saccharification were characterized filtered (0.22 μm) for quantifying sugars using Dionex Ion Exchange Chromatography 3000 equipped with an electrochemical detector (Dionex, Sunnyvale, USA) as previously described [39]. The sugar concentration from the IC reading was converted to milligrams of sugar per gram of dry matter or milligrams of sugar per plant. Sugar yield was measured from four replicates for each treatment.

2.10. Statistical Analysis. Experiments were conducted in a complete randomized design in at least three replicates, and each experiment was repeated twice. Data were analyzed using one-way ANOVA using the PROC GLM procedure [44]. After the significant F-tests, the Tukey multiple comparison procedure was used to separate the means ($P < 0.05$).

3. Results

3.1. Modulation of Vegetative Growth by Cg1. Ectopic expression of *Cg1* has been shown to alter growth and development in various plant species [25]. In this study, we investigated whether *Cg1* expressed in tobacco alone or with *glgC* affects plant growth and development. At least six independent lines were generated for each construct. Transgene insertion of at least two independent transgenic lines was validated by PCR using genomic DNA as template (Figure 2(a)).

After one month of growth, transgenic (T_1) lines coexpressing *Cg1* and *glgC* (*Cg1-glgC*) and *glgC* were not different from the empty vector and nontransgenic control lines (Figure 2(b)). Because *Cg1* lines did not produce seeds, we compared regenerated (T_0) *Cg1* plants with the WT, the empty vector control (NPTII), and *Cg1-glgC*. Interestingly, *Cg1* lines exhibited a distinct phenotype. The *Cg1* plants develop smaller leaves and grow slower than the controls and *Cg1-glgC* (Figure 2(c)). Likewise, in two-month-old plants (Figure 2(d)), transgenic lines expressing *Cg1-glgC* and *glgC* were not different from the nontransgenic (WT) and empty vector control lines, suggesting that coexpression of *Cg1* and *glgC* or *glgC* alone may not interfere with plant growth and development. The *Cg1* lines exhibited a bushy phenotype with increased branching and leaf number and reduced leaf size as compared to the WT and the other transgenic lines, which is more evident in two-month-old plants (Figure 2(c)) as compared to the one-month-old seedling, and is more pronounced in *Cg1*L2 (Figure 2(d)).

3.2. Cg1-Altered Flower Development Is Restored by glgC. In this study, expression of *Cg1* singly delayed flowering (Figures 2(c) and 3). Moreover, because the *Cg1* plants have more branching, the number of flowers per plants was higher than in the WT, and floral parts were reduced in size when visually observed (Figure 3). The flowers bore smaller petals as compared to WT, whereas there was no marked difference in flowering time (Figures 2(c) and 2(d)) and floral organ development (Figure 3) between the WT control, *Cg1-glgC*, or *glgC*. Moreover, the number of flowers per plant in *Cg1-glgC* and *glgC* does not appear to be different from that in WT. Intriguingly, none of the flowers of a total of six generated *Cg1* lines were fertile; therefore, no seed was recovered from these lines. On the contrary, the flowers of *Cg1-glgC* and *glgC* lines were fertile and produced normal seeds same as the WT or empty vector control lines. These findings suggest that coexpression of *glgC* with *Cg1* restores normal flower development and fertility.

3.3. Transgene Expression and Possible Regulation of Putative N. Benthamiana SPL Genes. Given that mR156 has been shown to modulate flowering via suppression of *SPL* transcription factors, we analyzed the transcript level of putative homologs of the *Arabidopsis SPL* genes including *SPL1*, *SPL9*, *SPL12*, and *SPL15* using quantitative PCR in WT and constitutively expressing *Cg1*, *Cg1-glgC*, or *glgC* lines. To distinguish from *Arabidopsis SPL* genes, the *SPL* genes analyzed in this study are denoted as *NbSPL* (for *N. benthamiana SPL* genes).

As shown in Figure 4, transcripts of both *Cg1* and *glgC* were accumulated at higher levels in the transgenic lines as compared to the nontransgenic control (WT). In *Cg1-glgC* lines, the transcript levels of *Cg1* increased by up to 8000-fold (Figure 4(a)) as compared to the WT while the level of *Cg1* transcript accumulation was over 120-fold higher than in WT. Interestingly, the transcripts of *Cg1* in *Cg1-glgC* lines were over 60-fold higher than in *Cg1* lines. Likewise, the transcripts of *GlgC* were abundantly accumulated in the *Cg1-glgC* and *glgC* lines (Figure 4(b). The increase in transcript abundance ranged from about 120- to 16,000-fold as compared to the WT. The highest increase in *glgC* transcript was detected in *glgCL5*.

On the other hand, transcript levels of all putative *NbSPL* genes analyzed were downregulated in most of the transgenic lines. This decrease in transcript levels was more severe in *Cg1* lines. Expression of *NbSPL1*, *NbSPL9*, *NbSPL12*, and *NbSPL15* was severely downregulated in the *Cg1* lines as compared to the WT, *Cg1-glgC*, or *glgC* (Figures 4(a)–4(d)). Expression of *NbSPL15* was more severely suppressed in the *Cg1* lines (Figures 4(c) and 4(e)). The expression of *NbSPL9* and *NbSPL15* was also downregulated in *Cg1-glgC*-coexpressing lines, but less severely compared to *Cg1* lines. Expression of *NbSPL1* and *NbSPL12* was not markedly affected in the *Cg1-glgC* lines (Figures 4(d) and 4(f)). Similarly, expression of the *SPL* genes in *glgC* lines was not markedly affected (Figure 4). These findings suggest that coexpression of *glgC* restores expression of *NbSPL* genes that was suppressed by *Cg1*.

3.4. Identification of Cg1 and Putative NbSPL Complementary Sites. Since the observed phenotype of the transgenic lines and the gene expression data suggest regulation of the *AtSPL* paralogue genes in tobacco by *Cg1*, we searched for the *Cg1*-binding sites in the putative *SPL* sequences using the targetfinder.pl software [41]. Putative *Cg1*-binding sites with high complementarity were identified in all the *SPL* genes (Supplementary Figure S1). While the complementary site was detected in the ORF of *NbSPL1*,

Figure 2: Transgenic expression of *Cg1* and *glgC* in tobacco. (a) Integration of *Cg1* and *glgC* into the tobacco genome as confirmed by PCR using genomic DNA as template. Primers specific to *NPTII*, *Cg1*, or *glgC* were used. (b) Phenotypes of transgenic (T1) and nontransformed tobacco lines after one month. (c) Phenotypes of transgenic (T0) lines regenerated from transgenic callus and nontransformed tobacco lines after one month. Arrows indicate initiation of flowering in WT, *NPTII*, and *Cg1-glgC* lines. (d) Phenotypes of transgenic (T1) and nontransformed tobacco lines obtained from seeds, and *Cg1* transgenic (T0) lines regenerated from transgenic callus (right panel) after two months of growth in the soil. Phenotypic alteration is observed in *Cg1*-expressing lines alone.

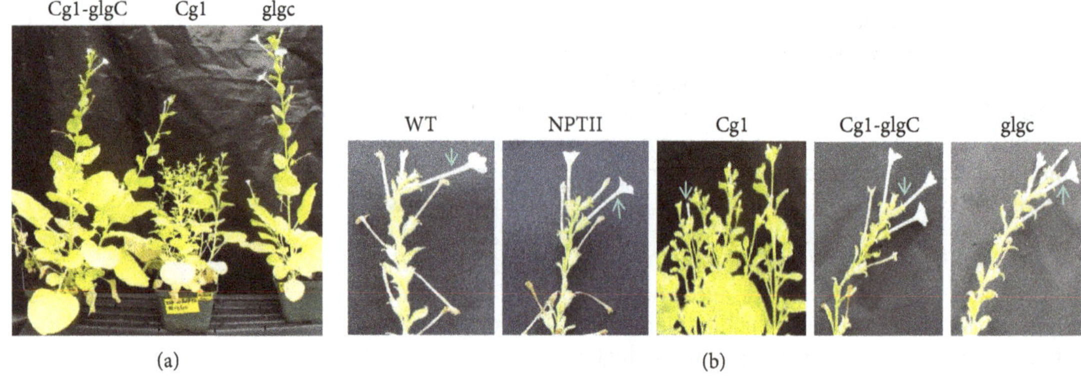

Figure 3: Comparison of T_0 transgenic lines after flowering. (a) *Cg1*-expressing lines appeared more bushy than did lines expressing *glgC* or coexpressing *Cg1* and *glgC*. (b) Close-up pictures to compare floral organs between transgenic and nontransgenic wild-type and empty vector control.

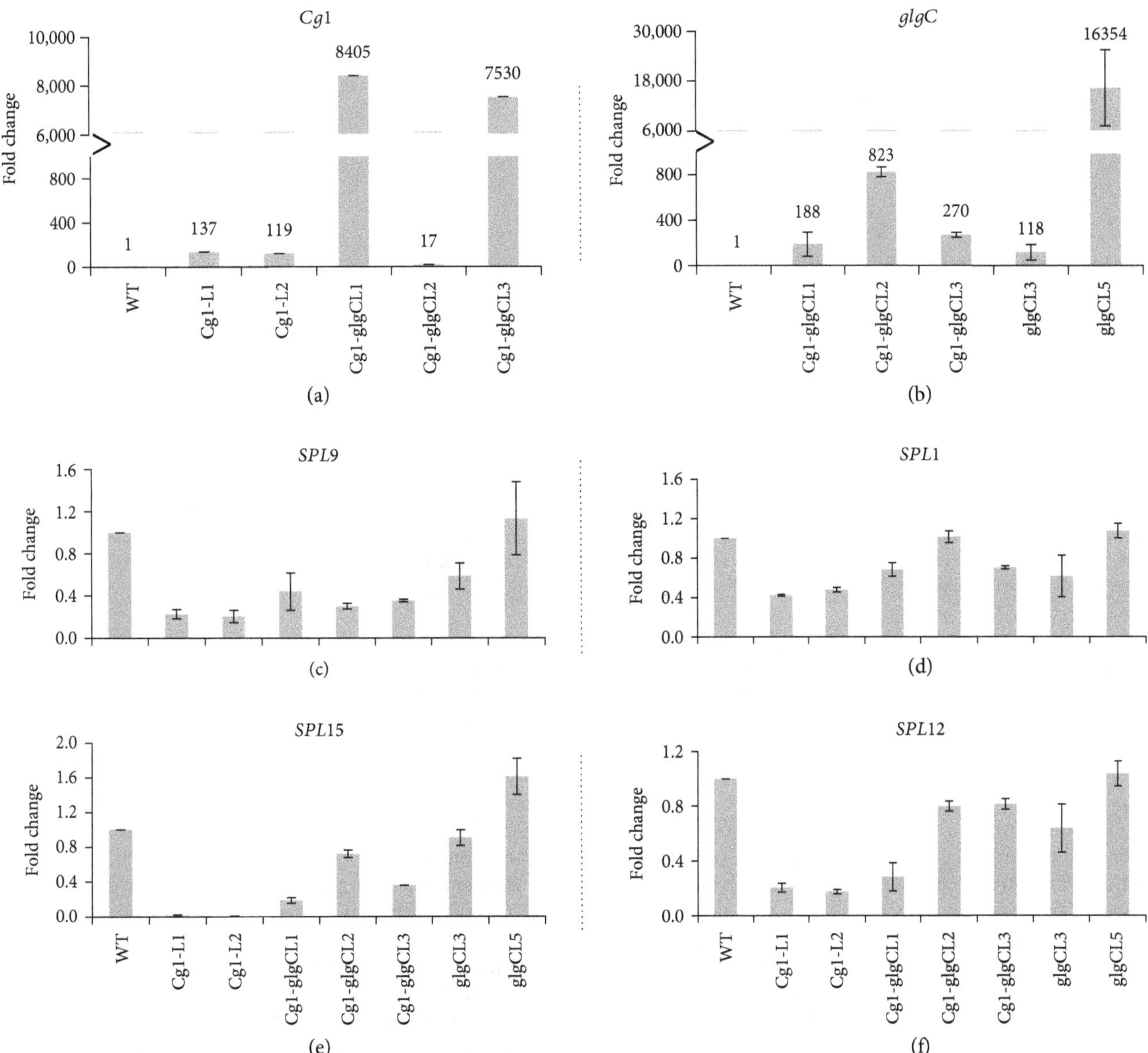

FIGURE 4: Analysis of gene expression in transgenic *N. benthamiana* using quantitative real-time PCR. First-strand cDNA was synthesized from total RNA isolated from two-month-old greenhouse established plants as described in the Material and Methods. Expression of *Cg1* (a), *glgC* (b), and putative *Cg1* targets *SPL9* (c), *SPL1* (d), *SPL15* (e), and *SPL12* (f) was determined by using transcript accumulation in the WT sample reference. Transcript level is expressed as fold change as compared to the WT. Bars represent means and standard error of four replicates. Experiments were repeated twice.

NbSPL9, and *NbSPL15*, it was detected in the 3′-UTR region of *NbSPL12* as previously reported for *Arabidopsis SPL3/4/5* [45, 46].

3.5. Overexpression of glgC with or without Cg1 Increases Shoot Dry Matter Yield. To see the effect of *glgC* on biomass accumulation, dry matter yield of *Cg1-glgC*- and *glgC*-expressing T1 lines and the WT was determined. Because the *Cg1* lines failed to produce seeds, we were not able to compare the biomass yield of *Cg1* lines with that of the WT and lines expressing *Cg1-glgC* and *glgC*. As shown in Figure 5(a), coexpressing lines *Cg1-glgCL1* and *glgCL5* showed between 9 and 48% increase in shoot biomass. In lines *Cg1-glgCL1* and *glgCL5*, biomass yield was increased by 48% and 42%, respectively, followed by *glgCL3* (28%) and *Cg1-glgCL2* (22%), while biomass yield of *Cg1-glgCL3* was only 9% higher than that of the WT.

3.6. Carbohydrate Content in WT and Transgenic Lines. To understand whether overexpression of *glgC* modulates carbohydrate content, we determined starch and sugar content in mature WT and transgenic lines. The analysis showed that the starch content at maturity was slightly reduced in the transgenic lines overexpressing *glgC* alone or significantly reduced in a line coexpressing *glgc* and *Cg1* as compared to the WT (Figure 5(b)). The starch content in the *Cg1-glgC*-

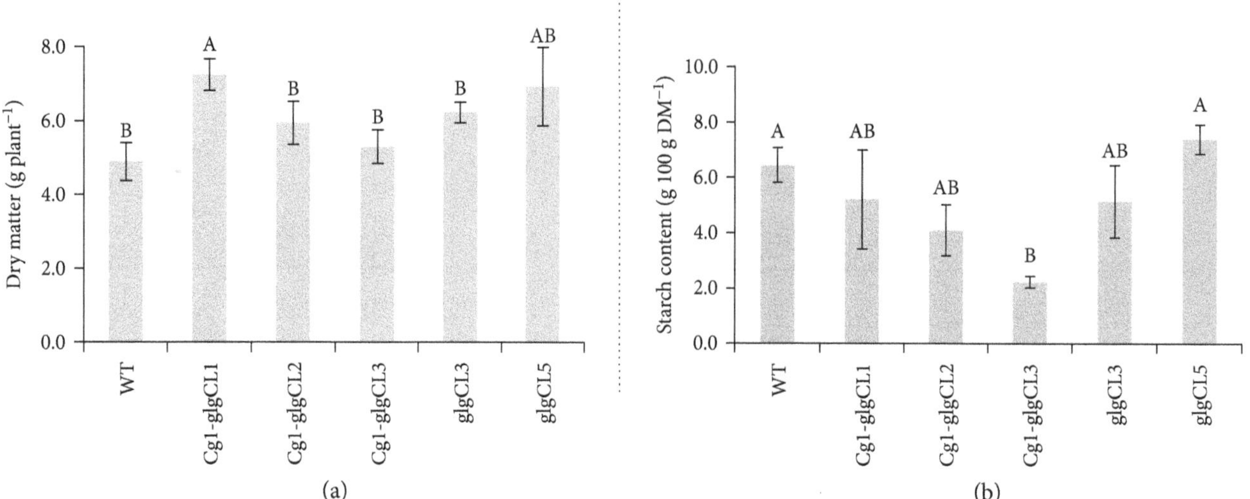

Figure 5: Determination of biomass and starch content. (a) Matured plants were harvested, and dry matter was determined after drying in an oven at 60°C for two days. (b) Starch content in dried biomass was determined according the Dairy One procedures as described in the Material and Methods. Bars represent means and standard error of three replicates. Experiments were repeated twice. Bars bearing the same letter are not significantly different ($P < 0.05$).

expressing line was reduced by up to threefold. This could be due to the age at which starch content was determined. However, whether decrease in starch content is due to age-dependent changes in carbohydrate dynamics or experimental procedures needs to be determined in detail through future studies.

The concentration of sugars was determined using IC-Dionex in the presence of standards of known sugar concentration. As shown in Figure 6 and Supplementary Table S2, glucose, fructose, and sucrose were the major sugars detected in the samples prior to saccharification by a cocktail of α-amylase, cellulase, and glucosidase. As compared to the WT, slightly more fructose and sucrose were released from the *glgC* lines (Figure 6(a), I). The amount of sugars released from *Cg1*-glgC lines was not different from that in the WT. However, the sugar content calculated based on total dry matter per plant in transgenic lines (*Cg1-glgCL1, GlgCL3,* and *glgCL5*) was slightly higher than in the WT (Figure 6(a), II) because these lines produced more biomass. Wild-type and transgenic biomasses were subjected to saccharification by a cocktail of α-amylase, cellulase, and glucosidase enzymes to release more sugars. After three days of saccharification, hexose (glucose, galactose, and mannose), and pentose (xylose and arabinose) sugars were released (Figure 6(b), Supplementary Table S3). Glucose was the dominant sugar released from all the tobacco lines (transgenic as well as control lines) (51–66 mg/g DM) accounting for 87% of the total sugars released. The release of glucose was enhanced by saccharification by about fivefold. There was no marked difference in the amount of sugars released from most of the transgenic lines and the WT control, while only *glg*CL3 released slightly more sugars (66 mg/g DM) than the WT did (51 mg/g DM), as well as the rest of the transgenic lines (Figure 6(b), I). Moreover, *GlgCL3* released at least 20% more glucose than the other transgenic lines did. The amount of glucose released from the *Cg1-glgC*-expressing lines was not significantly different from that in WT plants. There was no marked difference among the lines in the amount of galactose, mannose, xylose, and arabinose released. However, the total amount of sugars (mg/plant) released from most of the transgenic lines (*Cg-glgCL1, GlgCL2, glgCL3,* and *glgCL3*) was significantly higher than in the WT control (Figure 6(b), II). Moreover, the total amount of sugars produced in lines overexpressing *glgC* was slightly higher than in WT and *Cg1-glgC* lines.

3.7. Discussion. Overexpression of the maize *Cg1* in various plant species has been shown to enhance sugar and starch content [25, 26]. Likewise, expression of *glgC* has been shown to increase sink strength and starch accumulation in transgenic potato [47] and cassava [11]. Therefore, this study was conceived to see whether simultaneous expression of the two genes could modulate carbohydrate metabolism in the transgenic biomass.

3.8. Overexpression of Cg1 Alters Vegetative Growth. In this study, the growth of *Cg1*-expressing lines was significantly altered. At the early stage, the growth of *Cg1* lines was slower than that of the nontransgenic control (Figure 2(b)). Moreover, *Cg1* lines produced smaller and more leaves and branches as compared to the WT control. The observed increase in lateral growth could be due to a decrease in apical dominance [48] as reported previously in various plant species expressing *Cg1* [25, 26] or miRNA156 [49, 50]. The maize *Cg1* expressed in *Arabidopsis, Brachypodium,* switchgrass [25], and poplar [51] has been shown to produce plants with extra branches and leaves, while in corn, *Cg1* has been shown to increase the number of tillers [52]. Similarly, overexpression of the rice stem-loop fragment of the *OsmiR156b* precursor in switchgrass has been shown to increase tiller

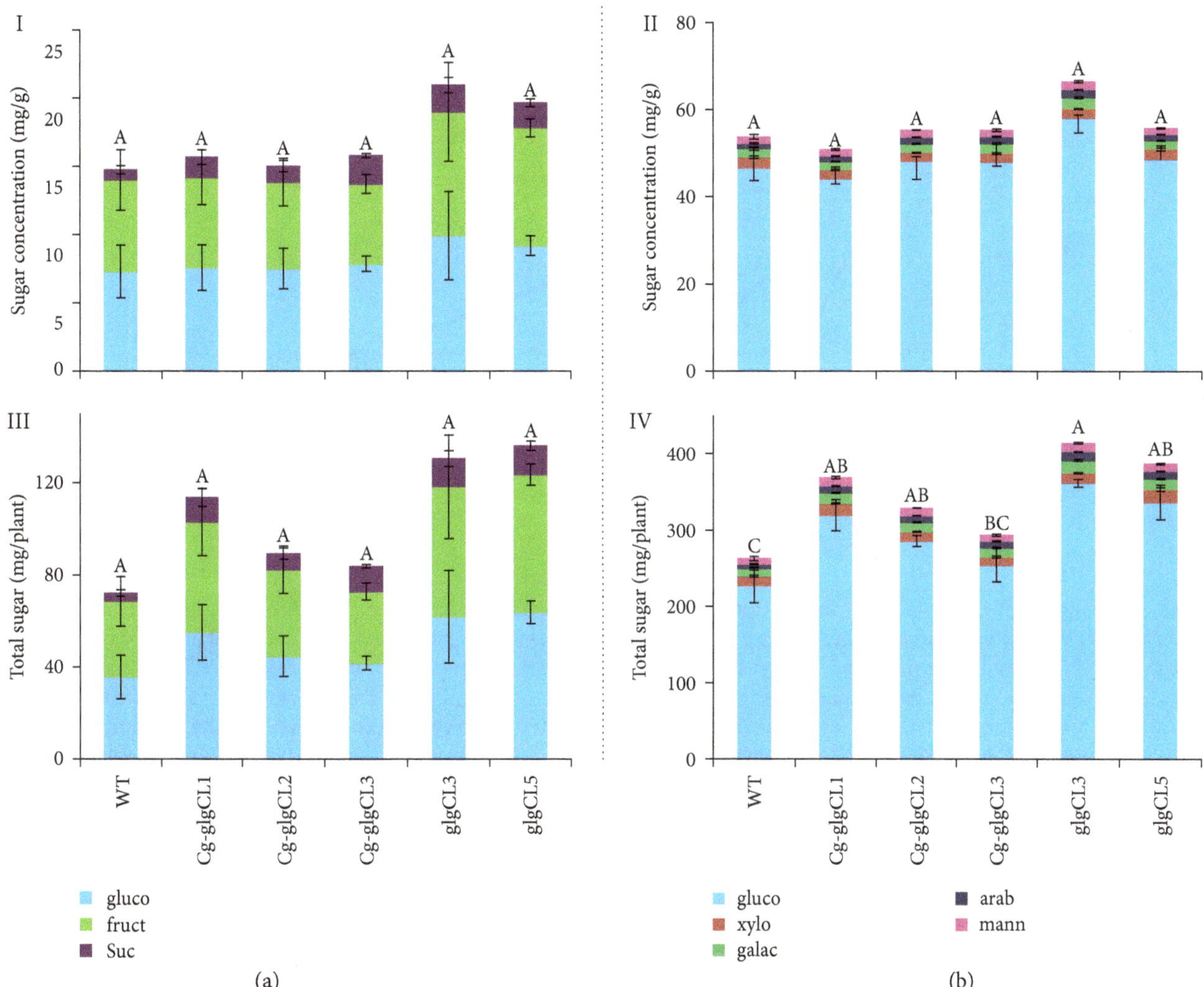

FIGURE 6: Biomass saccharification efficiency. Biomass was harvested at maturity and dried in an oven at 60°C for two days. One gram of ground biomass was used for the analysis as described in Material and Methods. Sugars released before (a) or after (b) saccharification with a cocktail of α-amylase, cellulase, and glucosidase were quantified using IC DIONEX. Bars represent means and standard error of four replicates, and measurement was repeated twice. Bars bearing the same letter are not significantly different ($P < 0.05$).

number [26]. Reduced leaf size and increased leaf number and alteration of other morphological traits have also been observed in tobacco (*N. tabacum*) overexpressing the *Arabidopsis* miR156A hairpin structure [50].

As reported by Muller and Leyser [53], variation in shoot architecture depends on the formation of axillary meristems and the subsequent regulation of their activation, which depends on the genotype, developmental stage, and environment, which in turn is mediated by hormonal signals. Axillary bud outgrowth and branching are mainly controlled by apical dominance and the crosstalk between plant hormones auxin, cytokinin, and strigolactone [53–55]. In this study, the *Cg1* lines exhibited a bushy phenotype with decreased plant height and increased branching (Figures 2 and 3). This suggests that *Cg1* may affect hormone balance, for example, decreasing the level of auxin and strigolactone or increasing the level of cytokinin; however, this remains to be studied.

3.9. Overexpression of Cg1 Alters Reproductive Development. In this study, we observed a delay in the transition from vegetative to reproductive phase in *Cg1* lines. As compared to the WT and transgenic lines (*NPTII*, *Cg1-glgC*, and *glgC*), flower initiation was delayed in *Cg1* lines. Initiation of flowering was observed in less than two months in the former while it was delayed for about two weeks in the latter (Figures 2(c) and 2(d)). This observation is consistent with previous reports on the prolonged juvenile phase in various plant species expressing *miRNAs* [48, 49, 51, 52].

It is well-established that members of the miRNA miR156 have been shown to prolong juvenile cell identities and delay flowering by targeting the transcripts of the *SPL* transcription factors, which in turn activate the expression of flowering regulators such as *LEAFY* and *APETALA1* [56] and a different microRNA, *miR172* [57]. Overexpression of *Cg1* in switchgrass has been shown to downregulate the expression of four *SPL* homologs [25] as compared to the nontransgenic control.

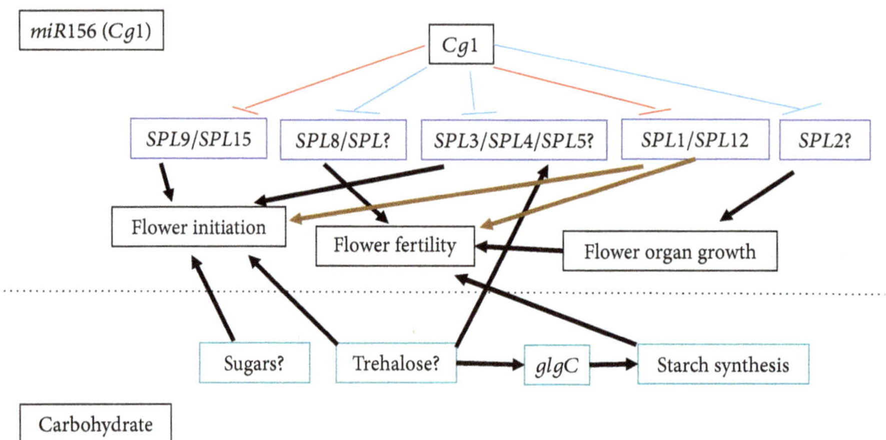

Figure 7: A simplified model proposing possible *miR156 (Cg1)* and carbohydrate-mediated pathways regulating flowering traits in *Cg1*- and *glgC*-expressing transgenic tobacco. Red lines, genes downregulated by *Cg1* in this study. Blue lines, miR156 target genes reported elsewhere and discussed here. Brown arrows, roles not yet reported. Black arrows, roles reported elsewhere and discussed here.

In this study, we analyzed the gene expression levels of *NbSPL1*, *NbSPL9*, *NbSPL12*, and *NbSPL15* (Figure 4). Expression of the *NbSPL* genes was more severely suppressed in *Cg1*- and *Cg1-glgC*-expressing lines as compared to the WT and *glgC* lines. The reduction in transcript level was more severe in *Cg1*, particularly for *NbSPL15*, which was downregulated by nearly a hundred fold, whereas the expression of closely related paralogue *NbSPL9* is suppressed by fivefold. In *Arabidopsis*, both *SPL9* and *SPL15* have been shown to redundantly regulate juvenile to adult phase transition [58]. *AtSPL15* has been implicated in the coordination of basal floral promotion pathways required for flowering in noninductive environments. *AtSPL15* has been shown to directly activate transcription of the MADS-box floral activator *FRUITFULL* (FUL) and *miR172b* in the shoot apical meristem and during floral induction, whereas *AtSPL9* is expressed later in flanks of the shoot apical meristem [59] and has also been implicated in the regulation of *miR172b* [60]. In contrast to miR156 which delays flowering by suppressing the expression of *SPLs*, *miR172* activates flowering by facilitating the degradation of its target transcription factors related to the *APETALA2* (AP2) gene, including *TARGET OF EAT1* (*TOE1*), *TOE2*, *TOE3*, *SCHLAFMUTZE* (*SMZ*), and *SCHNARCHZAPFEN* (*SNZ*), which are implicated in repressing the floral inducer *Flowering Locus T* (FT) [61], whereas SPL9 has been shown to induce flowering through activating MADS-box genes APETALA1 (AP1), FRUITFULL (FUL), and SUPPRESSOR OF OVEREXPRESSION OF CO1 (SOC1) [56, 62].

The expression level of *NbSPL1* and *NbSPL12* is also suppressed in *Cg1*-expressing lines as compared to the WT (Figures 4(d) and 4(f)), suggesting the presence of interaction between *Cg1* and *NbSPL1*/*SPL12* while the transcript level was not markedly affected in *Cg1-glgC* and *glgC* lines. *AtSPL1*, *AtSPL12*, and *AtSPL14* are expressed most strongly in cauline leaves (growing on the upper part of the stem), flowers, and latest-age shoot apices [63]. The role of *SPL1* and *SPL12*, both lacking negative regulation by *miR156* and *miR157* [2], in flowering is not well understood. However, recent reports suggest that *SPL1* and *SPL12* control the expression of many genes and regulate multiple biological processes in *Arabidopsis* inflorescence upon heat stress [30]. Given that miRNA156 has been implicated in various developmental processes including flowering time, flower fertility, alteration of cell wall composition, and biotic and abiotic stress responses [64], downregulation of *NbSPL1* and *NbSPL12* homologs in this study may suggest the presence of developmental processes regulated by the interaction of *Cg1-NbSPL1*/*SPL12* which needs to be identified in the future.

3.10. Overexpression of Cg1 Induces Flower Sterility. Seed production is a key step in the survival of flowering plants; however, its success depends on favorable genetic and environmental factors supporting optimum flower development and fertility. Our findings revealed that flowers of the *Cg1*-expressing lines were fully sterile, and no seed was recovered from these lines. Although the mechanism of observed sterility in tobacco expressing *Cg1* is yet to be understood, it is likely that flower fertility is regulated by *Cg1* and its target *SPL* paralogue genes as previously reported. In *Arabidopsis*, fully fertile flowers require the action of *AtSPL8* which functions redundantly with multiple miR156/7-targeted SPL genes including *AtSPL2*, *AtSPL9*, and *AtSPL15* [33]. In the current study, flower development and fertility were not affected in *Cg1-glgC*-coexpressing lines (Figures 2 and 3). The *Cg1-glgC* lines were fully fertile and produced normal seeds, which suggests that a pathway regulated by coexpression of *glgC* restores floral fertility that was suppressed by *Cg1*. *AtSPL8* is required for proper development of sporogenic tissues in *Arabidopsis* as early anther development has been shown to be affected in *AtSPL8* mutants (*spl8-1*) overexpressing miR156b, resulting in the development of small and fully sterile anthers [33]. Furthermore, *AtSPL8* and miR156-targeted *SPL* genes also regulate gynoecium development by interfering with auxin homeostasis and signaling [34]. Therefore, the absence of seed formation in *Cg1*-expressing lines could be due to suppression of a yet to be identified *AtSPL8* paralogue in tobacco. While it hampers plant fecundity, *Cg1*-induced pollen sterility has a significant biotechnological implication, for example, in eliminating the risk of transgene escape, which is one of the major concerns associated with GM crops [25].

3.11. Possible Role of Carbohydrates in Flower Initiation and Fertility. Our findings revealed that coexpression of *Cg1* and *glgC* does not have a marked effect on flowering time and flower fertility. The phenotype of *Cg1-glgC* was not different from the nontransgenic control, despite the up to sixtyfold higher *Cg1* transcript level in *Cg1-glgC* lines as compared to those expressing *Cg1* alone (Figure 4(b)). This suggests that besides *Cg1(miR156)-SPL* interaction, there is likely a *glgC*-sensitive pathway that is involved in the regulation of various aspects of flowering including flower initiation, flower organ development, and fertility. Therefore, since AGPase is a major enzyme controlling starch biosynthesis [13], reconstitution of normal flowering in *Cg1-glgC* lines as compared to *Cg1* lines could be due to *glgC*-mediated metabolic changes. Involvement of carbohydrates in developmental transitions has been reported before [64–69].

Trehalose-6-phosphate (T6P), which is an indicator of carbohydrate status in plants, has been implicated in the regulation of flowering [65]. A reduction in Tre6P by the loss-of-function mutation of TREHALOSE-6-PHOSPHATE SYNTHASE 1 (*TPS1*), an enzyme which converts UDP-glucose to Tre6P, has been shown to delay flowering in *Arabidopsis*, even under flower inductive environmental conditions [65, 70]. Wahl et al. [66] further showed that the Tre6P pathway controls expression of *SPL* genes via or independently of miR156. Tre6P has also been shown to regulate starch metabolism in plants [71–73]. For example, exogenous application of trehalose in *Arabidopsis* induced accumulation of starch by increasing the activity of AGPase [71], which has been suggested to be via a thioredoxin-mediated redox reaction [74]. Furthermore, rising sucrose levels in plants are accompanied by increases in the level of Tre6P, redox activation of AGPase, and stimulation of starch synthesis *in vivo* [71]. Although it remains to be studied, the lack of an abnormal phenotype in *glgC*-expressing tobacco lines in the current study could be due to Tre6P-mediated stimulation of AGPase activity and starch synthesis and alteration of carbohydrate dynamics during flower initiation and early floral development. Various genetic and physiological approaches have demonstrated the involvement of starch in the control of floral induction [75] and fertility [76–78].

3.12. Overexpression of glgC Increases Biomass and Sugar Yield. The biomass yield of transgenic lines expressing *Cg1-glgC* and *glgC* was higher than in the WT control. This increase in yield is likely due to *glgC*. *glgC* has been shown to increase biomass yield in cassava [11] and potato [46] by increasing sink strength for assimilates and releasing possible feedback inhibition on overall photosynthesis or carbon fixation [11]. On the contrary, starch content was not enhanced in *glgC*-expressing lines (Figure 5) as determined from biomass that was harvested after physiological maturity, which could be due to plant age. Therefore, a more detailed study is needed to determine the level of starch and sugars at different developmental stages. The amount of sugars released from the *Cg1-glgC* lines before and after saccharification was not different from the WT control. On the other hand, the amount of sugars released from *glgC* lines was slightly higher than in the WT (Figures 6(a-I) and 6(b-II)). However, the total sugar yield per plant was higher in both *Cg1-glgC* and *glgC* lines (Figures 6(a-III) and 6(b-IV)), suggesting that overexpression of *glgC* with or without *Cg1* has a potential to increase overall sugar production as biofuel feedstocks. Taken together, overexpression of *glgC* only increased total sugar production and sugar release while this was not observed in *Cg1-glgC* lines. Enhanced total sugar content per plant was observed for *glgC* and *Cg1-glgC* expressers as compared to the nontransgenic control, suggesting a potential of the transgenic approach to increase sugar production.

4. Conclusion

Cg1 expression in transgenic tobacco altered vegetative growth, delayed flowering, and led to loss of fertility. Coexpression of *glgC* with *Cg1* restored wild-type phenotype. *Cg1*-induced changes in vegetative and reproductive growth are likely regulated via suppression of its target *SPL* genes. The antagonistic effect of *glgC* in restoring the *Cg1* phenotype may suggest involvement of changes in carbohydrate dynamics in flower initiation and fertility. Based on the gene expression analysis and reviewed literature, we propose a model summarizing how *Cg1* could modulate flowering and fertility by downregulating the expression of its target *SPLs*, as well as possible involvement of carbohydrates in flower initiation and fertility (Figure 7). Overexpression of *Cg1* leads to downregulation of *SPL* transcription factors, which in turn regulates flower initiation, organ growth, and fertility. Likewise, carbohydrates including sugars, trehalose, and AGPase-mediated enhanced starch biosynthesis may be involved in the regulation of flower development and fertility. Future studies will focus on deciphering genetic and physiological mechanisms regulating the observed phenotype. Global transcriptome and metabolomic analysis as well as biomass composition analysis will help in identifying key genetic elements and pathways regulating the observed phenotypes.

Conflicts of Interest

The authors declare no competing interests.

Authors' Contributions

Ayalew Ligaba-Osena and Bertrand Hankoua conceived and designed the study. Ayalew Ligaba-Osena generated and characterized the transgenic lines, analyzed the data, and wrote the manuscript. Kay DiMarco and Tom L. Richard performed the sugar analysis. Tom L. Richard and Bertrand

Hankoua edited the manuscript. All authors have read and approved the final manuscript.

Acknowledgments

This research was supported by U.S. Department of Agriculture-National Institute of Food and Agriculture CBG (nos. 2011-38821-30974 and 2014-38821-22417), EPSCoR-National Science Foundation (no. 6635), and U.S. Department of Agriculture-National Institute of Food and Agriculture AFRI grant (no. 2012-68005-19703). The authors thank Dr. Michael J. Axtell (Pennsylvania State University) for the *SPL* sequence analysis, Dr. Blake Mayers (Donald Danforth Plant Science) for the helpful discussion, members of the USDA-ARS and Dr. Venugopal Kalavacharla for allowing Ayalew Ligaba-Osena to use lab facilities as needed, Dr. Tony Romeo (University of Florida) for providing the pO12 plasmid containing the *E. coli glgC*, and Drs. Dyremple Marsh and Dr. Samuel Besong for supporting this work with essential facilities and supplies as needed.

References

[1] International Energy Agency Secretariat, *World Energy Outlook*, CORLET, Paris, 2014.

[2] F. Birol, *World Energy Outlook. International Energy Agency Special Report*, IEA Publications, Paris, 2016.

[3] C. Mei and C. Rakhmatov, "Advances in the genetic manipulation of cellulosic bioenergy crops for bioethanol production," in *Biological Conversion of Biomass for Fuels and Chemicals. Explorations from Natural Utilization Systems*, RSC Energy and Environment Series, J. Z. Sun, S. Y. Ding, and J. Doran-Peterson, Eds., no. 10, pp. 53–82, Royal Society of Chemistry, London, 2014.

[4] V. Mendu, A. E. Harman-Ware, M. Crocker et al., "Identification and thermochemical analysis of high-lignin feedstocks for biofuel and biochemical production," *Biotechnology for Biofuels*, vol. 4, no. 1, p. 43, 2011.

[5] D. W. S. Wong, "Structure and action mechanism of ligninolytic enzymes," *Applied Biochemistry and Biotechnology*, vol. 157, no. 2, pp. 174–209, 2009.

[6] C. Petti, A. E. Harman-Ware, M. Tateno et al., "Sorghum mutant *RG* displays antithetic leaf shoot lignin accumulation resulting in improved stem saccharification properties," *Biotechnology for Biofuels*, vol. 6, no. 1, p. 146, 2013.

[7] C. I. Lacayo, M. S. Hwang, S. Y. Ding, and M. P. Thelen, "Lignin depletion enhances the digestibility of cellulose in cultured xylem cells," *PLoS ONE*, vol. 8, no. 7, article e68266, 2013.

[8] P. Alvira, E. Tomás-Pejó, M. Ballesteros, and M. J. Negro, "Pretreatment technologies for an efficient bioethanol production process based on enzymatic hydrolysis: a review," *Bioresource Technology*, vol. 101, no. 13, pp. 4851–4861, 2010.

[9] I. Farran, A. Fernandez-San Millan, M. Ancin, L. Larraya, and J. Veramendi, "Increased bioethanol production from commercial tobacco cultivars overexpressing thioredoxin f grown under field conditions," *Molecular Breeding*, vol. 34, no. 2, pp. 457–469, 2014.

[10] C. R. Meyer, J. A. Bork, S. Nadler, J. Yirsa, and J. Preiss, "Site-directed mutagenesis of a regulatory site of *Escherichia coli* ADP-glucose pyrophosphorylase: the role of residue 336 in allosteric behavior," *Archives of Biochemistry and Biophysics*, vol. 353, no. 1, pp. 152–159, 1998.

[11] U. Ihemere, D. Arias-Garzon, S. Lawrence, and R. Sayre, "Genetic modification of cassava for enhanced starch production," *Plant Biotechnology Journal*, vol. 4, no. 4, pp. 453–465, 2006.

[12] J. Preiss, "Biosynthesis of starch and its regulation," in *The Biochemistry of Plants*, J. Preiss, Ed., vol. 14, pp. 181–254, Academic Press, San Diego, 1988.

[13] B. Huang, T. A. Hennen-Bierwagen, and A. M. Myers, "Functions of multiple genes encoding ADP-glucose pyrophosphorylase subunits in maize endosperm, embryo, and leaf," *Plant Physiology*, vol. 164, no. 2, pp. 596–611, 2014.

[14] B. Müller-Röber, U. Sonnewald, and L. Willmitzer, "Inhibition of the ADP-glucose pyrophosphorylase in transgenic potatoes leads to sugar-storing tubers and influences tuber formation and expression of tuber storage protein genes," *The EMBO Journal*, vol. 11, no. 4, pp. 1229–1238, 1992.

[15] A. Tiessen, J. H. Hendriks, M. Stitt et al., "Starch synthesis in potato tubers is regulated by post-translational redox modification of ADP-glucose pyrophosphorylase: a novel regulatory mechanism linking starch synthesis to the sucrose supply," *The Plant Cell*, vol. 14, no. 9, pp. 2191–2213, 2002.

[16] O. Kötting, J. Kossmann, S. C. Zeeman, and J. R. Lloyd, "Regulation of starch metabolism: the age of enlightenment?," *Current Opinion in Plant Biology*, vol. 13, no. 3, pp. 320–328, 2010.

[17] M. A. Glaring, K. Skryhan, O. Kötting, S. C. Zeeman, and A. Blennow, "Comprehensive survey of redox sensitive starch metabolising enzymes in *Arabidopsis thaliana*," *Plant Physiology and Biochemistry*, vol. 58, pp. 89–97, 2012.

[18] R. Sanz-Barrio, P. Corral-Martinez, M. Ancin, J. M. Segui-Simarro, and I. Farran, "Overexpression of plastidial thioredoxin f leads to enhanced starch accumulation in tobacco leaves," *Plant Biotechnology Journal*, vol. 11, no. 5, pp. 618–627, 2013.

[19] G. Nölke, M. Houdelet, F. Kreuzaler, C. Peterhänsel, and S. Schillberg, "The expression of a recombinant glycolate dehydrogenase polyprotein in potato (*Solanum tuberosum*) plastids strongly enhances photosynthesis and tuber yield," *Plant Biotechnology Journal*, vol. 12, no. 6, pp. 734–742, 2014.

[20] C. Peterhansel, C. Blume, and S. Offermann, "Photorespiratory bypasses: how can they work?," *Journal of Experimental Botany*, vol. 64, no. 3, pp. 709–715, 2013.

[21] R. Kebeish, M. Niessen, K. Thiruveedhi et al., "Chloroplastic photorespiratory bypass increases photosynthesis and biomass production in *Arabidopsis thaliana*," *Nature Biotechnology*, vol. 25, no. 5, pp. 593–599, 2007.

[22] M. W. Rhoades, B. J. Reinhart, L. P. Lim, C. B. Burge, B. Bartel, and D. P. Bartel, "Prediction of plant microRNA targets," *Cell*, vol. 110, no. 4, pp. 513–520, 2002.

[23] K. Xie, C. Wu, and L. Xiong, "Genomic organization, differential expression, and interaction of SQUAMOSA promoter-binding-like transcription factors and microRNA156 in rice," *Plant Physiology*, vol. 142, no. 1, pp. 280–293, 2006.

[24] J. J. Kim, J. H. Lee, W. Kim, H. S. Jung, P. Huijser, and J. H. Ahn, "The *microRNA156-SQUAMOSA PROMOTER BINDING PROTEIN-LIKE3* module regulates ambient temperature-responsive flowering via *FLOWERING LOCUS T* in Arabidopsis," *Plant Physiology*, vol. 159, no. 1, pp. 461–478, 2012.

[25] G. S. Chuck, C. Tobias, L. Sun et al., "Overexpression of the maize *Corngrass1* microRNA prevents flowering, improves digestibility, and increases starch content of switchgrass," *Proceedings of the National Academy of Sciences of the United States of America*, vol. 108, no. 42, pp. 17550–17555, 2011.

[26] C. Fu, R. Sunkar, C. Zhou et al., "Overexpression of miR156 in switchgrass (Panicum virgatum L.) results in various morphological alterations and leads to improved biomass production," *Plant Biotechnology Journal*, vol. 10, no. 4, pp. 443–452, 2012.

[27] J. C. Preston and L. C. Hileman, "Functional evolution in the plant SQUAMOSA-PROMOTER BINDING PROTEIN-LIKE (SPL) gene family," *Frontiers in Plant Science*, vol. 4, no. 80, 2013.

[28] M. Xu, T. Hu, J. Zhao et al., "Developmental functions of miR156-regulated *SQUAMOSA PROMOTER BINDING PROTEIN-LIKE (SPL)* genes in *Arabidopsis thaliana*," *PLOS Genetics*, vol. 12, no. 8, article e1006263, 2016.

[29] U. S. Unte, A. M. Sorensen, P. Pesaresi et al., "*SPL8*, an SBP-box gene that affects pollen sac development in Arabidopsis," *The Plant Cell*, vol. 15, no. 4, pp. 1009–1019, 2003.

[30] Z. Wang, Y. Wang, S. E. Kohalmi, L. Amyot, and A. Hannoufa, "SQUAMOSA PROMOTER BINDING PROTEIN-LIKE 2 controls floral organ development and plant fertility by activating *ASYMMETRIC LEAVES* 2 in *Arabidopsis thaliana*," *Plant Molecular Biology*, vol. 92, no. 6, pp. 661–674, 2016.

[31] M. Riese, S. Höhmann, H. Saedler, T. Münster, and P. Huijser, "Comparative analysis of the SBP-box gene families in *P. patens* and seed plants," *Gene*, vol. 401, no. 1-2, pp. 28–37, 2007.

[32] Y. Zhang, S. Schwarz, H. Saedler, and P. Huijser, "SPL8, a local regulator in a subset of gibberellin-mediated developmental processes in Arabidopsis," *Plant Molecular Biology*, vol. 63, no. 3, pp. 429–439, 2007.

[33] S. Xing, M. Salinas, S. Höhmann, R. Berndtgen, and P. Huijser, "miR156-targeted and nontargeted SBP-box transcription factors act in concert to secure male fertility in *Arabidopsis*," *The Plant Cell*, vol. 22, no. 12, pp. 3935–3950, 2011.

[34] S. Xing, M. Salinas, A. Garcia-Molina, S. Höhmann, R. Berndtgen, and P. Huijser, "SPL8 and miR156-targeted SPL genes redundantly regulate Arabidopsis gynoecium differential patterning," *Plant Journal*, vol. 75, no. 4, pp. 566–577, 2013.

[35] T. Tzfira, G. W. Tian, B.°. Lacroix et al., "pSAT vectors: a modular series of plasmids for autofluorescent protein tagging and expression of multiple genes in plants," *Plant Molecular Biology*, vol. 57, no. 4, pp. 503–516, 2005.

[36] S. N. Ho, H. D. Hunt, R. M. Horton, J. K. Pullen, and L. R. Pease, "Site-directed mutagenesis by overlap extension using the polymerase chain reaction," *Gene*, vol. 77, no. 1, pp. 51–59, 1989.

[37] A. Ligaba, I. Dreyer, A. Margaryan, D. J. Schneider, L. Kochian, and M. Piñeros, "Functional, structural and phylogenetic analysis of domains underlying the Al sensitivity of the aluminum-activated malate/anion transporter, TaALMT1," *The Plant Journal*, vol. 76, no. 5, pp. 766–780, 2013.

[38] I. J. W. M. Goderis, M. F. C. de Bolle, I. E. J. A. François, P. F. J. Wouters, W. F. Broekaert, and B. P. A. Cammue, "A set of modular plant transformation vectors allowing flexible insertion of up to six expression units," *Plant Molecular Biology*, vol. 50, no. 1, pp. 17–27, 2002.

[39] A. Ligaba-Osena, B. Hankoua, K. DiMarco et al., "Reducing biomass recalcitrance by heterologous expression of a bacterial peroxidase in tobacco (*Nicotiana benthamiana*)," *Scientific Reports*, vol. 7, no. 1, p. 17104, 2017.

[40] Y. Tang, F. Wang, J. Zhao, K. Xie, Y. Hong, and Y. Liu, "Virus-based microRNA expression for gene functional analysis in plants," *Plant Physiology*, vol. 153, no. 2, pp. 632–641, 2010.

[41] N. Fahlgren and J. C. Carrington, "miRNA target prediction in plants," in *Plant MicroRNAs. Methods in Molecular Biology (Methods and Protocols)*, B. Meyers and P. Green, Eds., vol. 592, Humana Press, 2010.

[42] B. C. Saha and M. A. Cotta, "Comparison of pretreatment strategies for enzymatic saccharification and fermentation of barley straw to ethanol," *New Biotechnology*, vol. 27, no. 1, pp. 10–16, 2010.

[43] M. Selig, N. Weiss, and Y. Ji, "Enzymatic saccharification of lignocellulosic biomass," NREL Lab Anal Procedure NREL/TP-510-42629, Golden, 2008.

[44] P. H. Westfall, R. D. Tobia, D. Rom, R. D. Wolfinger, and H. Hochberg, *Multiple Comparisons of Multiple Tests Using the SAS System*, (SAS Institute Inc.), 1996.

[45] G. Wu and R. S. Poethig, "Temporal regulation of shoot development in Arabidopsis thaliana by miR156 and its target SPL3," *Development*, vol. 133, no. 18, pp. 3539–3547, 2006.

[46] M. Gandikota, R. P. Birkenbihl, S. Höhmann, G. H. Cardon, H. Saedler, and P. Huijser, "The miRNA156/157 recognition element in the 3′ UTR of the Arabidopsis SBP box gene *SPL3* prevents early flowering by translational inhibition in seedlings," *The Plant Journal*, vol. 49, no. 4, pp. 683–693, 2007.

[47] D. M. Stark, K. P. Timmerman, G. F. Barry, J. Preiss, and G. M. Kishore, "Regulation of the amount of starch in plant tissues by ADP glucose pyrophosphorylase," *Science*, vol. 258, no. 5080, pp. 287–292, 1992.

[48] R. Schwab, J. F. Palatnik, M. Riester, C. Schommer, M. Schmid, and D. Weigel, "Specific effects of microRNAs on the plant transcriptome," *Developmental Cell*, vol. 8, no. 4, pp. 517–527, 2005.

[49] B. Aung, M. Y. Gruber, L. Amyot, K. Omari, A. Bertrand, and A. Hannoufa, "MicroRNA156 as a promising tool for alfalfa improvement," *Plant Biotechnology Journal*, vol. 13, no. 6, pp. 779–790, 2015.

[50] S. Feng, Y. Xu, C. Guo et al., "Modulation of miR156 to identify traits associated with vegetative phase change in tobacco (Nicotiana tabacum)," *Journal of Experimental Botany*, vol. 67, no. 5, pp. 1493–1504, 2016.

[51] P. M. Rubinelli, G. Chuck, X. Li, and R. Meilan, "Constitutive expression of the *Corngrass1* microRNA in poplar affects plant architecture and stem lignin content and composition," *Biomass and Bioenergy*, vol. 54, pp. 312–321, 2013.

[52] G. Chuck, A. M. Cigan, K. Saeteurn, and S. Hake, "The heterochronic maize mutant *Corngrass1* results from overexpression of a tandem microRNA," *Nature Genetics*, vol. 39, no. 4, pp. 544–549, 2007.

[53] D. Muller and O. Leyser, "Auxin, cytokinin and the control of shoot branching," *Annals of Botany*, vol. 107, no. 7, pp. 1203–1212, 2011.

[54] M. Umehara, A. Hanada, S. Yoshida et al., "Inhibition of shoot branching by new terpenoid plant hormones," *Nature*, vol. 455, no. 7210, pp. 195–200, 2008.

[55] R. Dierck, E. de Keyser, J. de Riek et al., "Change in auxin and cytokinin levels coincides with altered expression of branching genes during axillary bud outgrowth in Chrysanthemum," *Plos One*, vol. 11, no. 8, article e0161732, 2016.

[56] A. Yamaguchi, M. F. Wu, L. Yang, G. Wu, R. S. Poethig, and D. Wagner, "The microRNA-regulated SBP-Box transcription factor SPL3 is a direct upstream activator of *LEAFY*, *FRUITFULL* and *APETALA1*," *Developmental Cell*, vol. 17, no. 2, pp. 268–278, 2009.

[57] G. Wu, M. Y. Park, S. R. Conway, J. W. Wang, D. Weigel, and R. S. Poethig, "The sequential action of miR156 and miR172 regulates developmental timing in Arabidopsis," *Cell*, vol. 138, no. 4, pp. 750–759, 2009.

[58] S. Schwarz, A. V. Grande, N. Bujdoso, H. Saedler, and P. Huijser, "The microRNA regulated SBP-box genes SPL9 and SPL15 control shoot maturation in Arabidopsis," *Plant Molecular Biology*, vol. 67, no. 1-2, pp. 183–195, 2008.

[59] Y. Hyun, R. Richter, C. Vincent, R. Martinez-Gallegos, A. Porri, and G. Coupland, "Multi-layered regulation of SPL15 and cooperation with SOC1 integrate endogenous flowering pathways at the Arabidopsis shoot meristem," *Developmental Cell*, vol. 37, no. 3, pp. 254–266, 2016.

[60] J. W. Wang, "Regulation of flowering time by the miR156-mediated age pathway," *Journal of Experimental Botany*, vol. 65, no. 17, pp. 4723–4730, 2014.

[61] J. Mathieu, L. J. Yant, F. Mürdter, F. Küttner, and M. Schmid, "Repression of flowering by the miR172 target SMZ," *PLoS Biology*, vol. 7, no. 7, article e1000148, 2009.

[62] J. W. Wang, B. Czech, and D. Weigel, "miR156-regulated SPL transcription factors define an endogenous flowering pathway in Arabidopsis thaliana," *Cell*, vol. 138, no. 4, pp. 738–749, 2009.

[63] L. M. Chao, Y. Q. Liu, D. Y. Chen, X. Y. Xue, Y. B. Mao, and X. Y. Chen, "Arabidopsis transcription factors SPL1 and SPL12 confer plant thermotolerance at reproductive stage," *Molecular Plant*, vol. 10, no. 5, pp. 735–748, 2017.

[64] V. Wahl, J. Ponnu, A. Schlereth et al., "Regulation of flowering by trehalose-6-phosphate signaling in Arabidopsis thaliana," *Science*, vol. 339, no. 6120, pp. 704–707, 2013.

[65] S. Yu, H. Lian, and J. W. Wang, "Plant developmental transitions: the role of microRNAs and sugars," *Current Opinion in Plant Biology*, vol. 27, pp. 1–7, 2015.

[66] A. Allsopp, "Juvenile stages of plants and the nutritional status of the shoot apex," *Nature*, vol. 173, no. 4413, pp. 1032–1035, 1954.

[67] G. Bernier, A. Havelange, C. Houssa, A. Petitjean, and P. Lejeune, "Physiological signals that induce flowering," *Plant Cell*, vol. 5, no. 10, pp. 1147–1155, 1993.

[68] L. Arrom and S. Munné-Bosch, "Sucrose accelerates flower opening and delays senescence through a hormonal effect in cut lily flowers," *Plant Science*, vol. 188-189, pp. 41–47, 2012.

[69] M. Buendía-Monreal and C. S. Gillmor, "Convergent repression of miR156 by sugar and the CDK8 module of Arabidopsis mediator," *Developmental Biology*, vol. 423, no. 1, pp. 19–23, 2017.

[70] H. Schluepmann, L. Berke, and G. F. Sanchez-Perez, "Metabolism control over growth: a case for trehalose-6-phosphate in plants," *Journal of Experimental Botany*, vol. 63, no. 9, pp. 3379–3390, 2012.

[71] A. Wingler, T. Fritzius, A. Wiemken, T. Boller, and R. A. Aeschbacher, "Trehalose induces the ADP-glucose pyrophosphorylase gene, ApL3, and starch synthesis in Arabidopsis," *Plant Physiology*, vol. 124, no. 1, pp. 105–114, 2000.

[72] J. E. Lunn, R. Feil, J. H. M. Hendriks et al., "Sugar-induced increases in trehalose 6-phosphate are correlated with redox activation of ADPglucose pyrophosphorylase and higher rates of starch synthesis in *Arabidopsis thaliana*," *Biochemical Journal*, vol. 397, no. 1, pp. 139–148, 2006.

[73] J. Ponnu, V. Wahl, and M. Schmid, "Trehalose-6-phosphate: connecting plant metabolism and development," *Frontiers in Plant Science*, vol. 2, 70 pages, 2011.

[74] A. Kolbe, A. Tiessen, H. Schluepmann, M. Paul, S. Ulrich, and P. Geigenberger, "Trehalose 6-phosphate regulates starch synthesis via posttranslational redox activation of ADP-glucose pyrophosphorylase," *Proceedings of the National Academy of Sciences of the United States of America*, vol. 102, no. 31, pp. 11118–11123, 2005.

[75] I. G. Matsoukas, A. J. Massiah, and B. Thomas, "Starch metabolism and antiflorigenic signals modulate the juvenile-to-adult phase transition in *Arabidopsis*," *Plant, Cell & Environment*, vol. 36, no. 10, pp. 1802–1811, 2013.

[76] R. Datta, K. C. Chamusco, and P. S. Chourey, "Starch biosynthesis during pollen maturation is associated with altered patterns of gene expression in maize," *Plant Physiology*, vol. 130, no. 4, pp. 1645–1656, 2002.

[77] G. Lebon, E. Duchêne, O. Brun, C. Magné, and C. Clément, "Flower abscission and inflorescence carbohydrates in sensitive and non-sensitive cultivars of grapevine," *Sexual Plant Reproduction*, vol. 17, no. 2, pp. 71–79, 2004.

[78] G. Lebon, E. Duchene, O. Brun, and C. Clement, "Phenology of flowering and starch accumulation in grape (Vitis vinifera L.) cuttings and vines," *Annals of Botany*, vol. 95, no. 6, pp. 943–948, 2005.

8

The Influence of Metabolic Syndrome and Sex on the DNA Methylome in Schizophrenia

Kyle J. Burghardt,[1] **Jacyln M. Goodrich**,[2] **Brittany N. Lines**,[1] **and Vicki L. Ellingrod**[3,4]

[1]*Department of Pharmacy Practice, Wayne State University Eugene Applebaum College of Pharmacy and Health Sciences, 259 Mack Avenue, Suite 2190, Detroit, MI 48201, USA*
[2]*Department of Environmental Health Sciences, University of Michigan School of Public Health, 1415 Washington Heights Ann Arbor, MI 48109, USA*
[3]*Department of Clinical Social and Administrative Sciences, College of Pharmacy, University of Michigan, 428 Church Street, Ann Arbor, MI 48109, USA*
[4]*Department of Psychiatry, School of Medicine, University of Michigan, 1301 Catherine, Ann Arbor, MI 48109, USA*

Correspondence should be addressed to Kyle J. Burghardt; kburg@wayne.edu

Academic Editor: Hieronim Jakubowski

Introduction. The mechanism by which metabolic syndrome occurs in schizophrenia is not completely known; however, previous work suggests that changes in DNA methylation may be involved which is further influenced by sex. Within this study, the DNA methylome was profiled to identify altered methylation associated with metabolic syndrome in a schizophrenia population on atypical antipsychotics. *Methods*. Peripheral blood from schizophrenia subjects was utilized for DNA methylation analyses. Discovery analyses ($n = 96$) were performed using an epigenome-wide analysis on the Illumina HumanMethylation450K BeadChip based on metabolic syndrome diagnosis. A secondary discovery analysis was conducted based on sex. The top hits from the discovery analyses were assessed in an additional validation set ($n = 166$) using site-specific methylation pyrosequencing. *Results*. A significant increase in *CDH22* gene methylation in subjects with metabolic syndrome was identified in the overall sample. Additionally, differential methylation was found within the *MAP3K13* gene in females and the *CCDC8* gene within males. Significant differences in methylation were again observed for the *CDH22* and *MAP3K13* genes, but not *CCDC8*, in the validation sample set. *Conclusions*. This study provides preliminary evidence that DNA methylation may be associated with metabolic syndrome and sex in schizophrenia.

1. Introduction

Antipsychotics, in particular second-generation or atypical antipsychotics (AAPs), increase the risk of metabolic syndrome 2-3-fold in patients with schizophrenia due to their effects on weight and insulin resistance [1–4]. The metabolic syndrome consists of a cluster of metabolic disorders that include obesity, dyslipidemia, hypertension, and insulin resistance [5]. Together these risk factors are predictive of cardiovascular disease, type 2 diabetes, and mortality [6, 7]. Despite the risk of metabolic syndrome and other adverse events, AAPs provide many beneficial, therapeutic benefits. Increased awareness including enhanced metabolic monitoring and pharmacologic treatment of metabolic disorders that arise during AAP use (e.g., blood pressure medication use) has helped to lower the risk of metabolic syndrome, yet this has not completely removed it. Additionally, the use of AAPs has expanded from schizophrenia to other populations such as pediatric which are particularly sensitive to these metabolic effects [8–10]. Therefore, a better understanding of the molecular mechanisms underlying metabolic syndrome in patients with schizophrenia is necessary so that personalized interventions and/or newer therapies can be developed that will minimize or remove the risk.

Previous work has linked aberrant genetic regulation of the folate cycle to metabolic syndrome risk in schizophrenia patients treated with AAPs [11]. Additionally, treatment of schizophrenia patients with folate may aid in improving some aspects of metabolic syndrome [12]. These findings suggest that a properly functioning folate system may be important to minimizing the risk for AAP-induced metabolic syndrome. Dysfunctional folate regulation could be causing metabolic syndrome for several reasons [13, 14]. At the molecular level, a product of the folate cycle is methyl molecules used for various cellular reactions including lipid, protein, and DNA methylation. Thus, it has been suggested that altered gene regulation through changes to DNA methylation may be responsible for AAP-associated metabolic syndrome.

DNA methylation at the global level may be altered in schizophrenia patients with metabolic syndrome in sex-specific ways [15]. Despite this previously identified association between global DNA methylation and AAP-induced metabolic syndrome, only one study has examined gene-specific methylation at the catechol-O-methyltransferase *(COMT)* gene and reported a negative finding [16]. Within the current study, we used an epigenome-wide strategy to identify and potentially validate candidate genes or regions that are associated with metabolic syndrome in a cohort of well-characterized schizophrenia subjects while also determining any sex-specific effects that may be present.

2. Methods

2.1. Subject Population Recruitment and Assessment. Potential subjects were recruited from mental health clinics and with public advertisements in the Southeastern Michigan and neighboring areas. A preliminary phone screening was used to assess for the following inclusion criteria: Diagnostic and Statistical Manual IV diagnosis with a schizophrenia-spectrum disorder, age 18 to 90 years, presently treated with an antipsychotic medication with no dosage changes in the past 6 months, and no known metabolic diseases such as dyslipidemia, hypertension, or diabetes prior to starting their antipsychotic treatment. Subjects were excluded if they were pregnant or unable to give blood. Potential subjects interested in participating were invited to the Michigan Clinical Research Center (MCRC) which is supported by the Michigan Institute for Clinical and Translational Research (MiCHR) to undergo full informed consent as approved by the University of Michigan Institutional Review Board. The study was registered with ClinicalTrials.gov (NCT00815854).

Following consent, subjects underwent a medical and medication history questionnaire which captured current and past medication use. Pharmacy and clinical records were used to verify dosages. For the purposes of description, antipsychotics were grouped according to their potential to cause metabolic side effects (i.e., high versus medium versus low) [17, 18]. Subjects on olanzapine and clozapine were placed in the high-risk metabolic group; subjects on quetiapine, paliperidone, and risperidone were placed in the medium-risk group; and subjects on aripiprazole and ziprasidone were placed in the low-risk group. We have employed this empirical categorization in previous metabolic studies [19]. Subjects underwent psychiatric screening by a trained clinical research assistant using the Structure Clinical Interview for DSM diagnoses (SCID-4) in order to confirm the diagnosis of schizophrenia [20]. Vital signs and anthropometric data including weight, height, hip circumference, and waist circumference were assessed by clinical research center nursing staff. All subjects underwent a fasting blood draw that was used for laboratory analyses (glucose and lipid panels) and genomic DNA extraction. Glucose and lipid levels were analyzed by the University of Michigan Hospital System (UMHS) laboratories. Samples for both the discovery and validation groups described in the results were collected using the above inclusion/exclusion criteria and protocol. Ninety-six samples were chosen for the discovery analysis. The discovery group was selected to include one-half with metabolic syndrome equally matched based on age, race, and sex (e.g., 48 subjects with metabolic syndrome matched with 48 subjects without metabolic syndrome). The remainder of the recruited subjects (166 additional samples) were used for validation of the discovery findings.

2.2. Genetics Analysis: Extraction and Preparation of Genomic DNA. Genomic DNA was extracted by the salt precipitation method [21] and cleaned using commercially available kits. DNA was quantified on a Qubit fluorimeter (Life Technologies) and 1 μg of bisulfite was converted using the Zymo EZ DNA Methylation-Gold kit (Zymo Technologies) according to manufacturer specifications.

2.3. Genetics Analysis: Discovery Analysis. For the discovery analysis, converted samples were submitted to the University of Michigan DNA Sequencing Core for analysis on the Illumina HumanMethylation450 BeadChip ("450K"). The core returned raw IDAT files for subsequent processing and statistical analysis by investigators. Discovery analyses were conducted in the combined samples (96 subjects) and within each sex (49 male, 47 female).

2.4. Genetics Analysis: Validation Analysis. For the validation analyses, primer sets were chosen based on the discovery findings where the goal was to choose locations within the same CpG island or the nearest CpG island to the top discovery finding for the combined analysis and the sex-specific analyses. Site-specific methylation was analyzed by the method of pyrosequencing on a PyroMark MD 96. Commercially available primer sets from Qiagen (Redwood City, CA, USA) were utilized for the *CDH22* (in a nearby CpG island ~300 base pairs away) and *CCDC8* genes (same CpG island). The *CDH22* primer set was designed to obtain an amplicon of approximately 115 base pairs that would analyze methylation in chromosome 20 at positions 44880277, 44880264, and 44880250 following pyrosequencing (genomic coordinates using GRCh37/hg19). The *CCDC8* primer set was designed to obtain an amplicon of 189 base pairs and analyze methylation at four chromosome 19 locations (46915716, 46915706, 46915704, and 46915701). Finally, a primer set for the *MAP3K13* gene was designed using the Qiagen Assay Design 2.0 program (in the same CpG island as the discovery finding). The resultant primers amplified a 146-base-pair region in the *MAP3K13* gene on Chromosome 3 for methylation analysis at positions 185000790, 185000779, 185000774, and 185000760. Primers for the self-designed *MAP3K13* assay are available upon request. All samples were performed in triplicate, and each batch was normalized by constructing a standard curve with samples of known methylation to account for bisulfite PCR bias [22]. No samples were removed due to excess variation amongst the replicates (defined as a coefficient of variation > 5%).

2.5. Statistical Analysis. Values are reported in mean ± standard deviations (s.d.). Student *t*-, chi-square, or Fisher's exact test was used for comparison of demographic and clinical variables between metabolic syndrome groups and discovery and validation groups. The epigenome-wide analysis for the discovery of differentially methylated genes associated with metabolic syndrome employed the use of RnBeads [23]. RnBeads is a comprehensive R statistical software package that enables users to utilize specific workflows for processing, normalizing, and analyzing DNA methylation data from the Illumina HumanMethylation450 BeadChip. Within RnBeads, our obtained data was loaded in raw IDAT file form where it was preprocessed and normalized according to published biostatistical and bioinformatics workflows which included correction for color bias, quantile normalization, probe-type bias, and batch adjustment [24, 25]. Preprocessed and normalized *M* values were then analyzed by CpG site (overall sample) or CpG island (sex-specific analysis) using linear regression with the *limma* package [26]. The CpG island analysis, employed through the RnBeads package in R, uses annotation data to group probes within the same CpG island and constructs a "combined" *p* value to assess that CpG island's overall association with metabolic syndrome by regression with the *limma* package. Three linear regressions were performed for discovery of differential methylation based on metabolic syndrome: (1) CpG sites (total of 393,193 sites) in the overall discovery sample, (2) CpG islands (total of 25,352 CpG islands) in males within the discovery samples, and (3) CpG islands in females within the discovery sample. Each regression used metabolic syndrome as the independent variable of interest while adjusting for other relevant variables. For the CpG site analysis in the overall discovery sample, regressions were adjusted for smoking status, antipsychotic type, and estimated cell types using the Houseman et al. method in R [27]. For the regional CpG island analyses based on sex, both the male and female regressions used age, smoking status, antipsychotic type, and estimated cell-type compositions. All regressions used the *sva* package within RnBeads to detect batch effects and control them by adding estimated surrogate variables as covariates to each model [28]. Top differentially methylated CpG sites or CpG islands were corrected for multiple testing using a false discovery rate (FDR, *q* value) cutoff of less than 0.05. [29].

Validation analyses were conducted in a separate sample of subjects to potentially replicate the top differentially methylated findings from the overall and sex-specific discovery analyses. Validation analyses used linear regressions, in a similar format to the epigenome-wide analysis, where each methylation site (within the three genes assessed) served as the dependent variable and metabolic syndrome status served as the independent variable of interest while adjusting for age, sex, race, smoking status, and antipsychotic type. A p value < 0.05 was considered statistically significant for the validation analyses.

2.6. Pathway Analysis. An exploratory pathway analysis utilizing the discovery analysis data was performed with Ingenuity Pathway Analysis (IPA) software build version: 430520M, content version: 31813283, release date: December 5, 2016 (Qiagen, Redwood City, CA, USA). Such an analysis may reveal pathway or network perturbations not captured by the top CpG site or CpG island approach employed in the discovery analysis due to various reasons including a lack of power. For the overall discovery analysis, the top 100 differentially methylated genes corresponding to the CpG sites were analyzed in IPA. For the sex-specific discovery analyses, the top 50 CpG islands genes were entered into IPA for analysis. The top 100 or 50 genes were chosen as an arbitrary cutoff that would include a representation of potentially influenced pathways by metabolic syndrome in the overall and sex-specific analyses. Alternate gene sets (e.g., top 1000) in the pathway analysis did not result in the identification of other pathways (data not shown). The IPA reference set chosen was the Ingenuity Knowledge Base, and the findings were restricted to humans only in the Core Analysis module. Top canonical pathways with an FDR-corrected *p* value below 0.05 were considered statistically significant for each analysis.

3. Results

3.1. Discovery and Validation Group Characteristics. The discovery sample, consisting of a total of 96 subjects, had an average age of 49.5 ± 8.4 years, 51% were male, 60% were Caucasian, and 35% were African-American. The distribution of antipsychotic type was similar between discovery and validation groups. As designed, approximately 50% of the discovery sample had a diagnosis of metabolic syndrome matched for age and race and split evenly for sex. There was a trend for a lower rate of smoking in females when compared to males ($p = 0.1$) in the discovery sample. The validation sample had a total of 166 subjects. The average age of the validation sample was 43.9 ± 12.0 years, 64% were male, and 46% had metabolic syndrome. The discovery and validation groups were similar except for a non-significant trend for more males in the validation group ($p = 0.08$). Both samples' demographic and clinical variables can be found in Table 1. Additionally, a breakdown by sex for each sample can be found in Table 2. Significant differences were not noted between males and females in the validation group.

3.2. Discovery: Differentially Methylated Sites Based on Metabolic Syndrome. As described in Methods, the included covariates in the final model to estimate the top differentially methylated CpG sites based on metabolic syndrome in the

TABLE 1: Demographic and clinical characteristics of discovery and validation groups.

	Discovery group ($n = 96$)	Validation group ($n = 166$)
Age (years ± s.d.)	49.8 ± 7.4	43.9 ± 12.0
Sex (% male)	51	64
Caucasian (%)/ African-American (%)	60/35	53/32
Metabolic syndrome (%)	50	46
% currently smoking	50	51
Olanzapine/clozapine (%)	29	29
Quetiapine/paliperidone/ risperidone (%)	38	39
Aripiprazole/ziprasidone (%)	33	32

The table depicts the mean ± s.d. or % values for the discovery and validation groups. No statistically significant differences were noted between the groups. A nonsignificant trend for more males in the validation group was observed ($p = 0.08$).

overall discovery sample were smoking status, antipsychotic type, and cell composition. A *Q-Q* plot with inflation factor was used to characterize the p value distributions and estimate the appropriateness of the model. We compared the model with and without the correction for batch effects through the addition of estimated surrogate variables with RnBeads. The comparisons can be found in Supplemental Figure 1. The model without batch effect correction had an inflation factor of 0.978 while correction for batch effects improved the model's visual fit and the associated inflation factor to 1.006. This suggests that there were small, but still present, sources of additional variation that were unaccounted for by our included covariates. From the *Q-Q* plot, the associations of methylation with respect to metabolic syndrome deviate from the null at higher p values as would be expected. Table 3 shows the top differentially methylated CpG sites associated with metabolic syndrome at a FDR less than 0.1. The top five CpG sites met a predefined FDR cutoff of <0.05 and were found in the following genes: cadherin-like 22 *(CDH22)*, family with sequence similarity 19 (chemokine- (C-C motif-) like), member A2 *(FAM19A2)*, cadherin-like 22 *(CDH5)*, casein kinase 1 *(CSNK1E)*, and Delta/notch-like EGF repeat *(DNER)*. An expanded table with proposed biological functions as well as previous correlations from the literature for each site's corresponding gene can be found in Supplementary Table 1. For further exploration, the top 100 differentially methylated CpG sites based on metabolic syndrome in the overall discovery sample can be found in Supplementary Table 2.

3.3. Discovery: Sex-Specific Differential Methylation in Metabolic Syndrome. Following our analyses in the overall discovery sample, we conducted a sex-specific analysis of differential methylation based on metabolic syndrome. For this secondary analysis, we chose to conduct analyses at the regional level of CpG islands. This was done to increase power in a limited sample size by decreasing the number of statistical tests being conducted.

For the male population, the model to identify top differentially methylated CpG islands based on metabolic syndrome included the following covariables: age, smoking status, antipsychotic type, and cell-type composition. The model that included the estimated surrogate variables had an improved lambda based on a *Q-Q* plot (lambda without surrogate variables = 0.889 versus lambda with surrogate variables = 1.001). The CpG islands associated with metabolic syndrome in males with an FDR p value < 0.1 can be found in Table 4. The top result, in the coiled-coil domain containing 8 *(CCDC8)* gene, was statistically significant after FDR correction.

We performed the same analysis using the same regression variables in females. Including the surrogate variables in the model improved the lambda from 0.893 to 1.021. Table 4 also contains the female analysis results. The top two CpG islands, found in the mitogen-activated protein kinase kinase kinase 13 *(MAP3K13)* and transmembrane phosphoinositide 3-phosphatase and tensin homolog 2 *(TPTE2)* genes, were statistically significant after FDR correction. An expanded table, for further exploration, showing the top 50 CpG islands for each sex can be found in Supplementary Table 3.

3.4. Discovery: Pathway Analysis. The exploratory pathway analysis of the overall sample results revealed that differential methylation related to metabolic syndrome in schizophrenia was enriched in the Wnt/β-catenin signaling pathway (FDR p value $= 6.21 \times 10^{-4}$). For the analysis in females only, the axonal guidance signaling pathway was the most enriched pathway (FDR p value $= 6.22 \times 10^{-4}$). Finally, the FAK signaling pathway was the top pathway for the male analysis of CpG islands associated with metabolic syndrome (FDR p value $= 1.32 \times 10^{-4}$). The top ten pathways for each analysis along with the identified genes that caused enrichment can be found in Supplementary Table 4.

3.5. Validation Analyses of Top Differentially Methylated Genes from Discovery. In the absence of access to a larger sample set for the discovery analyses, we sought to validate the top discovery findings in an additional sample of schizophrenia subjects from the same recruitment pool. For validation, methylation was analyzed by site-specific pyrosequencing at three sites: (1) the top differentially methylated CpG site in the overall discovery sample *(CDH22)* and (2) the top differentially methylated CpG islands for males *(CCDC8)* and (3) females *(MAP3K13)*. Significant associations at Chr20:44880277 and Chr20:44880264 in the *CDH22* gene were identified in the overall sample which held when adjusting for age, race, gender, smoking, and antipsychotic type (both $p = 0.04$). At both sites, higher methylation (hypermethylation) was observed in subjects with metabolic syndrome. The third site Chr20:44880250, assessed in the *CDH22* gene, did not reach statistical significance ($p = 0.7$).

Three sites within the *MAP3K13* gene at genomic locations Chr3:185000779, Chr3:185000774, and Chr3: 185000760 were associated with metabolic syndrome status in females after adjusting for age, race, smoking status, and antipsychotic type ($p = 0.01$, 0.04, and 0.01, resp.).

TABLE 2: Discovery and validation group broken down by sex.

	Discovery group		Validation group	
	Males ($n = 49$)	Females ($n = 47$)	Males ($n = 100$)	Females ($n = 66$)
Age (years ± s.d.)	49.4 ± 8.64	49.7 ± 8.29	42.9 ± 11.4	45.7 ± 13.1
Caucasian (%)/African-American (%)	55/40	65/30	53/36	53/25
Metabolic syndrome (%)	50	51	44	51
% currently smoking	57	43	55	56
Olanzapine/clozapine (%)	25	24	31	26
Quetiapine/paliperidone/risperidone (%)	35	38	44	31
Aripiprazole/ziprasidone (%)	40	38	25	43

The table depicts the mean ± s.d. or % values for the discovery and validation groups. No statistically significant differences were noted between males and females for either group. There was a trend for decreased smoking in females in the discovery group ($p = 0.1$).

TABLE 3: Top differentially methylated sites based on metabolic syndrome.

CpG probe ID	Gene	Chromosome	Position	CpG type	Fold change[a]	Raw p value	FDR-corrected p value
cg04640913	Cadherin-like 22 *(CDH22)*	20	44880515	South shore	0.123	9.26×10^{-07}	0.02*
cg12501957	Family with sequence similarity 19 (chemokine- (C-C motif-) like), member A2 *(FAM19A2)*	12	62629234	Open sea	−0.0266	1.05×10^{-06}	0.04*
cg05086443	Cadherin-like 22 *(CDH5)*	16	66437349	South shore	−0.0215	3.45×10^{-06}	0.04*
cg16653173	Casein kinase 1 *(CSNK1E)*	22	38713453	South shore	0.0675	3.83×10^{-06}	0.04*
cg16656316	Delta/notch-like EGF repeat *(DNER)*	2	230280621	Open sea	−0.0764	7.85×10^{-06}	0.04*
cg06378976	Transcription factor EB *(TFEB)*	6	41703613	South shore	0.135	1.05×10^{-05}	0.08
cg04457354	E2F transcription factor 3 *(E2F2)*	6	20447442	Open sea	−0.0221	2.03×10^{-05}	0.08
cg04953503	Melanophilin *(MLPH)*	2	238420656	Open sea	−0.00763	2.24×10^{-05}	0.08
Cg05434957	Islet autoantigen 1 *(ICA1)*	7	8301435	Island	0.118	2.26×10^{-05}	0.09
cg08464505	ATPase, class VI, type 11A *(ATP11A)*	13	113425982	South shore	−0.0113	2.87×10^{-05}	0.09
Cg22158175	Proteosome subunit, beta type, 8 *(PSMB8)*	6	32809475	North sea	−0.0162	3.10×10^{-05}	0.09
Cg17492940	Protein phosphatase 1, regulatory subunit 12B *(PPP1R12B)*	1	202407102	Open sea	−0.00380	3.30×10^{-05}	0.1
Cg04033559	Pyruvate dehydrogenase kinase, isozyme 1 *(PDK1)*	2	173461819	Open sea	−0.312	3.46×10^{-05}	0.1

Top differentially methylated sites based on metabolic syndrome status with an FDR p value < 0.1. Only FDR < 0.05 was considered statistically significant in this study. Columns 1 and 2 give the probe ID and associated gene name. Columns 3 and 4 give the genomic location (GRCh37/hg19) of the CpG site and CpG classification of the probe with respect to CpG islands (i.e., island versus shore versus sea). The final columns give the fold change with direction, unadjusted, and FDR-corrected p values. [a]Fold change calculated by log2 of the quotient in methylation in subjects with metabolic syndrome compared to subjects without metabolic syndrome. Positive fold change indicates an increase in methylation (hypermethylation) in the metabolic syndrome group. * indicates statistical significance based on an FDR cutoff below 0.05

Consistent with the discovery analysis, hypomethylation of *MAP3K13* was seen in female subjects with metabolic syndrome. *CCDC8* methylation did not show significant differences based on metabolic syndrome within males. The details of the unadjusted and adjusted validation analyses are found in Table 5.

4. Discussion

The purpose of this study was to identify areas of the DNA methylome that may be altered in subjects with AAP-associated metabolic syndrome [15]. To this end, we identified overall and sex-specific differentially methylated genes in a discovery sample of 96 schizophrenia subjects. Two of the three findings from the discovery group were validated in an additional group of schizophrenia subjects.

4.1. Differentially Methylated Genes Associated with Metabolic Syndrome. Differentially methylated CpG sites within the overall discovery sample were located within genes with either a known biological function in a cardiometabolic illness and/or a previously reported association with a metabolic phenotype or disease (Supplementary Table 1). The top site, located in the cadherin-like 22 *(CDH22)* gene, had increased

Table 4: Top differentially methylated CpG islands based on metabolic syndrome in males and females.

Female Chromosomal location (Chr:region)	Number of CpGs in island	Gene name	Fold change[a]	Raw p value	FDR-corrected p value	Male Chromosomal location (Chr:region)	Number of CpGs in island	Gene name	Fold change[a]	Raw p value	FDR-corrected p value
3: 185000558–185000896	33	Mitogen-activated protein kinase kinase kinase 13 *(MAP3K13)*	−0.196	0.000936	0.0423*	19: 46915312–46915802	44	Coiled-coil domain containing 8 *(CCDC8)*	0.0930	0.0000186	0.0325*
13: 20135400–20136041	53	Transmembrane phosphoinositide 3-phosphatase and tensin homolog 2 *(TPTE2)*	0.0267	0.00101	0.0423*	17: 40558006–40558274	19	Polymerase I and transcript release factor *(PTRF)*	−0.129	0.000278	0.102

Shows the CpG islands for the sex-specific methylation discovery analysis with an FDR < 0.1. FDR < 0.05 was considered significant. Column 1 provides the genomic location of the CpG island (GRCh37/hg19), column 2 provides the number of CpG sites located within the island according to annotation data, and column 3 provides the gene name where the CpG island is found. The remaining columns provide the fold change with direction, unadjusted, and FDR-corrected p values. [a]Fold change calculated by log2 of the quotient in methylation in subjects with metabolic syndrome compared to subjects without metabolic syndrome. Positive fold change indicates an increase in methylation (hypermethylation) in the metabolic syndrome group. * indicates statistical significance based on an FDR cutoff below 0.05

Table 5: Site-specific validation methylation analyses based on discovery findings.

Metabolic syndrome status		Validation population: males and females ($n = 166$)			Validation population: males only ($n = 100$)				Validation population: females only ($n = 66$)			
		CDH22 Chr20:44880277	*CDH22* Chr20:44880264	*CDH22* Chr20:44880250	*CCDC8* Chr19:46915716	*CCDC8* Chr19:46915706	*CCDC8* Chr19:46915704	*CCDC8* Chr19:46915701	*MAP3K13* Chr3:185000790	*MAP3K13* Chr3:185000779	*MAP3K13* Chr3:185000774	*MAP3K13* Chr3:185000760
Metabolic syndrome[a]	Crude beta ± s.e. (p value)	1.034 ± 0.520 ($p = 0.0488^*$)	1.14 ± 0.508 ($p = 0.264$)	0.334 ± 0.488 ($p = 0.5$)	−0.149 ± 1.09 ($p = 0.9$)	0.200 ± 1.18 ($p = 0.9$)	0.355 ± 1.03 ($p = 0.7$)	−0.063 ± 1.09 ($p = 0.9$)	−0.584 ± 0.491 ($p = 0.2$)	−0.969 ± 0.367 ($p = 0.0110^*$)	−0.705 ± 0.481 ($p = 0.1$)	−0.696 ± 0.517 ($p = 0.2$)
	Adjusted beta ± s.e.	1.27 ± 0.531 ($p = 0.0469^*$)	1.29 ± 0.625 ($p = 0.0414^*$)	0.222 ± 0.646 ($p = 0.7$)	0.235 ± 1.20 ($p = 0.8$)	0.573 ± 1.31 ($p = 0.7$)	0.547 ± 1.12 ($p = 0.6$)	0.711 ± 1.15 ($p = 0.5$)	−0.258 ± 0.575 ($p = 0.7$)	−1.12 ± 0.437 ($p = 0.0137^*$)	−1.17 ± 0.578 ($p = 0.0491^*$)	−1.42 ± 0.583 ($p = 0.0192^*$)

Gives beta values with standard error for unadjusted and adjusted regression model with methylation site as dependent variable and metabolic syndrome status as the independent variable. CDH22 validation regressions performed in overall validation population, CCDC8 validation regressions performed in males only, and MAP3K13 validation regressions performed in females only. CDH22 regressions adjusted for age, gender, race, smoking status, and antipsychotic type. CCDC8 and MAP3K13 regressions adjusted for age, race, smoking status, and antipsychotic type. [a]Reference group is subjects without metabolic syndrome; therefore, a positive beta value indicates hypermethylation in subjects with metabolic syndrome while a negative beta value indicates hypomethylation in subjects with metabolic syndrome. $*$ indicates statistical significance based on a p value below 0.05.

methylation in subjects with metabolic syndrome (i.e., hypermethylation). A member of the cadherin superfamily, this gene codes for a cell-adhesion protein that is predominately expressed in the brain and is important for tissue development and morphogenesis. This gene has been associated with type 2 diabetes in a previous genetic variation study [30]. Within this study, out of the top 5 most significant single nucleotide polymorphisms associated with type 2 diabetes, 3 were found in *CDH22*. It should be noted that within this study, this finding was not replicated in a separate data set. Nevertheless, it may be that *CDH22* regulation, through both genetic and epigenetic mechanisms, could point to a potentially important role for this gene in metabolic disease. Additionally, differential methylation was identified in the cadherin 5, type 2 *(CDH5)* gene, which is also in the cadherin superfamily. This particular cadherin isoform is highly important in the development of vascular endothelium which our group has shown to be influenced by AAP use, folate metabolism, and genetic variation [11, 12, 31]. Specifically, we previously have shown that genetic variation in the rate-limiting enzyme in folate metabolism, methylenetetrahydrofolate *(MTHFR)*, as well as endothelial nitric oxide synthetase *(eNOS)* is associated with a greater risk for endothelial dysfunction, a predictor of cardiovascular morbidity and mortality. Additional work may be needed to understand if genetic regulation at *CDH5* confers additional risk. Altogether, these findings may add further evidence of the complex links between altered folate regulation, DNA methylation, and AAP-associated metabolic syndrome and cardiovascular disease.

Within the top differentially methylated CpG sites in the overall population, several genes related to protein regulation and function were present (e.g., *CSNK1E, E2F2, PSMB8, PPP1R12B*, and *PDK1*). Control of protein function and action through phosphorylation and other modifications play a central role in several disease states including diabetes, lipid metabolism, insulin resistance, and metabolic syndrome [32, 33]. Additionally, altered basal and insulin-stimulated protein phosphorylation has been identified with AAP treatment in both preclinical models and patients [34–36]. In particular, the *CSNK1E* gene has been linked to the pathophysiology of schizophrenia and bipolar disorder which are the main conditions for which antipsychotics are used [37, 38]. Further work incorporating the power of DNA methylomics and proteomics in AAP treatment may yield further insight into psychiatric disease and its treatment.

4.2. Sex-Specific Methylation in Metabolic Syndrome: Females. In addition to looking at the associations between the DNA methylome and metabolic syndrome in schizophrenia in an overall manner, region-specific DNA methylation, at the CpG Island level, was performed within each sex based on our previous findings suggesting that sex may play a role [15]. The top differentially methylated CpG island associated with metabolic syndrome in females was in the gene encoding for the mitogen-activated protein kinase kinase kinase 13 *(MAP3K13)* protein. This protein, a member of the serine/threonine phosphatase kinase family, interacts with and regulates other proteins through its ability to phosphorylate specific mitogen-activated proteins including MAP2K7/MKK7 and MAPK8/JNK [39, 40]. Epigenetic regulation of this protein may play a role in MAPK and Jun amino terminal kinase (JNK) signaling pathways, both shown to play an important role in glucose homeostasis, a defining feature of the metabolic syndrome [41–43]. Overall, female subjects with metabolic syndrome had lower methylation in the investigated *MAP3K13* CpG island (both in the discovery and validation analyses) compared to female subjects without metabolic syndrome which may suggest higher expression and possibly activity of the kinase. Further work is needed to understand the effect of epigenetic regulation on *MAP3K13* expression and activity.

4.3. Sex-Specific Methylation in Metabolic Syndrome: Males. The top differentially methylated CpG island associated with metabolic syndrome in males was found in the coiled-coil domain containing 8 *(CCDC8)* gene, although this finding was not replicated in our validation sample. The *CCDC8* gene (alias name protein phosphatase 1, regulatory subunit 20 (PPP1R20)) encodes a protein involved in cell apoptosis following DNA damage as well as human growth and development and genomic integrity [44, 45]. Notably, this protein has been shown to modulate alternative splicing of the insulin receptor (INSR) [46], which may have downstream effects on MAPK/AKT signaling. Again, given that insulin resistance is a key feature of metabolic syndrome, further work with this gene and its effects on the MAPK/AKT pathway may be warranted.

4.4. Exploratory Pathway Analyses. The Wnt/β-catenin pathway, the most enriched pathway in the overall analysis, involves molecules from other pathways to regulate cell-specific processes including fate, proliferation, and migration. This pathway has been linked to insulin signaling and sensitivity and lipid metabolism which all have been known to be influenced by AAP treatment [47–52]. For females, the top canonical pathway was the axonal guidance signaling pathway which is involved in determining how nervous system axons reach their target. Such a pathway may be of importance in a psychiatric disorder where significant overlap is seen in metabolic and nervous system processes when considering both the disease itself as well as the medications used to treat the symptoms. Interactions between these pathways have been identified in other models of metabolic disease [53]. The FAK signaling pathways were the most enriched pathways in the male analysis. This pathway is involved in cell movement and adhesion and has also been linked to glucose dysregulation and insulin signaling [54, 55]. In summary, pathway analyses revealed sex-specific enriched pathways when considering DNA methylation in the context of metabolic syndrome. Despite the differences, an underlying theme of involvement in insulin signaling was present in the identified pathways (see Supplementary Table 4).

4.5. Strengths and Limitations. The current study utilized a sample of schizophrenia subjects who were stable in their antipsychotic therapy for 6 or more months to identify DNA methylation changes associated with metabolic syndrome. The sample was well characterized using detailed medication histories as well as anthropometric and metabolic assessments to diagnose metabolic syndrome. Some limitations should be considered when interpreting the findings of this study. DNA methylation was assessed in peripheral blood which is composed of multiple cell types. While we could use statistical techniques to estimate cell-type composition and control for it in the discovery analyses, we did not have access to cell-type composition in the gene-specific validation analyses. Validation analyses occurred by choosing methylation sites within the same CpG island as the discovery sites or, for the *CDH22* validation, the nearest CpG island which was within 300 base pairs of the original discovery site. We chose to focus validation work within CpG islands due to their known importance in gene regulation. Other sites may have stronger associations with metabolic syndrome, or in the case of the *CCDC8* gene which was not validated, other areas of the gene may have stronger associations with metabolic syndrome. Deeper, gene-specific methylation profiling within a tissue of interest (e.g., adipose, muscle, or brain) in various models should be considered to further understand the role of epigenetics in antipsychotic-induced metabolic syndrome. We did not have access to RNA samples to assess gene expression levels. Future work will need to functionally validate the effect of methylation on gene expression or other downstream products of the gene(s). Based on previous work establishing sample sizes and effect sizes in epigenome-wide studies, the discovery sample size was limited in its ability to detect smaller changes (e.g., smaller effect sizes) in methylation; however, DNA methylation at two of our genes was validated in an additional sample of schizophrenia subjects which does strengthen the findings in its present form [56]. The study includes a population on various AAPs. For the purposes of our study, we were interested in capturing a population on any AAP since all AAP increases cause weight gain and increase the risk of metabolic syndrome [57]. Our analyses (not shown) did not identify significant effects of AAP dosage; however, these analyses may have been underpowered to allow for appropriate interpretation. Future work may need to begin to analyze specific antipsychotics in mechanistic studies at specified dosages to better design interventions that prevent this side effect. Finally, this study utilized cross-sectional samples without a healthy control group. This makes determining cause and effect (e.g., if the medications are inducing changes in DNA methylation which subsequently causes metabolic syndrome or vice versa) and the effect of the psychiatric disease itself difficult. Evidence exists suggesting AAP effects on molecular features may be specific to specific psychiatric illnesses [58]. Nevertheless, the identification of gene methylation associated with AAP-associated metabolic syndrome will serve to direct further studies that look at DNA methylation changes before and after antipsychotic treatment coupled with a healthy control group to assess causation between metabolic syndrome and gene methylation.

5. Conclusion

Within our study, we identified gene methylation changes that are associated with antipsychotic-associated metabolic syndrome and changes that are specific to sex. The results here are preliminary, and future work is needed to understand the mechanistic role of these gene changes and possible therapies that could target and potentially prevent the negative consequences of antipsychotic-induced metabolic syndrome.

Conflicts of Interest

The authors have no conflicts of interest to disclose.

Authors' Contributions

Kyle J. Burghardt developed the project, obtained samples, ran the sample analyses and statistical analyses, conducted the interpretation, and wrote the manuscript. Jacyln M. Goodrich assisted in sample analysis, statistical analysis, and manuscript writing. Brittany N. Lines assisted in sample analysis and manuscript writing. Vicki L. Ellingrod assisted in obtaining samples from parent grant, the interpretation, and manuscript writing.

Acknowledgments

This work and the authors were supported by a supplement (S1) to the National Institute of Mental Health (R01 MH082784 to Vicki L. Ellingrod); National Institutes of Health and National Center for Research Resources, GCRC Program (UL1RR024986, UL1TR000433); the Chemistry Core of the Michigan Diabetes Research and Training Center (P30DK020572, P30DK092926); the Washtenaw Community Health Organization (WCHO, Ann Arbor, Michigan); the Brain and Behavior Research Foundation (formerly NARSAD, Great Neck, New York); the Rachael Upjohn Clinical Scholars Grant; a University of Michigan Environmental Health Sciences Core Center (P30 ES017885); an American College of Clinical Pharmacy Junior Investigator Grant (Kyle J. Burghardt); and a National Institutes of Health Loan Repayment Program Grant from the National Institute of Diabetes and Digestive and Kidney Diseases (L30 DK110823 to Kyle J. Burghardt). The authors would like to thank the Applied Genomics Technology Center at Wayne State University for allowing access to Ingenuity Pathway Analysis Software.

Supplementary Materials

Supplementary Figure 1: *Q-Q* plot of overall discovery sample epigenome-wide analysis. *Q-Q* plots for the model assessing the association between metabolic syndrome and methylation site using the Illumina HumanMethylation450 BeadChip for the overall population. Both models included smoking status, antipsychotic type, CD4T, CD8T, granulocytes, monocytes, and natural killer cell counts as covariates. Plot (a) depicts the *Q-Q* plot before performing surrogate variable adjustment (lambda = 0.978) and (b) after performing surrogate variable adjustment (lambda = 1.0006). Supplementary Table 1: annotated top differentially methylated CpG sites (FDR < 0.1) associated with metabolic syndrome in the overall sample. Top differentially methylated sites (FDR < 0.1) associated with metabolic syndrome annotated with biological function and previous links to cardiometabolic phenotypes in the literature. It expands Table 1 from the main manuscript including a description of the known or proposed biological function of the associated gene and previous literature references investigating its role in cardiometabolic outcomes (references listed at end of the supplementary file). Supplementary Table 2: top 100 differentially methylated CpG sites associated with metabolic syndrome in the overall sample. Top 100 CpG results with annotated genes from linear regression of methylation sites based on metabolic syndrome adjusted for smoking status, antipsychotic type, estimated cell types, and batch effects (components estimated using the *sva* package) in the overall sample. Supplementary Table 3: top 50 differentially methylated CpG islands associated with metabolic syndrome from the sex-specific analysis. Top 50 annotated results from sex-specific linear regression of CpG islands based on metabolic syndrome adjusted for smoking status, antipsychotic type, estimated cell types, and batch effects (components estimated using the *sva* package). Supplementary Table 4: top 10 enriched pathways for each discovery analysis. The 1000 CpG sites (or CpG islands for sex-specific analysis) with the smallest p values from the discovery analyses were entered into the Core Analysis module of Ingenuity Pathway Analysis (IPA) software. The top 10 canonical pathways are listed in the table for each analysis along with FDR-correct p values, ratios, and genes. *(Supplementary Materials)*

References

[1] M. De Hert, V. Schreurs, K. Sweers et al., "Typical and atypical antipsychotics differentially affect long-term incidence rates of the metabolic syndrome in first-episode patients with schizophrenia: a retrospective chart review," *Schizophrenia Research*, vol. 101, no. 1-3, pp. 295-303, 2008.

[2] M. De Hert, R. van Winkel, D. Van Eyck et al., "Prevalence of diabetes, metabolic syndrome and metabolic abnormalities in schizophrenia over the course of the illness: a cross-sectional study," *Clinical Practice and Epidemiology in Mental Health*, vol. 2, no. 1, p. 14, 2006.

[3] M. A. De Hert, R. van Winkel, D. Van Eyck et al., "Prevalence of the metabolic syndrome in patients with schizophrenia treated with antipsychotic medication," *Schizophrenia Research*, vol. 83, no. 1, pp. 87-93, 2006.

[4] K. J. Burghardt, B. Seyoum, A. Mallisho, P. R. Burghardt, R. A. Kowluru, and Z. Yi, "Atypical antipsychotics, insulin resistance and weight; a meta-analysis of healthy volunteer studies," *Progress in Neuro-Psychopharmacology & Biological Psychiatry*, vol. 83, pp. 55-63, 2018.

[5] S. M. Grundy, J. I. Cleeman, S. R. Daniels et al., "Diagnosis and management of the metabolic syndrome: an American Heart Association/National Heart, Lung, and Blood Institute scientific statement," *Circulation*, vol. 112, no. 17, pp. 2735-2752, 2005.

[6] E. S. Ford, "Risks for all-cause mortality, cardiovascular disease, and diabetes associated with the metabolic syndrome: a summary of the evidence," *Diabetes Care*, vol. 28, no. 7, pp. 1769-1778, 2005.

[7] A. Galassi, K. Reynolds, and J. He, "Metabolic syndrome and risk of cardiovascular disease: a meta-analysis," *The American Journal of Medicine*, vol. 119, no. 10, pp. 812-819, 2006.

[8] E. Longden and J. Read, "Assessing and reporting the adverse effects of antipsychotic medication: a systematic review of clinical studies, and prospective, retrospective, and cross-sectional research," *Clinical Neuropharmacology*, vol. 39, no. 1, pp. 29-39, 2016.

[9] X. Zhou, G. I. Keitner, B. Qin et al., "Atypical antipsychotic augmentation for treatment-resistant depression: a systematic review and network meta-analysis," *The International Journal of Neuropsychopharmacology*, vol. 18, no. 11, article pyv060, 2015.

[10] N. Ji and R. L. Findling, "An update on pharmacotherapy for autism spectrum disorder in children and adolescents," *Current Opinion in Psychiatry*, vol. 28, no. 2, pp. 91-101, 2015.

[11] V. L. Ellingrod, D. D. Miller, S. F. Taylor, J. Moline, T. Holman, and J. Kerr, "Metabolic syndrome and insulin resistance in schizophrenia patients receiving antipsychotics genotyped for the methylenetetrahydrofolate reductase (MTHFR) 677C/T and 1298A/C variants," *Schizophrenia Research*, vol. 98, no. 1-3, pp. 47-54, 2008.

[12] V. L. Ellingrod, T. B. Grove, K. J. Burghardt, S. F. Taylor, and G. Dalack, "The effect of folate supplementation and genotype on cardiovascular and epigenetic measures in schizophrenia subjects," *npj Schizophrenia*, vol. 1, no. 1, p. 15046, 2015.

[13] E. Setola, L. Monti, E. Galluccio et al., "Insulin resistance and endothelial function are improved after folate and vitamin B12 therapy in patients with metabolic syndrome: relationship between homocysteine levels and hyperinsulinemia," *European Journal of Endocrinology*, vol. 151, no. 4, pp. 483-489, 2004.

[14] M. R. Hayden and S. C. Tyagi, "Homocysteine and reactive oxygen species in metabolic syndrome, type 2 diabetes mellitus, and atheroscleropathy: the pleiotropic effects of folate supplementation," *Nutrition Journal*, vol. 3, no. 1, p. 4, 2004.

[15] K. J. Burghardt, J. R. Pilsner, M. J. Bly, and V. L. Ellingrod, "DNA methylation in schizophrenia subjects: gender and MTHFR 677C/T genotype differences," *Epigenomics*, vol. 4, no. 3, pp. 261-268, 2012.

[16] S. A. Lott, P. R. Burghardt, K. J. Burghardt, M. J. Bly, T. B. Grove, and V. L. Ellingrod, "The influence of metabolic syndrome, physical activity and genotype on catechol-O-methyl transferase promoter-region methylation in schizophrenia," *The Pharmacogenomics Journal*, vol. 13, no. 3, pp. 264-271, 2013.

[17] American Diabetes Association, American Psychiatric Association, American Association of Clinical Endocrinologists, and North American Association for the Study of Obesity, "Consensus development conference on antipsychotic drugs and obesity and diabetes," *Diabetes Care*, vol. 27, no. 2, pp. 596-601, 2004.

[18] M. De Hert, J. Detraux, R. van Winkel, W. Yu, and C. U. Correll, "Metabolic and cardiovascular adverse effects associated with antipsychotic drugs," *Nature Reviews Endocrinology*, vol. 8, no. 2, pp. 114-126, 2012.

[19] K. J. Burghardt, J. M. Goodrich, D. C. Dolinoy, and V. L. Ellingrod, "Gene-specific DNA methylation may mediate atypical antipsychotic-induced insulin resistance," *Bipolar Disorders*, vol. 18, no. 5, pp. 423-432, 2016.

[20] M. B. First, R. L. Spitzer, M. W. Gibbon, and J. B. Williams, "Structured Clinical Interview for DSM-IV Axis I Disorders," in *Research version, Non-Patient Edition*, New York State Psychiatric Institute, Biometrics Research, New York, 2002.

[21] D. K. Lahiri and J. I. Numberger Jr., "A rapid non-enzymatic method for the preparation of HMW DNA from blood for RFLP studies," *Nucleic Acids Research*, vol. 19, no. 19, p. 5444, 1991.

[22] J. M. Goodrich, B. N. Sánchez, D. C. Dolinoy et al., "Quality control and statistical modeling for environmental epigenetics: a study on in utero lead exposure and DNA methylation at birth," *Epigenetics*, vol. 10, no. 1, pp. 19–30, 2015.

[23] Y. Assenov, F. Müller, P. Lutsik, J. Walter, T. Lengauer, and C. Bock, "Comprehensive analysis of DNA methylation data with RnBeads," *Nature Methods*, vol. 11, no. 11, pp. 1138–1140, 2014.

[24] F. Marabita, M. Almgren, M. E. Lindholm et al., "An evaluation of analysis pipelines for DNA methylation profiling using the Illumina HumanMethylation450 BeadChip platform," *Epigenetics*, vol. 8, no. 3, pp. 333–346, 2014.

[25] T. J. Morris and S. Beck, "Analysis pipelines and packages for Infinium HumanMethylation450 BeadChip (450k) data," *Methods*, vol. 72, pp. 3–8, 2015.

[26] M. E. Ritchie, B. Phipson, D. Wu et al., "*limma* powers differential expression analyses for RNA-sequencing and microarray studies," *Nucleic Acids Research*, vol. 43, no. 7, article e47, 2015.

[27] E. Houseman, W. P. Accomando, D. C. Koestler et al., "DNA methylation arrays as surrogate measures of cell mixture distribution," *BMC Bioinformatics*, vol. 13, no. 1, p. 86, 2012.

[28] J. T. Leek, W. E. Johnson, H. S. Parker, A. E. Jaffe, and J. D. Storey, "The sva package for removing batch effects and other unwanted variation in high-throughput experiments," *Bioinformatics*, vol. 28, no. 6, pp. 882-883, 2012.

[29] Y. Benjamini and Y. Hochberg, "Controlling the false discovery rate: a practical and powerful approach to multiple testing," *Journal of the Royal Statistical Society. Series B (Methodological)*, vol. 57, no. 1, pp. 289–300, 1995.

[30] J. L. Bento, N. D. Palmer, M. Zhong et al., "Heterogeneity in gene loci associated with type 2 diabetes on human chromosome 20q13.1," *Genomics*, vol. 92, no. 4, pp. 226–234, 2008.

[31] K. Burghardt, T. Grove, and V. Ellingrod, "Endothelial nitric oxide synthetase genetic variants, metabolic syndrome and endothelial function in schizophrenia," *Journal of Psychopharmacology*, vol. 28, no. 4, pp. 349–356, 2013.

[32] M. Caruso, D. Ma, Z. Msallaty et al., "Increased interaction with insulin receptor substrate-1, a novel abnormality in insulin resistance and type 2 diabetes," *Diabetes*, vol. 63, no. 6, pp. 1933–1947, 2014.

[33] C. Terfve, E. Sabidó, Y. Wu et al., "System-wide quantitative proteomics of the metabolic syndrome in mice: genotypic and dietary effects," *Journal of Proteome Research*, vol. 16, no. 2, pp. 831–841, 2017.

[34] J. A. J. Jaros, D. Martins-de-Souza, H. Rahmoune et al., "Protein phosphorylation patterns in serum from schizophrenia patients and healthy controls," *Journal of Proteomics*, vol. 76, pp. 43–55, 2012.

[35] J. A. Jaros, H. Rahmoune, H. Wesseling et al., "Effects of olanzapine on serum protein phosphorylation patterns in patients with schizophrenia," *Proteomics Clinical Applications*, vol. 9, no. 9-10, pp. 907–916, 2015.

[36] L. Carboni and E. Domenici, "Proteome effects of antipsychotic drugs: learning from preclinical models," *Proteomics. Clinical Applications*, vol. 10, no. 4, pp. 430–441, 2016.

[37] S. Matsunaga, M. Ikeda, T. Kishi et al., "An evaluation of polymorphisms in casein kinase 1 delta and epsilon genes in major psychiatric disorders," *Neuroscience Letters*, vol. 529, no. 1, pp. 66–69, 2012.

[38] R. Pinacho, N. Villalmanzo, J. J. Meana et al., "Altered CSNK1E, FABP4 and NEFH protein levels in the dorsolateral prefrontal cortex in schizophrenia," *Schizophrenia Research*, vol. 177, no. 1–3, pp. 88–97, 2016.

[39] Z. Xu, A. C. Maroney, P. Dobrzanski, N. V. Kukekov, and L. A. Greene, "The MLK family mediates c-Jun N-terminal kinase activation in neuronal apoptosis," *Molecular and Cellular Biology*, vol. 21, no. 14, pp. 4713–4724, 2001.

[40] M. Masaki, A. Ikeda, E. Shiraki, S. Oka, and T. Kawasaki, "Mixed lineage kinase LZK and antioxidant protein-1 activate NF-kappaB synergistically," *European Journal of Biochemistry*, vol. 270, no. 1, pp. 76–83, 2003.

[41] S. Frojdo, H. Vidal, and L. Pirola, "Alterations of insulin signaling in type 2 diabetes: a review of the current evidence from humans," *Biochimica et Biophysica Acta (BBA) - Molecular Basis of Disease*, vol. 1792, no. 2, pp. 83–92, 2009.

[42] M. Fujishiro, Y. Gotoh, H. Katagiri et al., "Three mitogen-activated protein kinases inhibit insulin signaling by different mechanisms in 3T3-L1 adipocytes," *Molecular Endocrinology*, vol. 17, no. 3, pp. 487–497, 2003.

[43] H. Li and X. Yu, "Emerging role of JNK in insulin resistance," *Current Diabetes Reviews*, vol. 9, no. 5, pp. 422–428, 2013.

[44] D. Hanson, P. G. Murray, J. O'Sullivan et al., "Exome sequencing identifies CCDC8 mutations in 3-M syndrome, suggesting that CCDC8 contributes in a pathway with CUL7 and OBSL1 to control human growth," *American Journal of Human Genetics*, vol. 89, no. 1, pp. 148–153, 2011.

[45] J. Yan, F. Yan, Z. Li et al., "The 3M complex maintains microtubule and genome integrity," *Molecular Cell*, vol. 54, no. 5, pp. 791–804, 2014.

[46] D. Hanson, A. Stevens, P. G. Murray, G. C. M. Black, and P. E. Clayton, "Identifying biological pathways that underlie primordial short stature using network analysis," *Journal of Molecular Endocrinology*, vol. 52, no. 3, pp. 333–344, 2014.

[47] M. Abiola, M. Favier, E. Christodoulou-Vafeiadou, A. L. Pichard, I. Martelly, and I. Guillet-Deniau, "Activation of Wnt/beta-catenin signaling increases insulin sensitivity through a reciprocal regulation of Wnt10b and SREBP-1c in skeletal muscle cells," *PLoS One*, vol. 4, no. 12, article e8509, 2009.

[48] J. C. Yoon, A. Ng, B. H. Kim, A. Bianco, R. J. Xavier, and S. J. Elledge, "Wnt signaling regulates mitochondrial physiology and insulin sensitivity," *Genes & Development*, vol. 24, no. 14, pp. 1507–1518, 2010.

[49] J. K. Sethi and A. Vidal-Puig, "Wnt signalling and the control of cellular metabolism," *The Biochemical Journal*, vol. 427, no. 1, pp. 1–17, 2010.

[50] T. J. Vassas, K. J. Burghardt, and V. L. Ellingrod, "Pharmacogenomics of sterol synthesis and statin use in schizophrenia subjects treated with antipsychotics," *Pharmacogenomics*, vol. 15, no. 1, pp. 61–67, 2014.

[51] K. J. Burghardt, K. N. Gardner, J. W. Johnson, and V. L. Ellingrod, "Fatty acid desaturase gene polymorphisms and metabolic measures in schizophrenia and bipolar patients taking antipsychotics," *Cardiovascular Psychiatry and Neurology*, vol. 2013, Article ID 596945, 8 pages, 2013.

[52] J. S. Ballon, U. Pajvani, Z. Freyberg, R. L. Leibel, and J. A. Lieberman, "Molecular pathophysiology of metabolic effects of antipsychotic medications," *Trends in Endocrinology and Metabolism*, vol. 25, no. 11, pp. 593–600, 2014.

[53] T. Kelder, L. Eijssen, R. Kleemann, M. van Erk, T. Kooistra, and C. Evelo, "Exploring pathway interactions in insulin resistant mouse liver," *BMC Systems Biology*, vol. 5, no. 1, p. 127, 2011.

[54] D. Huang, A. T. Cheung, J. T. Parsons, and M. Bryer-Ash, "Focal adhesion kinase (FAK) regulates insulin-stimulated glycogen synthesis in hepatocytes," *The Journal of Biological Chemistry*, vol. 277, no. 20, pp. 18151–18160, 2002.

[55] M. Kruger, I. Kratchmarova, B. Blagoev, Y. H. Tseng, C. R. Kahn, and M. Mann, "Dissection of the insulin signaling pathway via quantitative phosphoproteomics," *Proceedings of the National Academy of Sciences of the United States of America*, vol. 105, no. 7, pp. 2451–2456, 2008.

[56] P. C. Tsai and J. T. Bell, "Power and sample size estimation for epigenome-wide association scans to detect differential DNA methylation," *International Journal of Epidemiology*, vol. 44, no. 4, pp. 1429–1441, 2015.

[57] M. Bak, A. Fransen, J. Janssen, J. van Os, and M. Drukker, "Almost all antipsychotics result in weight gain: a meta-analysis," *PLoS One*, vol. 9, no. 4, article e94112, 2014.

[58] K. J. Burghardt, S. J. Evans, K. M. Wiese, and V. L. Ellingrod, "An untargeted metabolomics analysis of antipsychotic use in bipolar disorder," *Clinical and Translational Science*, vol. 8, no. 5, pp. 432–440, 2015.

9

The Effect of Citalopram on Genome-Wide DNA Methylation of Human Cells

Riya R. Kanherkar,[1] Bruk Getachew,[2] Joseph Ben-Sheetrit,[3] Sudhir Varma,[4] Thomas Heinbockel,[1] Yousef Tizabi,[2] and Antonei B. Csoka[1]

[1]*Epigenetics Laboratory, Department of Anatomy, Howard University, 520 W St. NW, Washington, DC 20059, USA*
[2]*Department of Pharmacology, Howard University, 520 W St. NW, Washington, DC 20059, USA*
[3]*Tel-Aviv Brüll Community Mental Health Center, Clalit Health Services, 9 Hatzvi St., 6719709 Tel-Aviv, Israel*
[4]*HiThru Analytics LLC, 1001 Spring St. No. 219, Silver Spring, MD 20910, USA*

Correspondence should be addressed to Antonei B. Csoka; antonei.csoka@howard.edu

Academic Editor: Igor Koturbash

Commonly used pharmaceutical drugs might alter the epigenetic state of cells, leading to varying degrees of long-term repercussions to human health. To test this hypothesis, we cultured HEK-293 cells in the presence of 50 μM citalopram, a common antidepressant, for 30 days and performed whole-genome DNA methylation analysis using the NimbleGen Human DNA Methylation 3x720K Promoter Plus CpG Island Array. A total of 626 gene promoters, out of a total of 25,437 queried genes on the array (2.46%), showed significant differential methylation ($p < 0.01$); among these, 272 were hypomethylated and 354 were hypermethylated in treated versus control. Using Ingenuity Pathway Analysis, we found that the chief gene networks and signaling pathways that are differentially regulated include those involved in nervous system development and function and cellular growth and proliferation. Genes implicated in depression, as well as genetic networks involving nucleic acid metabolism, small molecule biochemistry, and cell cycle regulation were significantly modified. Involvement of upstream regulators such as BDNF, FSH, and NFκB was predicted based on differential methylation of their downstream targets. The study validates our hypothesis that pharmaceutical drugs can have off-target epigenetic effects and reveals affected networks and pathways. We view this study as a first step towards understanding the long-term epigenetic consequences of prescription drugs on human health.

1. Introduction

It has been hypothesized that pharmaceutical drugs can cause long-term epigenetic changes in the human genome [1, 2]. There is also evidence from animal models that antipsychotics can cause epigenetic changes [3] and that some drugs including antidepressants can interfere with the action of epigenetic enzymes, such as DNA methyltransferase 1 [4]. To test the hypothesis that pharmacological agents can change global DNA methylation in human cells, we chose a commonly used antidepressant, citalopram, and analyzed its effects on human cells by performing genome-wide DNA methylation analysis. Our hypothesis was that treatment with a typical pharmaceutical drug would cause widespread epigenetic changes. Confirmation of this hypothesis could have significant implications for the practice of medicine and for human health.

Citalopram belongs to the widely used class of antidepressant drugs called selective serotonin reuptake inhibitors (SSRIs) and is sold under the commercial name Celexa [5]. In terms of their mechanism of action, SSRIs prevent reuptake of the neurotransmitter serotonin (5-hydroxytryptamine (5-HT)) into the presynaptic cell, thereby increasing its bioavailability in the synaptic cleft, where it can bind to the postsynaptic receptors [6, 7]. Increasing the availability of serotonin in the synaptic cleft enhances serotonergic function and is believed to be responsible for alleviating depression-associated behavior [8]. By a different mechanism, SSRIs increase serotonin by downregulating presynaptic $5HT_{1B}$ autoreceptors (5-hydroxytryptamine or serotonin) that can otherwise inhibit serotonin release [9], thereby increasing synaptic serotonin availability [6]. While blocking reuptake of serotonin can increase its bioavailability and stimulate postsynaptic serotonin receptors to positively affect mood and anxiety, excessive firing of such serotonin-regulated neurons can negatively affect sleep, appetite, sexual function, and pain sensation, raising concerns regarding their adverse effects [6].

With the huge popularity of SSRIs and instances of controversial use in cases of "cosmetic psychopharmacology," i.e., by individuals without clinical diagnoses, such side effects are of significant clinical concern [7]. Although earlier in their developmental phase they were considered to have fewer adverse effects than their first-generation counterparts (viz., tricyclic antidepressants), postmarketing clinical trials documented adverse effects mostly in terms of sexual dysfunction including anorgasmia, erectile dysfunction, genital anesthesia, and diminished libido in almost 75% of treated patients [6, 10–12]. Interestingly, these side effects appear to endure after treatment in some cases [13, 14], which is hard to explain using a standard pharmacological model.

A plausible cause of these persistent side effects is changes to the epigenome [1–3]. The epigenome of a cell is a unique, dynamic entity consisting of distinct DNA methylation patterns across gene enhancers, promoters, and bodies along with histone modifications that do not involve any changes to the actual DNA sequence. Recently, the effects of environmental factors, developmental processes, or lifestyle habits, such as diet and drugs, on modulation of gene expression via epigenetic modification have been studied in detail [15]. Epigenetic changes resulting from environmental effects such as traumatic life events can rewire neural circuits and alter neurotransmitter and endocrine systems resulting in stress-related psychiatric disorders such as major depression or posttraumatic stress disorder [15]. Based on this evidence, it can be posited that potential unknown mechanisms of action of SSRIs, as well as side effects, could be through epigenetic modification of genes [1, 2, 15].

For these reasons, its long history of use in depression treatment as well as its well-documented side effects including sexual dysfunction, sleep disturbances, and weight gain [6], citalopram was tested to assess its effect on genome-wide DNA methylation of human cells, with additional analysis on affected gene networks and signaling pathways, including but not limited to those implicated in neuropsychological function.

2. Materials and Methods

2.1. Cell Culture. Human embryonic kidney (HEK-293) cells were chosen for this study because they are used broadly for biomedical research, ranging from signal transduction to protein interaction studies, and are hence a good candidate for studying epigenetics as well. Advantages of using these cells over primary neurons is that they can be easily proliferated, maintained, preserved, and studied. Also, they express significant amounts of protein and mRNA for neurofilament (NF) subunits, such as NF-L, NF-M, NF-H, and α-internexin, as well as other neuron-specific proteins, suggestive of their neuronal lineage [16].

The HEK-293 cell line was purchased from ATCC and cultured in growth medium containing Dulbecco's Modified Eagle Medium (DMEM) (Life Technologies, CA, USA) supplemented with 10% fetal bovine serum (FBS) (Life Technologies, CA, USA) and 1x penicillin-streptomycin solution (Life Technologies, CA, USA) in a humidified incubator with 5% CO_2 at 37°C. On reaching 90% confluence, cells were subcultured with a 1 : 6 split ratio in T25 flasks.

2.2. Cell Treatment. A toxicity curve was performed on the cells to determine the optimum concentration of citalopram hydrobromide (Sigma-Aldrich, MO, USA) that can be tolerated by the cells without changing their growth dynamics. Cells were cultured in growth media containing different concentrations of citalopram hydrobromide (10 μM, 50 μM, 90 μM, 120 μM, 160 μM, and 200 μM) for 48 hours. No effect was observed on cell growth kinetics or morphology below 120 μM, but at concentrations above 160 μM, an apoptotic-like cytotoxic effect was noted (Supplement 1). A 50 μM solution of citalopram hydrobromide was determined to be the maximum concentration that could be safely used without any possibility of inducing any change in growth kinetics. HEK-293 cells in the treatment group (in triplicates) were cultured with 50 μM citalopram hydrobromide for thirty days along with nontreated controls. All flasks were passaged and maintained under similar conditions as mentioned above for a period of thirty days.

2.3. DNA Extraction and MeDIP Chip Analysis. After a 30-day treatment, cells were lysed and genomic DNA was homogenized using QIAshredder (Qiagen) and extracted using the DNeasy kit (Qiagen) followed by sonication to generate fragments of about 200–1000 base pairs. Immunoprecipitation of methylated DNA was performed using Biomag™ magnetic beads coupled to a mouse monoclonal antibody against 5-methylcytidine. The immunoprecipitated DNA was eluted and purified by phenol-chloroform extraction and ethanol precipitation. The total input and immunoprecipitated DNA were labeled with Cy3- and Cy5-labeled random 9-mers, respectively, and hybridized to NimbleGen Human DNA Methylation 3x720K Promoter Plus CpG Island Arrays, which is a multiplex slide with 3 identical arrays per slide, and each array contains 27,728 CpG Islands annotated by UCSC and 22,532 well-characterized RefSeq promoter regions (from about −2440 bp to +610 bp of the Transcription Start Sites) totally covered by ~720,000 probes. Scanning was performed with the Axon GenePix 4000B Microarray Scanner by Arraystar Inc. (Rockville, MD, USA).

2.4. Data Normalization. Raw data was extracted as pair files by NimbleScan software. We performed median-centering, quantile normalization, and linear smoothing using NimbleScan by Nimblegen and R Bioconductor packages (Ringo, limma, and MEDME) [17]. The enrichment peaks and differentially methylated peaks were analyzed and annotated by NimbleScan software. The user guide and result data formats

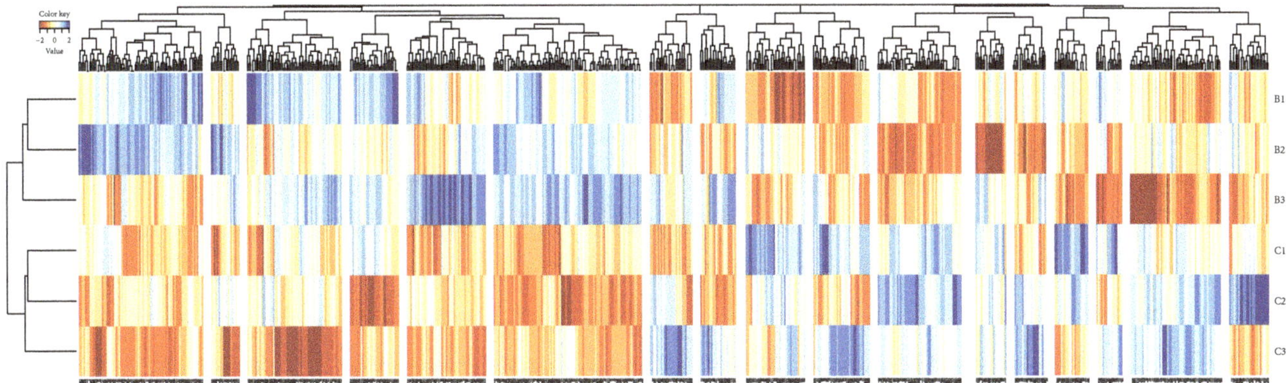

FIGURE 1: Heat map of hypermethylated and hypomethylated gene promoters. This heat map represents differentially methylated gene promoters between citalopram-treated (B1, B2, and B3) and control (C1, C2, and C3) samples with significant values from MeDIP chip analysis, grouped into clusters. The scale represents hypermethylated gene promoters (values 0 to +2) in blue and hypomethylated gene promoters (values 0 to −2) in red. Each column represents a gene, as specified on the gene axis at the bottom, that is either downregulated (hypermethylated promoters in blue) or upregulated (hypomethylated promoters in red) between samples represented in the six rows. The right axis represents overall methylation clustering between treated and control samples, and the top axis represents quantitative methylation clustering between significant genes.

can be found at http://www.nimblegen.com/downloads/support/NimbleScan_v2p6_UsersGuide.pdf. After normalization, a normalized log2-ratio data (*_ratio.gff file) was created for each sample. From the normalized log2-ratio data, a sliding-window peak-finding algorithm provided by NimbleScan v2.5 (Roche-NimbleGen) was applied to find the enriched peaks with specified parameters (sliding-window width: 750 bp; miniprobes per peak: 2; p-value minimum cutoff: 2; maximum spacing between nearby probes within peak: 500 bp). After obtaining the *_peaks.gff files, the identified peaks were mapped to genomic features: transcripts and CpG Islands.

2.5. Bioinformatics and Pathway Analysis of MeDIP Chip Results. t-tests and/or binomial tests were used to compute p values for differential methylation of CpG sites followed by multiple comparison correction of p values and computation of false detection ratio (FDR) using the Benjamini Hochberg method [18]. Genes that are significantly differentially methylated ($p < 0.01$) between the treated versus control groups were identified, and functional analysis of differentially methylated genes was performed using gene set enrichment analysis (GSEA). Gene promoters showing statistically significant changes in DNA methylation patterns were subjected to Ingenuity Pathway Analysis (IPA) (Ingenuity System Inc., CA, USA) for signaling pathway and gene network analysis. The z-scores predict activation states of transcriptional regulators and were calculated by an IPA-based algorithm (http://pages.ingenuity.com/rs/ingenuity/images/0812%20upstream_regulator_analysis_whitepaper.pdf).

3. Results and Discussion

3.1. Results. Genome-wide DNA methylation analysis revealed that citalopram causes significant differential methylation ($p < 0.01$) in 626 gene promoters (from about −2440 bp to +610 bp of the transcription start sites) compared to controls (2.46%). Overall, there were more gene promoters hypermethylated (354; 1.39%) than hypomethylated (272; 1.07%) (Supplement 2a). Means and standard deviations for all of the samples can be seen in Supplement 2b. A heat map (Figure 1) represents differential DNA methylation between treated (B1, B2, and B3) and control (C1, C2, and C3), grouped into clusters. Since our analysis only included significant gene promoters without intragenic and intergenic regions, we were able to translate our methylation data into gene expression data for IPA without complication; hypermethylated promoters representing downregulation and hypomethylated promoters representing upregulation of gene expression, by default. We assigned positive and negative values to peak differential methylation values to correlate to upregulation or downregulation of gene expression, respectively (Supplement 3). Hereafter, we refer to these gene promoters as genes for simplicity and refer to activation from gene induction as upregulation and inhibition from gene silencing as downregulation.

We especially wanted to analyse any differential methylation caused by citalopram at genes that are either a part of the epigenetic modifier groups or involved in depression-related behavior. We compared our dataset with a curated list of a total 601 genes and molecules implicated in psychological depression (IPA) and found 13 genes common showing significant differential methylation (Figure 2, Supplement 4), including BTG2, FABP6, GRIN1, HRH1, HSD17B1, MDFI, OXT, and TSPO. We also found six epigenetic enzymes with significant differential methylation, including HDAC6, SET, SETBP1, SETD82, SIRT1, and TDG (Supplement 5).

In a broad analysis of canonical signaling pathways, gene networks and biological functions using IPA's core analysis function, we found that significant genes from our dataset were enriched in canonical pathways including Hippo signaling (p value $= 9.14E-03$), PTEN signaling (p value $= 1.64E-02$), maturity-onset diabetes of the young (MODY)

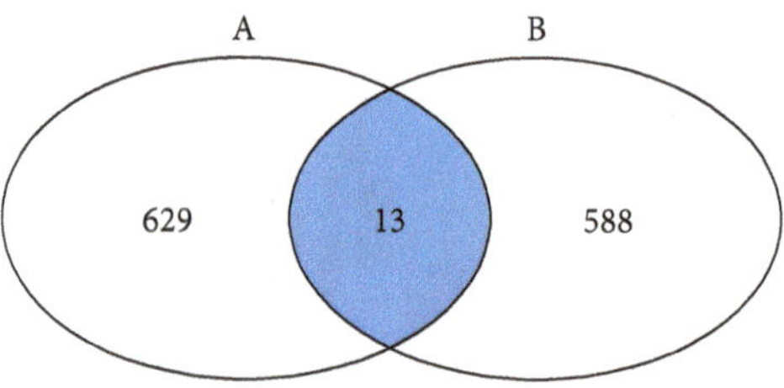

List of intersecting genes implicated in depression

Symbol	Entrez Gene Name	Gene symbol	Exp other	Exp p value	Location
AOC3	Amine oxidase, copper containing 3	AOC3	−0.218	0.002	Plasma membrane
APOE	Apolipoprotein E	APOE	−0.119	0.005	Extracellular space
BTG2	BTG family member 2	BTG2	−0.057	0.004	Nucleus
CACNA1D	Calcium voltage-gated channel subunit alpha1 D	CACNA1D	−0.463	0.003	Plasma membrane
CD3E	CD3e molecule	CD3E	0.162	0.004	Plasma membrane
CHAT	Choline O-acetyltransferase	CHAT	−0.455	0.006	Nucleus
FABP6	Fatty acid binding protein 6	FABP6	−0.748	0.003	Cytoplasm
GRIN1	Glutamate ionotropic receptor NMDA type subunit 1	GRIN1	−0.211	0.002	Plasma membrane
HRH1	Histamine receptor H1	HRH1	0.249	0.002	Plasma membrane
HSD17B1	Hydroxysteroid (17-beta) dehydrogenase 1	HSD17B1	0.163	0.003	Cytoplasm
MDFI	MyoD family inhibitor	MDFI	0.525	0.000	Cytoplasm
OXT	Oxytocin/neurophysin I prepropeptide	OXT	−0.297	0.002	Extracellular space
TSPO	Translocator protein	TSPO	−0.101	0.003	Cytoplasm

FIGURE 2: Venn diagram of genes involved in depression with significant differential methylation. This figure shows a set of 13 genes identified from our dataset with significant differential methylation resulting from citalopram treatment overlapping with a curated list of genes implicated in psychological depression according to the current IPA database, shown in Venn diagram form. Genes are classified according to their subcellular location, p value, and expression value where positive expression values indicate upregulation and negative expression values represent downregulation. A few important genes included in this list are OXT, GRIN1, CHAT, and CACNA1D which are potential regulators of neurophysiological processes and have a high degree of implication in psychological disorders.

signaling (p value = 1.87E − 02), and cyclins and cell cycle regulation signaling (p value = 1.98E − 02) (Table 1, Figures 3 and 4). These also included inflammation-related signaling pathways like TNFR2 (p value = 4.37E − 02) and TNFR1 (p value = 4.43E − 02) (Table 1). Many of these pathways show overlapping patterns (Supplement 6). The chief associated gene network functions of the canonical pathways include nucleic acid metabolism, small molecule biochemistry, and cell signaling associated with a number of diseases including cancer and nervous system dysfunction (Table 1). Genes are enriched for molecular and cellular functions including protein synthesis, cellular movement, and drug metabolism (Table 1). Novel regulatory networks involving CASZ1 in quality of metal ion and miR 199a-5p in growth of plasma membrane projections were identified (Supplement 7). Thus, a wide variety of gene networks and pathways were affected by the citalopram treatment.

Next, we analyzed the main upstream regulators predicted for differential regulation based on their downstream target states (hypermethylated or hypomethylated) and found that citalopram most importantly affected the NFκB complex (p value = 1.79E − 04) and L-dopa pathway (p value = 1.22E − 03) (Figure 5). Other significant upstream regulators with predicted differential regulation include FSH (p value = 4.22E − 02); BDNF (p value = 2.23E − 02); IL13 (p value = 8.92E − 03); PRKCD, a protein kinase C (p value = 4.64E − 01); and GLI1, a Kruppel family member of zinc finger proteins (p value = 3.99E − 01).

Finally, top physiological systems affected by citalopram identified by IPA included nervous system development diseases and function with 24 genes involved in neurotransmission (p value = 1.25E − 02), 23 genes related to outgrowth of neurites (p value = 1.59E − 02), 9 genes related to excitatory postsynaptic potential (p value = 1.48E − 02), 6 genes related to quantity of synapse (p value = 1.21E − 02), and 3 genes related to loss of dendritic spines (p value = 1.42E − 02). Additionally, 47 genes related to morphology of the nervous system (p value = 1.95E − 02), 35 genes related to development of the central nervous system (p value = 1.45E − 02), 21 genes related to sensation (p value = 3.67E − 03), 11 genes related to development of the cerebral cortex (p value = 1.20E − 02), 8 genes related to abnormal morphology of the hippocampus (p value = 1.72E − 02), 7 genes related to abnormal morphology of the synapse (p value = 2.96E − 03), and 3 genes related to development of the hypothalamus

Table 1: Top canonical pathways, upstream regulators, diseases and biofunctions, and networks.

(a)

Top canonical pathways Name	p value	Overlap
Hippo signaling	$9.14E-03$	8.1% 7/86
Hepatic cholestasis	$1.21E-02$	6.3% 10/159
PTEN signaling	$1.64E-02$	6.7% 8/119
Maturity-onset diabetes of the young (MODY)	$1.87E-02$	14.3% 3/21
Cyclins and cell cycle regulation	$1.98E-02$	7.7% 6/78

(b)

Top upstream regulators Upstream regulator	p value of overlap
NS-398	$5.54E-05$
NFκB (complex)	$1.79E-04$
ACKR3	$3.72E-04$
RP 73401	$8.44E-04$
L-Dopa	$1.22E-03$

(c)

Top diseases and biofunctions Name	p value	Number of molecules
Diseases and disorders		
Cancer	$2.00E-02$ to $1.20E-06$	511
Organismal injury and abnormalities	$2.00E-02$ to $1.20E-06$	515
Hypersensitivity response	$5.02E-04$ to $5.02E-04$	7
Dermatological diseases and conditions	$1.97E-02$ to $6.53E-04$	25
Immunological disease	$1.95E-02$ to $6.53E-04$	22
Physiological system development and function		
Lymphoid tissue structure and development	$2.12E-02$ to $7.93E-05$	56
Tissue morphology	$2.12E-02$ to $7.93E-05$	107
Humoral immune response	$1.37E-02$ to $1.84E-04$	28
Connective tissue development and function	$2.12E-02$ to $2.67E-04$	42
Nervous system development and function	$1.95E-02$ to $2.67E-04$	107

(d)

Top networks ID	Associated network functions	Score
(1)	Nucleic acid metabolism, small molecule biochemistry, cell signaling	42
(2)	Auditory disease, cancer, cardiovascular disease	42
(3)	Carbohydrate metabolism, drug metabolism, small molecule biochemistry	37
(4)	Cellular growth and proliferation, tissue development, cellular movement	37
(5)	Embryonic development, humoral immune response, lymphoid tissue structure and development	33

This is a comprehensive list of the top canonical pathways and top upstream regulators with predicted differential regulation as well as diseases, biofunctions, and networks with the highest enrichment involving significant genes identified in our dataset based on p values and other criteria set by IPA. Amongst the top canonical pathways, Hippo signaling has the highest overlap in genes from our dataset that are enriched in the pathway divided by the total number of genes enriched in the Hippo pathway, that is, 8.1% according to the current IPA database. NS-398 which is a cyclooxygenase inhibitor is amongst the top upstream regulator with a p value of $5.54E-05$. The top diseases associated with citalopram treatment include cancer with 511 molecules predicted to have differential regulation, whereas the top physiological system development and function predicted to be effected includes nervous system development and function with 107 molecules identified by IPA. The top associated gene network functions include nucleic acid metabolism, small molecule biochemistry, and cell signaling with a score of 42 generated by the IPA algorithm. A score of 50 is considered as high and below 20 is low. Predicted activation.

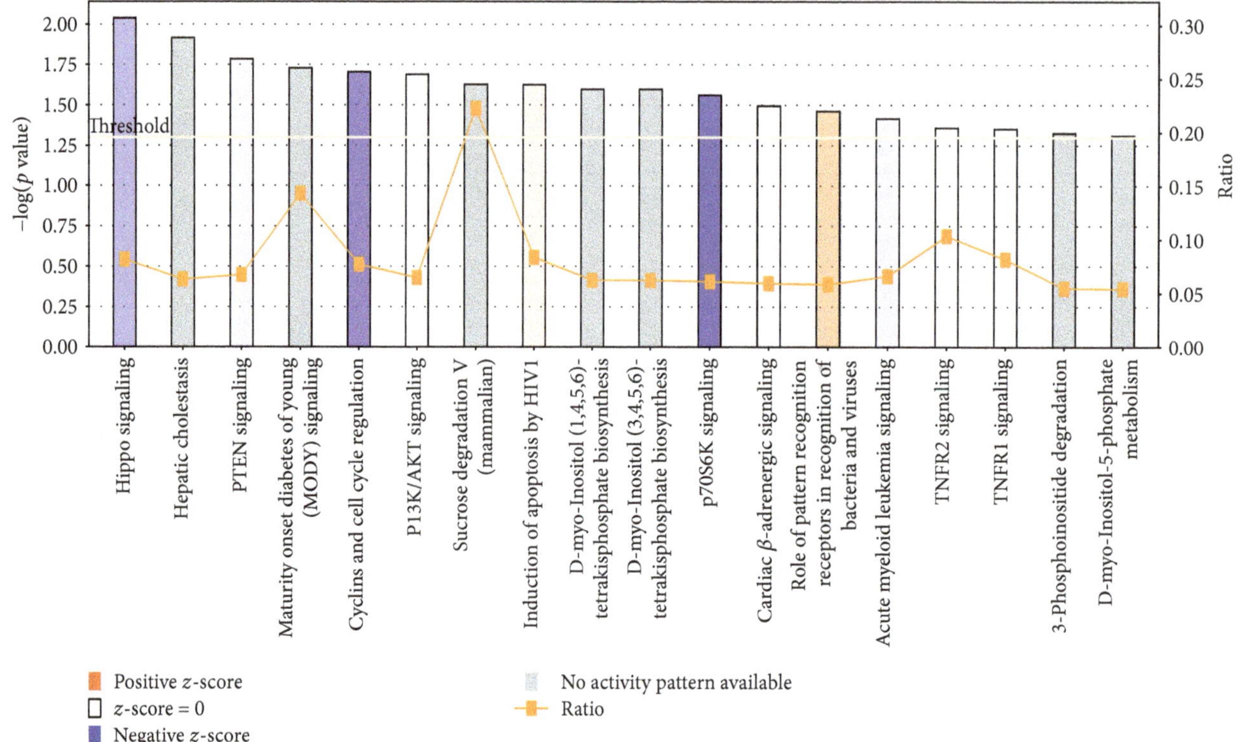

FIGURE 3: Top canonical pathways altered by citalopram. This bar graph enlists the top canonical pathways predicted to be altered by citalopram treatment using ingenuity pathway core analysis. Citalopram treatment results in differential methylation of significant genes from our dataset that are enriched in canonical pathways, like Hippo signaling, PTEN signaling, maturity-onset diabetes of the young (MODY) signaling, and cyclins and cell cycle regulation signaling based on their z-score, ratio, and −log (p value). A positive z-score (orange) denotes activation of pathway (e.g., role of pattern recognition receptors in recognition of bacteria and viruses), and a negative z-score (blue) denotes inhibition of a pathway (e.g., P70S6K signaling). The ratio (orange line with blocks) represents a ratio of genes from our dataset that is enriched in the pathway divided by the total number of genes enriched in the same pathway according to current IPA database [e.g., 22.5% sucrose degradation V (mammalian)]. Threshold is set at the lowest level of confidence that is acceptable statistically ($p < 0.05$).

(p value $= 1.50E-03$) were identified (Supplement 8). These results, in particular, were interesting because of the known mechanism of citalopram action on the nervous system-based serotonin transporter, along with unknown targets affected by epigenetic mechanisms, that can be further delineated in the future.

3.2. Discussion. We have previously outlined a potential mechanism for understanding the direct and indirect effects of environmental factors, including pharmaceutical drugs [1, 15] on the epigenome. Here, we attempted to confirm the hypothesis that pharmacological agents can cause permanent changes via epigenetic reprogramming. The results show that our first test drug, citalopram, can cause genome-wide DNA methylation alterations as revealed by significant differential methylation in hundreds of genes, as well as predicted impact on signaling pathways and/or physiological systems, some of which are described below.

3.2.1. Reproductive and Sexual Function. The OXT gene, producing oxytocin, is downregulated by citalopram. Since oxytocin plays a significant role in parturition and milk ejection and is also implicated in cognition, tolerance, adaptation, and complex sexual and maternal behavior, its downregulation by SSRIs may be one of the underlying causes of sexual dysfunction seen in many cases [12, 19]. In terms of upstream regulators, inhibition of the dopa pathway, involved in the synthesis of dopamine, also coincides with the numerous findings of negative effects of SSRIs on dopaminergic signaling including sexual dysfunction [20].

Amongst other upstream regulators, we saw predicted inhibition of follicle-stimulating hormone (FSH), which is responsible for maturation of ovarian follicles in females and spermatocytes in males. FSH is regulated by gonadotropin-releasing hormone (GnRH), also included in one of our gene networks, affecting functions including cell signaling, molecular transport, and vitamin and mineral metabolism (Supplement 8). Previous studies have confirmed the side effects of SSRIs on reproductive and neuroendocrine dysfunction in wildfish involving changes in ovarian and hypothalamic gene expression, spermatogenesis, and sex steroid production [21–23]. In a month-long treatment of male zebrafish with citalopram, different stages of spermatogenesis were inhibited, whereas short-term treatment downregulated the expression of GnRH and

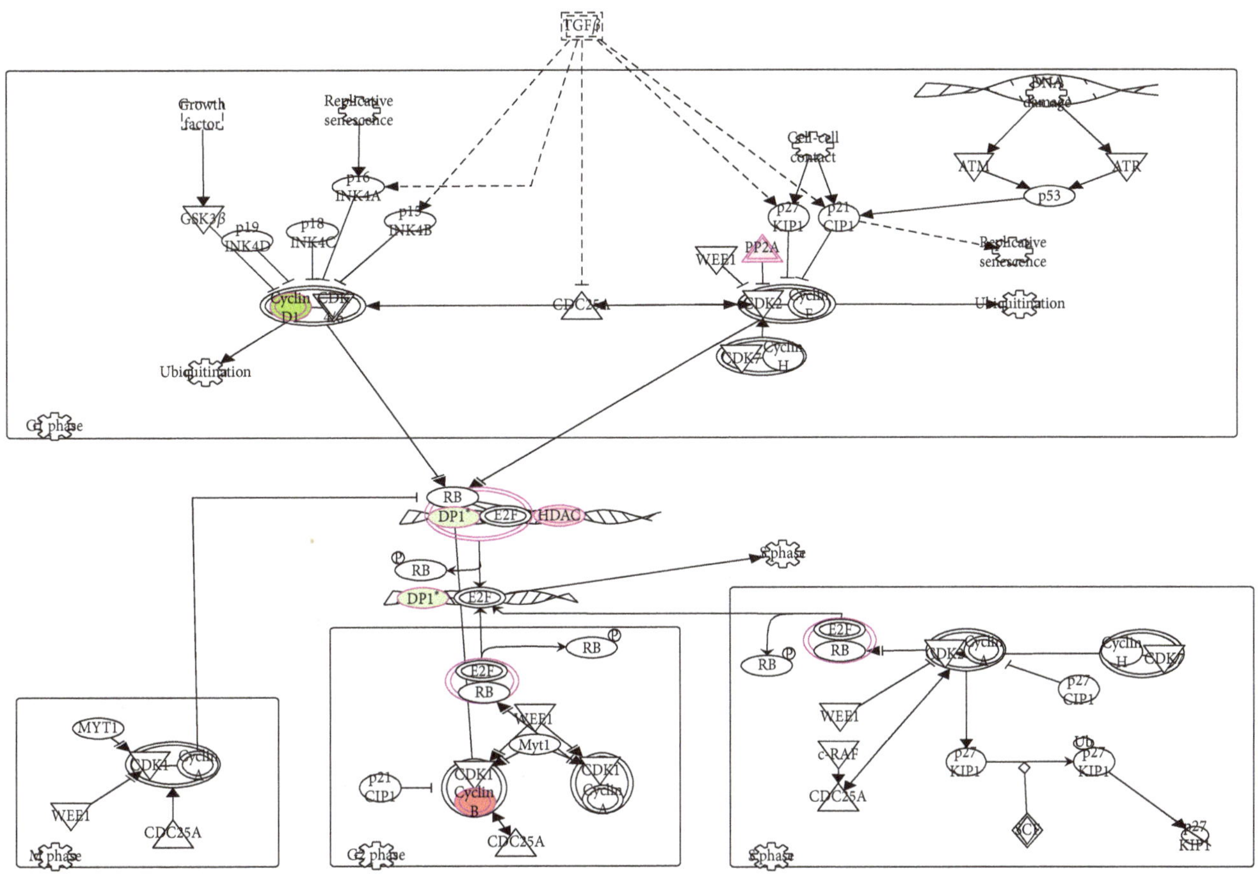

FIGURE 4: Cyclins and cell cycle regulation is the top pathway downregulated by citalopram treatment. This figure represents the cyclins and cell cycle regulation pathway as the top signaling pathway downregulated with a significant p value (p value = $1.98E-02$) due to citalopram treatment. IPA identifies this pathway as it involves differentially methylated genes from our dataset like cyclin B, HDAC, and P2A that are upregulated (red) and cyclin D1 and Dp1 that are downregulated (green). The color intensity is proportional to the extremity of upregulation or downregulation.

serotonin-related genes TPH2 and SERT [10]. Moreover, SSRIs affect the hypothalamic-pituitary-testis (HPT) axis in depressed male patients suffering from SSRI-induced sexual dysfunction due to significantly lower serum levels of luteinizing hormone (LH), FSH, and testosterone [24, 25]. These studies and our current data imply that the imbalances in GnRH, FSH, and LH production associated with abnormal serotonin levels might be epigenetic at source and at least partly responsible for SSRI-induced sexual and reproductive dysfunction.

3.2.2. Signaling Pathways: Molecular and Metabolic Interference. Primary pathways such as Hippo signaling, PTEN signaling, and cyclins and cell cycle regulation signaling were downregulated. The Hippo signaling pathway regulates organ size control, tumor suppression, tissue regeneration, and stem cell self-renewal [26]. Cyclins and cyclin-dependent kinase (CDK) family members are involved in a range of diverse functions including transcription, DNA damage repair, proteolytic degradation, epigenetic regulation, metabolism, stem cell self-renewal, neuronal functions, and spermatogenesis [27]. PTEN is a tumor suppressor, and modification of PTEN signaling networks results in manifestation of developmental defects and increased risk of cancer [28]. Thus, inhibition or dysregulation of signaling pathways may increase risk of cancer [29, 30]. Another interesting finding is the involvement of pathways for maturity-onset diabetes of the young (MODY). Previous studies report significant weight gain, insulin resistance, and worsening glycemic control as side effects of chronic SSRI usage [31].

3.2.3. Neurological and Psychiatric Pathways. The translation of early life stress into major depressive disorders in adulthood is possibly rooted in epigenetic alteration of candidate genes, including the serotonin transporter (SLC6A4), via DNA methylation, histone acetylation and methylation, and miRNAs, which also is a mode of therapeutic action of some antidepressant drugs [32–34]. We identified 13 genes associated with depression-related disorders that were differentially methylated by citalopram, which in some ways seems to be quite a low number considering the therapeutic target. In any case, B-cell translocation gene 2 (BTG2), reported to be upregulated (in the prefrontal cortex) in major depression, was downregulated [35]. Additionally, the MyoD family inhibitor (MDFI) that is downregulated in depression (dorsolateral prefrontal cortex) was upregulated [36].

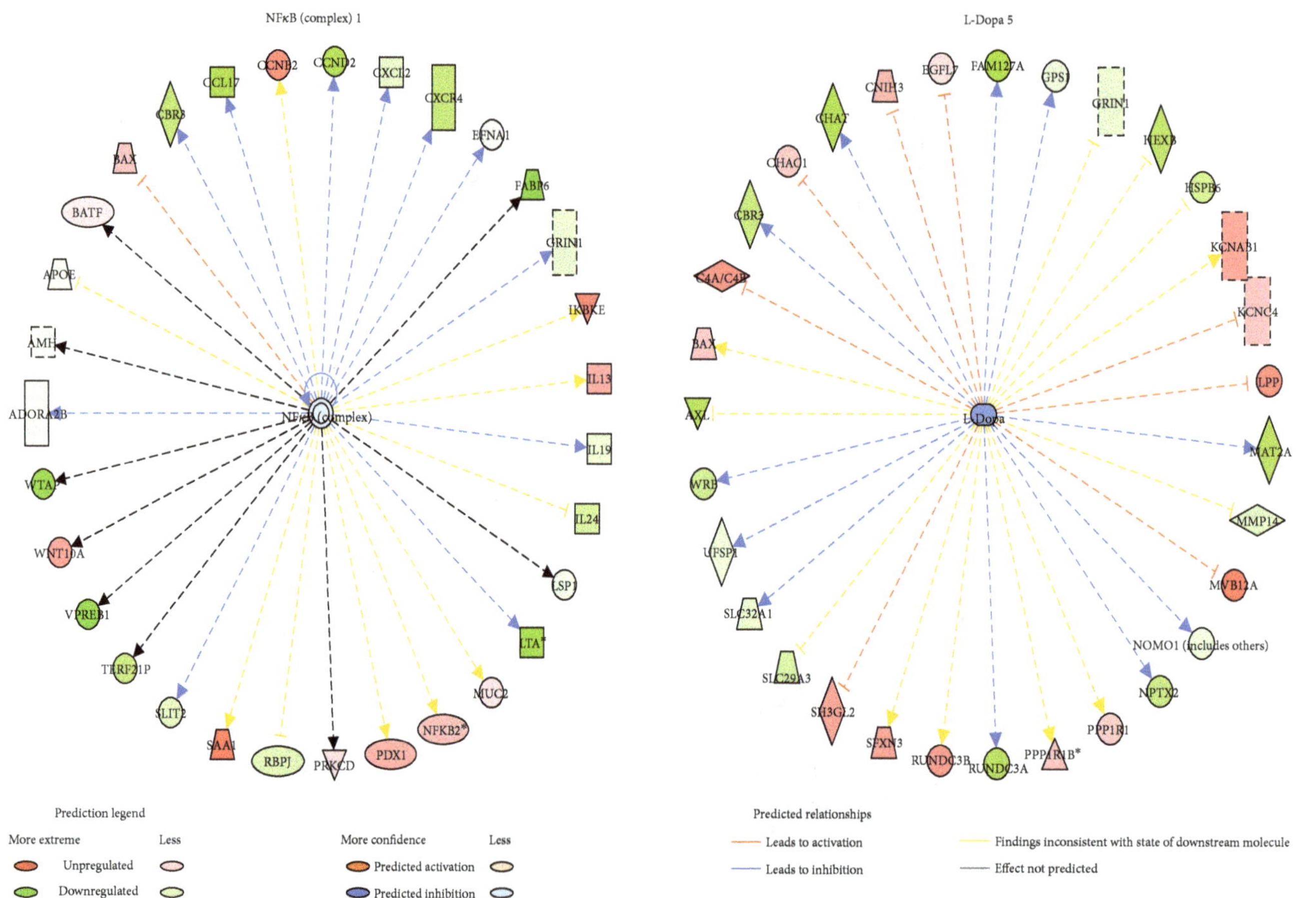

Figure 5: NFκB complex 1 and L-dopa 5 identified as upstream regulators with predicted inhibition. This figure represents the two upstream regulators with predicted significant differential regulation resulting from citalopram treatment based on the methylation states (hypermethylated or hypomethylated) of their downstream targets by ingenuity pathway upstream regulator analysis. NFκB complex 1 is a transcription factor predicted with a higher degree of inhibition (green) with p value = $1.79E-04$. L-Dopa, the initial substrate for neurotransmitter dopamine, is also predicted to be inhibited (green) at a lower degree with p value = $1.22E-03$. With upstream regulators NFκB complex 1 and L-dopa at the center, the dotted lines with arrow indicate downstream target genes that are upregulated (red) or downregulated (green) due to differential methylation as indicated in our dataset. A dashed line means indirect interaction, continuous line means direct interaction, line with arrow means "acts on" and line with bar at the end means "inhibits." These basic relationships between molecules represented in the figure are based on literature-reported effects, and the color coding represented in the legend is used for correlation of known relationships with observed gene expression effects resulting from treatment of citalopram. For example, based on literature, L-dopa indirectly acts on SLC32A1, a polyamine transporter. However, in the presence of citalopram, L-dopa bioavailability is decreased, and it is predicted to have an inhibitory effect on SLC32A1 indirectly such that it is downregulated.

Citalopram also downregulated translocator protein (TSPO), generally upregulated in depression [37]. In a study using a rat model involving long-term treatment of depression with escitalopram (a stereoisomer of citalopram), p11, a calcium-binding protein, generally downregulated in depression, was induced by specific hypomethylation of the p11 gene promoter, increasing gene expression and reversing depression-like behavior [38, 39]. Other genes including FABP6 (fatty acid binding protein 6), downregulated in the prefrontal cortex in major depression [35]; GRIN1 (glutamate ionotropic receptor NMDA subunit), implicated in stress-related psychiatric disorders [40]; HRH1 (histamine receptor H1), known to be blocked by TCAs [41]; and HSD17B1 (hydroxysteroid 17-beta dehydrogenase 1), associated with female depression [42], were likewise differentially methylated. These mechanisms indicate unique effects of SSRIs and suggest novel therapeutic targets for treatment of depression.

3.2.4. Inflammation. Inflammation-related upstream regulators like the NFκB complex are inhibited and IL13 activated. Inflammation plays an important role in the pathophysiology of depression as seen in many patients with elevated proinflammatory cytokine levels [43]. Modulation of inflammatory networks by antidepressants has previously been associated with decreased inflammation in male patients using SSRIs but, curiously, increased inflammation in patients using other types of antidepressants [44]. However, it should be noted that specific interactions between innate and adaptive immune systems and neurotransmitters and neuronal circuits may influence risk for depression and response to antidepressants [45, 46].

3.2.5. Nervous System Development and Function. Nervous system development and function (p value = $1.95E-02$ to $2.67E-04$) was one of the systems most significantly affected by the treatment (Table 1). 47 genes related to morphology of the nervous system (p value = $1.95E-02$), 35 genes related to development of the central nervous system (p value = $1.45E-02$), and 11 genes related to development of the cerebral cortex (p value = $1.20E-02$) were identified. These effects may be related to changes in autonomic functions (e.g., tachycardia), hypothermia, and changes in mental status (e.g., agitation, anxiety, and confusion) [47]. In mice, increased serotonergic activity postnatally can propagate abnormal neuroanatomical development of the somatosensory cortex along with functional response deficits [48]. Intrauterine antidepressant exposure can cause epigenetic changes affecting neonatal development and health [49] and lasting abnormal emotional behaviors [48, 49]. Thus, epigenetic changes at genetic loci involved in neuroanatomical development have major implications on the use of SSRIs to treat depressive behaviors.

One potential limitation with this pilot study is that HEK-293 cells have not been shown to express high levels of the serotonin transporter, SERT, nor been shown to synthesize high amounts of serotonin in the extracellular medium, compared to neurons. Hence, in this case, we argue that the effects of citalopram seen on DNA methylation in these cells are more likely to be 5HT-independent. HEK-293 cells *are* known to abundantly express a diverse repertoire of receptors such as β2-adrenergic, muscarinic acetylcholine, sphingosine-1-phosphate, P2Y1 and P2Y2, corticotropin-releasing factor type 1, and somatostatin- and thyrotropin-releasing hormone receptors, and citalopram has been shown to interact strongly with some of these receptors, in the concentration range used in this study, so it may well be eliciting epigenetic effects through these pathways. Moreover, the data is consistent with our initial hypothesis that the epigenetic effects of chemicals could be both direct (acting directly on DNA or DNA-modifying enzymes) and indirect (acting through receptors or signaling pathways) [1, 15], in which case a direct effect on SERT is not necessary to induce epigenetic changes. It is also possible that in the presence of 5-HT and SERT, we may see different epigenetic effects of citalopram from those observed in the HEK-293 cells. In any case, we intend to repeat this experiment using primary human neurons as the target cells, rather than a proliferating cell line, in order to gain greater insights into potential *in vivo* effects.

4. Conclusions

4.1. Whole-Genome Epigenetic Analysis as an Aspect of the Drug Development Process. In this study, we wanted to explore, in an initial investigative pilot experiment, the potential for a typical, widely used pharmaceutical drug to cause epigenetic changes in human cells, both beneficial and potentially harmful. We used human genome-wide promoter methylation analysis to delineate unique gene methylation profiles arising from short-term treatment with citalopram. These results could serve as proof of principle for such assays to become standard protocol during the toxicological analysis stage of drug development, from bench to bedside. Such epigenetic toxicological analysis could eventually revolutionize the safety of personalized medicine. We view this paper as an initial first step in a much broader inquiry into the epigenetic mechanisms of pharmacological agents.

4.2. Drugs and Waddington's Canal. We also wanted to explore the possibility that the magnitude of the epigenetic changes caused by a typical drug is enough to displace a cell from its normal "groove" in the "epigenetic landscape." This is a term derived from the original work by C. H. Waddington and represents a series of branching valleys depicting developmental pathways and ridges between valleys that are barriers to transitions between steady cellular states that reside in the valleys [50]. Waddington also coined the term "canalization," meaning that, up to a certain threshold, any genetic variation or environmental insult to a cell will be nullified and the cell will remain within its groove, but above this threshold, the cell would flip over into an adjacent pathway or "valley" [51]. A modern example of altering canalization is the phenomenon of reprogramming somatic cells to pluripotency, which is achieved by activating epigenetic switches and driving a cell back up its lineage to the highest point in the landscape via the reversal of differentiated gene expression to a fully embryonic-like state [52]. Interestingly, such total reprogramming of differentiated cells to pluripotency can now be achieved by the use of small molecules alone [53]. Therefore, we reasoned that if a chemical cocktail alone is capable of reversing a cell's lineage, then there is also a possibility that pharmaceutical drugs in isolation or in combination (as in polypharmacy) can alter cells' epigenetic profiles sufficiently that they are no longer in their original differentiated state. It is highly unlikely that this would represent a recanalization event per se but rather a slight "shift" in the groove causing marginal dysdifferentiation.

4.3. The Concept of "Pharmaceutical Reprogramming." As stated, it is likely that the epigenetic effects of citalopram are much too weak to induce phenotypic conversion or alter lineage but may be just robust enough to cause a partial dysdifferentiation event, whereby a cell's location in its epigenetic landscape is marginally altered. Such a differentiation "wobble" would result from all of the changes in DNA methylation altering the cell's normal biochemistry. We have termed this partial dysdifferentiation from pharmacological exposure "pharmaceutical reprogramming." Pharmaceutical reprogramming could affect cells and tissues at the submicroscopic level but might not be evident microscopically or macroscopically. It will be important to explore this hypothesis further in future studies, in order to better understand the epigenetic effects of drugs capable of affecting cellular function and integrity. The implications of these findings, if true, could have enormous importance for human health.

Conflicts of Interest

The authors declare that there are no conflicts of interest regarding the publication of this manuscript.

Acknowledgments

This project was supported by the National Institute of Health (NIH) R25 Resource Grant (1 R25 AG047843-01) to Antonei B. Csoka and by the Latham Trust Fund and NSF IOS-1355034 to Thomas Heinbockel. The authors thank Dr. William Southerland for providing them access to the Ingenuity Pathway Analysis Software.

Supplementary Materials

Supplementary 1. Supplement 1: images of HEK-293 cells in increasing concentrations of citalopram from 120 μM to 200 μM. No effect was observed on cell growth kinetics or morphology below 120 μM, but at a concentration above 160 μM, an apoptotic-like cytotoxic effect was noted.

Supplementary 2. Supplement 2a: table with a list of significantly differentially methylated peaks ($p < 0.01$) that lie within 2000 base pairs of the transcription start site of a gene, generated using the initial raw data. All coordinates were transformed from hg18 to hg38, and the genes were reannotated. Peaks were remapped to the latest genome (hg38) and refiltered according to distance from the nearest gene TSS. The "Direction" column indicates which way the differential methylation goes (hypermethylated in treated or hypomethylated). The "Peak differential methylation" column (col M) shows the average difference in methylation level between treated and untreated (positive values are higher methylation in treated compared to untreated; negatives are lower). There are more peaks hypermethylated in treated than hypomethylated (354 hypermethylated peaks versus 272 hypomethylated peaks). Each peak lies within 2000 base pairs of one or more genes. The list of genes is in column C and the distance of their TSS from the middle of the peak is in column L. Positive numbers indicate that the peak middle is upstream of the TSS; negative numbers indicate that the peak middle is downstream. Supplement 2b: means and standard deviations for all of the significant peaks.

Supplementary 3. Supplement 3: table with list of significantly differentially methylated peaks ($p < 0.01$) that lie within 2000 base pairs of the transcription start site of a gene, used as raw data for Ingenuity Pathway Analysis. Hypomethylated gene promoters with significantly differentially methylated peaks were assigned positive signs and hypermethylated gene promoters were assigned negative signs as opposed to the list in Supplement 1, to correlate with upregulation or downregulation of gene-expression, respectively.

Supplementary 4. Supplement 4: table listing all genes implicated in psychological depression curated by current IPA database. Information regarding gene/molecule and their respective Entrez Gene IDs for human, mouse, and rat are specified. This list was used for searching any common genes in our list of significant genes as obtained after MeDIP chip analysis on citalopram-treated human cells.

Supplementary 5. Supplement 5: table listing all epigenetic modifiers according to the current IPA database. This list was used for searching any common genes encoding epigenetic enzymes in our list of significant genes as obtained after MeDIP chip analysis on citalopram-treated human cells.

Supplementary 6. Supplement 6: overlapping between individual significant canonical pathways identified by IPA that are altered by citalopram treatment. Each node represents one canonical pathway, and each link represents a set of genes acting between two pathways determined by Fisher's exact test p value. Darker red shade of nodes represents highly significant pathways and lighter shade of red represents less significant ones. Line width of links corresponds to the number of molecules shared between two pathways where no line means no shared molecules between two pathways and blue line means strong overlap of molecules between canonical pathways.

Supplementary 7. Supplement 7: this figure represents a novel regulatory network identified by IPA. SET as predicted to be upregulated in our dataset can be involved in regulation of growth of plasma membrane projections in addition to miR 199a-5p, SIRT, and DDR1. Red color represents upregulation. Lighter red represents activation, and blue represents inhibition. Red line denotes activation, yellow line denotes finding inconsistencies, and black line denotes effect not predicted.

Supplementary 8. Supplement 8: an all-inclusive list of nervous system development-, function-, and disease-related genes based on significant genes from our dataset as identified by IPA analysis of citalopram-treated human cells. Each function with p value and activation z-score, gene/molecule names, and total number of genes/molecules with predicted differential regulation from citalopram treatment are listed.

References

[1] A. B. Csoka and M. Szyf, "Epigenetic side-effects of common pharmaceuticals: a potential new field in medicine and pharmacology," *Medical Hypotheses*, vol. 73, no. 5, pp. 770–780, 2009.

[2] J. Lotsch, G. Schneider, D. Reker et al., "Common non-epigenetic drugs as epigenetic modulators," *Trends in Molecular Medicine*, vol. 19, no. 12, pp. 742–753, 2013.

[3] M. G. Melka, B. I. Laufer, P. McDonald et al., "The effects of olanzapine on genome-wide DNA methylation in the hippocampus and cerebellum," *Clinical Epigenetics*, vol. 6, no. 1, p. 1, 2014.

[4] N. Zimmermann, J. Zschocke, T. Perisic, S. Yu, F. Holsboer, and T. Rein, "Antidepressants inhibit DNA methyltransferase

1 through reducing G9a levels," *The Biochemical Journal*, vol. 448, no. 1, pp. 93–102, 2012.

[5] N. G. Parker and C. S. Brown, "Citalopram in the treatment of depression," *The Annals of Pharmacotherapy*, vol. 34, no. 6, pp. 761–771, 2000.

[6] J. M. Ferguson, "SSRI antidepressant medications: adverse effects and tolerability," *The Primary Care Companion to The Journal of Clinical Psychiatry*, vol. 3, no. 1, pp. 22–27, 2001.

[7] Z. Lin, J. J. Canales, T. Bjorgvinsson et al., "Chapter 1—monoamine transporters: vulnerable and vital doorkeepers," *Progress in Molecular Biology and Translational Science*, vol. 98, pp. 1–46, 2011.

[8] L. Culpepper, "Escitalopram: a new SSRI for the treatment of depression in primary care," *The Primary Care Companion to The Journal of Clinical Psychiatry*, vol. 4, no. 6, pp. 209–214, 2002.

[9] J. F. Neumaier, D. C. Root, and M. W. Hamblin, "Chronic fluoxetine reduces serotonin transporter mRNA and 5-HT_{1B} mRNA in a sequential manner in the rat dorsal raphe nucleus," *Neuropsychopharmacology*, vol. 15, no. 5, pp. 515–522, 1996.

[10] A. Csoka, A. Bahrick, and O. P. Mehtonen, "Persistent sexual dysfunction after discontinuation of selective serotonin reuptake inhibitors," *The Journal of Sexual Medicine*, vol. 5, no. 1, pp. 227–233, 2008.

[11] P. Prasad, S. Ogawa, and I. S. Parhar, "Serotonin reuptake inhibitor citalopram inhibits GnRH synthesis and spermatogenesis in the male zebrafish," *Biology of Reproduction*, vol. 93, no. 4, p. 102, 2015.

[12] A. H. Clayton, H. A. Croft, and L. Handiwala, "Antidepressants and sexual dysfunction: mechanisms and clinical implications," *Postgraduate Medicine*, vol. 126, no. 2, pp. 91–99, 2014.

[13] J. Ben-Sheetrit, D. Aizenberg, A. B. Csoka, A. Weizman, and H. Hermesh, "Post-SSRI sexual dysfunction: clinical characterization and preliminary assessment of contributory factors and dose-response relationship," *Journal of Clinical Psychopharmacology*, vol. 35, no. 3, pp. 273–278, 2015.

[14] C. Hogan, J. Le Noury, D. Healy, and D. Mangin, "One hundred and twenty cases of enduring sexual dysfunction following treatment," *The International Journal of Risk & Safety in Medicine*, vol. 26, no. 2, pp. 109–116, 2014.

[15] R. R. Kanherkar, N. Bhatia-Dey, and A. B. Csoka, "Epigenetics across the human lifespan," *Frontiers in Cell and Developmental Biology*, vol. 2, p. 49, 2014.

[16] G. Shaw, S. Morse, M. Ararat, and F. L. Graham, "Preferential transformation of human neuronal cells by human adenoviruses and the origin of HEK 293 cells," *The FASEB Journal*, vol. 16, no. 8, pp. 869–871, 2002.

[17] K. D. Siegmund, "Statistical approaches for the analysis of DNA methylation microarray data," *Human Genetics*, vol. 129, no. 6, pp. 585–595, 2011.

[18] Y. Benjamini and Y. Hochberg, "Controlling the false discovery rate: a practical and powerful approach to multiple testing," *Journal of the Royal Statistical Society, Series B (Methodological)*, vol. 57, pp. 289–300, 1995.

[19] T. R. de Jong, J. G. Veening, B. Olivier, and M. D. Waldinger, "Oxytocin involvement in SSRI-induced delayed ejaculation: a review of animal studies," *The Journal of Sexual Medicine*, vol. 4, no. 1, pp. 14–28, 2007.

[20] E. Dremencov, M. El Mansari, and P. Blier, "Effects of sustained serotonin reuptake inhibition on the firing of dopamine neurons in the rat ventral tegmental area," *Journal of Psychiatry & Neuroscience*, vol. 34, no. 3, pp. 223–229, 2009.

[21] J. A. Mennigen, W. E. Lado, J. M. Zamora et al., "Waterborne fluoxetine disrupts the reproductive axis in sexually mature male goldfish, *Carassius auratus*," *Aquatic Toxicology*, vol. 100, no. 4, pp. 354–364, 2010.

[22] A. Pop, D. I. Lupu, J. Cherfan, B. Kiss, and F. Loghin, "Estrogenic/antiestrogenic activity of selected selective serotonin reuptake inhibitors," *Clujul Medical*, vol. 88, no. 3, pp. 381–385, 2015.

[23] A. Lister, C. Regan, J. Van Zwol, and G. Van Der Kraak, "Inhibition of egg production in zebrafish by fluoxetine and municipal effluents: a mechanistic evaluation," *Aquatic Toxicology*, vol. 95, no. 4, pp. 320–329, 2009.

[24] M. R. Safarinejad, "Evaluation of endocrine profile and hypothalamic-pituitary-testis axis in selective serotonin reuptake inhibitor-induced male sexual dysfunction," *Journal of Clinical Psychopharmacology*, vol. 28, no. 4, pp. 418–423, 2008.

[25] D. Prabhakar and R. Balon, "How do SSRIs cause sexual dysfunction? Understanding key mechanisms can help improve patient adherence, prognosis," *Current Psychiatry*, vol. 9, pp. 30–34, 2010.

[26] W. Juan and W. Hong, "Targeting the Hippo signaling pathway for tissue regeneration and cancer therapy," *Genes*, vol. 7, no. 9, 2016.

[27] S. Lim and P. Kaldis, "Cdks, cyclins and CKIs: roles beyond cell cycle regulation," *Development*, vol. 140, no. 15, pp. 3079–3093, 2013.

[28] M. Keniry and R. Parsons, "The role of PTEN signaling perturbations in cancer and in targeted therapy," *Oncogene*, vol. 27, no. 41, pp. 5477–5485, 2008.

[29] M. Cotterchio, N. Kreiger, G. Darlington, and A. Steingart, "Antidepressant medication use and breast cancer risk," *American Journal of Epidemiology*, vol. 151, no. 10, pp. 951–957, 2000.

[30] M. C. Casimiro, M. Crosariol, E. Loro, Z. Li, and R. G. Pestell, "Cyclins and cell cycle control in cancer and disease," *Genes & Cancer*, vol. 3, no. 11-12, pp. 649–657, 2012.

[31] K. Barnard, R. C. Peveler, and R. I. G. Holt, "Antidepressant medication as a risk factor for type 2 diabetes and impaired glucose regulation: systematic review," *Diabetes Care*, vol. 36, no. 10, pp. 3337–3345, 2013.

[32] A. Menke and E. B. Binder, "Epigenetic alterations in depression and antidepressant treatment," *Dialogues in Clinical Neuroscience*, vol. 16, no. 3, pp. 395–404, 2014.

[33] V. Vialou, J. Feng, A. J. Robison, and E. J. Nestler, "Epigenetic mechanisms of depression and antidepressant action," *Annual Review of Pharmacology and Toxicology*, vol. 53, no. 1, pp. 59–87, 2013.

[34] L. Booij, M. Szyf, A. Carballedo et al., "DNA methylation of the serotonin transporter gene in peripheral cells and stress-related changes in hippocampal volume: a study in depressed patients and healthy controls," *PLoS One*, vol. 10, no. 3, article e0119061, 2015.

[35] M. Tochigi, K. Iwamoto, M. Bundo, T. Sasaki, N. Kato, and T. Kato, "Gene expression profiling of major depression and suicide in the prefrontal cortex of postmortem brains," *Neuroscience Research*, vol. 60, no. 2, pp. 184–191, 2008.

[36] H. J. Kang, D. H. Adams, A. Simen et al., "Gene expression profiling in postmortem prefrontal cortex of major depressive

disorder," *The Journal of Neuroscience*, vol. 27, no. 48, pp. 13329–13340, 2007.

[37] R. Rupprecht, V. Papadopoulos, G. Rammes et al., "Translocator protein (18 kDa) (TSPO) as a therapeutic target for neurological and psychiatric disorders," *Nature Reviews Drug Discovery*, vol. 9, no. 12, pp. 971–988, 2010.

[38] P. A. Melas, M. Rogdaki, A. Lennartsson et al., "Antidepressant treatment is associated with epigenetic alterations in the promoter of P11 in a genetic model of depression," *International Journal of Neuropsychopharmacology*, vol. 15, no. 5, pp. 669–679, 2012.

[39] J. L. Warner-Schmidt, K. E. Vanover, E. Y. Chen, J. J. Marshall, and P. Greengard, "Antidepressant effects of selective serotonin reuptake inhibitors (SSRIs) are attenuated by antiinflammatory drugs in mice and humans," *Proceedings of the National Academy of Sciences of the United States of America*, vol. 108, no. 22, pp. 9262–9267, 2011.

[40] N. Weder, H. Zhang, K. Jensen et al., "Child abuse, depression, and methylation in genes involved with stress, neural plasticity, and brain circuitry," *Journal of the American Academy of Child and Adolescent Psychiatry*, vol. 53, no. 4, pp. 417–424.e5, 2014.

[41] F. Artigas, "Future directions for serotonin and antidepressants," *ACS Chemical Neuroscience*, vol. 4, no. 1, pp. 5–8, 2013.

[42] M. R. Sowers, A. L. Wilson, C. A. Karvonen-Gutierrez, and S. R. Kardia, "Sex steroid hormone pathway genes and health-related measures in women of 4 races/ethnicities: the Study of Women's Health Across the Nation (SWAN)," *The American Journal of Medicine*, vol. 119, no. 9, Supplement 1, pp. S103–S110, 2006.

[43] C. L. Raison, L. Capuron, and A. H. Miller, "Cytokines sing the blues: inflammation and the pathogenesis of depression," *Trends in Immunology*, vol. 27, no. 1, pp. 24–31, 2006.

[44] N. Vogelzangs, H. E. Duivis, A. T. F. Beekman et al., "Association of depressive disorders, depression characteristics and antidepressant medication with inflammation," *Translational Psychiatry*, vol. 2, no. 2, article e79, 2012.

[45] A. H. Miller and C. L. Raison, "The role of inflammation in depression: from evolutionary imperative to modern treatment target," *Nature Reviews Immunology*, vol. 16, no. 1, pp. 22–34, 2016.

[46] A. Cattaneo, F. Macchi, G. Plazzotta et al., "Inflammation and neuronal plasticity: a link between childhood trauma and depression pathogenesis," *Frontiers in Cellular Neuroscience*, vol. 9, p. 40, 2015.

[47] T. Richter, Z. Paluch, and S. Alusik, "The non-antidepressant effects of citalopram: a clinician's perspective," *Neuro Endocrinology Letters*, vol. 35, no. 1, pp. 7–12, 2014.

[48] T. Esaki, M. Cook, K. Shimoji, D. L. Murphy, L. Sokoloff, and A. Holmes, "Developmental disruption of serotonin transporter function impairs cerebral responses to whisker stimulation in mice," *Proceedings of the National Academy of Sciences of the United States of America*, vol. 102, no. 15, pp. 5582–5587, 2005.

[49] A. L. Non, A. M. Binder, L. D. Kubzansky, and K. B. Michels, "Genome-wide DNA methylation in neonates exposed to maternal depression, anxiety, or SSRI medication during pregnancy," *Epigenetics*, vol. 9, no. 7, pp. 964–972, 2014.

[50] S. Bhattacharya, Q. Zhang, and M. E. Andersen, "A deterministic map of Waddington's epigenetic landscape for cell fate specification," *BMC Systems Biology*, vol. 5, no. 1, p. 85, 2011.

[51] J. M. W. Slack, "Conrad Hal Waddington: the last renaissance biologist?," *Nature Reviews Genetics*, vol. 3, no. 11, pp. 889–895, 2002.

[52] B. D. MacArthur, A. Ma'ayan, and I. R. Lemischka, "Systems biology of stem cell fate and cellular reprogramming," *Nature Reviews Molecular Cell Biology*, vol. 10, no. 10, pp. 672–681, 2009.

[53] P. Hou, Y. Li, X. Zhang et al., "Pluripotent stem cells induced from mouse somatic cells by small-molecule compounds," *Science*, vol. 341, no. 6146, pp. 651–654, 2013.

Effect of Gegen Qinlian Decoction on Cardiac Gene Expression in Diabetic Mice

Jing Han,[1] Zhenglin Wang,[2] Wei Xing,[2] Yueying Yuan,[1] Yi Zhang,[3] Tiantian Lv,[2] Hongliang Wang,[2] Yonggang Liu,[4] and Yan Wu[1]

[1]*Institute of Chinese Medicine, Beijing University of Chinese Medicine, Beijing 100029, China*
[2]*College of Basic Medicine, Key Laboratory of Ministry of Education (Syndromes and Formulas), Key Laboratory of Beijing (Syndromes and Formulas), Beijing University of Chinese Medicine, Beijing 100029, China*
[3]*Modern Research Center for Traditional Chinese Medicine, Beijing University of Chinese Medicine, Beijing 100029, China*
[4]*College of Chinese Medicine, Beijing University of Chinese Medicine, Beijing 100029, China*

Correspondence should be addressed to Yan Wu; nayattmm@vip.sina.com

Academic Editor: Qin Feng

The aim of this research is to investigate the therapeutic effect of GGQL decoction on cardiac dysfunction and elucidate the pharmacological mechanisms. db/db mice were divided into DB group or GGQL group, and WT mice were used as control. All mice were accessed by echocardiography. And the total RNA of LV tissue samples was sequenced, then differential expression genes were analyzed. The RNA-seq results were validated by the results of RT-qPCR of 4 genes identified as differentially expressed. The content of pyruvate and ceramide in myocardial tissue was also measured. The results showed that GGQL decoction could significantly improve the diastolic dysfunction, increase the content of pyruvate, and had the trend to reduce the ceramide content. The results of RNA-seq showed that 2958 genes were differentially expressed when comparing the DB group with the WT group. Among them, compared with the DB group, 26 genes were differentially regulated in the GGQL group. The expression results of 4 genes were consistent with the RNA-seq results. Our study reveals that GGQL decoction has a therapeutic effect on diastolic dysfunction of the left ventricular and the effect may be related to its role in promoting myocardial glycolysis and decreasing the content of ceramide.

1. Introduction

There is a dramatically increasing epidemic of DM patients. The prevalence of DM is 4% in 1995, and this number is anticipated to reach 5.4% in 2025, amounting to 300 million DM patients [1]. Diabetic cardiomyopathy (DCM) is a major complication of diabetes, afflicting 12% of patients. The prevalence rate will reach 22% in people aged > 64 years old [2].

A very significant aspect in early period of DCM is left ventricular (LV) dysfunction, especially diastolic dysfunction. And it is characterized by LV hypertrophy and increased cardiac fibrosis [3].

Despite of intensive glycemic, lipidemic control and neurohormonal antagonists, the progress of DCM in diabetic patients has not impeded. Even worse, cardiovascular mortality has increased due to hypoglycemia [4]. In brief, the therapeutic rules for DCM come from the treatment of heart failure and DM. No specific medications for this disease have been put to clinical use. Therefore, more studies are required to explore new agents for this complex syndrome.

Gegen Qinlian (GGQL) decoction, which is composed of Radix Puerariae (ge gen), Radix Scutellariae (huang qin), *Coptis chinensis* Franch (huang lian), and Radix Glycyrrhizae Praeparata (zhi gan cao), has been addressed for its medicinal effect against DM for almost ten years [5]. It has been reported that GGQL decoction has the ability to improve hyperglycemia and hyperlipidemia [6, 7]. In addition, GGQL decoction has a positive inotropic, negative frequency effect on the isolated perfused rat heart [8]. Furthermore, R *Scutellariae baicalensis* reduced myocardial infarct size in myocardial ischemia-reperfusion injured rats [9]. Berberine, a compound from *C. chinensis,* has negative chronotropic, positive inotropic, antiarrhythmic, and vasodilator properties [10]. Thereafter, we speculate that GGQL decoction will have protective effect against DCM.

In recent years, transcriptomics has been used to determine the mechanism of traditional Chinese medicine [11]. Thus, the purpose of this study is to evaluate the effect of GGQL decoction on DCM and elucidate the pharmacological mechanisms by transcriptomics. First, we assessed the efficacy of GGQL decoction against injured cardiac functions using the DM mouse model. Then, we evaluate the influence of GGQL decoction on changes of transcriptomics. Some of the differently expressed genes were confirmed by RT-qPCR. Our extensive studies will determine the potential targets of GGQL decoction and provide new treatment strategies for DCM.

2. Materials and Methods

2.1. Preparation of Gegen Qinlian Decoction. To prepare the aqueous extract of Gegen Qinlian decoction, firstly, Radix Puerariae, Radix Scutellariae, *Coptis chinensis* Franch, and Radix Glycyrrhizae Praeparata, at the rate of 5 : 3 : 3 : 2, were soaked in 10 times of distilled water (v/w) for 30 minutes and then boiled for 1 h. The decoction was filtered and collected. Secondly, the residue was added into 10 times of distilled water (v/w) and boiled for 1 h, and the hot decoction was filtered. Thirdly, the filtrate was mixed and concentrated to the aqueous extract.

2.2. Animals and Grouping. Studies were performed following the Guide for the Care and Use of Laboratory Animals published by the National Institutes of Health and with the permission of the Care Committee of Beijing University of Chinese Medicine. Male C57BL/KsJ db/db mice and their similar genetic background age-matched C57BL/KsJ wild-type (WT) mice were obtained from the Nanjing Biomedical Research Institute of Nanjing University (Jiangsu, China). Mice were kept in the animal house with a 12 : 12 h light-dark cycle and controlled temperature of 22–25°C. At 8 weeks of age, db/db mice were randomly divided into the DB group or the GGQL group, and the age-matched WT mice were used as control. The aqueous extract of GGQL decoction was dissolved in distilled water, and the GGQL group was intragastrically administered at a dosage of 23.4 g crude drugs/kg/d for 8 weeks. The DB group and WT group were treated with an equal volume of distilled water. After 8 weeks' administration, all the mice were weighted and the blood glucose was measured. Then, all the mice were sacrificed, and the hearts were harvested. And the tissue of the left ventricle was collected and stored in the liquid nitrogen for subsequent mRNA isolation.

2.3. Echocardiographic Assessment. At the 8th week of administration, the mice were accessed by echocardiography. All the mice were taped on the heated procedure board and anesthetized with 1.5% isoflurane in oxygen. In the apical four-chamber view, the peak velocity of early diastolic mitral inflow velocity (*E*), the peak value of late diastolic mitral inflow velocity (*A*), and the E-to-A ratio (*E*/*A*) were measured by pulsed Doppler mode, using Vevo 2100 Imaging System (VisualSonics, Canada) with a 30 MHz high-frequency transducer. In tissue Doppler mode, the early diastolic velocity (*E*') and the late diastolic velocity (*A*') were measured, and ratio of early to late diastolic velocities (*E*'/*A*') was calculated. Data analysis was performed with the use of Vevo 2100 Analytic Software (VisualSonics, Canada).

Table 1: List of primers used in RT-qPCR.

Gene	Primer sequence (5′–3′)
Pgam1	F:AATTCAGGGAGGAACTGTGCT R:GGACAGGTTCCAGGGACAAAA
Acer2	F:GCTCTGTGAAAATACTGCCACC R:CAGTGTTGGCTCTGGGTAGG
Slc38a2	F:CAAACCTCCTGTGAGGGAGC R:GAATTGAGGTGACGGGACAGT
Ppp1r3c	F:AGAGCTCTTTCAGTGCCTCCA R:TGATGGCCCTCCTGATGATTTC

F: forward; R: reverse.

2.4. RNA Isolation and Quality Control. Total RNA was isolated from 15 LV tissue samples (5 samples each group) using Trizol reagent (Invitrogen, CA). Agarose gel electrophoresis was used to analyze RNA integrity and the presence of DNA contamination of samples. RNA purity (ratio of OD260/280 and OD260/230) was measured by NanoPhotometer spectrophotometer. RNA concentration was accurately quantified by Qubit 2.0 Fluorometer. And RNA integrity was accurately detected by Agilent 2100 bioanalyzer. Then, poly-A tail mRNA was enriched with Oligo(dT) beads, and the enriched mRNA was randomly fragmented by divalent cations in NEB Fragmentation Buffer. The fragmented mRNA was reverse transcripted, and the cDNA was synthesized. After end repair and addition of A tail and adaptor to the purified cDNA, cDNAs were selected about 200 bp with AMPure XP beads, PCR amplification was performed, and then the PCR product was purified. The insert size of library was detected by Agilent 2100 bioanalyzer, and the effective concentration of the library was accurately quantified.

2.5. RNA Sequencing. The mRNA was sequenced with Illumina HiSeq2500 platform. The read numbers mapped of each gene were counted with HTSeq v0.6.1. RPKM (reads per kilobase of exon model per million mapped reads) of each gene was calculated on the basis of the gene length and reads count mapped to the gene.

Gene Ontology (GO) and Kyoto Encyclopedia of Genes and Genomes (KEGG) enrichment was applied by the GOseq R package and KOBAS. The results of clustered analysis of differentially expressed genes were showed with the heatmap.

2.6. RT-qPCR Validation of RNA-seq. Four genes identified as differentially expressed were randomly selected to validate

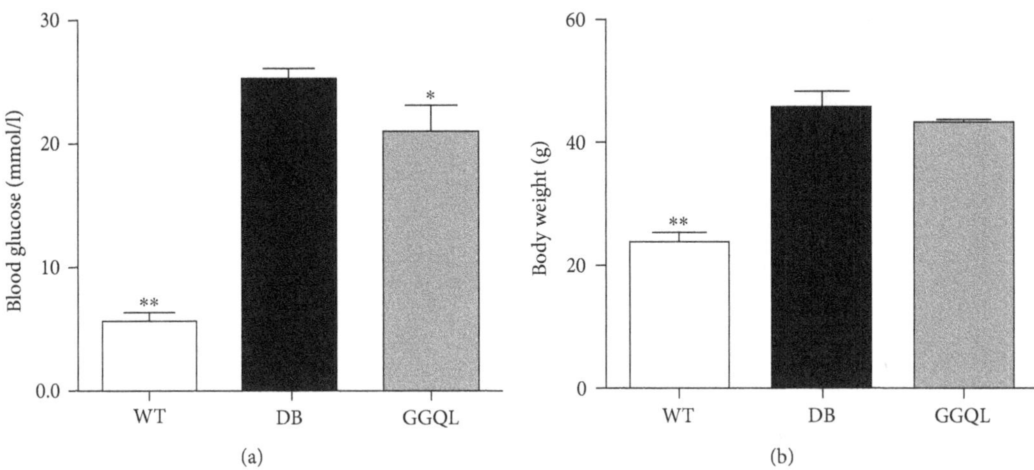

FIGURE 1: The results of blood glucose and bodyweight. (a) Blood glucose and (b) bodyweight in each group ($n = 8$). The DB group was the reference group to calculate P values, $^{*}P < 0.05$ and $^{**}P < 0.01$.

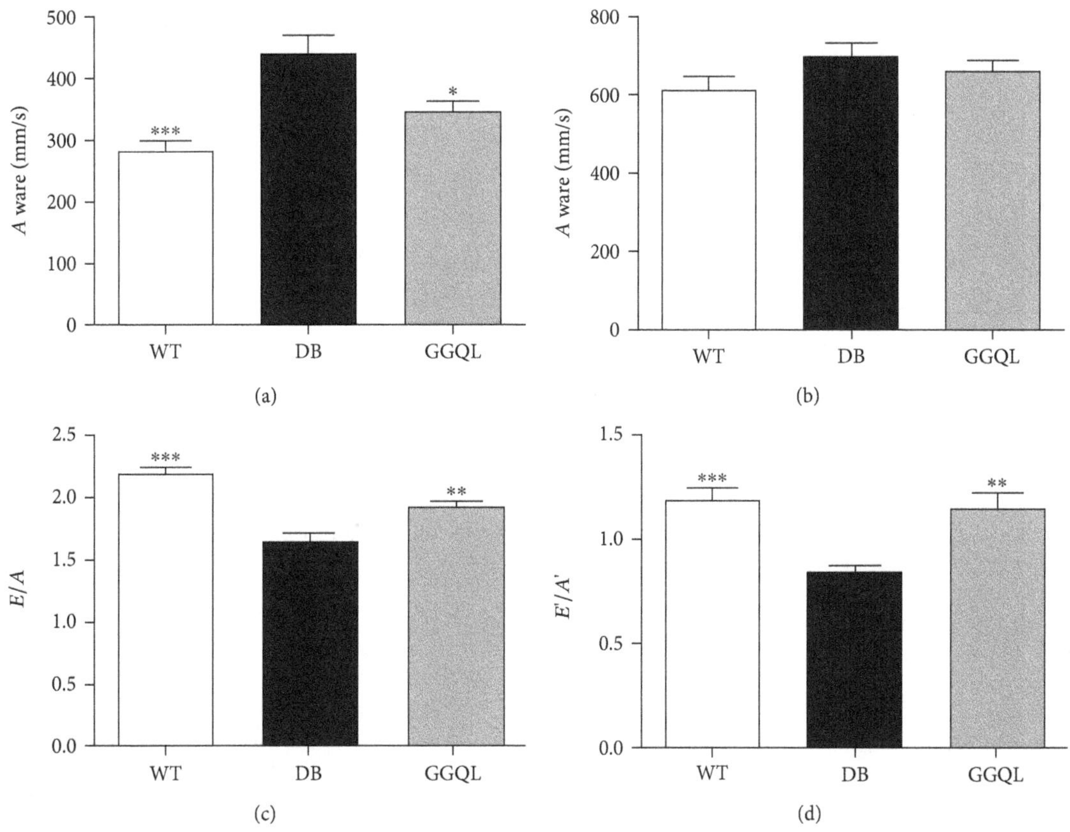

FIGURE 2: The results of cardiac function detected by echocardiograph. (a) E wave, (b) A wave, (c) E/A ratio, and (d) E'/A' ratio in each group ($n = 9$). The DB group was the reference group to calculate P values, $^{*}P < 0.05$, $^{**}P < 0.01$ and $^{***}P < 0.001$.

the results of RNA-seq by RT-qPCR. 1 μg of total RNA was reverse transcribed to cDNA in 20 μl reaction. Custom gene-specific primers for qRT-PCR were designed by Primer-BLAST, and the sequences of primers are listed in Table 1. Gene expression was measured in triplicates using the ABI StepOnePlus® instrument (ABI, USA) and SYBR GreenImaster mix (Roche, USA). The protocol of the reactions was 2 min at 50°C and 10 min at 95°C, followed by 40 cycles of denaturation at 95°C for 15 s and annealing at the corresponding melting temperatures for 30 s, and the melt curve was detected from 60°C to 95°C (0.5°C increments every 5 s). All the expression levels of mRNA were normalized with GAPDH as a housekeeping gene.

2.7. Measurement of Ceramide in Myocardial Tissue. On the day of extraction, the myocardial tissues were weighed

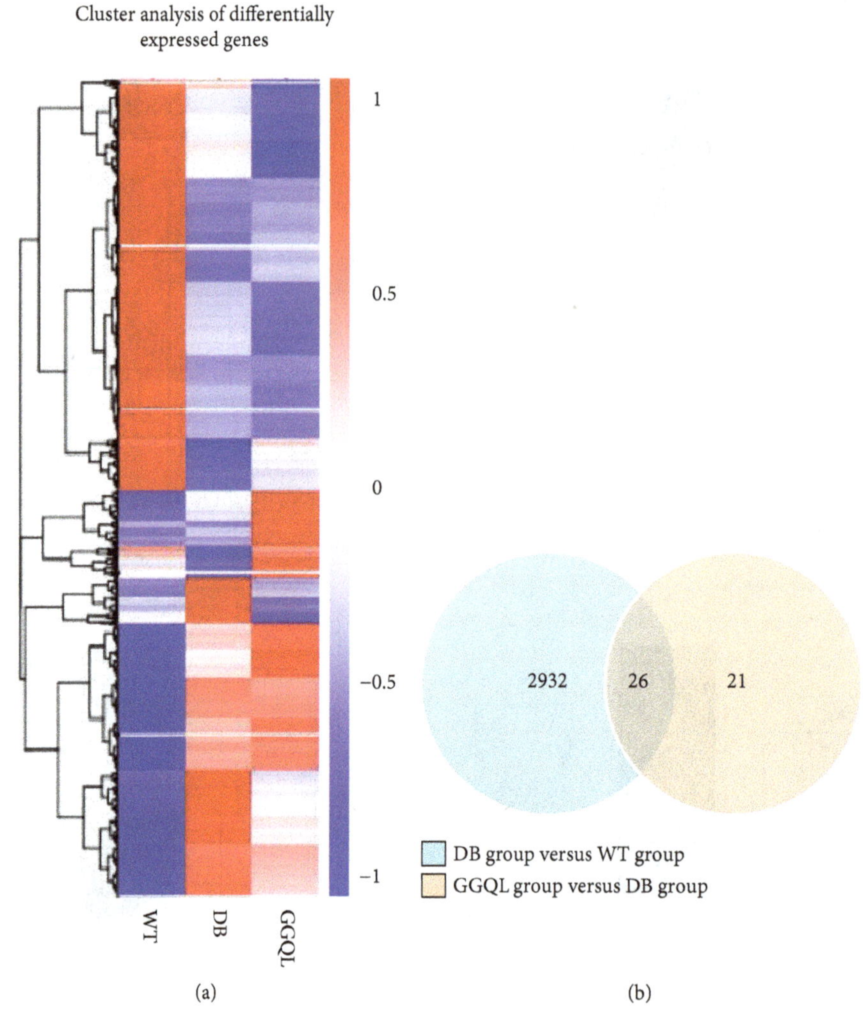

FIGURE 3: (a) Heatmap for cluster analysis of differentially expressed genes. (b) Venn diagram of differentially expressed genes.

with an analytical balance. The tissues were grinded, and cold 2: 1 chloroform: methanol was added. After vortexing, the samples were stored overnight at 4°C. After centrifugation, the organic phase was extracted and dried under nitrogen. The samples were reconstituted for UPLC/MS/MS analysis. The content of ceramide was measured with a UPLC/MS/MS. Ceramide were analyzed using an Agilent TQ-S triple quadrupole mass spectrometer with positive ion electrospray ionization (ESI) source by Daughters Scan. The chromatographic separation was performed by a Waters UPLC Acquity. A reverse-phase Acquity BEH C18 column (2.1 × 100 mm, 1.7 μm) was used as the analytical column. Chromatographic separation was carried out in binary gradient with 2 mM ammonium formate and 0.2% formic acid in water as solvent A and 1 mM ammonium formate and 0.2% formic acid in methanol as solvent B, and the flow rate was 0.4 ml/min. The column temperature was 40°C, and the injection volume was 2 μl.

2.8. Measurement of Pyruvate in Myocardial Tissue. The myocardial tissues were weighed about 20 mg with an analytical balance. The protein concentration and pyruvate content of each tissue homogenate were measured according to the test kits.

2.9. Statistical Analysis. Differential expression analysis between each two groups was measured by using the DESeq R package (1.10.1), which provides statistical routines to determine differential expression in digital gene expression data using a model based on the negative binomial distribution. The P values were adjusted with the Benjamini and Hochberg's approach for controlling the false discovery rate (FDR). Genes with an adjusted P value < 0.05 were considered to be differentially expressed.

All the data were presented as mean ± SEM. Statistical analysis was performed with one-way analysis of variance (ANOVA) by SPSS program package (version 17.0). It is considered to be statistical significant when P value was below 0.05.

3. Results

3.1. Blood Glucose and Bodyweight. The results showed that the blood glucose and bodyweight of db/db mice were significantly higher than those of WT mice. Gegen Qinlian

TABLE 2: Genes significantly regulated in the GGQL group versus the DB group and also significantly regulated in the DB group versus the WT group.

Gene name	Description	Adjusted P value (DB/WT)	Adjusted P value (GGQL/DB)
Pgam1	Phosphoglycerate mutase 1	<0.0001	0.0038
Igf1	Insulin-like growth factor 1	0.0060	0.0029
Cpeb4	Cytoplasmic polyadenylation element binding protein 4	0.0141	0.0283
Slc38a2	Solute carrier family 38, member 2	0.0000	<0.0001
Vegfa	Vascular endothelial growth factor A	0.0001	0.0005
Banp	BTG3-associated nuclear protein	<0.0001	0.0100
Rbm38	RNA-binding motif protein 38	0.0078	0.0380
Rasgef1b	RasGEF domain family, member 1B	0.0003	0.0037
Anxa3	Annexin A3	<0.0001	0.0309
Cyp1a1	Cytochrome P450, family 1, subfamily a, polypeptide 1	<0.0001	0.0016
Trib1	Tribbles homolog 1 (Drosophila)	0.0004	0.0118
Nphp3	Nephronophthisis 3 (adolescent)	<0.0001	0.0275
Fhl3	Four and a half LIM domains 3	0.0036	0.0045
Serpine1	Serine (or cysteine) peptidase inhibitor, clade E, member 1	<0.0001	0.0016
Acer2	Alkaline ceramidase 2	<0.0001	0.0016
Arhgap32	Rho GTPase activating protein 32	<0.0001	<0.0001
Atf4	Activating transcription factor 4	0.0044	0.0016
Zc3h6	Zinc finger CCCH type containing 6	<0.0001	0.0136
Tet1	Tet methylcytosine dioxygenase 1	0.0006	0.0016
Klk1b26	Kallikrein 1-related petidase b26	0.0146	0.0189
Ppp1r3c	Protein phosphatase 1, regulatory (inhibitor) subunit 3C	0.0000	0.0029
Ttc30b	Tetratricopeptide repeat domain 30B	<0.0001	0.0054
Fign	Fidgetin	<0.0001	0.0195
AI593442	Expressed sequence AI593442	<0.0001	0.0350
5730480H06Rik	RIKEN cDNA 5730480H06 gene	0.0019	0.0029
Gm26703	Predicted gene, 26703	<0.0001	<0.0001

Genes with an adjusted P value < 0.05 were considered to be differentially expressed.

decoction could significantly reduce the blood glucose of the diabetic mice, but it had no significant effect on the bodyweight of the diabetic mice (Figure 1).

3.2. Effect of GGQL Decoction on Cardiac Function. The results of echocardiography showed that A wave of the DB group increased significantly, and E/A ratio and E'/A' ratio decreased markedly compared to the WT group. The results indicated that the diastolic function of the left ventricle was impaired. After 8 weeks of treatment, compared to the DB group, E/A ratio and E'/A' ratio were recovered, and A wave was also downregulated, suggesting that GGQL decoction could improve the diastolic dysfunction caused by diabetes mellitus. In three groups, there was no significant difference in E wave (Figure 2).

3.3. Gene Expression Levels and DEGs. 15 samples from the WT group, DB group, and GGQL group were detected for DEGs, with 5 samples in each group. The hierarchal clustering results showed that 2958 genes were detected to be differentially expressed when comparing the DB group with the WT group, including 1450 upregulated genes and 1508 downregulated genes. Compared to the DB group, 47 genes were differentially regulated in the GGQL group (Figure 3). Among them, 26 genes were also differentially regulated in the DB group versus the WT group (Table 2), and 21 genes expressed no difference in the DB group versus the WT group (Table 3). The genes differentially expressed were related to anion binding, oxidoreductase activity, calcium-dependent phospholipid binding, ceramide metabolic process, and others.

3.4. Validation of Differentially Expressed Genes. Four differentially expressed genes were selected to validate by RT-qPCR, including Acer2, Slc38a2, Ppp1r3c, and Pgam1. Acer2 (alkaline ceramidase 2) regulates the hydrolysis of ceramides. Slc38a2 (solute carrier family 38 member 2) is a sodium-dependent amino acid transporter. Ppp1r3c (protein phosphatase 1 regulatory subunit 3C) is a glycogen targeting subunit of PP1 and regulates glycogen metabolism. Pgam1 (phosphoglycerate mutase 1) is an enzyme which catalyzes the reaction of 3-phosphoglycerate (3-PGA) to 2-phosphoglycerate (2-PGA).

Table 3: Genes significantly regulated in the GGQL group versus the DB group but no differential expression in the DB group versus the WT group.

Gene name	Description	Adjusted P value (DB/WT)	Adjusted P value (GGQL/DB)
Gatsl3	GATS protein-like 3	0.4265	<0.0001
Abca8b	ATP-binding cassette, subfamily A (ABC1), member 8b	1	0.0195
Apex2	Apurinic/apyrimidinic endonuclease 2	0.8103	0.0077
Dnajb4	DnaJ (Hsp40) homolog, subfamily B, member 4	0.1556	0.0071
Vgll4	Vestigial-like 4 (Drosophila)	0.9649	0.0136
Klhl36	Kelch-like 36	0.1950	0.0136
Amotl2	Angiomotin-like 2	0.0637	0.0030
Cish	Cytokine inducible SH2-containing protein	0.4114	<0.0001
Myo6	Myosin VI	1	0.0483
Caskin2	CASK-interacting protein 2	0.8910	0.0158
Baz1a	Bromodomain adjacent to zinc finger domain 1A	0.2177	0.0020
Fbxw17	F-box and WD-40 domain protein 17	0.6014	0.0480
Ddx60	DEAD (Asp-Glu-Ala-Asp) box polypeptide 60	0.9151	0.0097
Gck	Glucokinase	0.1227	0.0112
Zfp507	Zinc finger protein 507	0.6355	0.0201
Zfp518a	Zinc finger protein 518A	1	0.0416
Dusp7	Dual specificity phosphatase 7	0.7537	0.0416
Ifnlr1	Interferon lambda receptor 1	0.3676	0.0275
Rnasel	Ribonuclease L (2′,5′-oligoisoadenylate synthetase-dependent)	1	0.0275
Zfp763	Zinc finger protein 763	0.2005	0.0063
RP23-402K24.5	RIKEN cDNA 6230415J03 gene	0.4670	0.0350

Genes with an adjusted P value < 0.05 were considered to be differentially expressed.

The results of RNA-seq showed that, compared with the WT group, the expressions of Acer2, Slc38a2, Ppp1r3c, and Pgam1 were depressed in LV tissue of the DB group mice. And the expressions of all the four genes were significantly upregulated in the GGQL group. The results of RT-qPCR were consistent with the results of RNA-seq, and the results of RNA-seq were validated (Figure 4).

3.5. Content of Ceramide in Myocardial Tissue. The content of ceramide (d18:1/24:1) in the myocardial tissue of the model group was markedly higher than that of the wild-type group ($P < 0.05$). In the model group, the content of ceramide (d18:1/24:0) and ceramide (d17:1/24:1) appeared an increasing trend compared to that of the wild-type group. The levels of ceramide (d18:1/24:1), ceramide (d18:1/24:0), and ceramide (d17:1/24:1) in the myocardial tissue of the GGQL group had a reduced trend compared to that of the model group (Figure 5).

3.6. Content of Pyruvate in Myocardial Tissue. In the model group, the pyruvate content of myocardial tissue homogenate decreased significantly than that of the wild-type group ($P < 0.05$). Compared to the model group, the content of pyruvate in the myocardial tissue homogenate of the GGQL group increased obviously ($P < 0.05$) (Figure 5).

4. Discussion

DM is not only manifested as long-term and chronic elevated hyperglycemia but also accompanied with various complications. Diabetic cardiomyopathy (DCM) is one of the major causes of death in T2DM patients. DCM is a type of primary cardiomyopathy, independent of macrovascular and coronary artery diseases. The pathophysiology of DCM includes cardiomyocyte hypertrophy, focal necrosis, extracellular matrix accumulation, and interstitial fibrosis [12]. Early manifestations usually expressed as diastolic dysfunction with major pathological changes of declining in myocardial compliance and blocked diastolic filling [13]. Systolic dysfunction is the major pathological change of DCM in the late period of the disease, and this disturbance is prone to congestive heart failure.

db/db mice are the classic models for the research of T2DM-induced DCM. A number of investigators confirmed the existence of DCM of db/db mice with evaluation of echocardiography [14, 15]. According to the literature, under the detection of echocardiography, the E/A ratio and E'/A' ratio of the db/db mice decreased significantly compared with WT mice [14, 16, 17], suggesting impaired cardiac diastolic function.

In our study, compared with the WT mice, the A wave increased, and the left ventricular E/A and E'/A' ratio of the

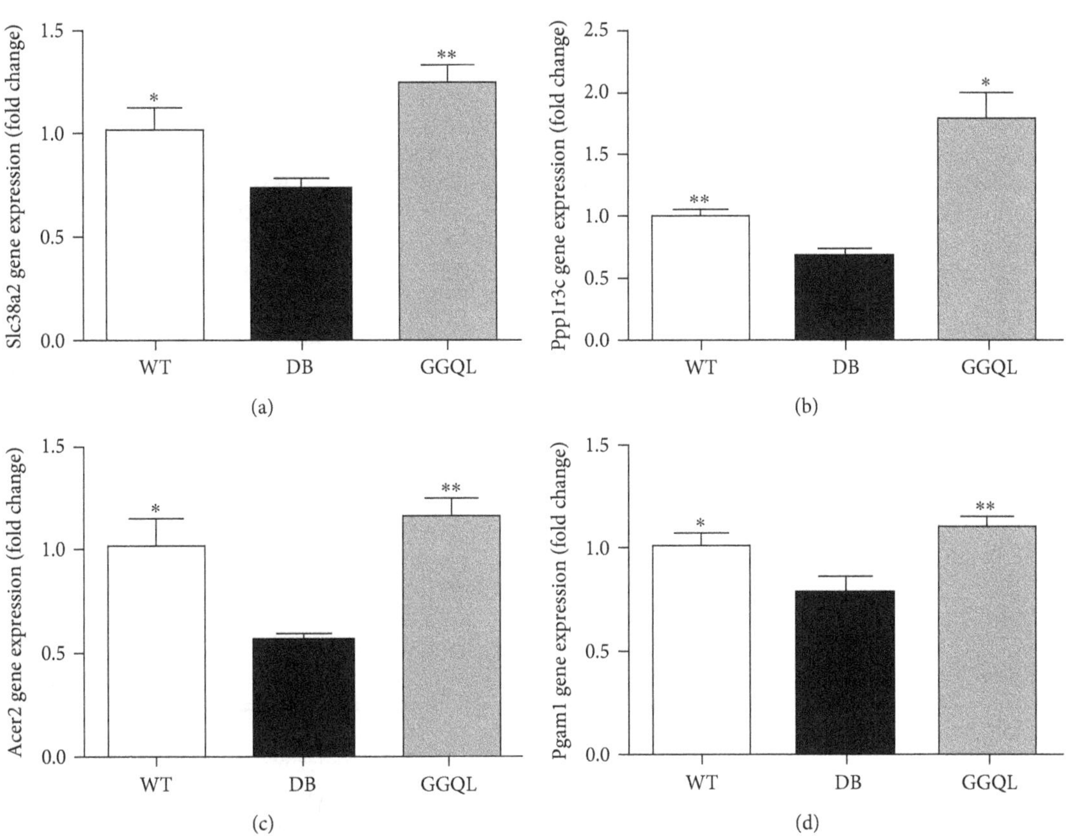

FIGURE 4: The results of RT-qPCR to validate differentially expressed genes. The mRNA expressions of (a) Slc38a2, (b) Ppp1r3c, (c) Acer2, and (d) Pgam1 in each group ($n = 4 - 5$). The DB group was the reference group to calculate P values, $^{*}P < 0.05$ and $^{**}P < 0.01$.

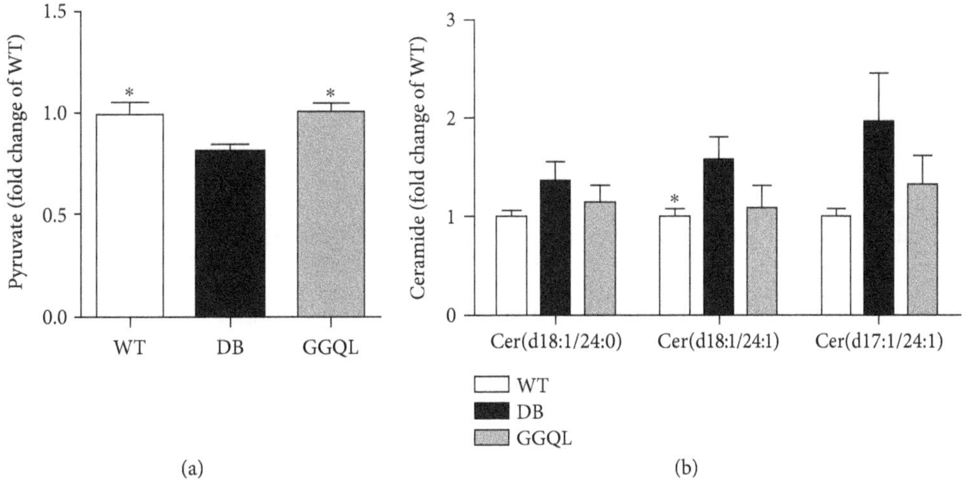

FIGURE 5: The results of pyruvate and ceramide content in the myocardial tissue. The content of (a) pyruvate and (b) ceramide in each group (n = 3–6). The DB group was the reference group to calculate P values, $^{*}P < 0.05$.

db/db mice markedly decreased, suggesting a diastolic dysfunction of the left ventricular; Gegen Qinlian decoction could significantly increase the left ventricle E/A and E'/A' ratio of the db/db mice and downregulate the A wave, suggesting that the drug can improve the function of the left ventricle.

Glycolysis is a process in which one molecule of glucose is converted to two molecules of pyruvate and produces two molecules of adenosine triphosphate (ATP), which provides energy for almost all the biological processes. Glycolysis is necessary for cardiac metabolism. Glucose utilization is impaired, meanwhile glycolysis and glucose oxidation are depressed in DCM. In diabetes, the capacity of cardiac glycolysis is chronically reduced [18]. The reduction of myocardial glycolysis may be associated with the development of heart failure and myocardial injury suffered by DM patients [19, 20].

Glycolysis consists of ten steps, and pyruvate is the product. Phosphoglycerate mutase-1 (Pgam1) is used to catalyze the interconversion of 3-phosphoglycerate (3-PGA) to 2-phosphoglycerate (2-PGA), which is the eighth step of glycolysis [21].

Ceramide is a main molecule of the sphingolipid metabolism, which mediates cell growth retardation and differentiation, and inhibits cell proliferation and promotes apoptosis [22–26]. The accumulation of tissue ceramides may be helpful to the development of insulin resistance [27]. Ceramide is an important mediator of myocardial lipid toxicity [28, 29]. Moreover, some studies showed that ceramide is related to cardiomyocyte apoptosis induced by ischemia/reperfusion injury [30, 31]. Alkaline ceramidase 2 (Acer2) is a type of ceramidases, which are key enzymes of the degradation of intracellular ceramide and hydrolyzed ceramide to sphingosine and then further phosphorylated to sphingosine-1-phosphate (S1P) by sphingosine kinase [32]. And the increased expression of Acer2 may cause the decrease in ceramide accumulation.

Previous research indicated that the cardiac glycolytic rate of db/db mice was reduced. Cardiac function, glycolysis, and glucose oxidation of isolated hearts of C57BL/KsJ (*db/db*) were damaged [19]. It is also reported that in the diabetic model animals, myocardial ceramide content was significantly increased accompanied with cardiac dysfunction [33, 34].

The results of this study showed that compared with the mice of the WT group, the mRNA content of Pgam1 and Acer2 in myocardium of the DB group mice decreased significantly. And the mRNA content of Pgam1 and Acer2 in myocardium of the GGQL group mice increased significantly, compared with the mice of the DB group. It indicated that the myocardial glycolysis and ceramide hydrolysis were depressed in the DB group mice, and potentially the treatment of Gegen Qinlian decoction protects the myocardium by promoting glycolysis and decreasing the content of ceramide. In this study, the results of the pyruvate content in the myocardial tissue suggested that the glycolysis decreased in the DB group compared to the WT group. And GGQL decoction has the effect to increase cardiac glycolysis in diabetic mice. The content of ceramide in the myocardial tissue of the DB group appeared an increasing trend compared to that of the wild-type group, and GGQL decoction trended to reduce the ceramide content in diabetic mice. The results were consistent with the results of RNA-seq and q-PCR.

Previous studies have shown that berberine has protective effect on the development of diabetic cardiomyopathy [35]. Liquiritigenin and liquiritin have a protective role in high fructose-induced myocardial fibrosis [36, 37]. Other investigators confirmed that various components of Gegen Qinlian decoction have moderating effect on glycolysis and ceramide levels. Berberine could significantly reduce serum levels of ceramide of nonalcoholic fatty liver disease patients [38] and stimulate the glycolysis in HepG2 hepatocytes and C2C12 myotubes [39]. Berberine compounds could upregulate the glycolysis which was depressed in hyperlipidemic rats [40]. Baicalin could decreased the blood glucose of diabetic rats by increasing the content of liver glycogen and promoting glycolysis [41]. It is concluded that the effect on improving diastolic function of the left ventricle and promoting the mRNA expression of Pgam1 and Acer2 of Gegen Qinlian decoction may be attributed to berberine, baicalin, liquiritigenin, and liquiritin.

In conclusion, the results of this study showed that Gegen Qinlian decoction has a therapeutic effect on diastolic dysfunction of the left ventricle in db/db mice, and the effect may be related to its role in promoting myocardial glycolysis and decreasing the content of ceramide.

Conflicts of Interest

The authors declare that they have no competing interests.

Authors' Contributions

Yan Wu conceived and designed the experiments. Jing Han, Zhenglin Wang, Wei Xing, Yueying Yuan, and Yonggang Liu performed the experiments. Tiantian Lv, Hongliang Wang, and Yi Zhang analyzed the data. Jing Han and Yan Wu contributed to the writing of the manuscript. All authors read and approved the final manuscript.

Acknowledgments

This work was supported by the National Nature Science Foundation of China (no. 81303083).

References

[1] W. Chen, Y. Xia, X. Zhao et al., "The critical role of Astragalus polysaccharides for the improvement of PPARα-mediated lipotoxicity in diabetic cardiomyopathy," *PLoS One*, vol. 7, no. 10, article e45541, 2012.

[2] K. Trachanas, S. Sideris, C. Aggeli et al., "Diabetic cardiomyopathy: from pathophysiology to treatment," *Hellenic Journal of Cardiology*, vol. 55, no. 5, pp. 411–421, 2014.

[3] K. Huynh, H. Kiriazis, L. J. E. Du XJ et al., "Coenzyme Q_{10} attenuates diastolic dysfunction, cardiomyocyte hypertrophy and cardiac fibrosis in the *db/db* mouse model of type 2 diabetes," *Diabetologia*, vol. 55, no. 5, pp. 1544–1553, 2012.

[4] J. Fuentes-Antras, B. Picatoste, E. Ramirez, J. Egido, J. Tunon, and O. Lorenzo, "Targeting metabolic disturbance in the diabetic heart," *Cardiovascular Diabetology*, vol. 14, no. 1, p. 17, 2015.

[5] Y. P. Zeng, Y. S. Huang, and Y. G. Hu, "Effect of Gegen Qinlian decoction combined with short-term intensive insulin treatment on patients with type 2 diabetes mellitus of dampness-heat syndrome," *Zhongguo Zhong Xi Yi Jie He Za Zhi*, vol. 26, no. 6, pp. 514–516, 2006, 520.

[6] Y. M. Li, X. M. Fan, Y. M. Wang, Q. L. Liang, and G. A. Luo, "Therapeutic effects of Gegen Qinlian decoction and its mechanism of action on type 2 diabetic rats," *Yao Xue Xue Bao*, vol. 48, no. 9, pp. 1415–1421, 2013.

[7] J. Li, T. Gao, Q. Song, H. Li, Y. Wang, and Q. Feng, "Study on mechanism of effect of Gegenqinlian decoction on lowering blood sugar and lipid of type 2 diabetic rats," *Hu Bei Zhong Yi Yao Da Xue Xue Bao*, vol. 17, pp. 7–9, 2015.

[8] Y. Tan, Y. Shi, L. Liu, and Y. Zhao, "Effects of Gegenqinlian decoction and effective components on the isolated perfused heart," *Journal of Harbin University of Commerce(Natural Sciences Edition)*, vol. 27, pp. 649–652, 2011.

[9] E. Chan, X. X. Liu, D. J. Guo et al., "Extract of *Scutellaria baicalensis* Georgi root exerts protection against myocardial ischemia-reperfusion injury in rats," *The American Journal of Chinese Medicine*, vol. 39, no. 4, pp. 693–704, 2011.

[10] C. W. Lau, X. Q. Yao, Z. Y. Chen, W. H. Ko, and Y. Huang, "Cardiovascular actions of berberine," *Cardiovascular Drug Reviews*, vol. 19, no. 3, pp. 234–244, 2001.

[11] Y. Wang, W. Lin, C. Li et al., "Multipronged therapeutic effects of Chinese herbal medicine Qishenyiqi in the treatment of acute myocardial infarction," *Frontiers in Pharmacology*, vol. 8, p. 98, 2017.

[12] R. B. Devereux, M. J. Roman, M. Paranicas et al., "Impact of diabetes on cardiac structure and function: the strong heart study," *Circulation*, vol. 101, no. 19, pp. 2271–2276, 2000.

[13] C. M. Schannwell, M. Schneppenheim, S. Perings, G. Plehn, and B. E. Strauer, "Left ventricular diastolic dysfunction as an early manifestation of diabetic cardiomyopathy," *Cardiology*, vol. 98, no. 1-2, pp. 33–39, 2002.

[14] L. M. Semeniuk, A. J. Kryski, and D. L. Severson, "Echocardiographic assessment of cardiac function in diabetic*db/db* and transgenic *db/db*-hGLUT4 mice," *American Journal of Physiology Heart and Circulatory Physiology*, vol. 283, no. 3, pp. H976–H982, 2002.

[15] K. Venardos, K. A. De Jong, M. Elkamie, T. Connor, and S. L. McGee, "The PKD inhibitor CID755673 enhances cardiac function in diabetic *db/db* mice," *PLoS One*, vol. 10, no. 3, article e0120934, 2015.

[16] M. van Bilsen, A. Daniels, O. Brouwers et al., "Hypertension is a conditional factor for the development of cardiac hypertrophy in type 2 diabetic mice," *PLoS One*, vol. 9, no. 1, article e85078, 2014.

[17] D. J. Stuckey, C. A. Carr, D. J. Tyler, E. Aasum, and K. Clarke, "Novel MRI method to detect altered left ventricular ejection and filling patterns in rodent models of disease," *Magnetic Resonance in Medicine*, vol. 60, no. 3, pp. 582–587, 2008.

[18] A. Avogaro, R. Nosadini, A. Doria et al., "Myocardial metabolism in insulin-deficient diabetic humans without coronary artery disease," *The American Journal of Physiology*, vol. 258, no. 4, Part 1, pp. E606–E618, 1990.

[19] D. D. Belke, T. S. Larsen, E. M. Gibbs, and D. L. Severson, "Altered metabolism causes cardiac dysfunction in perfused hearts from diabetic (*db/db*) mice," *American Journal of Physiology Endocrinology and Metabolism*, vol. 279, no. 5, pp. E1104–E1113, 2000.

[20] E. D. Abel, H. C. Kaulbach, R. Tian et al., "Cardiac hypertrophy with preserved contractile function after selective deletion of GLUT4 from the heart," *Journal of Clinical Investigation*, vol. 104, no. 12, pp. 1703–1714, 1999.

[21] L. A. Fothergill-Gilmore and H. C. Watson, "The phosphoglycerate mutases," *Advances in Enzymology and Related Areas of Molecular Biology*, vol. 62, pp. 227–313, 1989.

[22] Y. Mizutani, H. Sun, Y. Ohno et al., "Cooperative synthesis of ultra long-chain fatty acid and ceramide during keratinocyte differentiation," *PLoS One*, vol. 8, no. 6, article e67317, 2013.

[23] B. J. Pettus, C. E. Chalfant, and Y. A. Hannun, "Ceramide in apoptosis: an overview and current perspectives," *Biochimica et Biophysica Acta (BBA) - Molecular and Cell Biology of Lipids*, vol. 1585, no. 2-3, pp. 114–125, 2002.

[24] T. A. Taha, T. D. Mullen, and L. M. Obeid, "A house divided: ceramide, sphingosine, and sphingosine-1-phosphate in programmed cell death," *Biochimica et Biophysica Acta (BBA) - Biomembranes*, vol. 1758, no. 12, pp. 2027–2036, 2006.

[25] S. Furuya, J. Mitoma, A. Makino, and Y. Hirabayashi, "Ceramide and its interconvertible metabolite sphingosine function as indispensable lipid factors involved in survival and dendritic differentiation of cerebellar Purkinje cells," *Journal of Neurochemistry*, vol. 71, no. 1, pp. 366–377, 1998.

[26] Y. Uchida, A. D. Nardo, V. Collins, P. M. Elias, and W. M. Holleran, "*De novo* ceramide synthesis participates in the ultraviolet B irradiation-induced apoptosis in undifferentiated cultured human keratinocytes," *Journal of Investigative Dermatology*, vol. 120, no. 4, pp. 662–669, 2003.

[27] S. A. Summers, "Ceramides in insulin resistance and lipotoxicity," *Progress in Lipid Research*, vol. 45, no. 1, pp. 42–72, 2006.

[28] H. C. Chiu, A. Kovacs, D. A. Ford et al., "A novel mouse model of lipotoxic cardiomyopathy," *Journal of Clinical Investigation*, vol. 107, no. 7, pp. 813–822, 2001.

[29] D. L. Hickson-Bick, L. M. Buja, and J. B. McMillin, "Palmitate-mediated alterations in the fatty acid metabolism of rat neonatal cardiac myocytes," *Journal of Molecular and Cellular Cardiology*, vol. 32, no. 3, pp. 511–519, 2000.

[30] A. E. Bielawska, J. P. Shapiro, L. Jiang et al., "Ceramide is involved in triggering of cardiomyocyte apoptosis induced by ischemia and reperfusion," *American Journal of Pathology*, vol. 151, no. 5, pp. 1257–1263, 1997.

[31] G. A. Cordis, T. Yoshida, and D. K. Das, "HPTLC analysis of sphingomylein, ceramide and sphingosine in ischemic/reperfused rat heart," *Journal of Pharmaceutical and Biomedical Analysis*, vol. 16, no. 7, pp. 1189–1193, 1998.

[32] P. W. Wertz and D. T. Downing, "Ceramidase activity in porcine epidermis," *FEBS Letters*, vol. 268, no. 1, pp. 110–112, 1990.

[33] Y. T. Zhou, P. Grayburn, A. Karim et al., "Lipotoxic heart disease in obese rats: implications for human obesity," *Proceedings of the National Academy of Sciences of the United States of America*, vol. 97, no. 4, pp. 1784–1789, 2000.

[34] R. Basu, G. Y. Oudit, X. Wang et al., "Type 1 diabetic cardiomyopathy in the Akita (Ins2WT/C96Y) mouse model is characterized by lipotoxicity and diastolic dysfunction with preserved systolic function," *American Journal of Physiology. Heart and Circulatory Physiology*, vol. 297, no. 6, pp. H2096–H2108, 2009.

[35] W. Chang, M. Zhang, Z. Meng et al., "Berberine treatment prevents cardiac dysfunction and remodeling through activation of $5'$-adenosine monophosphate-activated protein kinase in type 2 diabetic rats and in palmitate-induced hypertrophic H9c2 cells," *European Journal of Pharmacology*, vol. 769, pp. 55–63, 2015.

[36] Y. Zhang, L. Zhang, Y. Zhang, J. J. Xu, L. L. Sun, and S. Z. Li, "The protective role of liquiritin in high fructose-induced myocardial fibrosis via inhibiting NF-κB and MAPK signaling pathway," *Biomedicine & Pharmacotherapy*, vol. 84, pp. 1337–1349, 2016.

[37] X. W. Xie, "Liquiritigenin attenuates cardiac injury induced by high fructose-feeding through fibrosis and inflammation suppression," *Biomedicine & Pharmacotherapy*, vol. 86, pp. 694–704, 2017.

[38] X. Chang, Z. Wang, J. Zhang et al., "Lipid profiling of the therapeutic effects of berberine in patients with nonalcoholic fatty liver disease," *Journal of Translational Medicine*, vol. 14, no. 1, p. 266, 2016.

[39] M. Xu, Y. Xiao, J. Yin et al., "Berberine promotes glucose consumption independently of AMP-activated protein kinase activation," *PLoS One*, vol. 9, no. 7, article e103702, 2014.

[40] M. Li, X. Shu, H. Xu et al., "Integrative analysis of metabolome and gut microbiota in diet-induced hyperlipidemic rats treated with berberine compounds," *Journal of Translational Medicine*, vol. 14, no. 1, p. 237, 2016.

[41] H. T. Li, X. D. Wu, A. K. Davey, and J. Wang, "Antihyperglycemic effects of baicalin on streptozotocin - nicotinamide induced diabetic rats," *Phytotherapy Research*, vol. 25, no. 2, pp. 189–194, 2011.

11

Topological Characterization of Human and Mouse m⁵C Epitranscriptome Revealed by Bisulfite Sequencing

Zhen Wei,[1,2] Subbarayalu Panneerdoss,[3,4] Santosh Timilsina,[3,4] Jingting Zhu,[1,5] Tabrez A. Mohammad,[3] Zhi-Liang Lu,[1,5] João Pedro de Magalhães,[2] Yidong Chen,[3,6] Rong Rong,[1,5] Yufei Huang,[6,7] Manjeet K. Rao,[3,4] and Jia Meng[1,5]

[1]*Department of Biological Sciences, Xi'an Jiaotong-Liverpool University, Suzhou, Jiangsu 215123, China*
[2]*Integrative Genomics of Ageing Group, Institute of Ageing and Chronic Disease, University of Liverpool, L7 8TX Liverpool, UK*
[3]*Greehey Children's Cancer Research Institute, University of Texas Health Science Center at San Antonio, San Antonio, TX 78229, USA*
[4]*Department of Cellular Structural Biology, University of Texas Health Science Center at San Antonio, San Antonio, TX 78229, USA*
[5]*Institute of Integrative Biology, University of Liverpool, L7 8TX Liverpool, UK*
[6]*Department of Epidemiology and Biostatistics, University of Texas Health Science Center at San Antonio, San Antonio, TX 78229, USA*
[7]*Department of Electrical and Computer Engineering, University of Texas at San Antonio, San Antonio, TX 78249, USA*

Correspondence should be addressed to Rong Rong; rong.rong@xjtlu.edu.cn, Yufei Huang; yhuang@utsa.edu, Manjeet K. Rao; raom@uthscsa.edu, and Jia Meng; jia.meng@xjtlu.edu.cn

Academic Editor: Yujing Li

Background. Compared with the well-studied 5-methylcytosine (m^5C) in DNA, the role and topology of epitranscriptome m^5C remain insufficiently characterized. *Results.* Through analyzing transcriptome-wide m^5C distribution in human and mouse, we show that the m^5C modification is significantly enriched at 5′ untranslated regions (5′UTRs) of mRNA in human and mouse. With a comparative analysis of the mRNA and DNA methylome, we demonstrate that, like DNA methylation, transcriptome m^5C methylation exhibits a strong clustering effect. Surprisingly, an inverse correlation between mRNA and DNA m^5C methylation is observed at CpG sites. Further analysis reveals that RNA m^5C methylation level is positively correlated with both RNA expression and RNA half-life. We also observed that the methylation level of mitochondrial RNAs is significantly higher than RNAs transcribed from the nuclear genome. *Conclusions.* This study provides an in-depth topological characterization of transcriptome-wide m^5C modification by associating RNA m^5C methylation patterns with transcriptional expression, DNA methylations, RNA stabilities, and mitochondrial genome.

1. Introduction

DNA methylation is a well-established and extensively studied epigenetic phenomenon [1–4]. In contrast, mRNA methylation is still relatively an uncharted territory [5]. Although the presence of the chemical modifications to tRNA has been established in the 1970s [6–8], little is known about the epigenetic modifications to mRNA and other noncoding RNAs. Even less was known about their abundance, role, and mode of regulation until recently when several studies showed that N^6-methyladenosine (m^6A) is the most abundant messenger RNA (mRNA) modification in eukaryotes [9], and suggested to regulate a number of biological processes including translation efficiency [10], circadian clock [11], microRNA processing [12], RNA-protein interaction [13], RNA stability [14], heat shock response [15], and differentiation [16].

Compared to m^6A, even little is known about the abundance and role of transcriptome 5-methylcytosine (m^5C) modification. Existing studies of m^5C in cellular RNA have been largely confined to rRNA and tRNA [17]. For example,

RNA m^5C modification in plant rRNA and tRNA is reported to be conserved [18] and is shown to affect the stability of synthetic RNA [19, 20]. In the mammalian system, cytosine-5 methylation in tRNA has been shown to regulate Mg^{2+} binding, anticodon stem-loop conformation, and secondary structure stabilization [21, 22]. In addition, m^5C in tRNAs is reported to regulate protein translation in stress response, tissue differentiation, and neurodevelopment disorders [23–29]. In rRNA, m^5C is shown to regulate the translation process [30]. A recent study also showed that hm^5C, the intermediate of RNA m^5C demethylation, is enriched in poly(A)-tailed RNA and the coding sequences of the mRNA transcript, and it is associated with brain development and the active transcription of mRNA [11].

A recent advancement of the RNA bisulfite-sequencing (BS-Seq) technique [31–34] has enabled the transcriptome-wide m^5C profiling at single-base resolution and confirmed its widespread existence in the human transcriptome [34, 35]. Intriguing differences with respect to the degree of transcriptome m^5C methylation, functional classification, and position bias were reported with this technique [36], and it was recently shown that transcriptome m^5C promotes mRNA export through methyltransferase NSUN2 and reader ALYREF [37].

It is observed that m^5C modification may account for 20% of the total internal methylations on poly(A) RNA in the BHK21 cell line [38, 39]. However, it is not clear whether the transcriptome m^5C modification is differentially enriched in different cell types, and the topological relationship between RNA methylation and DNA methylation under the same cell lines has not been investigated.

In this study, using the BS-Seq approach, we identified transcriptome-wide mRNA m^5C methylome in mouse and human cells. Our results revealed that transcriptome m^5C is enriched and conserved at the 5′UTRs of target transcripts in both human and mouse cells. Interestingly, under all the examined cell lines, we observed a negative correlation of the methylation patterns between RNA m^5C methylation and DNA m^5C under the CpG context, and the RNA m^5C methylations are enriched on mitochondrial transcriptome.

2. Material and Methods

2.1. Sample Preparation and RNA Bisulfite Sequencing. MCF10A normal mammary epithelial cells and MDA-MB-468 breast cancer cells were obtained from ATCC. MCF10A cells were cultured and maintained in DMEM/F12 (Life Technologies, USA) supplemented with 5% horse serum, EGF (20 ng/ml), hydrocortisone (0.5 μg/ml), insulin (10 μg/ml), and anti-anti (Life Technologies, USA). Likewise, MDA-MB-468 cells were cultured in RPMI (Life Technologies, USA) supplemented with 10% FBS and anti-anti (Life Technologies, USA). For BS-Seq, total RNA was isolated from MCF10A and MDA-MB-468 cells and enriched for poly(A)+ RNA using poly(A) selection kits. The purified RNA is subjected to sodium bisulfite treatment at 60 degrees for 8 hours. The bisulfite-treated RNA was then reverse transcribed and subjected to deep sequencing using the Illumina RNA-Seq protocol. The data has been deposited under Gene Expression Ominous (GEO) with Accession Number GSE84230. To replenish the transcriptome BS-Seq data of the aforementioned human samples (MCF10A and MDA-MB-468), additional datasets are obtained from public resources, including DNA BS-Seq data from MCF10A (GEO GSM659628) [40], transcriptome m^5C methylation data from mouse embryo stem cells (ESCs) and mouse whole brain profiled by RNA BS-Seq (GEO GSE83432) [36], and mouse ESC DNA methylation data (GSM1873374) [7, 41].

2.2. Quality Control and Alignment of BS-Seq Data. The FASTQ files from BS-Seq samples are trimmed with Trim Galore [42], it removes low-quality 3′ ends with a Phred score threshold of 20, and it can remove potential adaptor contamination. Then, the reads are aligned to the reference genomes of mouse and human (mm10 and hg19) with MeRanGs in MeRanTK [43]. The methylation is called using MeRanCall, and regions of the 5′ ends and 3′ ends of the reads are ignored based on the threshold cutoff suggested by the M-bias plot generated by MeRanGs. The minimum read coverage for the methylation report was set at 10, and the minimum read base quality (Phred score) for methylation call is filtered at 30. The maximum read duplication level is set at 10 to prevent the PCR artefacts; the minimum nonconversion rate to report is set at 0 to include the nonmethylated sites as background control for further analysis.

For DNA bisulfite samples, the trimmed reads are aligned using Bismark under the following alignment setting: –score_min L,0,-0.6. The SAM files are filtered by Samtools using -F 1540 and -q 30 to remove reads that are duplicated and quality scores that are lower than 30. The methylation status of genome-wide cytosine sites is reported from the filtered SAM files with the Bismark methylation extractor using the following argument: –cytosine report. Also, the conversion rate biased ends are also ignored during methylation call based on the M-bias plots. The minimum read coverage was filtered at 10 as well.

2.3. Filtering False Positive m^5C Sites due to RNA Secondary Structure. It is known that secondary structures on RNAs prohibit bisulfite conversion and thus can result in false positive detection of transcriptome m^5C sites. As shown in Figure S1, the detected m^5C sites from MeRanTK are enriched with double-stranded regions of RNA, which are likely to be false positive errors due to a secondary structure. For this reason, an R package rBS2ndStructure was created to facilitate the elimination of the false positive methylation calls due to RNA secondary structures. Specifically, the RNA secondary structure is predicted with RNAfold from the Vienna RNA package [44] as it was performed by Amort et al. [36]. The transcriptome-wide full-length transcripts are extracted from UCSC gene annotation for both mm10 and hg19. Then, the double-stranded structures are predicted with the MEA method under alpha = 0.1. The folding temperature is set at 70 degrees, and the maximum pairing distance is set at 150 bp. For the mitochondrial chromosome and transcripts longer than 8000 bp, the structures are predicted using sliding windows of 2000 bp and step size of 1000 bp. For both the RNA and the DNA methylation reports, the methylation sites overlapped with the predicted regions of secondary structures are filtered. Due to the lack of computational resources to predict structures on large intronic sequences, the cytosine sites that do not locate on the exons of known transcripts or the mitochondrial chromosome are filtered. The resulting methylation reports are then analyzed under the R environment using primarily GenomicFeatures [45], Guitar [46], and ggplot2 [47] packages.

The rBS2ndStructure package is publically available at Github (https://github.com/ZhenWei10/rBS2ndStructure) with precomputed RNA secondary structures of genome assembly mm10 and hg19 for convenient processing of RNA BS-Seq result.

2.4. Quantitative Analysis of Methylation Status. The methylation ratio (mRatio) of a specific cytosine site is calculated by

$$\text{Methylation ratio} = \frac{\#\text{of unconverted Cs}}{\#\text{of unconverted Cs} + \#\text{of converted Cs}}, \tag{1}$$

where "# of unconverted Cs" and "# of converted Cs" indicates the count of methylated (unconverted) Cs and unmodified Cs (converted Cs) at a specific cytosine site, respectively. The methylation rate is conceptually similar to the well-adapted concept of "beta value" in DNA methylation analysis [48], which indicates the percentage of methylated Cs among all Cs. Also, it is not difficult to show that

$$\begin{aligned}\text{Methylation ratio} &= \frac{\#\text{of unconverted C}}{\#\text{of unconverted C} + \#\text{of converted C}}\\ &= 1 - \frac{\#\text{of converted C}}{\#\text{of unconverted C} + \#\text{of converted C}}\\ &= 1 - \text{convertion rate},\end{aligned} \tag{2}$$

where the conversion rate has been previously defined in [35] and a smaller value suggests a higher percentage of RNA m^5C methylation.

To differentiate a set of statistically significantly methylated cytosine sites against potential technical randomness due to incomplete bisulfite conversion, the p values for the methylation state of both the DNA and RNA methylation are calculated by Fisher's exact test against the background conversion odds after the filtering of the sites mapped to introns and secondary structures. The adjusted p values (FDR) are then adjusted by the Benjamin & Hochberg method. The positive methylation states were decided when FDR < 0.05.

For the mouse samples containing 3 biological replicates, the methylated sites are judged as FDR < 0.05 among all 3 replicates. For other insignificant methylated sites to be kept in the analysis, the sites should be reproduced 3 times with coverage > 10. The converted reads and nonconverted reads are added on each site when combining the biological replicates.

The background bisulfite nonconversion rate is 2.75%, 2.74%, 1.18%, and 0.81% for MCF10A, MDA468, mouse ESC, and mouse brain samples, respectively (taking the average for samples with more than one biological replicate). The difference among nonconversion rates might be due to the biological difference of cell lines, batch variation, and different BS-Seq protocols.

2.5. Differential Methylation Analysis. The odds ratio (OR) or methylation fold change from differential analysis is defined as

$$\text{Odds ratio from differential methylation} = \frac{(\#\text{of unconverted Cs under cond\_1}/\#\text{of converted Cs under cond\_1})}{(\#\text{of unconverted Cs under cond\_2}/\#\text{of converted Cs under cond\_2})}. \tag{3}$$

Odds ratio (or methylation fold change) indicates whether the methylation is enriched under one condition compared with another condition. A value greater than 1 suggests increased methylation level, where as a value less than 1 suggests decreased methylation level. The statistical significance of the odds ratio is evaluated by the QNB method, which tests the homogeneity of association between methylated and unmodified molecules under two experimental conditions with the within-group variability assessed through 4 cross-linked negative binomial distributions [49].

Similar to the odds ratio from differential methylation analysis, the enrichment odds ratio of m^5C sites within a specific region can be defined as

$$\text{Enrichment odds ratio} = \frac{(\#\text{of } m^5\text{C sites within a region}/\#\text{of C sites within a region})}{(\text{total}\#\text{of } m^5\text{C sites}/\text{total}\#\text{of C sites})}. \tag{4}$$

A value greater than 1 suggests that methylation sites are enriched within the tested region, and the statistical significance of enrichment can be evaluated by Fisher's exact test. Please note that, in this analysis, we used the total number of cytosine sites reported from MeRanTK rather than the total number of all 4 types of nucleotides.

2.6. Assessing the Distribution of m^5C Sites on mRNA. The distribution pattern of m^5C sites on mRNA is assessed with the Guitar R/Bioconductor [46]. Compared with other software tools and methods, the Guitar package provides an improved resolution by relying on only the mRNA transcripts that simultaneously have sufficient long (more than 100 bp) 5′ UTRs, CDSs, and 3′ UTRs. For instance, transcripts without annotated 5′UTRs will be excluded from the analysis. Additionally, Guitar does not rely on only the primary transcript (often defined as the longest transcript among all isoforms in practice) when solving an ambiguous association between a m^5C site and the isoform transcripts of a gene; instead, all ambiguous associations are considered with the weight of association evenly divided. For example, if a single m^5C site locates on the 3′UTR of a transcript and CDS of another isoform transcript of that gene, it is counted as if half of the m^5C site is located on the 3′ UTR and the other half located on 5′ UTR. In this way, the isoform information is largely retained. To our knowledge, the Guitar package should provide the most accurate assessment of a transcriptomic distribution pattern.

2.7. Differential Expression Analysis. Differential expression analysis was performed with the DESeq2 package [50] and the aligned RNA BS-Seq data.

2.8. Cell Culture and Viral Infection. Jurkat T lymphocytes were maintained in RPMI 1640 medium (Hyclone) supplemented with 5% (*v*/*v*) FBS (Gibco) and 100 U/ml penicillin/streptomycin (Hyclone). For infection, Jurkat cells were infected with known amounts (3 × 108 genome copies per 2 × 105 cells) of SRV for 18 hours at 37°C, followed by washing three times with PBS (Hyclone). Infected cells were incubated in completed culture medium for the indicated time. Successful infection was identified as the appearance of cytopathic effects in infected cells at 8 to 10 days postinfection.

2.9. Reverse Transcription and Real-Time PCR. SRV genome in culture medium was extracted by viral RNA extraction kit (TIANGEN) and reverse transcripted into cDNA by a reverse transcriptase PCR kit (TaKaRa). Cellular genome was extracted by a TIANamp Genomic DNA Kit (TaKaRa). Real-time PCR was performed in a 7500 Fast Real-Time PCR System (Applied Biosystems) by using a Premix Ex Taq (Probe qPCR) kit (TaKaRa). SRV genome positive control, primers, and probe, as well as GAPDH primers and probe were kindly provided by VRL China Ltd. [51].

2.10. Immunofluorescence Assay. Cells were seeded on poly-L-lysine (Sigma) coated slides, fixed with 4% paraformaldehyde for 15 minutes, permeabilized with precold pure methanol for 20 min at −20°C, and blocked with 5% BSA for 1 hour. Cells were then stained with the serum from an SRV-infected monkey (1 : 25 diluted in blocking buffer) overnight and visualized with DyLight™ 488-Labeled Anti-Human antibody (KPL). Cells were counterstained with Hoechst (Life Technologies) for 10 minutes and mounted on microscopy slides. Samples were imaged with a ZEISS LSM 880 Confocal Laser Scanning Microscope.

3. Results

3.1. Overview of mRNA m^5C Methylome Revealed by BS-Seq. After successful processing of the RNA BS-Seq datasets, a total of 3440 (0.40%), 1915 (0.29%), 35,246 (0.757%), and 25,301 (0.50%) RNA cytosine sites were identified as m^5C methylation (FDR < 0.05) sites in MCF10A, MDA-MB-468, mouse embryonic stem cell (ESC), and mouse whole brain, respectively. The overall transcriptome m^5C methylation level was much lower than the DNA m^5C methylation level (Figure S2). Importantly, we found that m^5C was widespread in different RNA families, where more than 50% of them were located on mRNA (Figure 1(a)). In MCF10A cells, 7131 protein-coding genes had sites reported after the filtering, of which 225 (3.15%) mRNAs contained m^5C sites. In MDA-MB-468 cells, 6320 protein-coding genes had reads aligned, of which 128 (2.06%) contained m^5C sites. In ESC and brain samples, the methylation status was available for 11,325 and 13,108 protein-coding genes, of which 3579 (31.6%) and 3065 (23.4%) contained m^5C sites. The difference in number of m^5C sites between different conditions is mostly due to different sequencing depth.

3.2. mRNA m^5C Is Enriched in 5′UTRs of Human and Mouse. To study the spatial organization of m^5C sites in the transcriptome, we first analyzed the relative enrichment (see Materials and Methods for more details) of m^5C sites on different types of RNA and at different regions (shown in Figure 1(b)) by compensating for the cytosine sites that do not carry m^5C modification. Our results showed that m^5C sites were consistently and significantly enriched at 5′UTRs in human and mouse with enrichment odds ratio of 3.138, 4.802, 2.744, and 1.601 (please see Table S1 for more details). The similar topology was already reported by previous studies [35, 36], and our observation further confirmed their conclusions. Also, we did observe a slight enrichment of m^5C sites in 3′UTR in mouse brain (enrichment odds ratio = 1.013 and 1.19*E* − 02), which is also reported in the study of Amort et al. [36]. 3′UTR enrichment was not observed in the other samples (odds ratio = 0.964, 0.971, and 0.617).

To further substantiate these findings, we plotted the distribution of the methylated and unmethylated cytosine sites located on mRNAs with the Guitar package [46]. In order to improve the resolution of this analysis and differentiate the distribution of m^5C sites on usually short 5′UTRs, only the mRNAs with a 5′UTR longer than 100 bp are used. As shown in Figure 2(a), the methylated cytosine sites were consistently enriched at 5′UTRs across all 4 samples when compared to unmethylated groups. Interestingly, this trend is also supported by the cytosine methylation sites reported

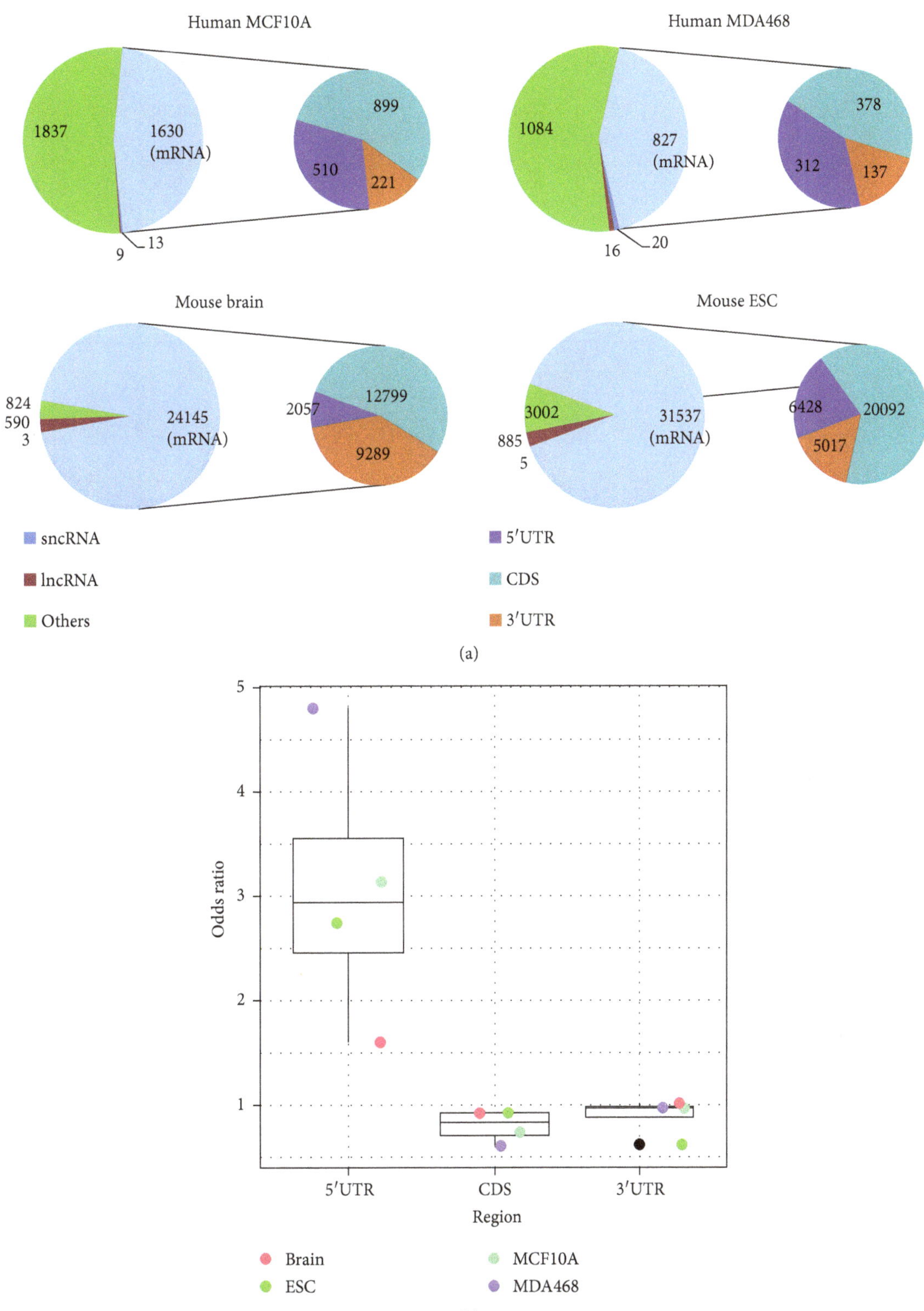

FIGURE 1: Distribution of transcriptome m^5C modification sites in human and mouse. (a) The pie chart shows transcriptome-wide distribution of m^5C sites in MCF10A, MDA-MB-468, mouse embryonic stem cell (ESC), and whole brain. The majority of the identified m^5C sites are located on mRNAs. (b) Graph showing status of m^5C frequency in different regions of mRNA. The result indicates that detected cytosine sites are consistently enriched at the 5′UTR on mRNA compared with the CDS and 3′UTR.

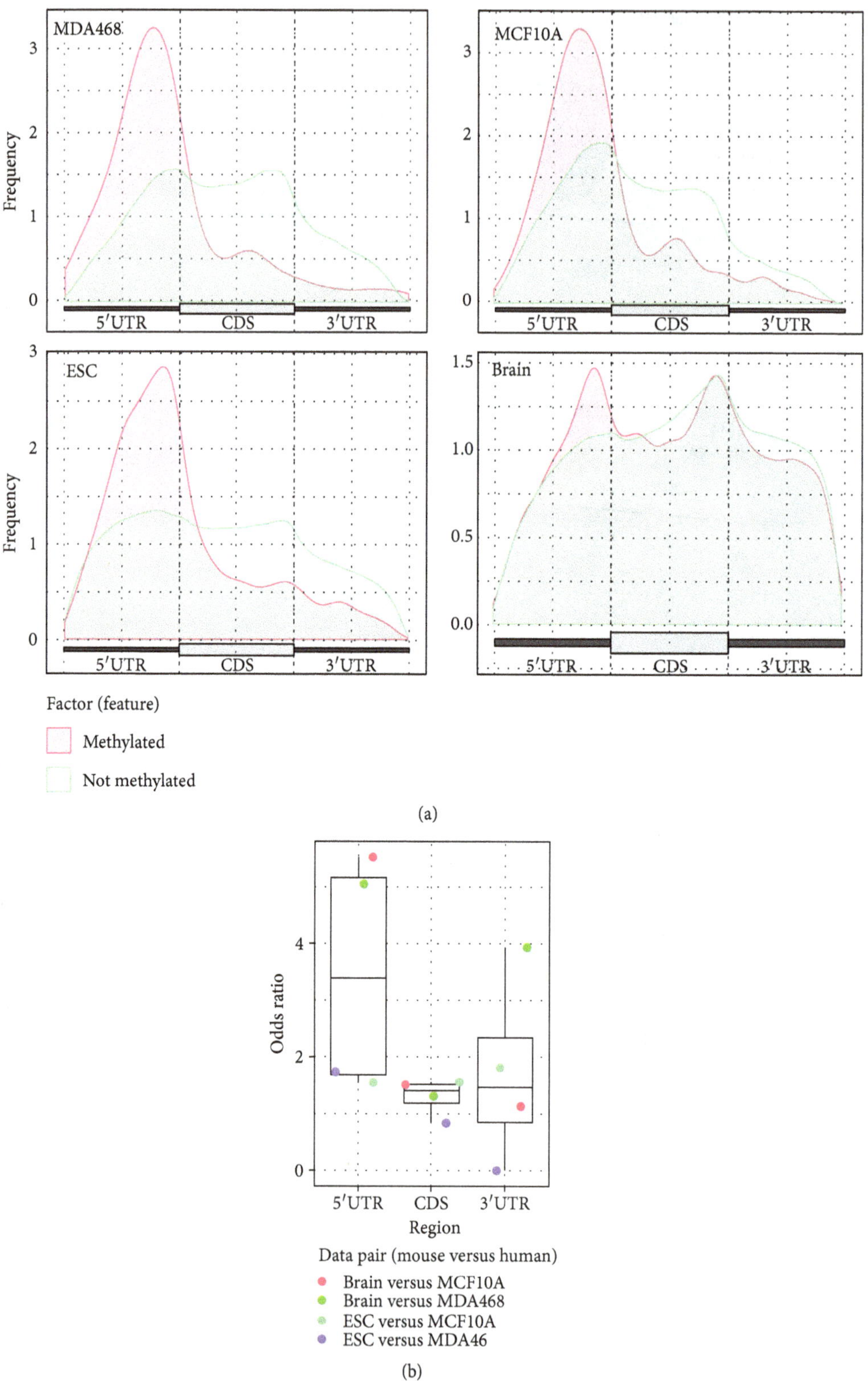

Figure 2: Conservation of m^{5}C in different mRNA regions. (b) Graph showing the status of m^{5}C frequency in different regions of the transcripts. We divided all the detected cytosine sites into 2 groups based on whether it is methylated. The result indicates that cytosine sites with significant methylation levels are consistently enriched at the 5′UTRs and near the start codon in all 4 samples. (b) A correlated methylation pattern is observed on 5′UTRs between different cell lines/tissues in human and mouse. The conserved cytosine residuals were retrieved with liftOver utility (http://genome.ucsc.edu/cgi-bin/hgLiftOver), and the correlation analysis is performed with Fisher's exact test. It is important to note that, although we failed to observe a correlated m^{5}C methylation pattern on CDSs and 3′UTRs of mRNA, it is possible that such pattern may emerge on strictly matched cell lines/tissues.

by Squires et al. [35], and there is no significant enrichment of m^5C sites observed in 3′UTR when all cytosine methylated sites were used as background (Figure S3).

When we further compared the methylation status of the conserved loci in human and mouse between different cell lines/tissues, we observed that, although the cell types/tissues we used were not strictly matched, a strong correlated methylation pattern was observed on the 5′UTR region (Figure 2(b) and Table S2). However, unlike the 5′UTR, the correlated pattern of m^5C sites were not consistently observed in CDSs or 3′UTRs in our study; the observed heterogeneity of the m^5C methylome in different transcript regions suggests that the m^5C mapped to the 5′UTR of the transcripts are more likely to be functionally important.

3.3. m^5C Site Exists under Different Nucleotide Contexts. Because RNA methyltransferase Dnmt2 shares strong sequence homology to all DNA DNMT methyltransferases [52], we reason that exploring the relationship between transcriptome and DNA m^5C methylation profiles may unravel interesting interplay between the two kinds of reversible chemical modifications. In mammalian cells, DNA methylation occurs mainly at CG dinucleotides (including ACG, CCG, TCG, and GCG, see Figure 3(b)). To study whether, like DNA methylation, transcriptome m^5C methylation also occurs at the similar position, we analyzed methylated cytosine in the transcriptome. For this purpose, we examined all the possible C-centered trinucleotide combinations. Unlike DNA, transcriptome m^5C occurs at all C-centered trinucleotides (Figure 3(a)) and was observed to be specifically enriched at GCA, ACG, CCG, GCG, CCC, and GCC. These results were found to be consistent within the same species (Pearson correlation = 0.96 and 0.92, Figure 3(c)) and between different species (Pearson correlation = 0.72, 0.75, 0.45, and 0.48, Figure 3(c)).

3.4. Negative Correlation in Methylation Level Is Observed between mRNA and the Corresponding Exonic Region of DNA. We next examined whether there exists any correlation between m^5C methylated/nonmethylated (m^5C methylation ratio) in the transcriptome and corresponding DNA exonic regions at each C-centered trinucleotide sites. Because DNA methylation occurs mainly at CG dinucleotides, as expected, we observed no strong correlation at non-CG trinucleotides. However, we observed significant negative correlation in methylation ratios between RNA and DNA at all four CG-containing trinucleotides. As a higher percentage of m^5C in mRNA is detected, the corresponding DNA exonic CG dinucleotide was less likely to be methylated (Figure 4(a)). Next, we grouped m^5C methylated at all CG sites according to their methylation ratio (methylated and unmethylated) and investigated their distributions in mRNA and the corresponding exonic regions of DNA. Consistent with our previous finding, we observed a significant negative correlation in both human and mouse cells. In particular, 5′UTR in mRNA showed a high methylation ratio, whereas the corresponding DNA region showed a significantly low methylation ratio (Figure 4(b)).

3.5. Transcriptome m^5C Sites Exhibit a Clustering Effect. In DNA methylation, it has been shown that the correlation of methylation rates between two CpG sites is related to the distance (see Figure S4), and the clustering effect can be as high as 0.7 for probes within 200 bp [53]. To address whether the mRNA m^5C methylation also exhibits a clustering effect, we examined the proportion of m^5C sites that are within 10 bp distance of other m^5C sites and compared this proportion with that from 1000 times of random permutation. Our analysis revealed that m^5C showed an obvious clustering effect in both mRNA and DNA (Figures 5(a) and 5(b). In the ESC cell line, more than 76.7% of the mRNA methylation sites had at least one methylation site mapped within the 10 nt-flanked region, compared with 7.7% of such event by random permutation of methylation states on insignificant methylation sites of the methylated genes. In mouse ESC and brain cells, more than 43.02% and 30.06% of mRNA m^5C methylation sites existed within the m^5C-p-m^5C dimmers, compared with expected rate of 1.02% and 0.77% of such dimmers by the random permutation.

To further elucidate the clustering effect, we calculated the correlation of the methylation ratio between two cytosine sites with a specific distance. To our surprise, mRNA methylation exhibited a stronger clustering effect compared with DNA (Figure 5(c)). In addition, the correlation of the methylation ratio was consistently stronger within 1–3 nt distance as revealed by the higher correlation of the methylation ratio (0.76 in MCF10A and 0.79 in ESC). These results indicated that most CpC dimers are comethylated; the correlation of the methylation ratio can be as high as 0.58 in MCF10A and 0.47 in ESC for cytosine sites with a distance of 4–10 nt. Though the overall clustering effect of DNA methylation was not as strong as mRNA methylation, when only the CpG dinucleotide was considered, DNA methylation exhibited a stronger clustering effect than mRNA methylation (see Figure S4).

3.6. Transcriptome m^5C Is Strongly Enriched in Mitochondrial Transcripts. To further establish a physiological relevance of m^5C distribution, we examined the methylation level of RNAs encoded in different chromosomes. Surprisingly, m^5C modification was strongly enriched in RNAs transcribed specifically from mitochondrial DNA in normal and breast cancer cells as well as in mouse ESC and brain as revealed by enrichment odds ratios of 818.42949, 634.72723, 1028.52065, and 67.28553, respectively. In contrast, the enrichment odds ratios of RNA methylation for transcripts from other chromosomes were found to be roughly the same (Figure 6(a)). The RNA transcripts of all the major genes located on a mitochondrial chromosome were significantly methylated (Figure 6(b)). Previously, it was reported that methyltransferase NSUN5 can regulate mitochondrial gene expression [54], and we speculate that RNA m^5C may play a more vital regulatory role in mitochondria-related biological processes.

3.7. Dysregulation of RNA Methylome in Breast Cancer. Comparison of normal (MCF10A) and breast cancer (MDA-BM-468) m^5C epitranscriptomes identified 162 significant differential methylation sites (DMSs) located on

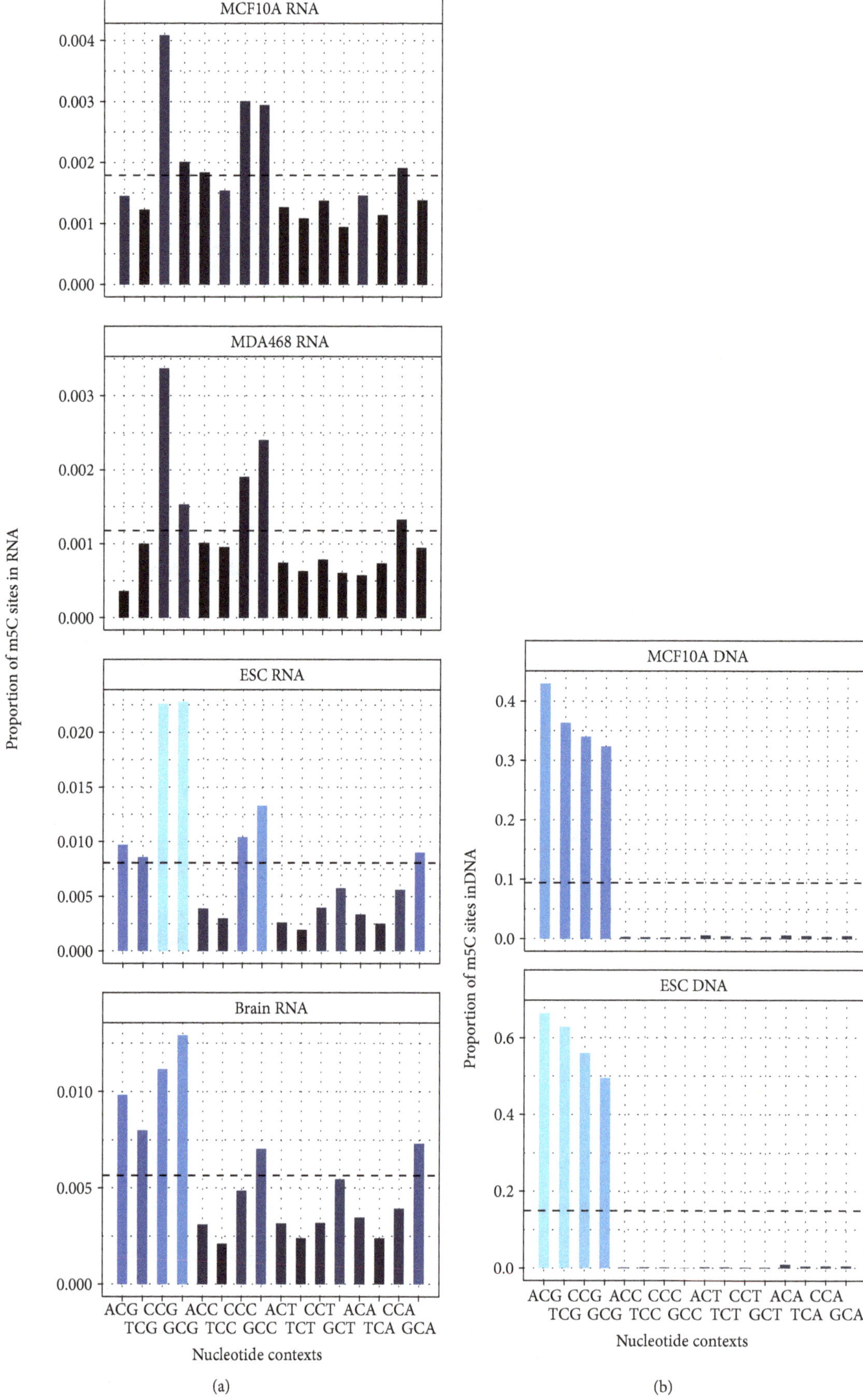

Figure 3: Continued.

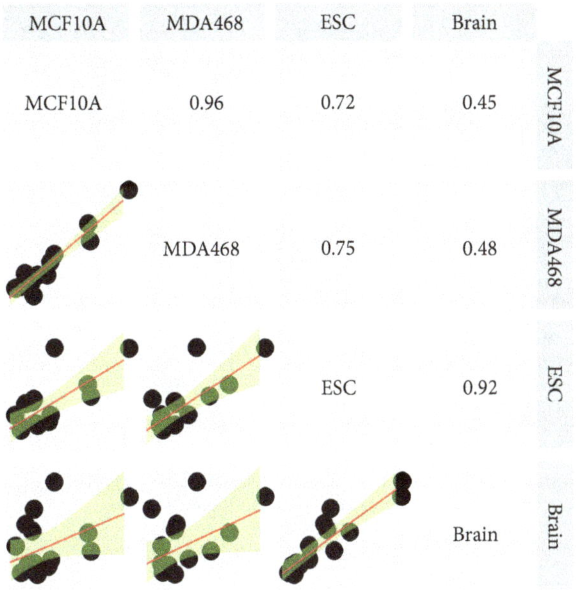

(c)

FIGURE 3: Comparative distributions of mRNA and DNA m^5C methylation. (a) Bar graph shows the proportion of mRNA m^5C sites under different combinations of C-centered trinucleotides in mouse and human cells. The dotted line shows the average percentage of methylation under all trinucleotide contexts within the entire transcriptome. We observed that RNA m^5C occurs under all trinucleotide contexts and is slightly enriched in sequences containing CCG, GCG, GCC, GCU, and GCA. (b) Bar graph showing proportion of DNA m^5C sites in mouse and human cells. DNA cytosine sites were enriched exclusively in sequences containing CG dinucleotides (ACG, CCG, CCG, and TCG). (c) The coefficient of correlation between RNA methylation and trinucleotide sequences was found to be consistent between samples from the same species (Pearson correlation = 0.96 for human and 0.92 for mouse) and also between human and mouse cells (Pearson correlation = 0.72, 0.75, 0.45, and 0.48).

47 annotated genes at a significance level of 0.05. Among the 47 differentially methylated genes, 35 shows hypomethylation and 12 shows hypermethylation in cancer cells compared with the normal control cell line. The majority of the differential methylation sites show hypomethylation (Excel Sheet S1 and Figure 7(a)), and the m^5C hypomethylations are mostly located in the CDS and 3′UTR region of mRNA but not in the 5′UTR region (Figure 7(b)). We then investigated whether different m^5C mRNA methylation levels in normal and breast cancer cells have any functional correlation. We performed functional gene set enrichment analysis on genes containing DMS using the DAVID web server and found that many of the 47 differentially methylated genes are related to important biological functions of cancer, for example, regulation of apoptosis and programmed cell death with RTN4, NME2, CASP14, HSPB1, RPL11, and RPS3 differentially methylated (Excel Sheet S1).

Interestingly, like the difference between the breast cancer cell line MDA-MB-468 and the normal epicelial cell line MCF10A, similar mechanistic mouse stem cells [55] also exhibit dominant hypomethylation in the m^5C epitranscriptome when compared with mouse brain cells with 2513 genes hypomethylated and 767 genes hypermethylated (Figure 7(c) and Excel Sheet S2). Also similar to the previous case, the hypomethylations are mostly located in the CDS and 3′ UTR regions of mRNA, but not in the 5′UTR region (Figure 7(d)). Using DAVID, we found that hypermethylated genes in ESC cells are mostly enriched with the regulation of cell cycle (FZR1, E2F5, BOP1, TRRAP, CDK4, JUNB, etc.), cell death (SIVA1, MCL1, YPEL3, ARF6, UBQLN1, SHF, CIAPIN1, APLP1, GPX1, CASP3, etc.), and mRNA metabolic process (SCAF1, FIP1L1, STRAP, RBM15B, CWC15, XAB2, YBX1, AUH, SF3B2, APLP1, HNRNPL, etc.); the hypomethylated genes are enriched with functions related to ATP synthesis (ATP6V1F, ATP6V1C1, ATP6V0C, ATP6V1A, ATP6V0E, ATP6V1E1, ATP5C1, etc.) and mitochondrial ribosome (MRPL15, MRPL27, MRPL16, MRPL36, MRPL39, MRPL34, DAP3, etc.) (Excel Sheet S2). These results may suggest that the m^5C methylations are selectively methylate transcripts having functions.

3.8. *Positive Correlation between m^5C mRNA Methylation and Expression Changes.*

In our data, as the gene expression is also estimated from RNA bisulfite-sequencing data, a direct comparison of expression and m^5C methylation changes may be problematic due to dependent noise. To eliminate the interference of dependent noise between expression and methylation data, the samples are further divided for different purposes. Specifically, the 3 biological replicates are divided into 2 groups, with 1 sample used for the estimation of expression changes and the other 2 samples for estimation of methylation changes. The expression changes and methylation changes are then compared. This procedure was repeated for 3 times using different grouping combinations.

A consistent and significantly positive correlation is observed (0.274, 0.303, and 0.254) between $\log_2$ expression fold change and $\log_2$ methylation fold change when comparing mouse embryo stem cells with brain cells (Figure S5),

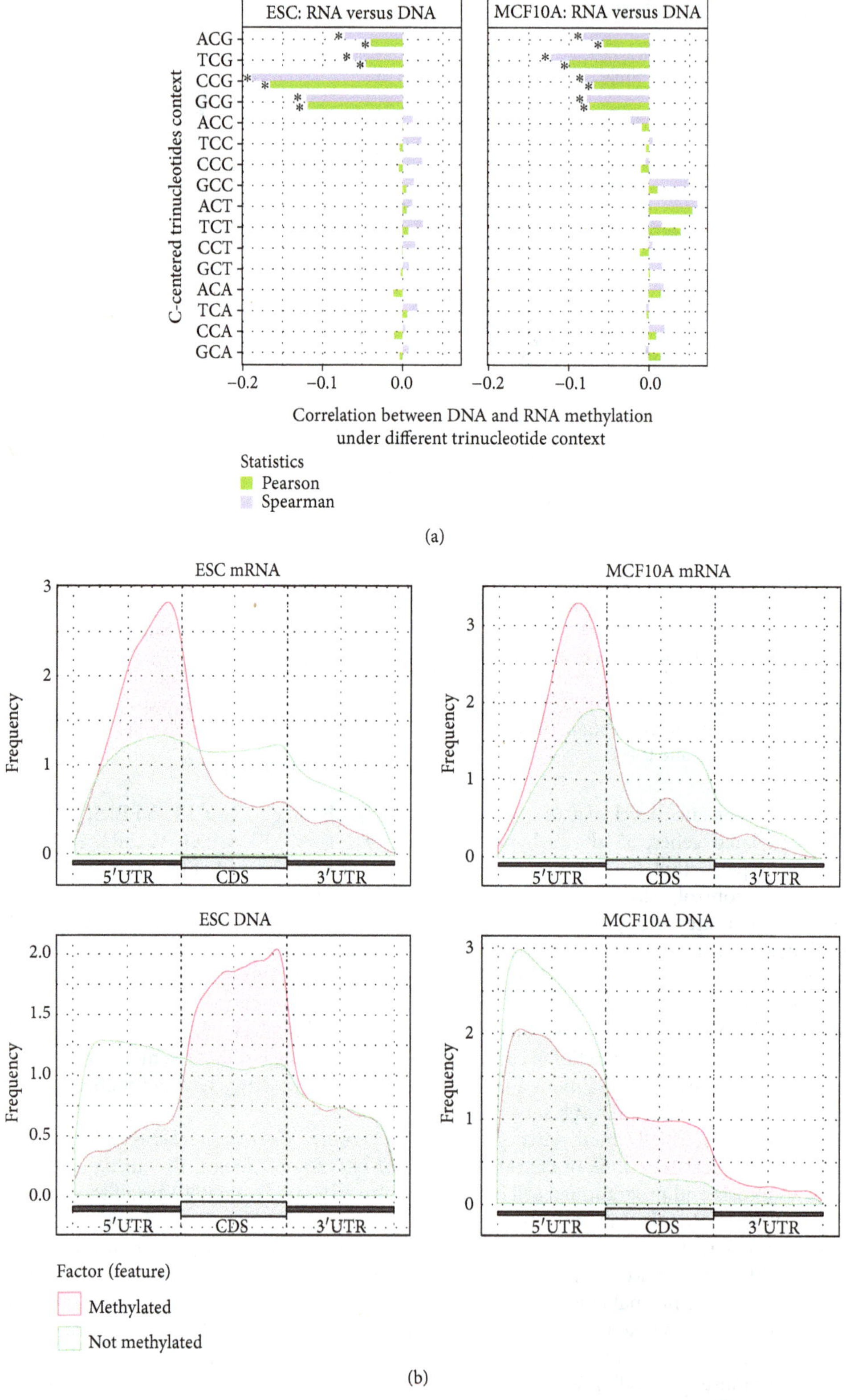

Figure 4: The methylation ratio of corresponding m^{5}C DNA and mRNA CpG islands shows negative correlation. (a) Negative correlation is observed between DNA and mRNA methylation ratio consistently under all four CG containing trinucleotides (ACG, TCG, CCG, and GCG) in both human and mouse, that is, if a specific CG dinucleotide in DNA is methylated, the corresponding dinucleotide in mRNA is significantly less likely to be methylated. *The top 4 nucleotide contexts under which the strongest correlation between DNA and RNA methylation level exists. (b) Comparative distributions of m^{5}C methylated CG sites in DNA and RNA show an enrichment of sites with a high methylation ratio in mRNA 5′UTR as opposed to an enrichment of low-methylation-ratio sites in DNA 5′UTR. The pattern is consistent in both the human MCF10A cell line and mouse embryo stem cells.

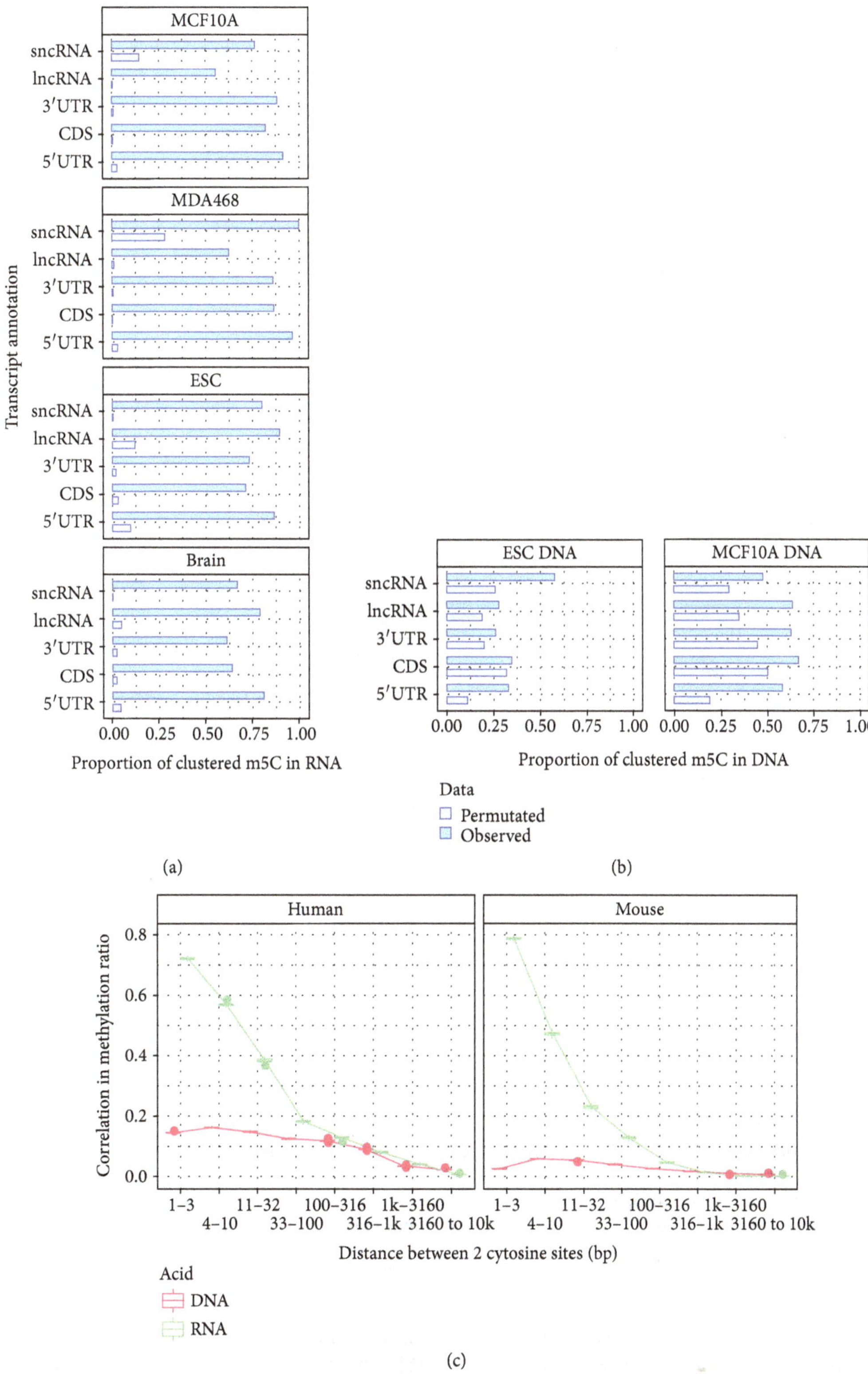

FIGURE 5: RNA m^{5}C modification exhibits a clustering effect. (a) Bar graph shows the proportion of clustered m^{5}C sites within 10 nt flanked regions. To evaluate the statistical significance, we generated 1000 permutated results as a comparison with the bars indicating a 99% confidence interval. Using these criteria, m^{5}C methylation showed a strong clustering effect consistently on different RNA families and on different regions of mRNA in human and mouse. Around 50% of the m^{5}C sites were clustered with each other within a 10 bp region. (b) DNA methylation also exhibited a clustering effect. However, the pattern is not that strong when all nucleotide contexts are considered. (c) Line graph showing correlation between RNA/DNA m^{5}C methylation and distance between cytosine sites. RNA m^{5}C methylation showed strong correlation with cytosine sites that are immediately close to each other. The clustering effect of DNA methylation is strong when only CpG context is considered (Figure S4).

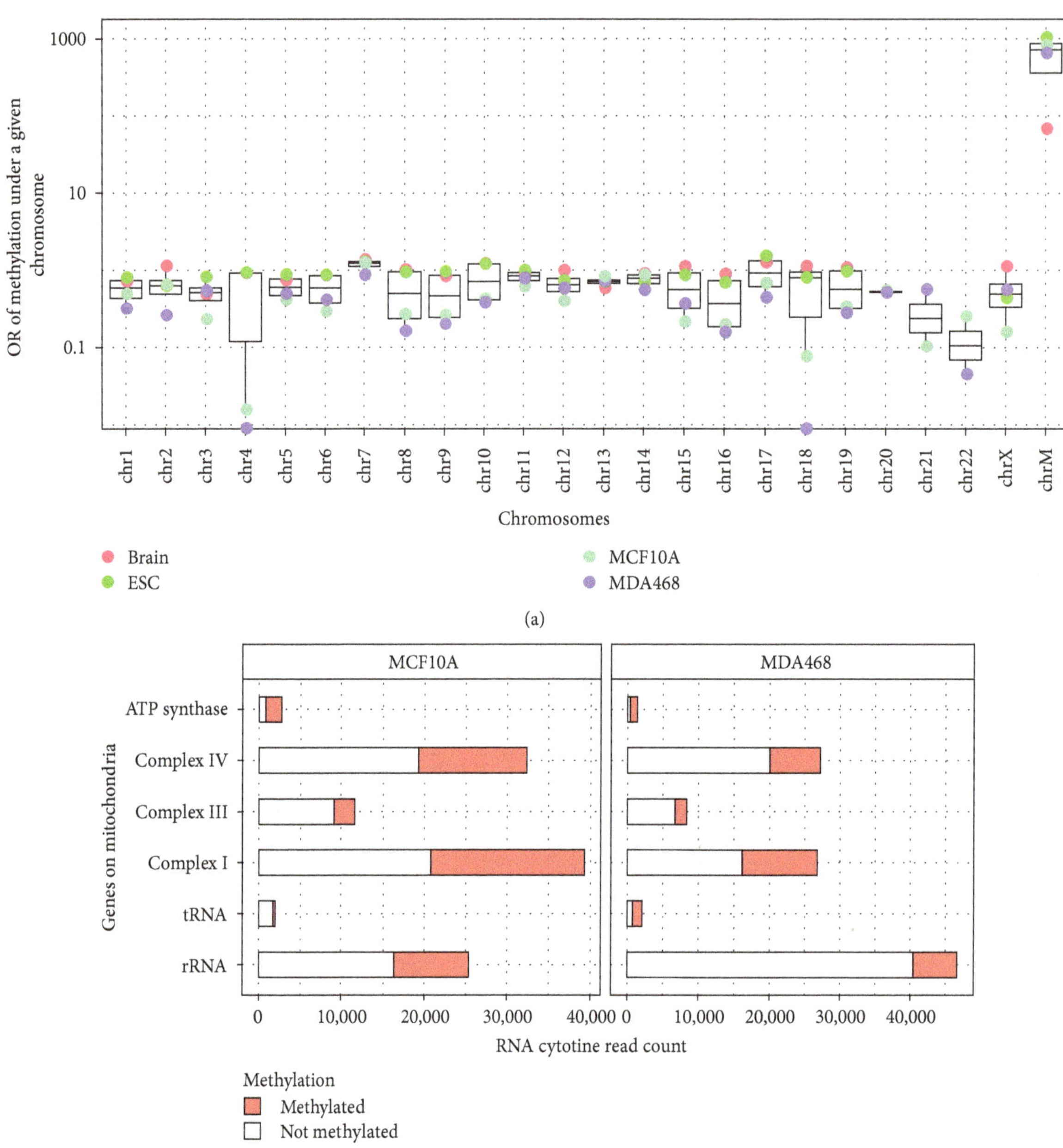

FIGURE 6: m^5C is enriched on mRNAs transcribed from mitochondrial DNA. (a) Bar graph depicting m^5C mRNA methylation sites on different chromosomes. RNAs transcribed from mitochondrial DNA (M) showed drastically increased frequency of m^5C sites (enrichment odds ratio of 818.42949, 634.72723, 1028.52065, and 67.28553). (b) Bar graph showing the number of methylated cytosine reads stacked with unmodified cytosine reads generated from 6 major classes of mitochondrial genes. The RNA transcripts of all the major genes located on a mitochondrial chromosome were significantly methylated.

suggesting that increased methylation level is likely to be associated with increased expression level. Although the specific molecular mechanism is not yet clear, the observed positive correlation between RNA m^5C and RNA expression confirmed our previous observed anticorrelation between DNA and RNA m^5C methylation (see Figure 4) from a different perspective.

To explain the positive correlation between expression and transcriptome m^5C methylation, we compared the methylation status of all the genes and their half-life, where the half-life of mouse genes were obtained from a previous study [56]. The mRNAs are classified into two groups based on whether they have at least one m^5C site or not. To exclude the confounding factor (effective size in methylation site calling), a generalized linear model of the binomial family was fitted to the half-life with both expression and methylation information. Our result suggests that there exists a significant positive correlation (p value $= 2.23e-12$) between the mRNA half-life and its m^5C methylation status in mouse embryo stem cells, and the positive association is also

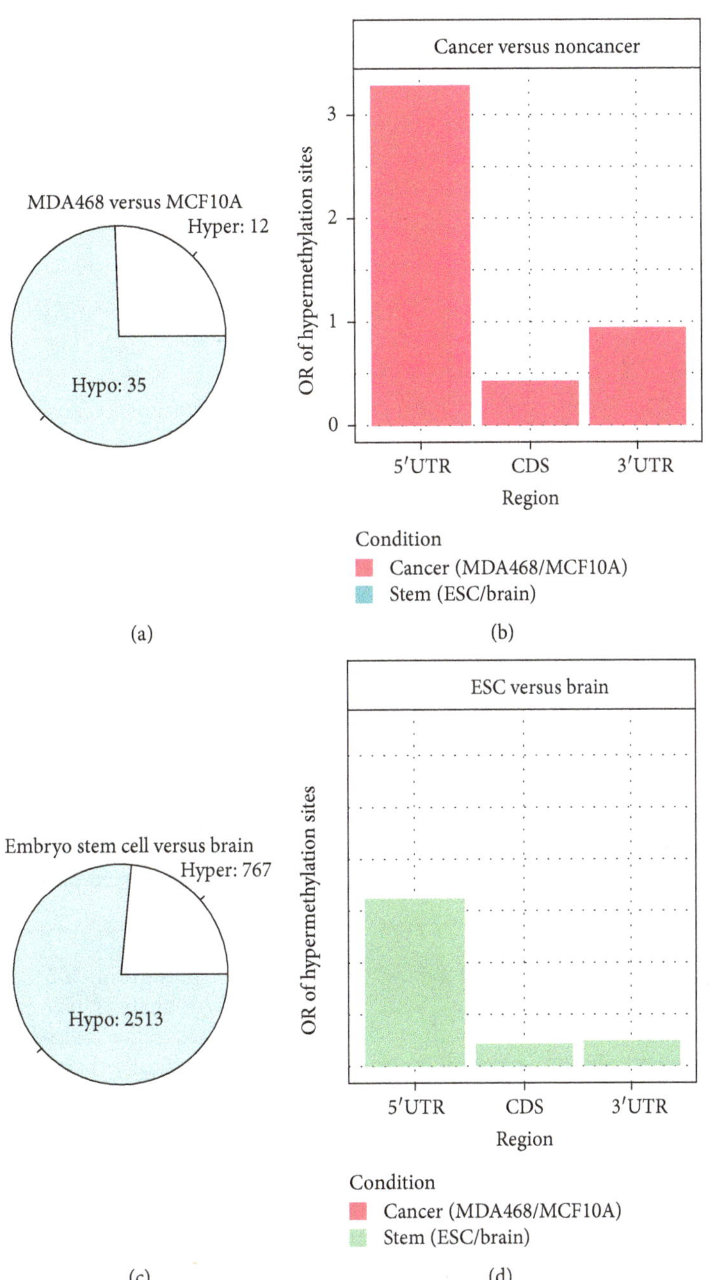

FIGURE 7: Differential m^{5}C mRNA methylation in different tissues. (a) Pie-diagram showing hypo- and hypermethylation in MDA468 when compared to MCF10A. A total of 47 differential methylated genes were identified between the breast cancer (MDA-MB-468) and normal control cell lines (MCF10A) with primary hypomethylation under cancer condition. (b) Bar graph showing odds ratio of hypermethylation sites with respect to all differentially methylated sites on different regions of mRNA. Hypermethylated sites were strongly enriched in 5′UTRs. (c) Pie diagram showing hypermethylation in mouse embryo stem cells when compared to whole brain cells. (d) Bar graph showing odds ratio of hypermethylation sites with respect to all differentially methylated sites on different regions of mRNA in the mouse experiment. Hypermethylated sites were strongly enriched in 5′UTRs.

confirmed on mouse whole brain dataset (p value = 0.0374). To further exclude the impact of mRNA expression in calling methylation status, we also extracted the genes whose $\log_2$ expression levels fall between 7 and 11, and then fit their mRNA half-life with a local regression. As shown in Figure 8, compared with the genes of a similar expression level but without an m^{5}C site, the half-life of the mRNAs that carry m^{5}C sites is clearly longer and the pattern is consistent in both mouse brain and ESC.

3.9. Dysregulation of RNA Methylome after Simian Retrovirus Infection. Simian retrovirus (SRV) infection of Jurkat T lymphocytes (Jurkat cells) was confirmed by syncytia formation, of which the membrane of the neighboring cells fused to one

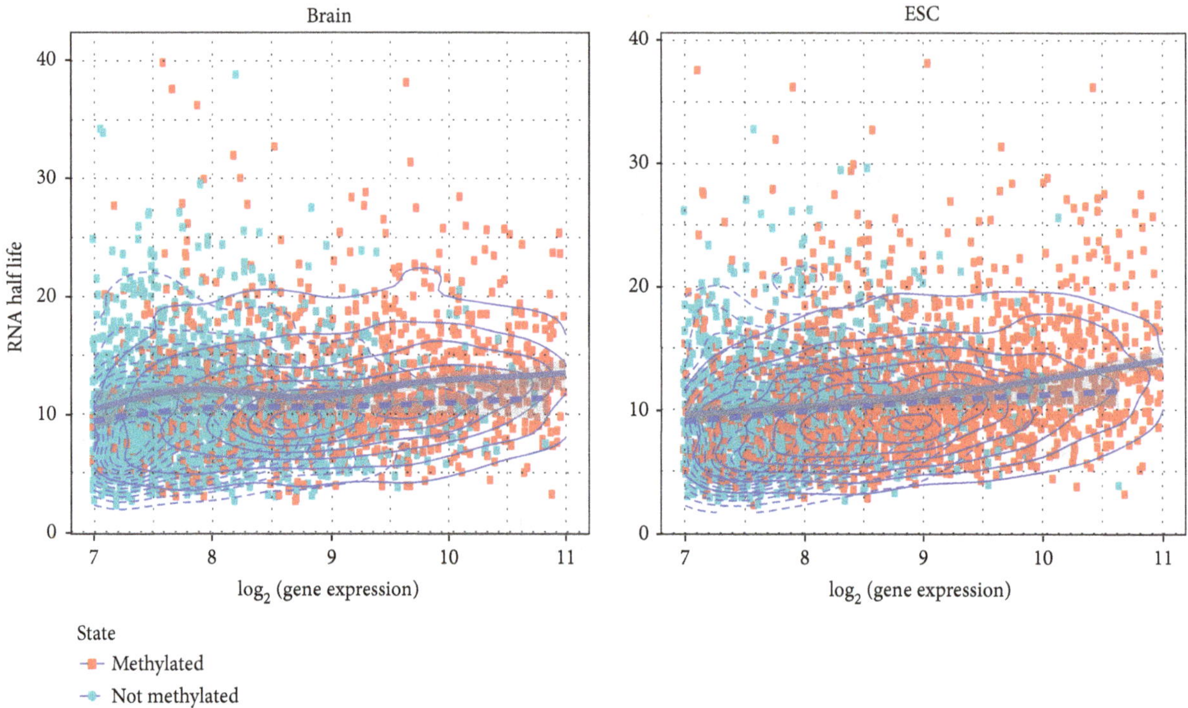

Figure 8: RNA m^5C status is positively correlated with RNA half-life. In the above figure, each red dot represents a gene that carries m^5C sites, and each blue dot represents a gene that does not carry an m^5C site. When comparing the methylated and unmethylated genes of similar expression, the genes that carry an m^5C site have longer RNA half-life than those that do not carry m^5C sites. (a) Positive correlation between RNA methylation status and RNA half-life is observed in mouse brain (p value = 0.0374, generalized linear model of binomial family). (b) Positive correlation between RNA methylation status and RNA half-life is observed in mouse embryo stem cells (p value = $2.23E-12$, generalized linear model of binomial family).

another. At 10 days postinfection, the formation of syncytium was observed among the Jurkat cells incubated with SRV (Figure 9(a)). The syncytium of Jurkat cells contains multiple nuclei and its size is dramatically larger than a single cell. SRV long terminal repeats (LTRs), which are reverse-transcribed from the RNA genome during the infection, contain critical sequences necessary for the integration, synthesis, and expression of viral DNA [1]. Therefore, the extent of SRV infection was assayed by monitoring SRV-LTR expression in Jurkat cells through quantitative real-time PCR. As shown in Figure 9(b), the copy number of SRV-LTR was gradually increased from 2 days to 10 days postinfection and then tended to be stable afterwards. Taken together, these results indicated that SRV was able to infect Jurkat cells and the infection reached maximum level after 10 days postinfection.

In order to investigate whether SRV could replicate in Jurkat cells, the SRV virions released in the culture medium were determined by measuring viral genome copy number through quantitative real-time PCR. As shown in Figure 9(c), the copy number of the SRV genome was gradually increased from 2 days to 14 days postinfection, suggesting that SRV was able to replicate in Jurkat cells.

We then measured the RNA methylome with bisulfite sequencing. A total of 2475 m^5C sites located on 517 genes are reported as differentially methylated 10 days postinfection of SRV with QNB p value < 0.05. Among them, 389 sites located on 158 genes are hypomethylated, while 2086 sites from 382 genes are hypermethylated. A gene ontology analysis using the DAVID website suggests that the differentially methylated genes are related to virus infection, specifically, hypermethylated genes are enriched with DNA replication (p value = $6.07E-5$), mitotic nuclear division (p value = $4.37E-4$), DNA replication initiation (p value = $3.48E-3$), autophagosome assembly (p value = $8.54E-3$), strand displacement (p value = $1.42E-2$), double-strand break repair via homologous recombination (p value = $3.42E-3$), and so on, while hypomethylated genes are enriched with the following biological processes including negative regulation of epidermal growth factor receptor signaling pathway (p value = $3.24E-3$), DNA damage checkpoint (p value = $2.54E-2$), cell migration (p value = $1.42E-2$), and so on (see Figure 9 and Excel Sheet S3). Similar to before, a positive correlation (0.07) is observed between RNA methylation level and expression level; however, as there are 23 genes that carry hyper- and hypomethylated sites simultaneously, it is expected that RNA m^5C carries more complicated biomolecular functions.

4. Discussion and Conclusion

The distribution of m^5C methylation in mRNA has been mysterious with inconsistent evidence reported from previous studies [35, 37]. Here, we profiled the human and mouse m^5C epitranscriptome using RNA BS-Seq data in human

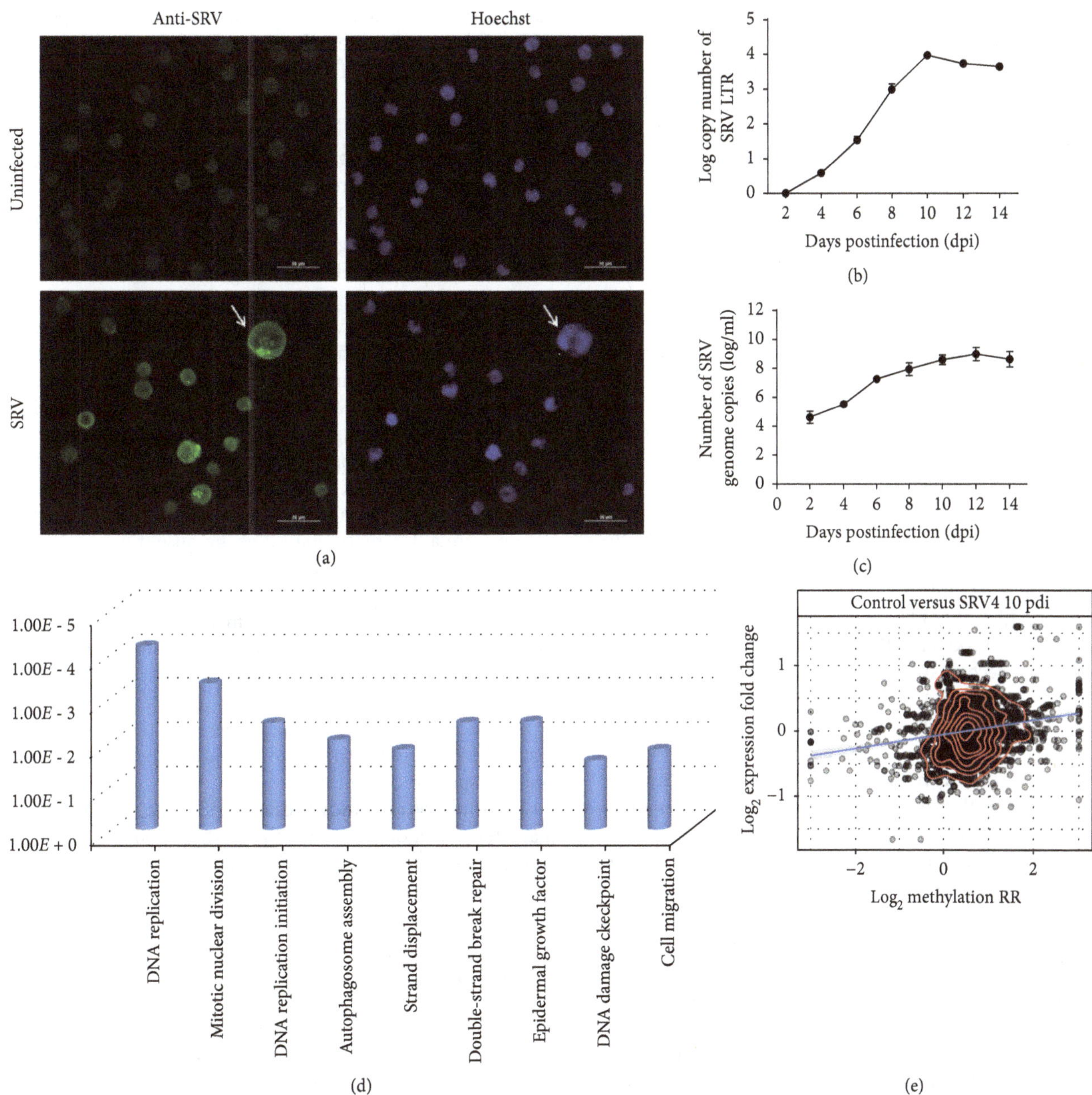

Figure 9: Dysregulation of RNA methylome after SRV infection of Jurkat cell. (a) At 10 days postinfection, uninfected or SRV-infected Jurkat cells were stained with SRV antibodies (green). Nuclei were visualized by Hoechst staining (blue). Arrows indicate the syncytium of infected cells. Scale bar: 50 μM. (b) The relative level of SRV-LTR in infected Jurkat cells was measured every two days by real-time PCR. GAPDH was used as the internal control. The relative level of SRV-LTR at each time point was normalized to the data at 2 dpi; mean ± SD, $n = 3$. (c) The absolute copy number of the SRV genome in culture medium was measured every two days by real-time PCR. SRV-LTR and SRV genome were not detected in all uninfected cells and culture medium, respectively; mean ± SD, $n = 3$. (d) The differentially methylated genes are enriched with the following functions including DNA replication (p value = 6.07E – 5), mitotic nuclear division (p value = 4.37E – 4), DNA replication initiation (p value = 3.48E – 3), autophagosome assembly (p value = 8.54E – 3), strand displacement (p value = 1.42E – 2), double-strand break repair via homologous recombination (p value = 3.42E – 3), and so on, while hypomethylated genes are enriched with the following biological processes including negative regulation of epidermal growth factor receptor signaling pathway (p value = 3.24E – 3), DNA damage checkpoint (p value = 2.54E – 2), cell migration (p value = 1.42E – 2), and so on (see Figure 9 and Excel Sheet S5). (e) A weak but positive correlation (Pearson correlation = 0.07) is observed between RNA methylation level and expression level, which is consistent with our previous result; however, there are 23 genes that carry hyper- and hypomethylated sites simultaneously, which implies that RNA m^{5}C carries more complicated biomolecular functions.

MCF10A, human MDA468, mouse ESC, and mouse whole brain cells. To eliminate the data sample bias, we employed a rigorous quality control procedure by filtering false positive m^{5}C sites due to the secondary structure and performed a comprehensive comparative analysis on cross-species conserved locus, cross-sample comparison of topological transcriptome distributions of m^5C, and differential m^5C analysis. Our analysis clearly shows that m^5C is enriched at the 5′UTR in human and mouse cells, confirming the discovery of a few independent studies [35, 36, 46]. Additionally, an unambiguous correlated methylation pattern is observed on 5′UTRs, but not on CDS and 3′UTR, in different mouse and human cell lines/tissues, suggesting a more complex aggregation pattern of m^5C that may be further characterized. Together, these observations strongly imply the functional relevance of m^5C RNA methylation and 5′UTR of mRNA. It is important to note that, although we failed to observe a correlated m^5C methylation pattern on CDS and 3′UTR regions of mRNA, it is still possible that such pattern may emerge on strictly matched cell lines/tissues.

When comparing the DNA and RNA methylome in matched cell lines in human and mouse, a negative correlation in the methylation level is observed on matched locus on DNA and RNA, which is quite surprising given that the methyltransferase of DNA and RNA may share strong sequence homology [52]. This anticorrelation pattern is consistent at all four CG-containing trinucleotide contexts and ruled out the possibility of sample contamination or off-target effect, which should both lead to false positive correlation in data. It is possible that there exists an underlying biomolecular mechanism that functions on the matched locus of DNA and RNA in parallel to ensure their orchestrated methylation status.

Similar to DNA methylation, a clustering effect of m^5C on mRNA is also observed in both human and mouse. The local dependency, that is, the adjacent cytosine locus often exhibits a similar methylation status, has been widely used in DNA methylation data analysis for more robust and accurate quantification of epigenetics status [57–59]. It is reasonable to expect that similar statistical approaches may be carried over into the field of single-base resolution RNA methylation data to enhance the analysis of bisulfite RNA methylation sequencing data. It is worth mentioning that, around 30%–43% of m^5C residuals exist in pairs in our results after filtering potential secondary structures that may lead to incomplete conversion and false positive m^5C sites. The number may be over- or underestimated because of the unfiltered secondary structure, which leads to an overestimation of the clustering effect, and structured regions excluded from the analysis, which may affect the estimation in both directions. It is necessary to develop a more sensitive unbiased approach that can eliminate the impact of the RNA structure to more accurately assess the distribution of transcriptome m^5C modification.

Intriguingly, we observed a strong enrichment of m^5C methylation on mitochondrial transcripts with more than 50 folds of enrichment. Previously, it was reported that methyltransferase NSUN5 can regulate mitochondrial gene expression [54], and we speculate that RNA m^5C methylation may play a more vital regulatory role in mitochondria-related biological processes.

Additionally, in order to have a glimpse of the dynamics of m^5C on mRNA, differential RNA methylation analysis was performed between breast cancer cell line MDA-MB-468 and the control cell line MCF10A; a total of 47 genes are reported to be differentially methylated, including RTN4, NME2, CASP14, HSPB1, RPL11, and RPS3, which are related to apoptosis and programmed cell death. Although we showed previously that m^5C on mRNA are more likely to be linked to the 5′UTR function, it is observed that the differential methylation sites between the breast cancer cell line and normal control cell lines are mostly located on the CDS and 3′UTR. These observations together implied a profound role of m^5C methylation on different regions of mRNA and in cancer pathology.

Interestingly, an overall positive correlation between RNA m^5C methylation and RNA expression level is observed in our mouse and human datasets, which added to the growing importance of mRNA m^5C methylation in regulating gene expression. Although the specific molecular mechanism is not yet clear, the observed positive correlation between RNA m^5C and RNA expression echoes our previous observed anticorrelation between DNA and RNA m^5C methylation from a different perspective, because it has been well established that DNA methylation is anticorrelated with RNA expression. However, as it is known that the most abundant RNA modification m^6A methylation may enhance or reduce the stability of the RNA molecule through interaction with different m^6A readers [14, 60] or regulate RN-protein interaction [13], it is reasonable to assume that RNA m^5C may have versatile functionalities, and may get dominated by a distinct mechanism under a specific condition.

In summary, our study presented an in-depth topological characterization of the m^5C RNA methylome in human and mouse. There are interesting patterns depicted and quantified, which call for further studies to explain novel biomolecular mechanisms.

Abbreviations

m^6A: N^6-Methyladenosine
hm^5C: Hydroxymethylcytosine
Ψ: Pseudouridine
m^1A: N^1-Methyladenosine
m^5C: 5-Methylcytosine
BS-Seq: Bisulfite sequencing
MDA-MB-468: MDA468
MEF: Mouse embryo fibroblast
MEF-Dnmt2$^-$: Mouse embryo fibroblast with Dnmt2 knockdown
WGBS: Whole genome bisulfite sequencing
RRBS: Reduced representation bisulfite sequencing
mRatio: Methylation ratio
OR: Odds ratio
DMS: Differential methylation site
lncRNA: Long noncoding RNA
sncRNA: Small noncoding RNA.

Conflicts of Interest

The authors declare that they have no competing interests.

Authors' Contributions

Jia Meng, João Pedro de Magalhães, and Yufei Huang conceived the project. Zhen Wei and Jia Meng carried out the computational analysis. Subbarayalu Panneerdoss, Santosh Timilsina, and Manjeet K. Rao performed the wet experiments, and Tabrez A. Mohammad and Yidong Chen performed bioinformatics support for BS-Seq. All authors provided valuable discussion and helped draft the manuscript. All authors read and approved the final manuscript. Zhen Wei, Subbarayalu Panneerdoss, Santosh Timilsina, and Jingting Zhu contributed equally to this work and are co-first author.

Acknowledgments

This work was supported by the National Natural Science Foundation of China (61401370 and 31671373 to Jia Meng, 81373469 to Zhi-Liang Lu), the Jiangsu Natural Science Foundation (BK20140403 to Jia Meng), the Jiangsu University Natural Science Program (16KJB180027 to Jia Meng), the US National Institutes of Health (R01GM113245 to Yufei Huang), the IIMS Translational Technology Resource (TTR) Award to Manjeet K. Rao and Yidong Chen, and NCI Cancer Center Shared Resources NCI P30CA54174 to Yidong Chen. Part of the BS-Seq experiment was performed by the Genome Sequencing Facility of the Greehey Children's Cancer Research Institute, UTHSCSA. The authors thank the computational support from the UTSA Computational Systems Biology Core, funded by the National Institute on Minority Health and Health Disparities (G12MD007591) from the National Institutes of Health.

References

[1] M. M. Suzuki and A. Bird, "DNA methylation landscapes: provocative insights from epigenomics," *Nature Reviews Genetics*, vol. 9, no. 6, pp. 465–476, 2008.

[2] K. D. Robertson, "DNA methylation and human disease," *Nature Reviews Genetics*, vol. 6, no. 8, pp. 597–610, 2005.

[3] L. E. Reinius, N. Acevedo, M. Joerink et al., "Differential DNA methylation in purified human blood cells: implications for cell lineage and studies on disease susceptibility," *PLoS One*, vol. 7, no. 7, article e41361, 2012.

[4] W. Xie, C. L. Barr, A. Kim et al., "Base-resolution analyses of sequence and parent-of-origin dependent DNA methylation in the mouse genome," *Cell*, vol. 148, no. 4, pp. 816–831, 2012.

[5] C. He, "Grand challenge commentary: RNA epigenetics?," *Nature Chemical Biology*, vol. 6, no. 12, pp. 863–865, 2010.

[6] U. Schibler, D. E. Kelley, and R. P. Perry, "Comparison of methylated sequences in messenger RNA and heterogeneous nuclear RNA from mouse L cells," *Journal of Molecular Biology*, vol. 115, no. 4, pp. 695–714, 1977.

[7] R. Desrosiers, K. Friderici, and F. Rottman, "Identification of methylated nucleosides in messenger RNA from Novikoff hepatoma cells," *Proceedings of the National Academy of Sciences of the United States of America*, vol. 71, no. 10, pp. 3971–3975, 1974.

[8] D. T. Dubin and R. H. Taylor, "The methylation state of poly A-containing-messenger RNA from cultured hamster cells," *Nucleic Acids Research*, vol. 2, no. 10, pp. 1653–1668, 1975.

[9] H. Grosjean, *Fine-Tuning of RNA Functions by Modification and Editing*, Springer, Berlin, Heidelberg, 2005.

[10] X. Wang, B. S. Zhao, I. A. Roundtree et al., "N^6-Methyladenosine modulates messenger RNA translation efficiency," *Cell*, vol. 161, no. 6, pp. 1388–1399, 2015.

[11] J. M. Fustin, M. Doi, Y. Yamaguchi et al., "RNA-methylation-dependent RNA processing controls the speed of the circadian clock," *Cell*, vol. 155, no. 4, pp. 793–806, 2013.

[12] C. R. Alarcon, H. Lee, H. Goodarzi, N. Halberg, and S. F. Tavazoie, "N^6-Methyladenosine marks primary microRNAs for processing," *Nature*, vol. 519, no. 7544, pp. 482–485, 2015.

[13] N. Liu, Q. Dai, G. Zheng, C. He, M. Parisien, and T. Pan, "N^6-Methyladenosine-dependent RNA structural switches regulate RNA-protein interactions," *Nature*, vol. 518, no. 7540, pp. 560–564, 2015.

[14] X. Wang, Z. Lu, A. Gomez et al., "N^6-Methyladenosine-dependent regulation of messenger RNA stability," *Nature*, vol. 505, no. 7481, pp. 117–120, 2014.

[15] J. Zhou, J. Wan, X. Gao, X. Zhang, S. R. Jaffrey, and S.-B. Qian, "Dynamic m^6A mRNA methylation directs translational control of heat shock response," *Nature*, vol. 526, no. 7574, pp. 591–594, 2015.

[16] S. Geula, S. Moshitch-Moshkovitz, D. Dominissini et al., "m^6A mRNA methylation facilitates resolution of naïve pluripotency toward differentiation," *Science*, vol. 347, no. 6225, pp. 1002–1006, 2015.

[17] S. Hussain, J. Aleksic, S. Blanco, S. Dietmann, and M. Frye, "Characterizing 5-methylcytosine in the mammalian epitranscriptome," *Genome Biology*, vol. 14, no. 11, p. 215, 2013.

[18] A. L. Burgess, R. David, and I. R. Searle, "Conservation of tRNA and rRNA 5-methylcytosine in the kingdom Plantae," *BMC Plant Biology*, vol. 15, no. 1, pp. 199–117, 2015.

[19] L. Warren, P. D. Manos, T. Ahfeldt et al., "Highly efficient reprogramming to pluripotency and directed differentiation of human cells with synthetic modified mRNA," *Cell Stem Cell*, vol. 7, no. 5, pp. 618–630, 2010.

[20] X. Zhang, Z. Liu, J. Yi et al., "The tRNA methyltransferase NSun2 stabilizes $p16^{INK4}$ mRNA by methylating the 3′ -untranslated region of p16," *Nature Communications*, vol. 3, no. 1, p. 712, 2012.

[21] Y. Chen, H. Sierzputowska-Gracz, R. Guenther, K. Everett, and P. F. Agris, "5-Methylcytidine is required for cooperative binding of magnesium(2+) and a conformational transition at the anticodon stem-loop of yeast phenylalanine tRNA," *Biochemistry*, vol. 32, no. 38, pp. 10249–10253, 1993.

[22] Y. Motorin and M. Helm, "tRNA stabilization by modified nucleotides," *Biochemistry*, vol. 49, no. 24, pp. 4934–4944, 2010.

[23] F. Tuorto, R. Liebers, T. Musch et al., "RNA cytosine methylation by Dnmt2 and NSun2 promotes tRNA stability and protein synthesis," *Nature Structural & Molecular Biology*, vol. 19, no. 9, pp. 900–905, 2012.

[24] M. Schaefer, T. Pollex, K. Hanna et al., "RNA methylation by Dnmt2 protects transfer RNAs against stress-induced cleavage," *Genes & Development*, vol. 24, no. 15, pp. 1590–1595, 2010.

[25] C. T. Y. Chan, Y. L. J. Pang, W. Deng et al., "Reprogramming of tRNA modifications controls the oxidative stress response by codon-biased translation of proteins," *Nature Communications*, vol. 3, no. 1, p. 937, 2012.

[26] K. Rai, S. Chidester, C. V. Zavala et al., "Dnmt2 functions in the cytoplasm to promote liver, brain, and retina development in zebrafish," *Genes & Development*, vol. 21, no. 3, pp. 261–266, 2007.

[27] C.-M. Wei, A. Gershowitz, and B. Moss, "Methylated nucleotides block 5′ terminus of HeLa cell messenger RNA," *Cell*, vol. 4, no. 4, pp. 379–386, 1975.

[28] S. Hussain, F. Tuorto, S. Menon et al., "The mouse cytosine-5 RNA methyltransferase NSun2 is a component of the chromatoid body and required for testis differentiation," *Molecular and Cellular Biology*, vol. 33, no. 8, pp. 1561–1570, 2013.

[29] S. Blanco, S. Dietmann, J. V. Flores et al., "Aberrant methylation of tRNAs links cellular stress to neuro-developmental disorders," *The EMBO Journal*, vol. 33, no. 18, pp. 2020–2039, 2014.

[30] C. S. Chow, T. N. Lamichhane, and S. K. Mahto, "Expanding the nucleotide repertoire of the ribosome with post-transcriptional modifications," *ACS Chemical Biology*, vol. 2, no. 9, pp. 610–619, 2007.

[31] M. Schaefer, "Chapter fourteen. RNA 5-methylcytosine analysis by bisulfite sequencing," *Methods in Enzymology*, vol. 560, pp. 297–329, 2015.

[32] M. Schaefer, T. Pollex, K. Hanna, and F. Lyko, "RNA cytosine methylation analysis by bisulfite sequencing," *Nucleic Acids Research*, vol. 37, no. 2, article e12, 2009.

[33] Y. Motorin, F. Lyko, and M. Helm, "5-Methylcytosine in RNA: detection, enzymatic formation and biological functions," *Nucleic Acids Research*, vol. 38, no. 5, pp. 1415–1430, 2010.

[34] V. Khoddami and B. R. Cairns, "Identification of direct targets and modified bases of RNA cytosine methyltransferases," *Nature Biotechnology*, vol. 31, no. 5, pp. 458–464, 2013.

[35] J. E. Squires, H. R. Patel, M. Nousch et al., "Widespread occurrence of 5-methylcytosine in human coding and non-coding RNA," *Nucleic Acids Research*, vol. 40, no. 11, pp. 5023–5033, 2012.

[36] T. Amort, D. Rieder, A. Wille et al., "Distinct 5-methylcytosine profiles in poly(A) RNA from mouse embryonic stem cells and brain," *Genome Biology*, vol. 18, no. 1, p. 1, 2017.

[37] X. Yang, Y. Yang, B. F. Sun et al., "5-Methylcytosine promotes mRNA export—NSUN2 as the methyltransferase and ALYREF as an m^5C reader," *Cell Research*, vol. 27, no. 5, pp. 606–625, 2017.

[38] J. M. Adams and S. Cory, "Modified nucleosides and bizarre 5′-termini in mouse myeloma mRNA," *Nature*, vol. 255, no. 5503, pp. 28–33, 1975.

[39] M. Salditt-Georgieff, W. Jelinek, J. E. Darnell, Y. Furuichi, M. Morgan, and A. Shatkin, "Methyl labeling of HeLa cell hnRNA: a comparison with mRNA," *Cell*, vol. 7, no. 2, pp. 227–237, 1976.

[40] E.-J. Lee, L. Pei, G. Srivastava et al., "Targeted bisulfite sequencing by solution hybrid selection and massively parallel sequencing," *Nucleic Acids Research*, vol. 39, no. 19, article e127, 2011.

[41] F. Neri, S. Rapelli, A. Krepelova et al., "Intragenic DNA methylation prevents spurious transcription initiation," *Nature*, vol. 543, no. 7643, pp. 72–77, 2017.

[42] F. Krueger, "Trim Galore. A wrapper tool around Cutadapt and FastQC to consistently apply quality and adapter trimming to FastQ files," 2015.

[43] D. Rieder, T. Amort, E. Kugler, A. Lusser, and Z. Trajanoski, "meRanTK: methylated RNA analysis ToolKit," *Bioinformatics*, vol. 32, no. 5, pp. 782–785, 2016.

[44] I. L. Hofacker, "Vienna RNA secondary structure server," *Nucleic Acids Research*, vol. 31, no. 13, pp. 3429–3431, 2003.

[45] M. Lawrence, W. Huber, H. Pagès et al., "Software for computing and annotating genomic ranges," *PLoS Computational Biology*, vol. 9, no. 8, article e1003118, 2013.

[46] X. Cui, Z. Wei, L. Zhang et al., "Guitar: an R/Bioconductor package for gene annotation guided transcriptomic analysis of RNA-related genomic features," *BioMed Research International*, vol. 2016, Article ID 8367534, 8 pages, 2016.

[47] H. Wickham, *ggplot2: Elegant Graphics for Data Analysis*, Springer Science & Business Media, New York, NY, USA, 2009.

[48] P. Du, X. Zhang, C.-C. Huang et al., "Comparison of beta-value and M-value methods for quantifying methylation levels by microarray analysis," *BMC Bioinformatics*, vol. 11, no. 1, p. 587, 2010.

[49] L. Liu, S. W. Zhang, Y. Huang, and J. Meng, "QNB: differential RNA methylation analysis for count-based small-sample sequencing data with a quad-negative binomial model," *BMC Bioinformatics*, vol. 18, no. 1, p. 387, 2017.

[50] M. I. Love, W. Huber, and S. Anders, "Moderated estimation of fold change and dispersion for RNA-seq data with DESeq2," *Genome Biology*, vol. 15, no. 12, p. 550, 2014.

[51] C. L. Zao, J. A. Ward, L. Tomanek, A. Cooke, R. Berger, and K. Armstrong, "Virological and serological characterization of SRV-4 infection in cynomolgus macaques," *Archives of Virology*, vol. 156, no. 11, pp. 2053–2056, 2011.

[52] M. G. Goll, F. Kirpekar, K. A. Maggert et al., "Methylation of $tRNA^{Asp}$ by the DNA methyltransferase homolog Dnmt2," *Science*, vol. 311, no. 5759, pp. 395–398, 2006.

[53] P. Qiu and L. Zhang, "Identification of markers associated with global changes in DNA methylation regulation in cancers," *BMC Bioinformatics*, vol. 13, Supplement 13, p. S7, 2012.

[54] Y. Cámara, J. Asin-Cayuela, C. B. Park et al., "MTERF4 regulates translation by targeting the methyltransferase NSUN4 to the mammalian mitochondrial ribosome," *Cell Metabolism*, vol. 13, no. 5, pp. 527–539, 2011.

[55] R. Pardal, M. F. Clarke, and S. J. Morrison, "Applying the principles of stem-cell biology to cancer," *Nature Reviews Cancer*, vol. 3, no. 12, pp. 895–902, 2003.

[56] B. Schwanhausser, D. Busse, N. Li et al., "Global quantification of mammalian gene expression control," *Nature*, vol. 473, no. 7347, pp. 337–342, 2011.

[57] X. Wang, J. Gu, L. Hilakivi-Clarke, R. Clarke, and J. Xuan, "DM-BLD: differential methylation detection using a hierarchical Bayesian model exploiting local dependency," *Bioinformatics*, vol. 33, no. 2, pp. 161–168, 2017.

[58] H.-U. Klein and K. Hebestreit, "An evaluation of methods to test predefined genomic regions for differential methylation in bisulfite sequencing data," *Briefings in Bioinformatics*, vol. 17, no. 5, pp. 796–807, 2016.

[59] S. Li, F. E. Garrett-Bakelman, A. Akalin et al., "An optimized algorithm for detecting and annotating regional differential methylation," *BMC Bioinformatics*, vol. 14, Supplement 5, p. S10, 2013.

[60] H. Huang, H. Weng, W. Sun et al., "Recognition of RNA N^6-methyladenosine by IGF2BP proteins enhances mRNA stability and translation," *Nature Cell Biology*, vol. 20, no. 3, pp. 285–295, 2018.

12

Changes in miRNA Gene Expression during Wound Repair in Differentiated Normal Human Bronchial Epithelium

Beata Narożna[1] **Wojciech Langwiński,**[1] **Claire Jackson,**[2] **Peter M. Lackie,**[2] **John W. Holloway**[2,3] **Zuzanna Stachowiak,**[1] **Monika Dmitrzak-Węglarz**[4] **and Aleksandra Szczepankiewicz**[1]

[1] *Laboratory of Molecular and Cell Biology, Department of Pediatric Pulmonology, Allergy and Clinical Immunology, Poznan University of Medical Sciences, Poznan, Poland*
[2] *Clinical and Experimental Sciences, Faculty of Medicine, University of Southampton, Southampton, UK*
[3] *Human Development and Health, Faculty of Medicine, University of Southampton, Southampton, UK*
[4] *Department of Psychiatric Genetics, Poznan University of Medical Sciences, Poznan, Poland*

Correspondence should be addressed to Aleksandra Szczepankiewicz; alszczep@gmail.com

Academic Editor: Mohamed Salem

Purpose. Airway epithelium acts as a protective barrier against the particles from the inhaled air. Damage to the epithelium may result in loss of the barrier function. Epithelial repair in response to injury requires complex mechanisms, such as microRNA, small noncoding molecules, to regulate the processes involved in wound repair. We aimed to establish if the microRNA gene expression profile is altered during the airway epithelial repair in differentiated cells. *Methods.* miRNA gene expression profile during the wound closure of differentiated normal human bronchial epithelium (NHBE) from one donor was analysed using quantitative real-time PCR. We have analysed the expression of 754 genes at five time points during a 48-hour period of epithelium repair using TaqMan Low Density Array. *Results.* We found out that 233 miRNA genes were expressed in normal human bronchial epithelium. Twenty miRNAs were differentially expressed during the wound repair process, but only one (miR-455-3p) showed significance after FDR adjustment ($p = 0.02$). Using STEM, we have identified two clusters of several miRNA genes with similar expression profile. Pathway enrichment analysis showed several significant signaling pathways altered during repair, mainly involved in cell cycle regulation, proliferation, migration, adhesion, and transcription regulation. *Conclusions.* miRNA expression profile is altered during airway epithelial repair of differentiated cells from one donor in response to mechanical injury *in vitro*, suggesting their potential role in wound repair.

1. Introduction

Upper respiratory airways are lined by pseudostratified columnar epithelium, composed of ciliated cells, goblet cells, and basal cells. The airway epithelium functions as a barrier that protects the lungs from inhaled pathogens and environmental particles [1]. The combined function of ciliated and secretory cells is responsible for maintaining efficient mucociliary clearance. Respiratory epithelium also plays a crucial role in the integration of innate and adaptive immune responses [2].

Damage to the airway epithelium might result in loss of barrier function and mucosal activation [3]. Physiological repair of the epithelium occurs in a series of overlapping processes. Studies in animals have shown that damage to the epithelium is rapidly restored by migration of remaining cells into the wound, followed by proliferation and differentiation [4, 5]. Epithelial injury and impaired repair underlie remodelling of the airways and underlie the development of respiratory diseases such as asthma [6]. Studies on cell cultures from asthmatic patients have shown that increased Th2 cell numbers and higher production of IL4 and IL14 cytokines result in decreased barrier integrity [7, 8].

MicroRNAs are small, noncoding RNA molecules that regulate gene expression either by degrading the target mRNA or by acting as translational enhancers or repressors. Several studies have shown their importance as regulators in development [9], proliferation [10], differentiation [11], apoptosis [12], immune response [13], and stem cell division [14]. However, the exact role of miRNA genes in the process of airway epithelial wound repair is still unclear. We hypothesized that miRNAs might coordinate the regulation of genes involved in the wound repair process. Our previous study has shown that microRNA profile was altered during bronchial epithelial repair in cells grown in monolayer (16HBE14o- cell line) [15]. The purpose of this study was to investigate whether changes in miRNA expression profile are also observed in the differentiated epithelium during repair. Since the epithelium injury and its abnormal repair are crucial in the pathogenesis of chronic airway diseases, such as asthma, the development of targeted therapy using inhibitors or synthetic miRNAs could be a breakthrough in the treatment of these diseases.

2. Materials and Methods

2.1. Cell Culture and Wounding Assays. NHBE cell line (Lonza) from one nonsmoking donor was cultured in a bronchial epithelium growth medium (BEBM medium with SingleQuot Kit Supplements and Growth Factors, Lonza) on 75 cm^2 until 85% confluent. Subsequently, the cells were passaged (3×10^5 cells on 1.2 cm^5) onto collagen-coated transwell inserts in a 12-well culture plate (Costar, Corning) and were cultivated until confluent. Afterwards, the basolateral chamber medium was switched to the air-liquid interface (ALI) (1 : 1, BEBM : DMEM, 3.5 g/L D-glucose with SingleQuots), supplemented with 100 nM retinoic acid (Sigma-Aldrich). The apical surface of the cell culture was then exposed to air. The medium was replaced three times per week. Any mucus that has appeared on the top of cells was systematically removed. Cilia were observed 3–4 weeks post-transition to ALI. Differentiated cells were scratched by P200 Gilson pipette tip. Cell debris was removed, and a fresh medium was added to the basolateral chamber.

2.2. Time-Lapse Microscopy. Time-lapse images were collected for 48 hours until complete wound closure at 15-minute intervals on the Olympus IX81 microscope, using xCellence software. The chamber was maintained at 36 ± 1°C and 5% CO_2 atmosphere. The wound area and the ongoing repair process were analysed using ImageJ software [16]. Wound recovery was calculated by tracing the new wound edge at each time interval and comparing the wound width to that of the original wound edge at the beginning of the experiment.

2.3. RNA Isolation. RNA was isolated from NHBE cells with the use of miRCURY RNA Isolation Kit—Cell and Plant (Exiqon), according to the manufacturer's instructions. Three biological replicates were collected at five consecutive time points: baseline (before wounding), 8, 16, 24, and 48 h after wounding. The amount of starting material was around 0.7×10^5 cells per well. Samples were stored at −70°C until the microarray experiment could be performed. The total RNA concentration was measured using NanoDrop 2000 spectrophotometer.

Table 1: Sequences of the primers used for target gene expression analysis.

Gene	Direction	Sequence
TGFB1	F	TTCAACACATCAGAGCTCC
	R	GCTGTATTTCTGGTACAGCT
TGFB3	F	CAAATTCAAAGGCGTGGAC
	R	ATTAGATGAGGGTTGTGGTG
TGFBR1	F	GAATCCTTCAAACGTGCTG
	R	TCATGAATTCCACCAATGGA
TGFBR2	F	GCTGTATGGAGAAAGAATGAC
	R	CAGAATAAAGTCATGGTAGGG
TGFBR3	F	TGATAATGGATTTCCGGGAG
	R	CTGCAATTAAACACCACGA
PPIA	F	AGACAAGGTCCCAAAGAC
	R	ACCACCCTGACACATAAA

2.4. MicroRNA Profiling. For profiling, we used the TaqMan Array Human MicroRNA Cards A and B that contain 754 human microRNAs. Reverse transcription was done using Megaplex Primer Pools (Human Pools A v.2.1 and B v.3.0, Thermo Fisher Scientific) and TaqMan® MicroRNA Reverse Transcription Kit (Thermo Fisher Scientific) according to the manufacturer's protocol. TaqMan Universal PCR Master Mix, No AmpErase UNG (Thermo Fisher Scientific) was combined with diluted cDNA and loaded into TaqMan Array Human MicroRNA Card A v.2.0 or Card B v.3.0 and centrifuged. Quantitative real-time PCR was conducted in the 7900HT Fast Real-Time PCR System (Applied Biosystems). The reaction was performed in triplicate for each sample. Raw expression data were acquired from SDS 2.4 software (Applied Biosystems) and further analysed with RQ Manager 1.2.1 (Applied Biosystems). The comparative analysis of obtained datasets between baseline and each time point was accomplished in DataAssist v.3.01 software (Applied Biosystems). Undetermined values were considered as equal to the maximum allowable Ct value (37). In order to reduce background noise, we have excluded miRNAs that were not expressed in 90% of the samples. Outliers were removed from the analysis after applying a refined Grubbs' outlier test. Each Ct value of target miRNA was normalized against the mean of the selected endogenous control, U6 snRNA-001973. Normalized miRNA expression was assessed against the baseline using the $2^{-\Delta Ct}$ method. All up- or downregulated miRNAs with a fold expression ≥2 and p value < 0.05 were considered to be differentially expressed. The p values were adjusted for multiple tests using the Benjamini-Hochberg false discovery rate (FDR).

2.5. Cluster Analysis. The clusters of miRNAs with similar expression profile over time were identified by cluster analysis in STEM (Short Time series Expression Miner) software available at http://www.cs.cmu.edu/~jernst/stem/ [17].

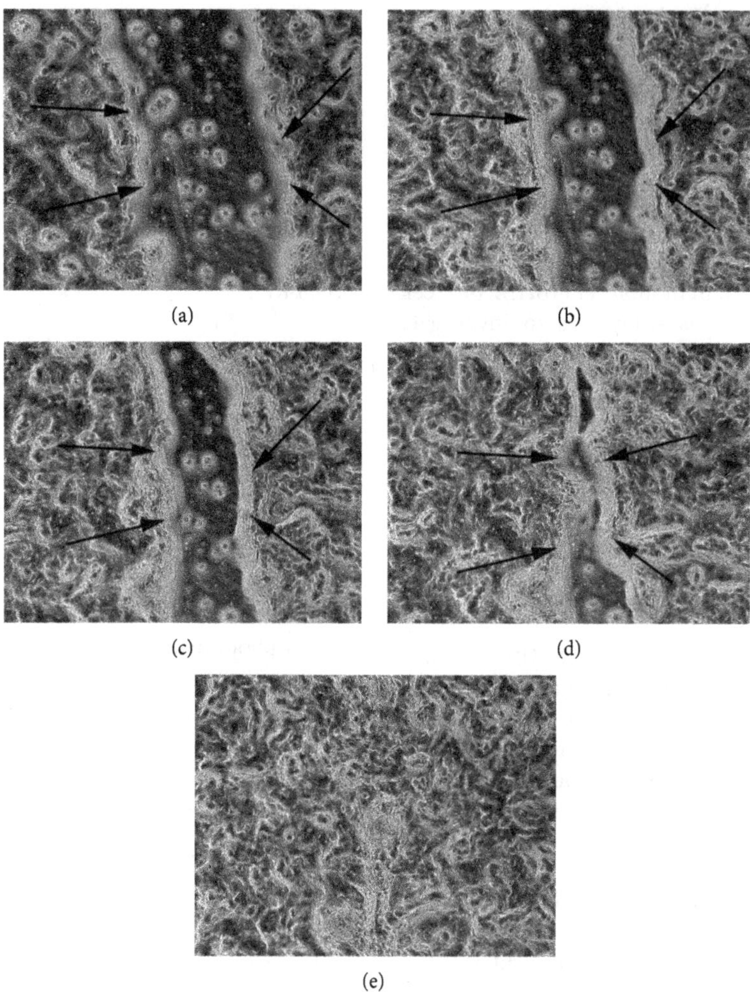

FIGURE 1: Representative images of wound repair at different time points: (a) 0 hrs, (b) 8 hrs, (c) 16 hrs, (d) 24 hrs, and (e) 48 hrs postwounding. $n = 3$ wells for each time point.

2.6. Target Genes and Pathways Prediction. We have performed pathway enrichment analysis to identify common biological pathways for miRNAs with similar expression profile. For each miRNA, we have identified the best predicted mRNA target genes using miRNA BodyMap tool (http://www.mirnabodymap.org). All available prediction algorithms were used: DIANA, PITA, TargetScan, RNA22 (3'UTR), RNA22 (5'UTR), TargetScan_cons, MicroCosm, miRDB, TarBase, and miRecords. However, to minimise the target prediction noise, we included the target genes predicted by at least four of these algorithms.

The list containing the best predictions was then analysed with the use of the Database for Annotation, Visualization and Integrated Discovery (DAVID) v.6.7 [18, 19], which allowed identifying BioCarta and KEGG pathways [20], enriched functional-related gene groups, and biological themes.

2.7. Mir-455-3p Targets. Based on available literature, target prediction results, and gene function, we have chosen the genes encoding isoforms of transforming growth factor β (*TGFβ1*, *TGFβ2*, and *TGFβ3*) and their receptors (*TGFβR1*, *TGFβR2*, and *TGFβR3*) for the gene expression analysis. Reverse transcription was done with the use of GoScript™ Reverse Transcription System (Promega), and resulting cDNA was analysed in quantitative real-time PCR using dye-based qPCR Master Mix (Promega) and set of specific primers spanning exon-exon junction (sequences are presented in Table 1). Each Ct value of target mRNA was normalized against an endogenous control, *PPIA*, which had shown to have the most stable expression in all samples tested (out of 10 potential endogenous controls).

3. Results

3.1. Epithelial Wound Repair Model. Based on the representative images from time-lapse microscopy, we have selected the following time points for miRNA profile analysis: the baseline immediately before injury (Figure 1(a)); 8 hours after wounding: the cells adjacent to the wound have initiated a response and have started to migrate, covering around 15% of the original wound area (SD 4.50, SEM 2.0) (Figure 1(b)); 16 hours after wounding: 35% of the wounded area has been covered by the cells (SD 4.24, SEM 1.89)

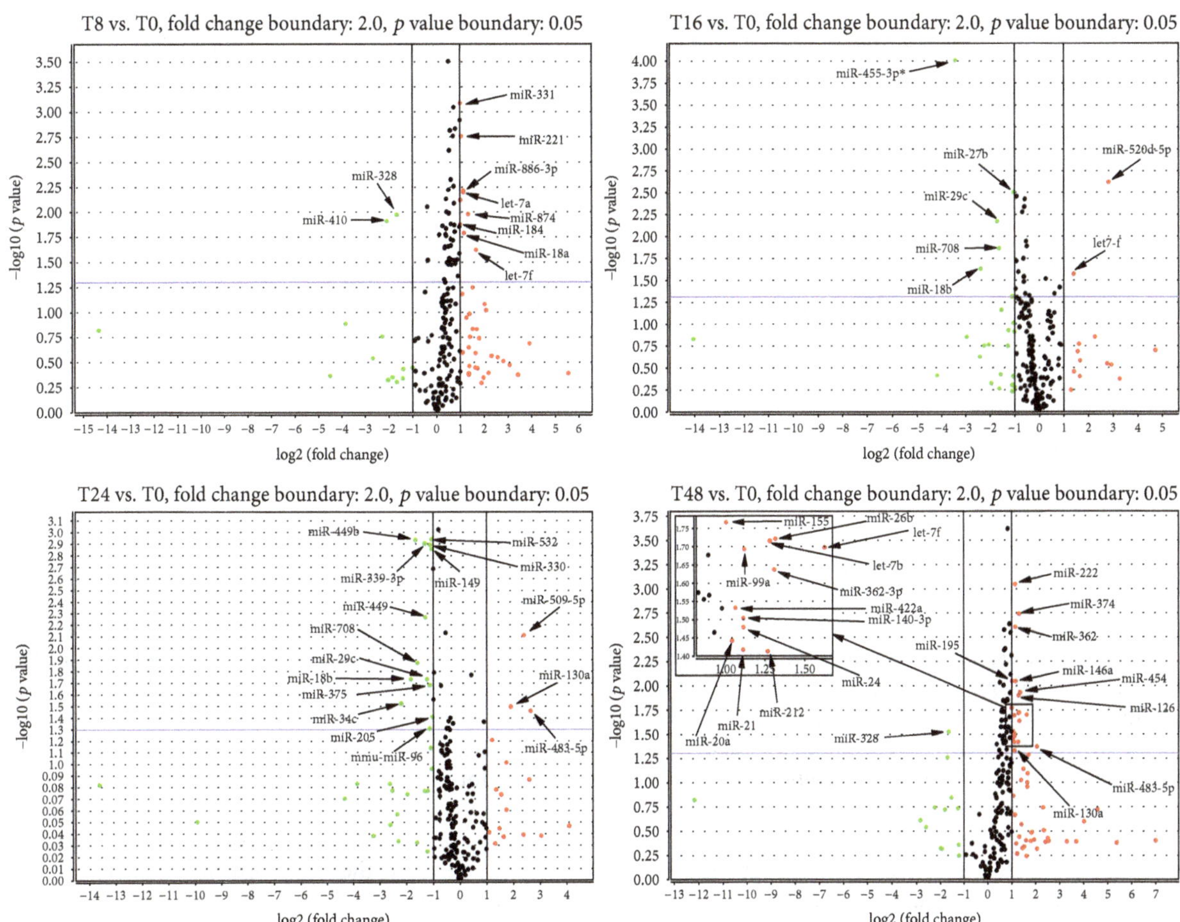

FIGURE 2: Volcano plots for different time points. T8: 8 h after wounding; T16: 16 h after wounding; T24: 24 h after wounding; T48: 48 h after wounding (reference: baseline; fold-change boundary: 2.0; p value boundary: 0.05; arrows are pointing out statistically significant results; *results remaining significant after FDR correction).

(Figure 1(c)); 24 hours after wounding: 50% of the wounded area covered by cells (SD 3.43, SEM 1.54) (Figure 1(d)); the wounded area entirely covered by cells (Figure 1(e)). Cell death was not observed in the repairing areas, with the exception of the cells damaged by the mechanical scratch.

3.2. Altered miRNA Expression Profile during Epithelial Wound Repair. After normalization, we found out that 230 miRNAs from Card A and 3 miRNAs from Card B were expressed in normal bronchial epithelium. Analysis of miRNA expression has revealed a large number of genes with significantly increased or decreased expression at different time points (fold change above 2.0, $p < 0.05$). Figure 2 shows volcano plots for each time point of wound repair (8, 16, 24, and 48 hours postwounding) compared to the baseline. After multiple testing correction (false discovery rate p value less than 0.05), we found significant changes in the expression for miR-455-3p ($p = 0.02$) 16 hours postwounding as compared to the baseline.

3.3. Cluster Analysis. To investigate if miRNA genes share a common expression profile during epithelial repair, we performed cluster analysis using the STEM algorithm. This calculation revealed that, out of 40 model profiles, two profiles (profiles 9 and 17) showed significant enrichment during repair ($p = 2.4E-21$ and $p = 5.2E-7$, respectively) (Figures 3 and 4). The list of miRNA genes for each profile can be found in Supplementary Materials (available here). Profile 9 groups 42 miRNA genes (12 of them differed significantly before FDR correction) and is characterised by a gradual increase in miRNA expression until 8 hours after wounding and decrease until 16 hours postwounding, followed by a significant increase at 24 hours of wound repair and decrease at 48 hours approximately to baseline. Profile 17, consisting of 11 genes (3 of them differed significantly before FDR correction), has a similar pattern until 24 hours, after which the decrease in expression is much smaller.

3.4. Pathway Enrichment Analysis. Our next step was to investigate whether miRNAs with the same expression patterns during airway epithelial repair may regulate target genes from the same biological pathways. Firstly, we have created a list of the best target mRNAs for each miRNA gene (8292 genes in total for profile 9 and 173939 genes for profile 17) and analysed these genes with DAVID online database. We have found several significantly enriched pathways for

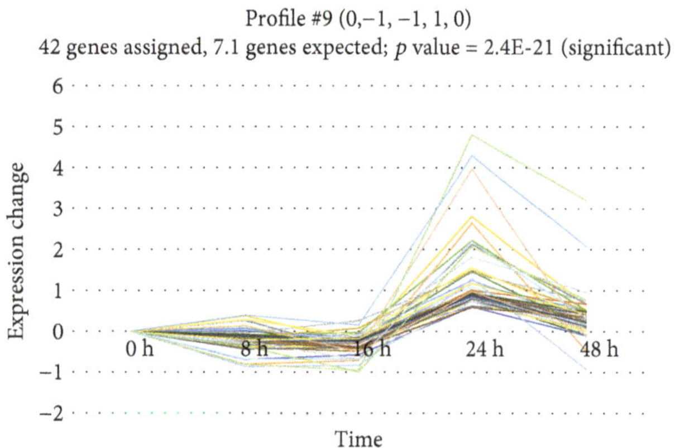

Figure 3: Profile 9 of differentially expressed miRNAs with similar expression pattern during wound repair.

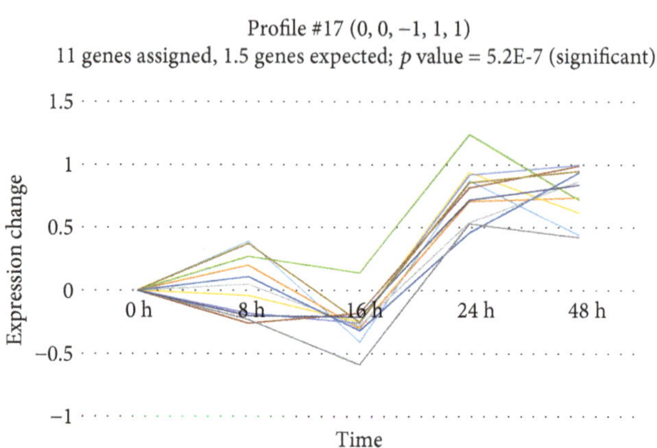

Figure 4: Profile 17 of differentially expressed miRNAs with similar expression pattern during wound repair.

each profile that remained significant after correction for multiple testing (Tables 2 and 3), including the MAPK signaling pathway (regulation of transcription and translation, inflammation, cell stress response, differentiation, division, proliferation, metabolism, motility, and apoptosis), PI3K-Akt signaling pathway (metabolism, growth, proliferation, survival, transcription, and protein synthesis), focal adhesion (cell motility, survival, proliferation, differentiation, and regulation of gene expression), regulation of actin cytoskeleton (cell motility and shape regulation, proliferation, secretion, phagocytosis, and cell communication), ErbB signaling pathway (cell proliferation, differentiation, motility, and survival), and neurotrophin signaling pathway.

3.5. Mir-455-3p Targets. Analysis of six targets from TGFβ family showed that the mRNA expression of two of them, *TGF-β1* and *TGF-βR3*, showed significantly altered expression upon repair (Figure 5). Their expression profiles during wound repair suggest negative correlation with the expression profile of miR-455-3p: the expression of these two genes significantly increased 8 hours postwounding, while the miRNA expression decreased, whereas between 8 and 24 hours of repair, the expression of *TGF-β1* and *TGF-βR3* decreased and miR-455-3p expression increased. The expression of the other four genes (*TGFβ2*, *TGF-β3*, *TGF-βR1*, and *TGF-βR2)* was not significantly changed during repair ($p > 0.05$).

4. Discussion

The most important finding of our study is the involvement of miRNA genes in the *in vitro* wound repair of the differentiated airway epithelium from one donor. That corresponds to the results of our previous research [15], reporting changes in miRNA expression in undifferentiated cell line (16HBE14o-). Our recent findings [21] have also confirmed the importance of miRNA during this process: silencing of *DICER* and *DROSHA*, the enzymes crucial for RNA interference, resulted in significantly delayed wound repair. Profiling analysis of both cell types (undifferentiated and pseudostratified epithelium) showed some similarities between the expressions of several miRNA genes, suggesting their importance in the fundamental repair processes.

Epithelial wound repair in the airways *in vivo* consists of cell spreading and migration within 12–24 hours postwounding. The proliferation process begins by 15–24 hours and can continue for days to weeks. While *in vitro* model mimics *in vivo* situation, it happens in a much shorter time frame. What is interesting, the NHBE cells require more time than undifferentiated 16HBE14o- until complete wound closure. Since most of the cells are ciliated, they need to dedifferentiate first. Therefore, the migration and proliferation observed in the first hours of the repair are most likely caused by the involvement of basal cells, suggesting that the ALI model may be more appropriate to study the wound repair.

The cluster analysis of time series miRNA expression data from one donor revealed distinct expression patterns of miRNA gene clusters during wound repair. The relationship between these miRNAs, their putative targets, and changes in individual protein levels during airway epithelial wound repair needs to be validated in future studies. However, the observed changes in the expression of the whole cluster of miRNA genes seem to be a consequence of the injury. For both profiles, we identified several signaling pathways responsible for the regulation of wound repair, some of which were also significant during the repair in undifferentiated 16HBE14o- cell line: the ErbB signaling pathway, MAPK signaling pathway, pathways in cancer, and neurotrophin signaling pathway [15].

Only the expression of one miRNA gene, miR-455-3p, was found to be significantly altered during the wound repair. MiR-455-3p was previously described as a tumour suppressor in several human cancers [22–24]. Gao et al. [25] found downregulated expression of this gene in non-small-cell lung cancer (NSCLC) tissues that correlated with poor prognosis of NSCLC patients. Furthermore, they revealed that miR-455-3p inhibited cell proliferation and migration *in vitro* via direct targeting *HOXB5*, a member of the HOX gene family. Decreased expression of miR-455-3p in our study complies with the previous findings and possibly enables to switch on genes involved in cell proliferation and migration required during wound repair.

Table 2: The results of pathway analysis of predicted target genes for profile 9 in DAVID database.

Category	Pathway	Enrichment score	No. of genes*	p	p (FDR corrected)
KEGG	Pathways in cancer	1.4	259	$5.70E-17$	$1.40E-13$
KEGG	PI3K-Akt signaling pathway	1.4	221	$1.90E-12$	$2.60E-09$
KEGG	MAPK signaling pathway	1.5	170	$7.80E-12$	$1.00E-08$
KEGG	Focal adhesion	1.5	141	$3.40E-11$	$4.50E-08$
KEGG	Axon guidance	1.6	92	$1.50E-09$	$2.00E-06$
KEGG	Ras signaling pathway	1.4	146	$7.30E-09$	$9.70E-06$
KEGG	cGMP-PKG signaling pathway	1.5	111	$3.90E-08$	$5.20E-05$
KEGG	Rap1 signaling pathway	1.4	135	$4.50E-08$	$6.00E-05$
KEGG	Proteoglycans in cancer	1.4	129	$7.00E-08$	$9.40E-05$
KEGG	FoxO signaling pathway	1.5	92	$1.00E-07$	$1.40E-04$
KEGG	Regulation of actin cytoskeleton	1.4	133	$3.10E-07$	$4.10E-04$
KEGG	Signaling pathways regulating pluripotency of stem cells	1.5	94	$3.70E-07$	$4.90E-04$
KEGG	T cell receptor signaling pathway	1.6	73	$3.90E-07$	$5.30E-04$
KEGG	Adrenergic signaling in cardiomyocytes	1.5	97	$5.00E-07$	$6.60E-04$
KEGG	Melanoma	1.7	54	$5.00E-07$	$6.70E-04$
KEGG	cAMP signaling pathway	1.4	125	$6.50E-07$	$8.70E-04$
KEGG	Glutamatergic synapse	1.5	78	$1.40E-06$	$1.90E-03$
KEGG	Wnt signaling pathway	1.4	91	$1.90E-06$	$2.60E-03$
KEGG	HTLV-I infection	1.3	154	$2.10E-06$	$2.80E-03$
KEGG	Dopaminergic synapse	1.5	85	$2.90E-06$	$3.90E-03$
KEGG	Renal cell carcinoma	1.7	49	$3.00E-06$	$3.90E-03$
KEGG	Neurotrophin signaling pathway	1.5	80	$4.70E-06$	$6.30E-03$
KEGG	TNF signaling pathway	1.5	72	$5.70E-06$	$7.60E-03$
KEGG	ErbB signaling pathway	1.5	61	$7.30E-06$	$9.80E-03$
KEGG	Sphingolipid signaling pathway	1.4	79	$1.10E-05$	$1.50E-02$
KEGG	Prostate cancer	1.5	61	$1.30E-05$	$1.70E-02$
KEGG	Insulin signaling pathway	1.4	88	$2.10E-05$	$2.80E-02$
KEGG	Inflammatory mediator regulation of TRP channels	1.5	66	$2.30E-05$	$3.00E-02$
KEGG	Dorsoventral axis formation	2	24	$2.30E-05$	$3.10E-02$
KEGG	Pancreatic cancer	1.6	47	$3.00E-05$	$4.10E-02$
KEGG	Glioma	1.6	47	$3.00E-05$	$4.10E-02$
KEGG	Oxytocin signaling pathway	1.4	98	$3.50E-05$	$4.70E-02$
KEGG	Platelet activation	1.4	83	$3.50E-05$	$4.70E-02$
KEGG	Insulin resistance	1.4	71	$3.60E-05$	$4.80E-02$

*Number of predicted target genes that are involved in the particular pathway.

Table 3: The results of pathway analysis of predicted target genes for profile 17 in DAVID database.

Category	Pathway	Enrichment score	No. of genes*	p	p (FDR corrected)
KEGG	Pathways in cancer	1.5	128	$4.40E-07$	$5.90E-04$
KEGG	Proteoglycans in cancer	1.7	74	$1.10E-06$	$1.40E-03$
KEGG	MAPK signaling pathway	1.6	88	$2.80E-06$	$3.70E-03$
KEGG	Inflammatory mediator regulation of TRP channels	2	42	$5.80E-06$	$7.80E-03$
KEGG	PI3K-Akt signaling pathway	1.5	110	$9.60E-06$	$1.30E-02$
KEGG	GnRH signaling pathway	2	39	$1.30E-05$	$1.80E-02$
KEGG	Rap1 signaling pathway	1.6	72	$3.20E-05$	$4.30E-02$

*Number of predicted target genes that are involved in the particular pathway.

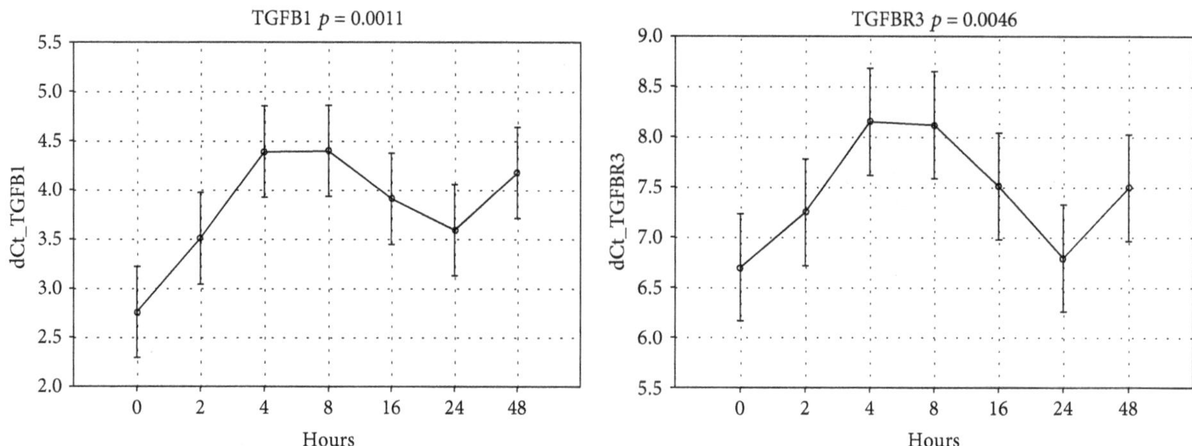

FIGURE 5: ANOVA analysis of TGF-β1, TGF-β3, TGF-βR1, TGF-βR2, and TGF-βR3 changes in mRNA expression.

Martinez-Anton et al. [11] have studied the differences in miRNA expression changes during differentiation of normal human bronchial cells. MiR-455-3p expression was significantly downregulated during differentiation; we have also observed its decreased expression during the epithelial wound repair. Their confirmatory experiments also showed that MUC1 mRNA is a target gene of miR-455-3p. Since MUC1 encodes mucin responsible for protecting the cells from pathogens, decreased expression of miR-455 in our study suggests that the injury results in the loss of barrier function.

TGF-β is a multifunctional cytokine, involved in cell growth, inflammation, and repair [26]. Ong et al. have discovered that miR-455-3p was induced by TGF-β in lung fibroblast, and immunoprecipitation of Ago2 revealed enrichment of miR-455-3p targets related to TGF-β and/or the Wnt pathway [27]. Our results showed that the expression profiles of TGF-β1 and TGF-βR3 correspond to the changes in the expression of miR-455-3p, suggesting this miRNA is possibly involved in posttranscriptional regulation of these two TGF genes.

The main limitation of our study was the use of biological replicates from only one donor. While changes in miRNA expression during wound repair were observed in cells from one donor and in other airway epithelium cell line, we cannot state for certain that this will be the case for other donors, due to donor-to-donor variability. Additional experiments including more donors, as well as the confirmation of the miR-455-3p involvement in *TGF-β1* and *TGF-βR3* expression regulation, are planned in future research.

In conclusion, we showed that the expression of multiple miRNAs is altered during airway epithelial repair in differentiated cells from one donor, suggesting their importance in the regulation of this process. We also observed two common expression profiles for several miRNA genes, and in silico analysis of predicted mRNA target genes has shown that they coordinate signaling pathways involved in the wound repair. We also found out that the *TGF* gene family is a possible target of miRNA genes altered upon epithelial repair, in particular miR-455-3p. However, further experiments on cells from more donors are required to investigate the exact role of identified miRNAs and their targets in epithelial repair.

Conflicts of Interest

The authors declare that they have no conflict of interest.

Acknowledgments

BN was a recipient of an EAACI Research Fellowship 2015. We thank the personnel from the Biomedical Imaging Unit and Primary Ciliary Dyskinesia (PCD) Diagnostic and Research Team, University of Southampton, for the assistance and technical support. Target gene expression analysis was supported by Poznan University of Medical Sciences grant no. 502-14-01105122-10347; cluster expression analysis and pathways prediction were financed by Polish National Science Centre grant no. 2017/25/N/NZ3/00332, whereas the rest of the study was funded by Polish National Science Centre grant no. 2011/01/M/NZ3/02906.

References

[1] R. W. A. Godfrey, "Human airway epithelial tight junctions," *Microscopy Research & Technique*, vol. 38, no. 5, pp. 488–499, 1997.

[2] A. Kato and R. P. Schleimer, "Beyond inflammation: airway epithelial cells are at the interface of innate and adaptive immunity," *Current Opinion in Immunology*, vol. 19, no. 6, pp. 711–720, 2007.

[3] Q. Sha, A. Q. Truong-Tran, J. R. Plitt, L. A. Beck, and R. P. Schleimer, "Activation of airway epithelial cells by toll-like receptor agonists," *American Journal of Respiratory Cell and Molecular Biology*, vol. 31, no. 3, pp. 358–364, 2004.

[4] K. P. Keenan, T. S. Wilson, and E. M. McDowell, "Regeneration of hamster tracheal epithelium after mechanical injury," *Virchows Archiv B Cell Pathology Including Molecular Pathology*, vol. 43, no. 1, pp. 213–240, 1983.

[5] J. M. Zahm, H. Kaplan, A. L. Herard et al., "Cell migration and proliferation during the in vitro wound repair of the respiratory epithelium," *Cell Motility and the Cytoskeleton*, vol. 37, no. 1, pp. 33–43, 1997.

[6] S. T. Holgate, "The sentinel role of the airway epithelium in asthma pathogenesis," *Immunological Reviews*, vol. 242, no. 1, pp. 205–219, 2011.

[7] P. Wawrzyniak, M. Wawrzyniak, K. Wanke et al., "Regulation of bronchial epithelial barrier integrity by type 2 cytokines and histone deacetylases in asthmatic patients," *The Journal of Allergy and Clinical Immunology*, vol. 139, no. 1, pp. 93–103, 2017.

[8] B. Jakiela, A. Gielicz, H. Plutecka et al., "Th2-type cytokine-induced mucus metaplasia decreases susceptibility of human bronchial epithelium to rhinovirus infection," *American Journal of Respiratory Cell and Molecular Biology*, vol. 51, no. 2, pp. 229–241, 2014.

[9] X. Karp and V. Ambros, "Developmental biology. Encountering microRNAs in cell fate signaling," *Science*, vol. 310, no. 5752, pp. 1288-1289, 2005.

[10] M. J. Bueno, I. P. de Castro, and M. Malumbres, "Control of cell proliferation pathways by microRNAs," *Cell Cycle*, vol. 7, no. 20, pp. 3143–3148, 2008.

[11] A. Martinez-Anton, M. Sokolowska, S. Kern et al., "Changes in microRNA and mRNA expression with differentiation of human bronchial epithelial cells," *American Journal of Respiratory Cell and Molecular Biology*, vol. 49, no. 3, pp. 384–395, 2013.

[12] P. Xu, M. Guo, and B. A. Hay, "MicroRNAs and the regulation of cell death," *Trends in Genetics*, vol. 20, no. 12, pp. 617–624, 2004.

[13] A. Liston, M. Linterman, and L. F. Lu, "MicroRNA in the adaptive immune system, in sickness and in health," *Journal of Clinical Immunology*, vol. 30, no. 3, pp. 339–346, 2010.

[14] S. D. Hatfield, H. R. Shcherbata, K. A. Fischer, K. Nakahara, R. W. Carthew, and H. Ruohola-Baker, "Stem cell division is regulated by the microRNA pathway," *Nature*, vol. 435, no. 7044, pp. 974–978, 2005.

[15] A. Szczepankiewicz, P. M. Lackie, and J. W. Holloway, "Altered microRNA expression profile during epithelial wound repair in bronchial epithelial cells," *BMC Pulmonary Medicine*, vol. 13, no. 1, p. 63, 2013.

[16] C. A. Schneider, W. S. Rasband, and K. W. Eliceiri, "NIH image to ImageJ: 25 years of image analysis," *Nature Methods*, vol. 9, no. 7, pp. 671–675, 2012.

[17] J. Ernst and Z. Bar-Joseph, "STEM: a tool for the analysis of short time series gene expression data," *BMC Bioinformatics*, vol. 7, no. 1, p. 191, 2006.

[18] D. W. Huang, B. T. Sherman, and R. A. Lempicki, "Systematic and integrative analysis of large gene lists using DAVID bioinformatics resources," *Nature Protocols*, vol. 4, no. 1, pp. 44–57, 2009.

[19] D. W. Huang, B. T. Sherman, X. Zheng et al., "Extracting biological meaning from large gene lists with DAVID," *Current Protocols in Bioinformatics*, vol. 27, no. 1, pp. 13.11.1–13.11.13, 2009.

[20] M. Kanehisa, S. Goto, S. Kawashima, Y. Okuno, and M. Hattori, "The KEGG resource for deciphering the genome," *Nucleic Acids Research*, vol. 32, pp. D277–D280, 2004.

[21] W. Langwinski, B. Narozna, P. M. Lackie, J. W. Holloway, and A. Szczepankiewicz, "Comparison of miRNA profiling during airway epithelial repair in undifferentiated and differentiated cells in vitro," *Journal of Applied Genetics*, vol. 58, no. 2, pp. 205–212, 2017.

[22] H. Yang, Y. N. Wei, J. Zhou, T. T. Hao, and X. L. Liu, "MiR-455-3p acts as a prognostic marker and inhibits the proliferation and invasion of esophageal squamous cell carcinoma by targeting FAM83F," *European Review for Medical and Pharmacological Sciences*, vol. 21, no. 14, pp. 3200–3206, 2017.

[23] J. Zheng, Z. Lin, L. Zhang, and H. Chen, "MicroRNA-455-3p inhibits tumor cell proliferation and induces apoptosis in HCT116 human colon cancer cells," *Medical Science Monitor*, vol. 22, pp. 4431–4437, 2016.

[24] Y. Zhao, M. Yan, Y. Yun et al., "MicroRNA-455-3p functions as a tumor suppressor by targeting eIF4E in prostate cancer," *Oncology Reports*, vol. 37, no. 4, pp. 2449–2458, 2017.

[25] X. Gao, H. Zhao, C. Diao et al., "miR-455-3p serves as prognostic factor and regulates the proliferation and migration of non-small cell lung cancer through targeting HOXB5," *Biochemical and Biophysical Research Communications*, vol. 495, no. 1, pp. 1074–1080, 2018.

[26] D. A. Clark and R. Coker, "Molecules in focus transforming growth factor-beta (TGF-β)," *The International Journal of Biochemistry & Cell Biology*, vol. 30, no. 3, pp. 293–298, 1998.

[27] J. Ong, W. Timens, V. Rajendran et al., "Identification of transforming growth factor-beta-regulated microRNAs and the microRNA-targetomes in primary lung fibroblasts," *PLoS One*, vol. 12, no. 9, article e0183815, 2017.

13

Differential Expression Profiling of Long Noncoding RNA and mRNA during Osteoblast Differentiation in Mouse

Minjung Kim,[1] Youngseok Yu,[1] Ji-Hoi Moon,[1,2] InSong Koh,[3,4] and Jae-Hyung Lee[1,2]

[1]*Department of Life and Nanopharmaceutical Sciences, Kyung Hee University, Seoul, Republic of Korea*
[2]*Department of Maxillofacial Biomedical Engineering, School of Dentistry, and Institute of Oral Biology, Kyung Hee University, Seoul, Republic of Korea*
[3]*Department of Physiology, College of Medicine, Hanyang University, Seoul, Republic of Korea*
[4]*Department of Biomedical Informatics, Hanyang University, Seoul, Republic of Korea*

Correspondence should be addressed to InSong Koh; insong@hanyang.ac.kr and Jae-Hyung Lee; jaehlee@khu.ac.kr

Academic Editor: Ignazio Piras

Long noncoding RNAs (lncRNAs) are emerging as an important controller affecting metabolic tissue development, signaling, and function. However, little is known about the function and profile of lncRNAs in osteoblastic differentiation in mice. Here, we analyzed the RNA-sequencing (RNA-Seq) datasets obtained for 18 days in two-day intervals from neonatal mouse calvarial pre-osteoblast-like cells. Over the course of osteoblast differentiation, 4058 mRNAs and 3948 lncRNAs were differentially expressed, and they were grouped into 12 clusters according to the expression pattern by fuzzy c-means clustering. Using weighted gene coexpression network analysis, we identified 9 modules related to the early differentiation stage (days 2–8) and 7 modules related to the late differentiation stage (days 10–18). Gene ontology and KEGG pathway enrichment analysis revealed that the mRNA and lncRNA upregulated in the late differentiation stage are highly associated with osteogenesis. We also identified 72 mRNA and 89 lncRNAs as potential markers including several novel markers for osteoblast differentiation and activation. Our findings provide a valuable resource for mouse lncRNA study and improves our understanding of the biology of osteoblastic differentiation in mice.

1. Introduction

Ossification is a tightly regulated process which is performed by specialized cells called osteoblasts differentiated from mesenchymal progenitors. The osteoblast differentiation process is regulated by several key factors and signaling pathways. Runt-related transcription factor 2 (Runx2) is the master switch in the commitment of mesenchymal progenitors to osteoblast lineage [1]. Runx2 is affected by several upstream regulators such as the Wnt/Notch system, Sox9, Msx2, and hedgehog signaling as well as by cofactors such as Osx and Atf4 [1–4]. A few paracrine and endocrine factors, including bone morphogenetic proteins (BMP) and parathyroid hormone, serve as coactivators. Vitamin D and histone deacetylase enzymes coordinate this process more finely [1].

Long noncoding RNAs (lncRNAs) are a class of RNA transcripts longer than 200 nucleotides, lacking open reading frames and protein-coding possibilities [5]. Tens of thousands of lncRNAs have been identified in mammalian genomes in recent decades [6]. Recently, in many studies, lncRNA has emerged as an important regulator in a variety of biological processes, such as epigenetic regulation, chromatin remodeling, genomic imprinting, transcriptional control, and pre-/posttranslational mRNA processing [7–11]. In terms of osteogenesis, several lncRNAs have been found to act as key regulators. One such example is maternal expression gene 3 (MEG3) regulating the expression of Bmp4, Runx2, and Osx in human mesenchymal stem cells [12, 13]. Antidifferentiation ncRNA (ANCR) inhibits Runx2 expression in association with the enhancer of zeste homolog 2

(EZH2); thus, downregulation of ANCR promotes osteoblast differentiation through modulation of EZH2/Runx2 [14].

In order to fully understand the lncRNA biology including its role in osteogenesis, it is necessary to characterize the expression pattern of lncRNA during osteoblast differentiation. In this study, we analyzed RNA-sequencing datasets obtained at nine different time points in the osteoblast differentiation of preosteoblasts isolated from neonatal mouse calvaria, using various bioinformatic approaches. We focused on identifying differentially expressed lncRNAs and mRNAs throughout the process and finding potential markers that exhibited significant changes in expression during the osteoblast differentiation.

2. Materials and Methods

2.1. RNA-Seq Data Processing and Analysis of Differential Gene Expression. We analyzed the RNA-Seq data generated by Kemp et al. [15] (GSE54461, nine time points: 2, 4, 6, 8, 10, 12, 14, 16, and 18 days of osteoblast differentiation). All sequencing reads were aligned to the mouse genome reference (mm10) using the GSNAP alignment tool [16]. Ensembl release 74 annotations were used to measure gene expression. In the case of lncRNAs, NONCODE v4 (http://www.noncode.org) annotations were used. Since NONCODE v4 annotations were based on mm9 mouse genome assembly, the positions of lncRNAs of NONCODE were converted from mm9 to mm10 using the LiftOver utility in UCSC. To assess gene expression, RPKM (reads per kilobase of exon per million mapped reads) values were calculated [17]. Hierarchical clustering of genes expressed in samples at the nine time points was performed using the flashClust R package. To establish the differences in gene expression patterns among nine time points of osteoblast differentiation, we performed differential gene expression analysis using the R package DESeq [18]. The false discovery rate (FDR) was controlled by adjusting p values using the Benjamini–Hochberg algorithm. Differentially expressed genes were defined as those with FDR less than 5% with an absolute value of fold change ≥ 2. Similar differential expression analyses were performed for lncRNAs.

2.2. Time Series Analysis of Differential Gene Expression. R package DESeq [18] and edgeR [19] were employed to identify genes that were differentially expressed across the differentiation time period to a significant extent, designated as time series genes. We selected time series genes that displayed significant differential expression with FDR < 5% in both DESeq and edgeR, absolute fold change ≥ 2 (between day 2 and at least one other time point), and maximum RPKM ≥ 3 across the time series. Details of the methods for each package are described in the supplementary methods. Similarly, time series expression tests were conducted for lncRNAs. Using the RPKM of each time series gene and lncRNA, principal component analysis (PCA) was performed with the aid of the "prcomp" module in R.

2.3. Time Series Gene Clustering. The time series genes identified were clustered using the R package Mfuzz [20] that performs soft clustering based on the fuzzy c-means algorithm. The advantage of soft clustering is that the algorithm clearly reflects the strength of association of an individual gene with a cluster. Average RPKM values (triplicates at each time point) of individual genes were employed as input values for Mfuzz clustering. The number of clusters was set to 12 and the fuzzifier coefficient, M, to 1.5. Heat maps of the clusters were drawn using the R module "heatmap.2" in the "gplots" package [21].

2.4. Weighted Gene Coexpression Network Analysis. Gene coexpression network analysis was performed using the R package "WGCNA" [22]. Details of the methods for constructing gene coexpression network analysis are described in the supplementary methods.

2.5. Gene Ontology (GO) Term and KEGG Pathway Enrichment Analysis. GO terms of each gene were obtained from Ensembl BioMart and KEGG pathways from the KEGG PATHWAY database. Details of the procedure to conduct the enrichment analysis are described in the supplementary methods.

2.6. Analysis of Motif Enrichment. MEME Suite 4.12.20 [23] and the HOmo sapiens COmprehensive MOdel COllection (HOCOMOCO) v11 mouse transcription factor database [24] were used for the identification of motifs in the promoter regions on the lncRNAs identified as potential markers. Details of the procedure to perform the motif enrichment analysis are described in the supplementary methods.

3. Results and Discussion

For convenience, the analysis results of mRNA and lncRNAs were described separately. The overall expression pattern was described first, followed by time series analysis, generation of modules by weighted gene coexpression network analysis, and identification of markers for osteoblast differentiation (Supplementary Figure 1).

3.1. Expression Profile of mRNA during Osteoblast Differentiation. A total of 12 to 33 million reads (at each time point) were processed and mapped against mouse genome reference (mm10) sequence, and uniquely mapped reads (89.01–90.74% of the total reads) were used for further analysis (Table 1). Based on Ensembl release 74 annotations, a total of 28,582 genes had at least one read for whole RNA-Seq datasets. By hierarchical clustering [25], two distinct clusters were generated for the early (2, 4, 6, and 8 days) and late (10, 12, 14, 16, and 18 days) differentiation stages (Figure 1(a)). As expected, the number of differentially expressed genes increased with time (Figure 1(b)).

Comparing the expression of individual genes at different time points using all possible combinations, 46% of the mapped genes (13,130 out of 28,582) were differentially expressed over time (fold-change difference ≥ 2, FDR < 0.05). For example, a comparison of gene expression between the two time points selected in the early and late stages (day 2 and day 18) is shown in Figure 1(c). A total of 7238

TABLE 1: RNA-Seq read mapping summary.

SRA ID	Sample	Total reads	Uniquely mapped reads	% of uniquely mapped reads
SRR1146385	B6_2_rep1	20,605,665	18,353,730	89.07
SRR1146386	B6_2_rep2	21,204,511	18,881,462	89.04
SRR1146387	B6_2_rep3	21,196,674	18,867,579	89.01
SRR1146388	B6_4_rep1	21,042,912	18,976,744	90.18
SRR1146389	B6_4_rep2	21,636,813	19,507,642	90.16
SRR1146390	B6_4_rep3	21,628,526	19,496,233	90.14
SRR1146391	B6_6_rep1	22,012,993	19,750,118	89.72
SRR1146392	B6_6_rep2	22,652,935	20,316,973	89.69
SRR1146393	B6_6_rep3	22,656,580	20,317,497	89.68
SRR1146394	B6_8_rep1	16,068,544	14,395,620	89.59
SRR1146395	B6_8_rep2	16,509,407	14,785,805	89.56
SRR1146396	B6_8_rep3	16,498,155	14,769,672	89.52
SRR1146397	B6_10_rep1	11,685,016	10,532,491	90.14
SRR1146398	B6_10_rep2	12,016,157	10,826,863	90.10
SRR1146399	B6_10_rep3	12,016,476	10,827,771	90.11
SRR1146400	B6_12_rep1	32,109,724	29,035,678	90.43
SRR1146401	B6_12_rep2	33,017,225	29,851,897	90.41
SRR1146402	B6_12_rep3	32,970,446	29,805,326	90.40
SRR1146403	B6_14_rep1	22,996,425	20,867,104	90.74
SRR1146404	B6_14_rep2	23,630,602	21,437,833	90.72
SRR1146405	B6_14_rep3	23,616,942	21,422,044	90.71
SRR1146406	B6_16_rep1	26,956,212	24,189,458	89.74
SRR1146407	B6_16_rep2	27,696,210	24,846,901	89.71
SRR1146408	B6_16_rep3	27,659,416	24,807,073	89.69
SRR1146409	B6_18_rep1	22,811,796	20,549,338	90.08
SRR1146410	B6_18_rep2	23,478,940	21,148,934	90.08
SRR1146411	B6_18_rep3	23,452,037	21,114,152	90.03

genes were differentially expressed between days 2 and 18 (4431 upregulated and 2807 downregulated on day 18). GO and KEGG pathway analyses revealed that genes upregulated on day 18 were strongly associated with osteogenesis processes, such as "extracellular matrix binding," "positive regulation of bone mineralization," "collagen type I," "elevation of cytosolic calcium ion concentration," "bone mineralization involved in bone maturation," and "positive regulation of Wnt receptor signaling pathway" (Supplementary Table 1). On the other hand, genes downregulated on day 18 were associated with cell proliferation processes, such as "cell division," "G1/S transition of mitotic cell cycle," "regulation of cell cycle," "M phase of mitotic cell cycle," and "DNA replication and response to DNA damage stimulus" (Supplementary Table 2).

3.2. Dynamic Changes of mRNA Expression over Time of Osteoblast Differentiation. During the differentiation process, several genes are dynamically expressed via complex regulatory mechanisms. To characterize temporal gene expression changes and patterns, we identified genes that were differentially expressed across the time course of osteoblast differentiation. A number of rules were applied to establish significant differential expression of genes at different time points: (1) statistical significance of temporal gene expression changes assessed via edgeR and DESeq methods (FDR < 0.05), (2) absolute fold change between day 2 and at least one other time point ≥ 2, and (3) maximum RPKM across the time series ≥ 3.

In total, 4058 genes were significantly differentially expressed during osteoblast differentiation. They were associated with osteogenesis processes, such as "extracellular matrix organization," "osteoblast proliferation," "osteoblast differentiation," "positive regulation of osteoblast differentiation," and "actin cytoskeleton and bone mineralization." Principal component analysis (PCA) of these genes revealed that 95% of the variations in gene expression could be explained by the first two principal components (PCs) (Supplementary Figure 2) and that the PCs dominantly separate the datasets according to differentiation stage (days 2–8 versus days 10–18) (Figure 1(d)). To evaluate the osteoblast differentiation stages in the current dataset, we have checked the expression level of the twelve mature osteoblast markers (Bglap, Ibsp, Dmp1, Col13a1, Pthr1, Lifr, Bambi, Dlx3, Hnf1a, Phex, Ptgis, and Cdo1) described in Kalajzic et al. [26]. The nine genes, except three genes (Pthr1, Bambi, and Hnf1a), exhibited very similar expression patterns and showed the highest expression at day 18, which indicate that the cells at day 18 were mature osteoblasts (Supplementary Figure 3).

Osteoblasts undergo several stages before maturation and mineralization of bone matrix. Differentiation of osteoblasts, both in vitro and in vivo, is divided into three stages: (1) cell proliferation, (2) matrix maturation, and (3) matrix mineralization [27]. Here, we examined differences in gene expression patterns at the early and late osteoblast differentiation stages. Based on the current study, gene expression profile analysis demonstrated that osteoblast differentiation could be clearly subdivided into two stages. The early stage corresponded to "cell proliferation" based on the Stein and Jane definition [27]. However, we could not separate the late differentiation stage into "matrix maturation" and "matrix mineralization" stages based on the hierarchical clustering and PCA analysis. In addition, the division of osteoblast differentiation into two stages (early and late stages) is observed in the time series gene clustering analysis and weighted gene coexpression network analysis described below.

3.3. Clustering Analysis of Time Series Genes. Differentially expressed genes identified by time series analysis were clustered according to their expression profiles (RPKM values) using the fuzzy c-means algorithm implemented in R Mfuzz package. Genes with similar time-specific expression patterns were clustered into 12 groups, each containing 191–570 genes (Figure 2(a)). Eight among the 12 clusters (clusters 2, 3, 4, 5, 6, 8, 11, and 12) showed high expression patterns at the early osteoblast differentiation stage. Functional enrichment analysis for each cluster showed that these 8 clusters were highly enriched for genes related to cell proliferation processes. On the other hand, genes included in

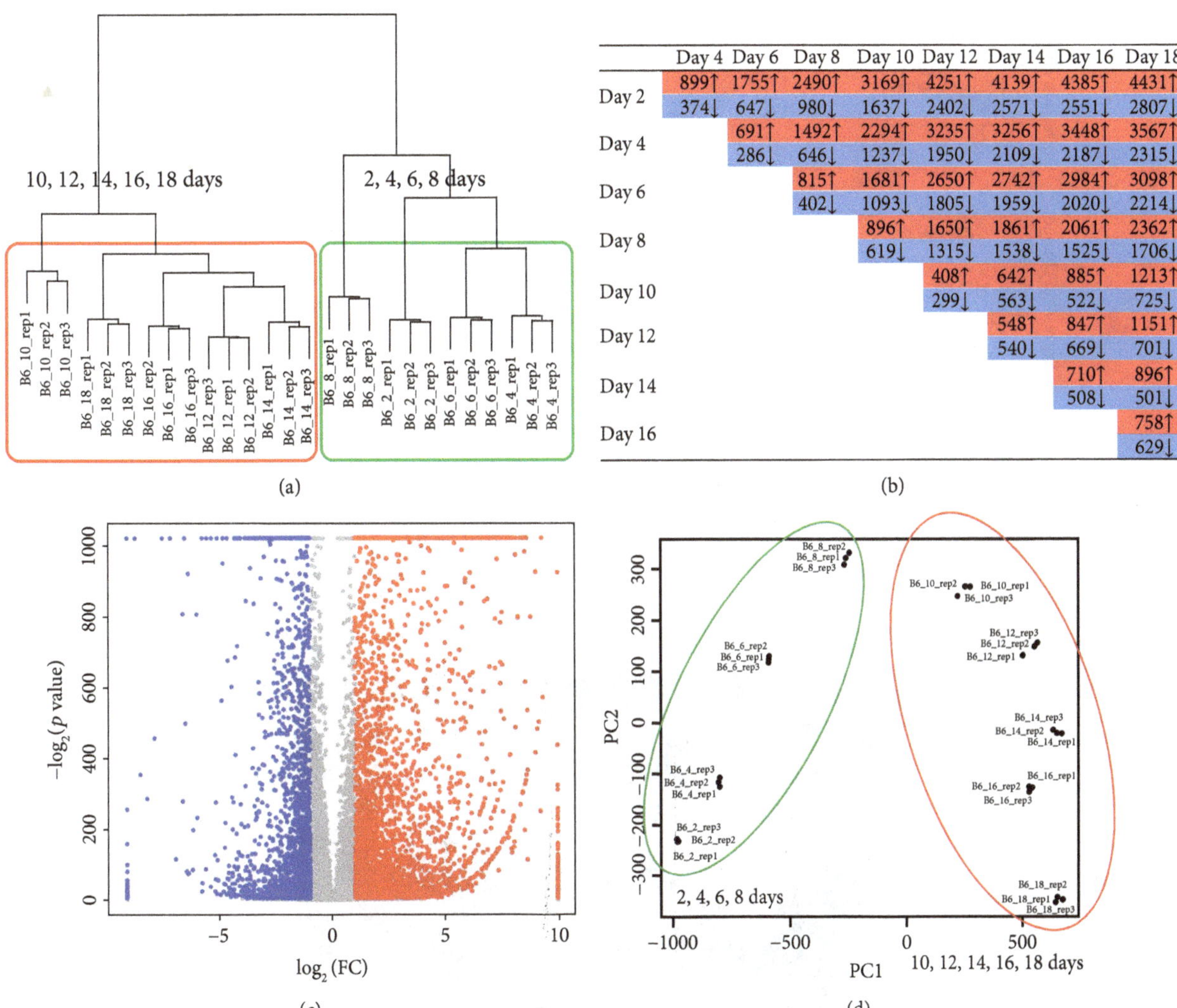

FIGURE 1: Global expression patterns of genes at the nine time points of osteoblast differentiation. (a) Hierarchical clustering of the transcriptome over the time period of osteoblast differentiation. All known genes with RPKM values ≥ 3 were used for analysis. (b) Number of genes showing up- or downregulation during osteoblast differentiation (fold change ≥ 2 or ≤0.5, FDR < 0.05). (c) Volcano plot showing differentially expressed genes in red and blue. The x- and y-axes represent the magnitude of fold changes ($\log_2$ transformed) and adjusted p value ($-\log_2$) by Benjamini-Hochberg correction, respectively. FC = fold change (d18/d2). (d) Principal component analysis (PCA) of 4058 time series gene expression profiles for different samples (three replicates at each time point of osteoblast differentiation).

the remaining 4 clusters (1, 7, 9, and 10) showed high expression patterns at the late osteoblast differentiation stage. These clusters included several genes associated with functional roles in osteogenesis, such as "extracellular matrix organization," "osteoblast differentiation," "metabolic process," "collagen fibril organization," "Wnt-protein binding," and "skeletal system development."

We focused on clusters 1, 7, 9, and 10, each containing 511, 570, 243, and 297 genes with high expression patterns in the late differentiation stage (Figure 2(b)). Assessment of individual genes within the clusters showed that several are involved in osteogenesis. For example, cluster 1 contained Fgfr2 that plays an essential role in skeletal development [28], cluster 7 contained Bmp4 that promotes formation of the bone and cartilage by stimulating differentiation processes of osteoblasts [29], and cluster 9 contained Fgf18 that has been shown to stimulate the proliferation of cultured mouse primary osteoblasts [30]. Expression levels of Fgfr2, Bmp4, and Fgf18 were particularly high between 12 and 18 days of the osteoblast differentiation process.

We determined the common function among the three clusters through GO and KEGG enrichment analysis. As shown in Table 2, the statistically significant GO terms in the clusters 1, 7, and 9 were "collagen metabolism," "subset of extracellular matrix," "regulation of osteoblast differentiation," and "Wnt signaling" which is known to be involved in the regulation of osteoblast lineage cells [31]. The enriched pathways in the KEGG analysis in the clusters 1, 7, and 9, were "cell adhesion," "cytokine-cytokine receptor interaction," and "PPAR signaling pathway" which is functionally associated with bone metabolism [32]. In the case of the cluster 10, "extracellular matrix" is the only statistically significant GO/KEGG term. This indicates that the genes belonging to the three clusters

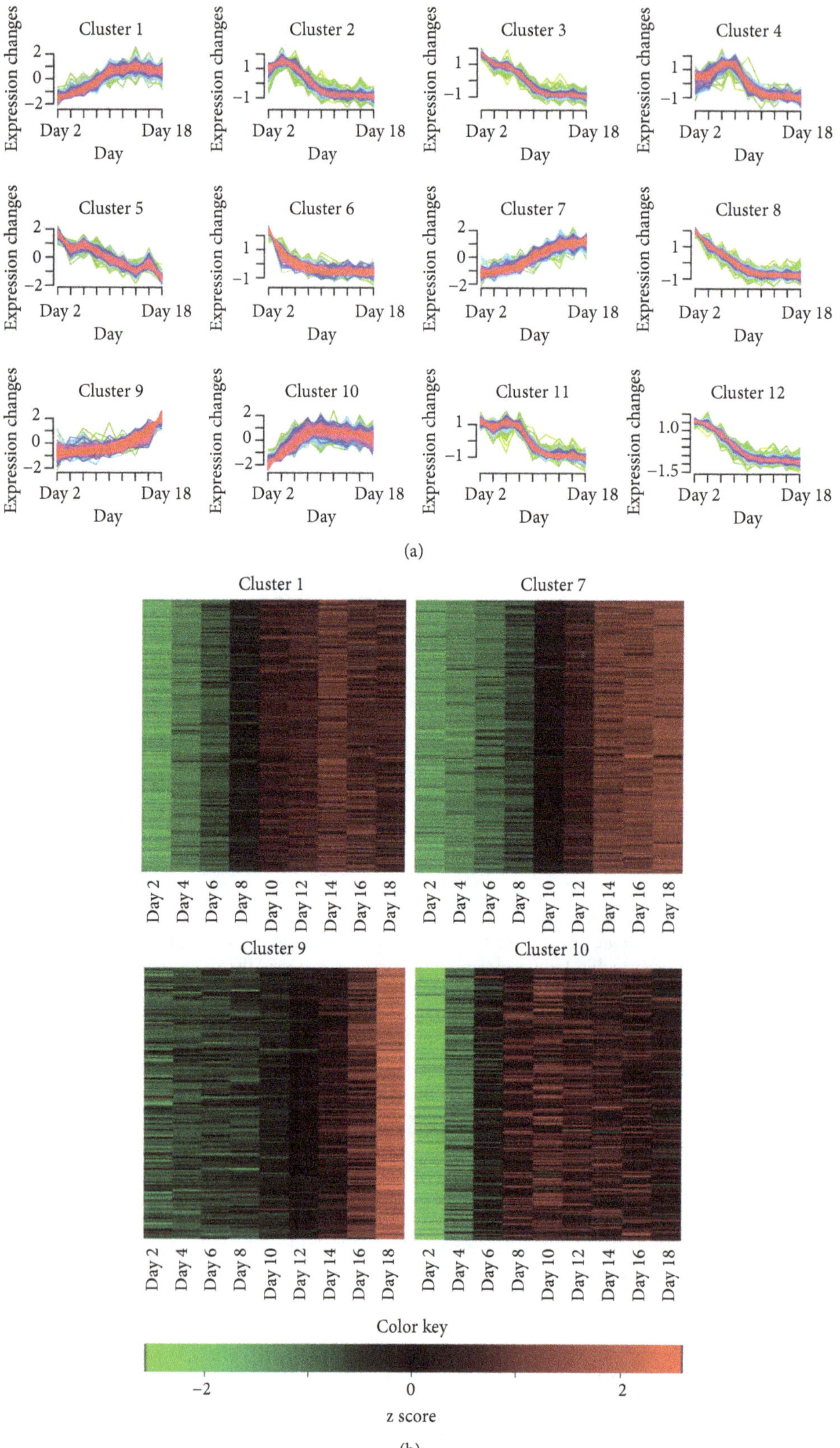

Figure 2: Clustering analysis of time series genes during osteoblast differentiation. (a) Soft clusters of 4058 time series gene expression data using Mfuzz. The numbers on the x-axis (time, 1~9) correspond to the nine time points of osteoblast differentiation (days 2, 4, 6, 8, 10, 12, 14, 16, and 18). (b) Heat maps and representative gene expression profiles (RNA-Seq) of four major clusters. The color key indicates gene expression values.

Table 2: Enriched GO terms and KEGG pathways of cluster 1, 7, and 9 genes.

Cluster	Gene ontology	p value	KEGG pathway	p value
Cluster 1	Wnt-activated receptor activity	<0.0001	ECM-receptor interactions	0.0007
	Collagen type I	<0.0001	Focal adhesion	0.0022
	Calcium ion binding	0.0001		
	Osteoblast differentiation	0.0003		
	Wnt-protein binding	0.0003		
Cluster 7	Collagen binding	<0.0001	Cytokine-cytokine receptor interactions	0.004
	Positive regulation of cell-substrate adhesion	<0.0001		
	Extracellular region	<0.0001		
	Extracellular matrix	<0.0001		
	Positive regulation of osteoblast differentiation	0.0001		
Cluster 9	Collagen	<0.0001	PPAR signaling pathway	<0.0001
	Biomineral tissue development	<0.0001	Osteoclast differentiation	0.001
	Regulation of bone mineralization	<0.0001	Adipocytokine signaling pathway	0.006
	Extracellular region	<0.0001	Cytokine-cytokine receptor interactions	0.0089
	Extracellular space	<0.0001		
Cluster 10	Extracellular matrix	<0.0001		

(1, 7, and 9) are highly correlated with the late osteoblast stage but the cluster 10 is less correlated with the late osteoblast stage. Hence, together with the coexpression network analysis described below, we used the genes belonging to the three clusters (1, 7, and 9 except 10) to find biomarkers of osteoblast differentiation.

3.4. Construction of a Gene Coexpression Network. Gene coexpression patterns may provide information on gene networks or pathways related to different biological phenomena. We constructed a coexpression network of genes using time series gene expression data. The weighted gene coexpression network analysis (WGCNA) generated a total of 16 network modules (Figure 3(a)). The eigengenes for all 16 modules were calculated and the correlations between eigengenes and differentiation stages evaluated (Figure 3(b)). Genes belonging to 9 modules (midnight blue, purple, red, tan, brown, turquoise, black, blue, and green) were enriched with GO and KEGG pathways associated with the early osteoblast stage (days 2–8), such as "cell proliferation" and "osteoblast differentiation processes." Meanwhile, genes belonging to the remaining 7 modules (cyan, salmon, magenta, green yellow, pink, yellow, and gray) were enriched with GO and KEGG pathways associated with the late osteoblast stage (days 10–18), such as "cell adhesion" and "osteoclast differentiation."

Notably, the genes belonging to the cyan and salmon modules exhibited high trait association correlation ($r > 0.6$), and they were enriched with the following KEGG pathways and GO terms, besides "cell adhesion" and "osteoclast differentiation," as mentioned above: PPAR signaling, subset of extracellular matrix, calcium metabolism, glucose homeostasis, cell signaling, and regulation of bone mineralization. This indicates that these two modules are strongly associated with late osteoblast differentiation (day 18). Thus, from the two modules (cyanide and salmon), we extracted 36 and 62 hub genes, respectively, with high connectivity (MM > 0.9), and these genes were used to find biomarkers for osteoblast differentiation as described below.

3.5. Identification of Markers Associated with Osteoblast Differentiation. We combined the results from gene clustering analysis and gene coexpression networks to determine novel biomarkers for osteoblast differentiation. Focusing on the late osteoblast differentiation stage, we considered genes belonging to clusters 1, 7, and 9 as well as the hub genes belonging to the cyan and salmon modules, as described above. We set the following cut-off values for biomarker candidates: absolute value of fold changes between early differentiation (average expression from samples of days 2–8) and late differentiation (average expression from samples of days 10–18) must be >2 and at least one of the samples should have an RPKM value > 5 for the gene.

We identified a total of 72 genes as potentially useful marker candidates for osteoblast differentiation (Supplementary Table 3). Functions associated with these genes include "regulation of bone mineralization (GO:0030500)," "extracellular matrix (GO:0031012)," "collagen (GO:0005581)," "collagen binding (GO:0005518)," and "cell adhesion (GO:0007155)" (Supplementary Table 4). Extracellular matrix, including collagen, is important for bone matrix construction [33], and cell adhesion affects cell-matrix interactions and further cell differentiation [34, 35].

Examples of genes associated with each function are as follows. Potential markers associated with "regulation of bone mineralization" are Ifitm5, Bglap, and Bglap2. Potential markers that are functionally associated with extracellular matrix included Abi3bp, Mfap4, Rarres2, Dcn, Dpt, Cilp, and Sod3 as well as the well-known gene Itgbl1 (integrin, beta-like 1; with EGF-like repeat domains) (Figure 4(a)). Dcn (decorin) is also known as a skeletal-related gene [36]. We identified C1qa, Marco, C1qc, Adipoq, and C1qb as

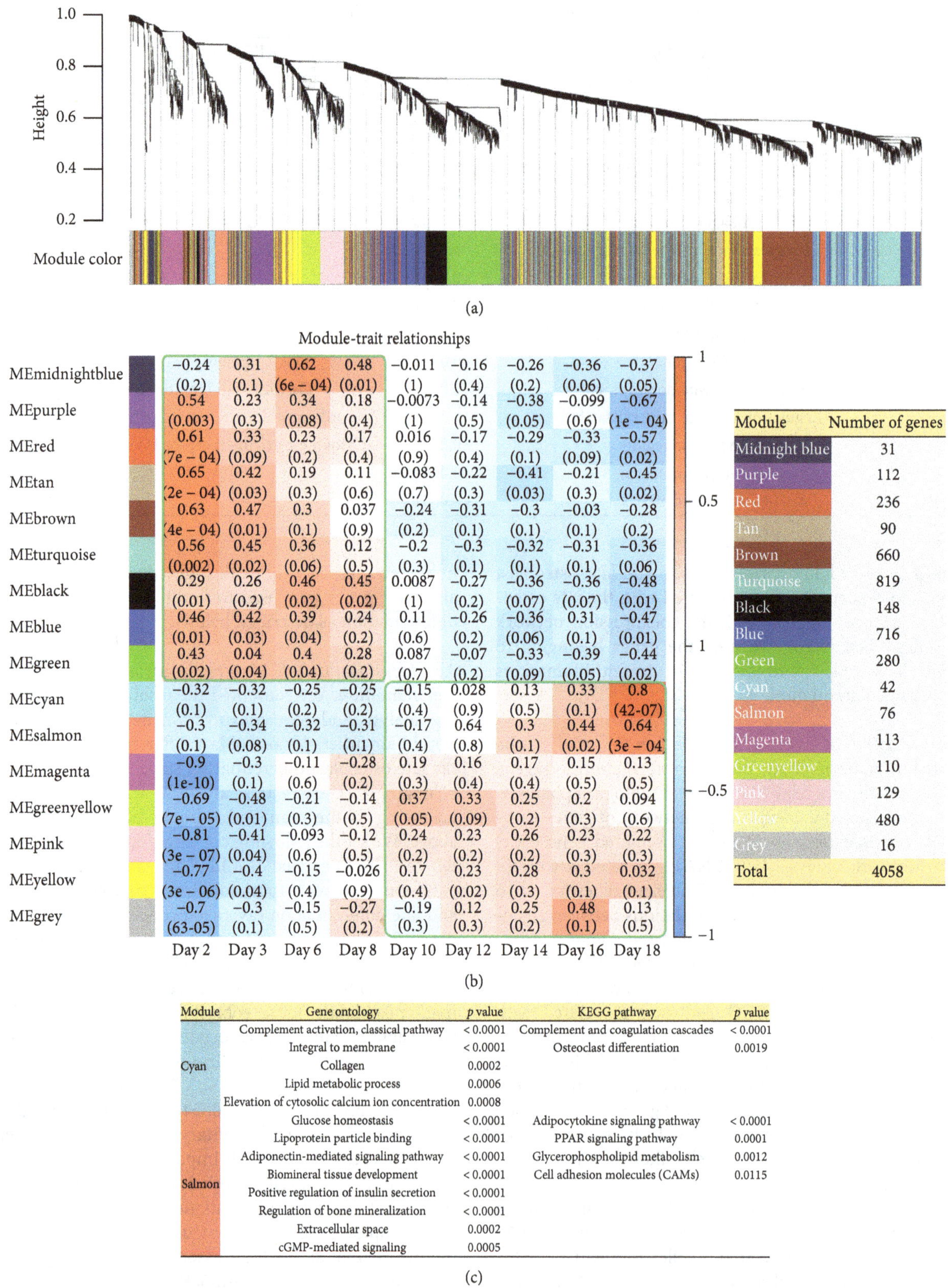

Module	Number of genes
Midnight blue	31
Purple	112
Red	236
Tan	90
Brown	660
Turquoise	819
Black	148
Blue	716
Green	280
Cyan	42
Salmon	76
Magenta	113
Greenyellow	110
Pink	129
Yellow	480
Grey	16
Total	4058

Module	Gene ontology	p value	KEGG pathway	p value
Cyan	Complement activation, classical pathway	< 0.0001	Complement and coagulation cascades	< 0.0001
	Integral to membrane	< 0.0001	Osteoclast differentiation	0.0019
	Collagen	0.0002		
	Lipid metabolic process	0.0006		
	Elevation of cytosolic calcium ion concentration	0.0008		
Salmon	Glucose homeostasis	< 0.0001	Adipocytokine signaling pathway	< 0.0001
	Lipoprotein particle binding	< 0.0001	PPAR signaling pathway	0.0001
	Adiponectin-mediated signaling pathway	< 0.0001	Glycerophospholipid metabolism	0.0012
	Biomineral tissue development	< 0.0001	Cell adhesion molecules (CAMs)	0.0115
	Positive regulation of insulin secretion	< 0.0001		
	Regulation of bone mineralization	< 0.0001		
	Extracellular space	0.0002		
	cGMP-mediated signaling	0.0005		

FIGURE 3: WGCNA analysis of 4058 time series gene expression data at the nine time points of osteoblast differentiation. (a) Hierarchical cluster trees depict coexpression modules identified using WGCNA. (b) A heat map plot of module-trait association. Each row corresponds to the module eigengene, the first principal component of a module. Each column corresponds to trait, the nine time points of osteoblast differentiation (days 2, 4, 6, 8, 10, 12, 14, 16, and 18). Each cell contains the corresponding correlation value and p value (left panel). Each of the sixteen modules contains between 16 and 819 genes (right panel). (c) Functional annotations of cyan and salmon modules. Gene ontology (GO) and KEGG pathway terms and corresponding p values are shown.

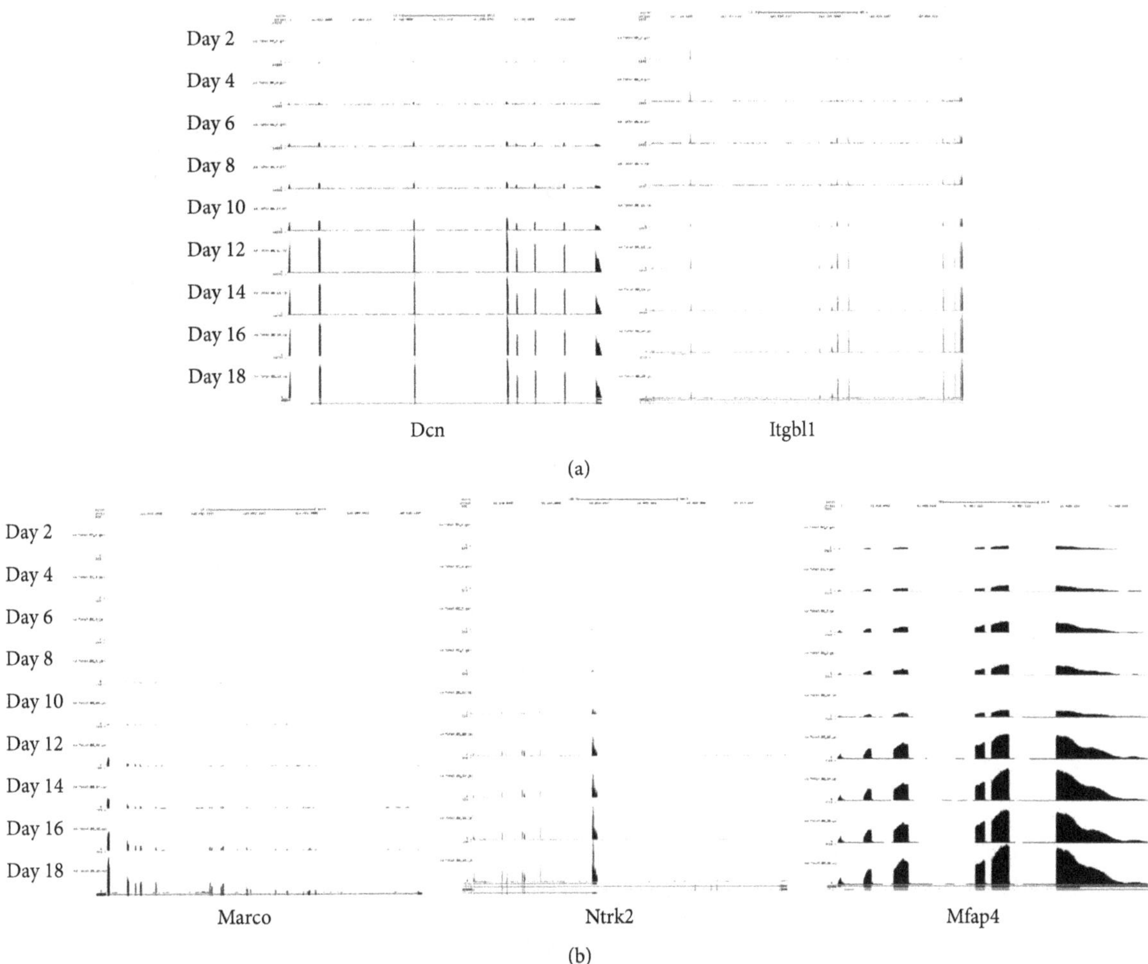

FIGURE 4: Expression patterns of five genes upregulated during late osteoblast differentiation stages. (a) Expression profiles of two known genes, Dcn (left panel) and Itgbl1 (right panel). (b) Expression profiles of three novel biomarker candidates, Marco (left panel), Ntrk2 (middle panel), Mfap4 (right panel).

collagen-related genes and identified Abi3bp, Srgn, and Dcn as genes related to "collagen binding." Markers associated with "cell adhesion" included genes such as Cd36, Fbln7, Mfap4, Ibsp, and Dpt. In addition, Marco (macrophage receptor with collagen structure), Ntrk2 (neurotrophic tyrosine kinase receptor type 2), and Mfap4 (microfibrillar-associated protein 4) were identified as novel potential markers (Figure 4(b)). Although we identified potentially useful marker candidates for osteoblast differentiation, our study does have limitations. Since we rigorously filtered out 4058 time series genes based on the clustering analysis and WGCNA analysis (trait correlation and hub genes), the current method did not identify the known biomarkers (Fgfr2, Bmp4, and Fgf18) described in the clustering analysis. In addition, because the candidates were mainly identified based on the bioinformatic observations, their biological relevance would need to be further investigated at the cellular and molecular levels experimentally.

3.6. Expression Profiles of lncRNAs during Osteoblast Differentiation. The NONCODE (v4) database was used for lncRNA annotations, and at least one read was mapped onto 37,431 lncRNAs during osteoblast differentiation. Hierarchical clustering was performed based on the expression profiles in each sample. Upon application of similar approaches for known Ensembl genes to lncRNAs, we identified that 54% of the total lncRNAs (20,167 out of 37,431) were differentially expressed across the stages of osteoblast differentiation. Similar to the expression profile of mRNA described above, the expression patterns of lncRNAs were different between the early differentiation stage (2–8 days) and the late differentiation stage (10–18 days) (Figure 5(a)). The number of differentially expressed lncRNAs tended to increase with time (Figure 5(b)).

3.7. Dynamic Changes in lncRNAs during Osteoblast Differentiation. Examination of the temporal expression changes of lncRNAs revealed that levels of 3948 out of 37,431 lncRNAs were dynamically altered over the course of osteoblast differentiation. In PCA analysis using 3948 time series lncRNAs, 97% of the variations in lncRNA expression patterns could be explained by the first two principal components (PCs) (Figure 5(c)). Some of the 3948 time series lncRNAs highlighted were associated with osteogenic

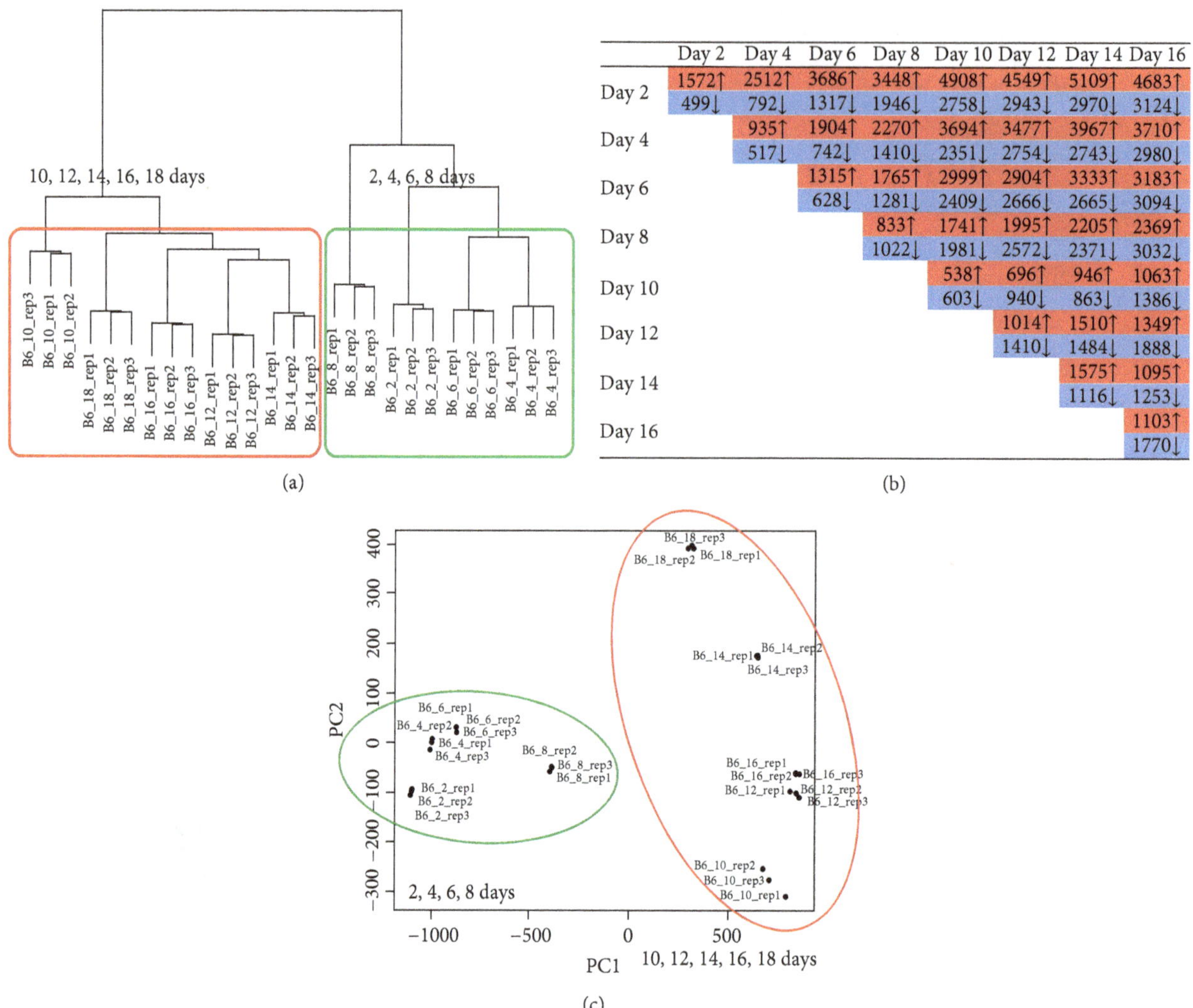

	Day 2	Day 4	Day 6	Day 8	Day 10	Day 12	Day 14	Day 16
Day 2	1572↑	2512↑	3686↑	3448↑	4908↑	4549↑	5109↑	4683↑
	499↓	792↓	1317↓	1946↓	2758↓	2943↓	2970↓	3124↓
Day 4		935↑	1904↑	2270↑	3694↑	3477↑	3967↑	3710↑
		517↓	742↓	1410↓	2351↓	2754↓	2743↓	2980↓
Day 6			1315↑	1765↑	2999↑	2904↑	3333↑	3183↑
			628↓	1281↓	2409↓	2666↓	2665↓	3094↓
Day 8				833↑	1741↑	1995↑	2205↑	2369↑
				1022↓	1981↓	2572↓	2371↓	3032↓
Day 10					538↑	696↑	946↑	1063↑
					603↓	940↓	863↓	1386↓
Day 12						1014↑	1510↑	1349↑
						1410↓	1484↓	1888↓
Day 14							1575↑	1095↑
							1116↓	1253↓
Day 16								1103↑
								1770↓

FIGURE 5: Global expression patterns of lncRNAs at the nine time points of osteoblast differentiation. (a) Hierarchical clustering of lncRNAs across the stages of osteoblast differentiation. (b) Number of lncRNAs showing up- or downregulation during osteoblast differentiation (fold change ≥ 2 or ≤ 0.5, FDR < 0.05). (c) Principal component analysis (PCA) of 3948 time series lncRNA expression profiles for different samples (three replicates for each time point of osteoblast differentiation).

function such as "actin cytoskeleton organization," "extracellular matrix organization," "Wnt signaling pathway," "embryonic skeletal system morphogenesis," "Notch signaling pathway," and "collagen fibril organization."

3.8. Clustering Analysis of Time Series lncRNAs. Using R package Mfuzz [20], differentially expressed 3948 time series lncRNAs were divided into 12 clusters, and each cluster contained 142–595 lncRNAs (Figure 6(a)). The lncRNAs contained in 5 clusters (2, 3, 6, 9, and 11) were highly expressed in early osteoblast differentiation stage. Functional annotation analysis showed that these clusters are associated with "regulation of cell proliferation," "actin cytoskeleton organization," and "cell-cycle-related processes." The remaining seven clusters (1, 4, 5, 7, 8, 10, and 12) included lncRNAs that displayed a higher expression in the late osteoblast differentiation stage and related to "cell adhesion," "cell differentiation," "collagen fibril organization," "Wnt signaling pathway," and "phosphorylation processes."

3.9. Identification of lncRNAs Associated with Osteoblast Differentiation. Expression patterns of lncRNAs belonging to cluster 5 were not only very similar but also significantly increased over time, especially between 10 and 18 days. Thus, we focused on cluster 5 to find the lncRNAs markers associated with late differentiation of osteoblasts. Cut-off values were set for lncRNAs (Figure 6(b), upper panel), as in the procedures in novel biomarker identification for known genes.

In total, 89 lncRNAs were identified as potentially useful markers for osteoblast differentiation (Supplementary Table 5). GO enrichment analysis for these lncRNAs disclosed significant enrichment of "cell adhesion (GO:0007155)" and "Wnt signaling pathway (GO:0016055)," which is important in the differentiation and/or function of osteoblasts [31]. For examples, NONMMUG016555, NONMMUG038646, NONMMUG007704, NONMMUG001799, NONMMUG 005875, NONMMUG002249, NONMMUG012067, NONMMUG035551, NONMMUG015487, and NONMMUG

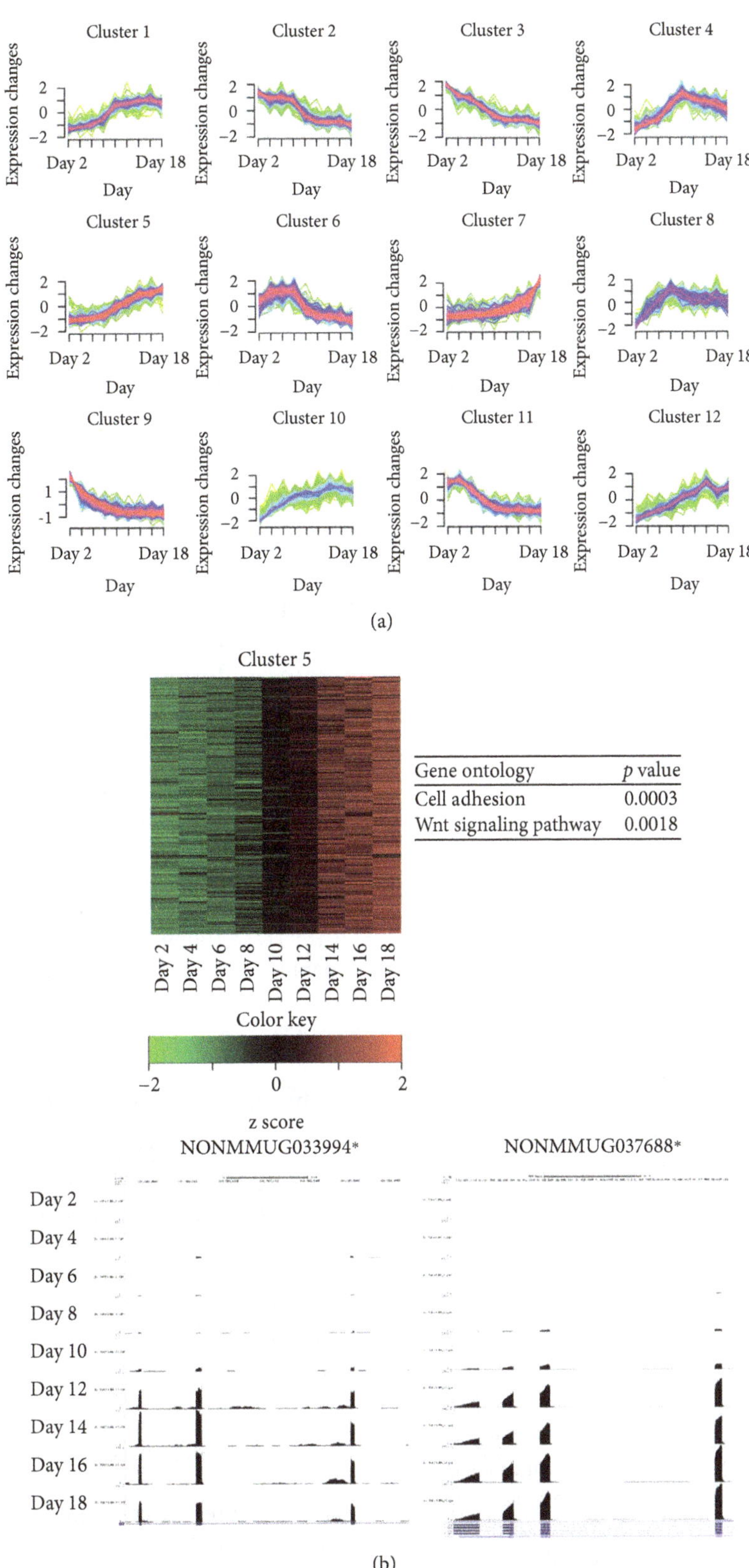

Gene ontology	p value
Cell adhesion	0.0003
Wnt signaling pathway	0.0018

Figure 6: Clustering analysis of time series lncRNAs during osteoblast differentiation. (a) Soft clusters of 3948 time series lncRNAs generated by Mfuzz. The numbers on the x-axis (time, 1–9) correspond to the nine time points of osteoblast differentiation (days 2, 4, 6, 8, 10, 12, 14, 16, and 18). (b) Heat map of cluster 5. The color key indicates gene expression values (top left panel). Enriched GO terms and corresponding p values in cluster 5 are shown (top right panel). Gene expression profiles of two representative lncRNAs, NONMMUG033994 (bottom left panel) and NONMMUG037688 (bottom right panel), visualized in the bottom panel. *NONCODE v4 database ID.

Table 3: Enriched transcriptional binding motifs in the 89 lncRNA promoter regions.

Motif ID	Transcription factor	p value	Adjusted p value
WT1_MOUSE.H11MO.1.A	Wt1	8.014E-06	0.004246
EGR2_MOUSE.H11MO.0.A	Egr2	5.517E-05	0.02887
ZN281_MOUSE.H11MO.0.A	Zn281	5.737E-05	0.03001
ZIC1_MOUSE.H11MO.0.B	Zic1	7.868E-05	0.04092
SP5_MOUSE.H11MO.1.C	Sp5	9.138E-05	0.04737

035553 are related to "cell adhesion." NONMMUG038646, NONMMUG016590, and NONMMUG001799 are related to "Wnt signaling pathway (GO:0016055)." In addition, we identified novel candidate lncRNA markers, and two examples are presented in Figure 6(b) (bottom panel).

Next, to see whether the identified lncRNAs could be regulated by common regulation factors, we searched the sequence motifs in the promoter regions (the upstream 1000 bp nucleotide sequences of lncRNAs) of the 89 lncRNAs with background controls generated from all NONCODE lncRNA promoter regions. We performed the motif enrichment analysis using MEME Suite package [23] and the HOCOMOCO v11 mouse transcription factor database [24] to detect known transcription factor binding motifs that are significantly enriched within the promoter region. A total of five significant motifs were identified (Table 3). The most significant enriched motif (WT1_MOUSE.H11MO.1.A) was the Wilms tumor protein (Wt1) transcription factor binding motif. Recently, it was shown that Wt1 was expressed during development in the limb tissue of E11.5 to E16.5 mice and the loss of the Wt1 expression affected a reduction of non-haematopoetic MSC cells in the E18.5 hindlimb [37]. These findings suggest that Wt1 may play a functional role in the bone development. The second most significant enriched motif (EGR2_MOUSE.H11MO.0.A) was early growth response 2 (Egr2) transcription factor binding motif. Chandra et al. previously suggested that EGFR-induced Egr2 expression is crucial for osteoprogenitor maintenance and new bone formation [38]. Further research on the multiple layers of regulatory mechanisms, including lncRNA expression and transcription regulations, will be required to fully understand the role of these lncRNAs during bone development.

4. Conclusions

We employed public RNA-Seq data (NCBI GEO accession number GSE54461) to profile mRNA and lncRNA expression and identified a series of biomarkers potentially associated with bone differentiation. Data from the current study showed that a combination of bioinformatic approaches, including (1) time series, (2) clustering, and (3) coexpression network analyses, provides an effective means to identify novel candidate markers associated with osteoblast differentiation. Expression patterns of mRNA and lncRNA during osteoblast differentiation were defined as two distinct stages (early and late osteoblast differentiation). Functional annotation showed that the members in the late stage are involved in osteogenesis processes. Notably, common transcription factor binding motifs were enriched in the identified lncRNA markers. Our findings provide a valuable resource for lncRNA study and understanding of the mechanism of mouse osteoblast differentiation. Further research is needed to determine the functions of lncRNA and mRNA identified as potential markers in this study.

Conflicts of Interest

The authors declare that they have no conflicts of interest.

Acknowledgments

This work was supported by a grant of the Korea Health Technology R&D project through KHIDI funded by the Ministry of Health and Welfare, Republic of Korea (HI14C0175), the Collaborative Genome Program for Fostering New Post-Genome Industry of the National Research Foundation (NRF) funded by the Ministry of Science and ICT (MSIT) (2017M3C9A6047623), and a grant (2016R1D1A1B03930275) of the Basic Science Research Program through the National Research Foundation of Korea (NRF) funded by the Ministry of Education, Republic of Korea.

Supplementary Materials

Supplementary Table 1: functional annotation of upregulated genes on day 18 (day 2 versus day 18). Supplementary Table 2: functional annotation of downregulated genes on day 18 (day 2 versus day 18). Supplementary Table 3: known Ensembl genes potentially associated with osteoblast differentiation. Supplementary Table 4: enriched GO terms of known Ensembl genes potentially associated with osteoblast differentiation. Supplementary Table 5: lncRNAs (NONCODE v4) potentially associated with osteoblast differentiation. Supplementary Figure 1: Overview of the methods and procedures to identify biomarkers. Supplementary Figure 2: proportion of variance for principal components. Supplementary Figure 3: expression profiles of known mature osteoblast markers. *(Supplementary Materials)*

References

[1] A. Rutkovskiy, K. O. Stensløkken, and I. J. Vaage, "Osteoblast differentiation at a glance," *Medical Science Monitor Basic Research*, vol. 22, pp. 95–106, 2016.

[2] K. Nakashima, X. Zhou, G. Kunkel et al., "The novel zinc finger-containing transcription factor osterix is required for

osteoblast differentiation and bone formation," *Cell*, vol. 108, no. 1, pp. 17–29, 2002.

[3] F. Otto, A. P. Thornell, T. Crompton et al., "*Cbfa1*, a candidate gene for cleidocranial dysplasia syndrome, is essential for osteoblast differentiation and bone development," *Cell*, vol. 89, no. 5, pp. 765–771, 1997.

[4] T. Komori, H. Yagi, S. Nomura et al., "Targeted disruption of *Cbfa1* results in a complete lack of bone formation owing to maturational arrest of osteoblasts," *Cell*, vol. 89, no. 5, pp. 755–764, 1997.

[5] X. Y. Zhao and J. D. Lin, "Long noncoding RNAs: a new regulatory code in metabolic control," *Trends in Biochemical Sciences*, vol. 40, no. 10, pp. 586–596, 2015.

[6] W. Zhang, R. Dong, S. Diao, J. Du, Z. Fan, and F. Wang, "Differential long noncoding RNA/mRNA expression profiling and functional network analysis during osteogenic differentiation of human bone marrow mesenchymal stem cells," *Stem Cell Research & Therapy*, vol. 8, no. 1, p. 30, 2017.

[7] P. Carninci, T. Kasukawa, S. Katayama et al., "The transcriptional landscape of the mammalian genome," *Science*, vol. 309, no. 5740, pp. 1559–1563, 2005.

[8] T. Hung and H. Y. Chang, "Long noncoding RNA in genome regulation: prospects and mechanisms," *RNA Biology*, vol. 7, no. 5, pp. 582–585, 2010.

[9] J. E. Wilusz, H. Sunwoo, and D. L. Spector, "Long noncoding RNAs: functional surprises from the RNA world," *Genes & Development*, vol. 23, no. 13, pp. 1494–1504, 2009.

[10] T. Zhang, X. Zhang, K. Han et al., "Analysis of long noncoding RNA and mRNA using RNA sequencing during the differentiation of intramuscular preadipocytes in chicken," *PLoS One*, vol. 12, no. 2, article e0172389, 2017.

[11] M. Kretz, D. E. Webster, R. J. Flockhart et al., "Suppression of progenitor differentiation requires the long noncoding RNA ANCR," *Genes & Development*, vol. 26, no. 4, pp. 338–343, 2012.

[12] L. Wang, Y. Wang, Z. Li, Z. Li, and B. Yu, "Differential expression of long noncoding ribonucleic acids during osteogenic differentiation of human bone marrow mesenchymal stem cells," *International Orthopaedics*, vol. 39, no. 5, pp. 1013–1019, 2015.

[13] W. Zhuang, X. Ge, S. Yang et al., "Upregulation of lncRNA MEG3 promotes osteogenic differentiation of mesenchymal stem cells from multiple myeloma patients by targeting BMP4 transcription," *Stem Cells*, vol. 33, no. 6, pp. 1985–1997, 2015.

[14] L. Zhu and P. C. Xu, "Downregulated LncRNA-ANCR promotes osteoblast differentiation by targeting EZH2 and regulating Runx2 expression," *Biochemical and Biophysical Research Communications*, vol. 432, no. 4, pp. 612–617, 2013.

[15] J. P. Kemp, C. Medina-Gomez, K. Estrada et al., "Phenotypic dissection of bone mineral density reveals skeletal site specificity and facilitates the identification of novel loci in the genetic regulation of bone mass attainment," *PLoS Genetics*, vol. 10, no. 6, article e1004423, 2014.

[16] T. D. Wu and S. Nacu, "Fast and SNP-tolerant detection of complex variants and splicing in short reads," *Bioinformatics*, vol. 26, no. 7, pp. 873–881, 2010.

[17] A. Mortazavi, B. A. Williams, K. McCue, L. Schaeffer, and B. Wold, "Mapping and quantifying mammalian transcriptomes by RNA-Seq," *Nature Methods*, vol. 5, no. 7, pp. 621–628, 2008.

[18] S. Anders and W. Huber, "Differential expression analysis for sequence count data," *Genome Biology*, vol. 11, no. 10, article R106, 2010.

[19] M. D. Robinson, D. J. McCarthy, and G. K. Smyth, "edgeR: a bioconductor package for differential expression analysis of digital gene expression data," *Bioinformatics*, vol. 26, no. 1, pp. 139-140, 2010.

[20] L. Kumar and M. E. Futschik, "Mfuzz: a software package for soft clustering of microarray data," *Bioinformation*, vol. 2, no. 1, pp. 5–7, 2007.

[21] G. R. Warnes, B. Bolker, L. Bonebakker et al., "gplots: various R programming tools for plotting data. 2013," *R Package Version 2.1*, 2014.

[22] P. Langfelder and S. Horvath, "WGCNA: an R package for weighted correlation network analysis," *BMC Bioinformatics*, vol. 9, no. 1, p. 559, 2008.

[23] R. C. McLeay and T. L. Bailey, "Motif enrichment analysis: a unified framework and an evaluation on ChIP data," *BMC Bioinformatics*, vol. 11, no. 1, p. 165, 2010.

[24] I. V. Kulakovskiy, I. E. Vorontsov, I. S. Yevshin et al., "HOCOMOCO: towards a complete collection of transcription factor binding models for human and mouse via large-scale ChIP-Seq analysis," *Nucleic Acids Research*, vol. 46, no. D1, pp. D252–D259, 2018.

[25] M. B. Eisen, P. T. Spellman, P. O. Brown, and D. Botstein, "Cluster analysis and display of genome-wide expression patterns," *Proceedings of the National Academy of Sciences of the United States of America*, vol. 95, no. 25, pp. 14863–14868, 1998.

[26] I. Kalajzic, A. Staal, W. P. Yang et al., "Expression profile of osteoblast lineage at defined stages of differentiation," *The Journal of Biological Chemistry*, vol. 280, no. 26, pp. 24618–26, 2005.

[27] G. S. Stein and B. L. Jane, *Molecular Mechanisms Mediating Developmental and Hormone-Regulated Expression of Genes in Osteoblasts in Cellular and Molecular Biology of Bone*, Academic Press, San Diego, CA, USA, 1993.

[28] K. Yu, J. Xu, Z. Liu et al., "Conditional inactivation of FGF receptor 2 reveals an essential role for FGF signaling in the regulation of osteoblast function and bone growth," *Development*, vol. 130, no. 13, pp. 3063–3074, 2003.

[29] P. Ducy and G. Karsenty, "The family of bone morphogenetic proteins," *Kidney International*, vol. 57, no. 6, pp. 2207–2214, 2000.

[30] T. Shimoaka, T. Ogasawara, A. Yonamine et al., "Regulation of osteoblast, chondrocyte, and osteoclast functions by fibroblast growth factor (FGF)-18 in comparison with FGF-2 and FGF-10," *The Journal of Biological Chemistry*, vol. 277, no. 9, pp. 7493–7500, 2002.

[31] F. Long, "Building strong bones: molecular regulation of the osteoblast lineage," *Nature Reviews Molecular Cell Biology*, vol. 13, no. 1, pp. 27–38, 2012.

[32] M. Takano, F. Otsuka, Y. Matsumoto et al., "Peroxisome proliferator-activated receptor activity is involved in the osteoblastic differentiation regulated by bone morphogenetic proteins and tumor necrosis factor-α," *Molecular and Cellular Endocrinology*, vol. 348, no. 1, pp. 224–232, 2012.

[33] A. Matsugaki, G. Aramoto, T. Ninomiya, H. Sawada, S. Hata, and T. Nakano, "Abnormal arrangement of a collagen/apatite extracellular matrix orthogonal to osteoblast alignment is

constructed by a nanoscale periodic surface structure," *Biomaterials*, vol. 37, pp. 134–143, 2015.

[34] Y. S. Lee and C. M. Chuong, "Adhesion molecules in skeletogenesis: I. Transient expression of neural cell adhesion molecules (NCAM) in osteoblasts during endochondral and intramembranous ossification," *Journal of Bone and Mineral Research*, vol. 7, no. 12, pp. 1435–1446, 1992.

[35] B. Chen, B. Ji, and H. Gao, "Modeling active mechanosensing in cell-matrix interactions," *Annual Review of Biophysics*, vol. 44, no. 1, pp. 1–32, 2015.

[36] N. C. Ho, L. Jia, C. C. Driscoll, E. M. Gutter, and C. A. Francomano, "A skeletal gene database," *Journal of Bone and Mineral Research*, vol. 15, no. 11, pp. 2095–2122, 2000.

[37] S. L. McHaffie, *Investigating the Role of Wt1 in Bone and Marrow Biology. Doctoral Dissertation*, University of Edinburgh, UK, 2014.

[38] A. Chandra, S. Lan, J. Zhu, V. A. Siclari, and L. Qin, "Epidermal growth factor receptor (EGFR) signaling promotes proliferation and survival in osteoprogenitors by increasing early growth response 2 (EGR2) expression," *The Journal of Biological Chemistry*, vol. 288, no. 28, pp. 20488–20498, 2013.

14

Exploring Long Noncoding RNAs in Glioblastoma: Regulatory Mechanisms and Clinical Potentials

Tao Zeng,[1] Lei Li,[1] Yan Zhou,[2] and Liang Gao[1]

[1]*Department of Neurosurgery, Shanghai Tenth People's Hospital, Tongji University School of Medicine, No. 301 Middle Yanchang Road, Shanghai 200072, China*
[2]*Medical Research Institute, College of Life Sciences, Wuhan University, Wuhan 430071, China*

Correspondence should be addressed to Liang Gao; lianggaoh@126.com

Academic Editor: Margarita Hadzopoulou-Cladaras

Gliomas are primary brain tumors presumably derived from glial cells. The WHO grade IV glioblastoma (GBM), characterized by rapid cell proliferation, easily recrudescent, high morbidity, and mortality, is the most common, devastating, and lethal gliomas. Molecular mechanisms underlying the pathogenesis and progression of GBMs with potential diagnostic and therapeutic value have been explored industriously. With the advent of high-throughput technologies, numerous long noncoding RNAs (lncRNAs) aberrantly expressed in GBMs were discovered recently, some of them probably involved in GBM initiation, malignant progression, relapse and resistant to therapy, or showing diagnostic and prognostic value. In this review, we summarized the profile of lncRNAs that has been extensively investigated in glioma research, with a focus on their regulatory mechanisms. Then, their diagnostic, prognostic, and therapeutic implications were also discussed.

1. Introduction

The WHO grade IV glioblastoma (GBM), characterized by high recurrence rate, high morbidity, and mortality, is the most common, aggressive, and deadly primary intracranial tumor in adults, reflecting the urgent need to develop new diagnostic and therapeutic targets for this devastating disease [1–3]. The standard therapy for GBM is the combination of maximal surgical tumor resection, radiotherapy, and chemotherapy [1, 4]. However, the average life expectancy for GBM patients is only approximately 15 months after initial diagnosis even though optimal treatment has been received [5, 6]. GBM malignancy and poor prognosis are closely correlated with the deregulation of signaling pathways controlling tumor cell proliferation, resistance to apoptosis, invasion, angiogenesis, and immune evasion [6, 7]. Thus, investigations revealing essential molecular mechanisms governing these features of glioblastoma with potential diagnostic and therapeutic value have drawn remarkable attention [8]. With the advent of high-throughput technologies, a wide variety of noncoding RNAs, including microRNAs and long noncoding RNAs (lncRNAs), has been identified in glioma tissues and cell lines, some of which show strong functional indications [9]. Interestingly, the so-called glioma stem cells (GSCs), which are believed to be responsible for GBM initiation, therapy-resistant, and relapse, carry diverse molecular and genetic changes, including aberrant lncRNA expression [10, 11]. Additionally, the competitive regulatory network formed among lncRNAs, protein-coding transcripts, and microRNAs seems to play a crucial regulatory role in the proliferation, metastasis, and resistance to apoptosis of GSCs [12–14]. Although researches of lncRNAs in GBMs are still in infancy, exploring their roles and mechanisms would not only deepen our understandings on molecular features of GBMs but also open new windows for unveiling novel diagnostic and therapeutic targets [15]. Here, we summarize recent progress regarding GBM-associated lncRNAs that have been under intensive investigations, with emphases on their regulatory mechanisms and clinical relevance.

2. Overview of lncRNAs

LncRNAs, a class of RNAs greater than 200 nucleotides (nt) without significant protein-coding capacity, were initially documented in the epigenetic regulation of X chromosome inactivation during embryogenesis [16, 17]. Occasionally, functional short peptides can be derived from lncRNAs [18, 19]. Until now, the NONCODE database has annotated 87,774 and 96,308 lncRNA genes in the mouse and human genome, respectively, which are far more than protein-coding genes (PCGs) [20]. LncRNA genes are largely categorized according to their locations and transcription orientations relative to the closest protein-coding genes. Thus, they can be antisense (partially or fully overlapped with PCGs), sense, divergent, convergent, intronic, and intergenic (no PCGs within a 5-kilobase range) [21]. Compared to protein-coding mRNAs, lncRNAs transcribed from intergenic regions (lincRNAs) are less spliced and largely nonpolyadenylated and are mostly attached to chromatin [22]. LncRNAs are usually expressed at lower levels than protein-coding mRNAs and display more cell type- and tissue-specific expression patterns. Notably, about 40% of lncRNAs are mainly expressed in the brain, reflecting the cellular and functional complexity of the brain [23]. Furthermore, lncRNA expression is spatiotemporally regulated during neural development [24–27] and upon neural activity [28, 29]. Since many lncRNAs are capable of forming a complex with DNA, RNA, and proteins dynamically, they regulate gene expression at multiple levels including chromatin remodeling, histone and DNA modification, and the process of transcription, as well as RNA splicing, transport, and stability [6, 30–33]. Particularly, lncRNAs could play *cis*-regulatory roles to positively or negatively control the transcription of neighboring PCGs [34]. Although functions of most lincRNAs have yet to be unveiled, some have been found to facilitate the chromatin structure and histone modifications, to act as a coactivator or corepressor in the nucleus [35, 36] or to modulate signal transduction in the cytosol [37]. In addition, some lncRNAs bear complementary sites (also known as miRNA response elements, MRE) for microRNAs (miRNAs). miRNAs have been widely recognized to be involved in almost every facet in the development and malignant progression of gliomas, including the maintenance of the stemness of GSCs, invasiveness, angiogenesis, epigenetic regulation, and signaling pathways [38–42]. The so-called competitive endogenous RNAs (ceRNAs) function as molecular "sponges" for miRNAs via their MREs, thus derepressing the target genes of the respective miRNAs [43–45]. Circular RNAs or transcripts of pseudogenes might also behave as ceRNAs. Some studies further indicated that a few lncRNAs could modulate gene expression and/or cell signaling at multiple molecular levels simultaneously [46, 47]. Given the broad involvement of lncRNAs in cellular events, it is not surprising to reveal that a number of lncRNAs play pivotal roles in embryogenesis, tissue homeostasis, and the development and progression of various diseases [48–51].

3. Aberrant lncRNA Expression in Glioblastomas

Dysregulated expression of lncRNAs is associated with human diseases such as cardiovascular diseases, neurodegenerative disorders, and malignant tumors, which are also evident in brain tumors [52, 53]. Moreover, several studies showed that abnormal expression profiles of lncRNA in clinical glioma specimens are correlated with histological differentiation and malignancy grades, which may have clinical implications in the diagnosis and prognostication [54, 55]. A high-throughput screen study by Han et al. identified 1308 differentially expressed lncRNAs in GBMs compared to normal brain tissue. Among them, *ASLNC22381* and *ASLNC2081* were predicted to be involved in the recurrence and malignant progression of GBM by upregulating the expression of *IGF-1* (insulin-like growth factor 1) [56]. Similarly, a transcriptome comparison study identified differentially expressed lncRNAs and mRNAs between GBM and normal brain samples. Gene ontology (GO) and pathway analysis-predicted genes involved in GBM pathogenesis could be modulated by lncRNAs, such as the *HOX* cluster-associated lncRNAs [57]. In another study, Zhang et al. identified 129 differentially expressed lncRNAs in glioma tissues. Their analysis revealed that the levels of lncRNA *MEG3* were significantly downregulated in GBMs, whereas those of *HOTAIRM1* (HOX antisense intergenic RNA myeloid 1) and *CRNDE* were upregulated [58]. Importantly, the expression pattern of these lncRNAs correlates with histological classification and malignancy grades of gliomas [58]. Although mechanistic evidence remains to be explored, differentially expressed lncRNAs could be a start point for digging out novel biomarkers for diagnosis or targets for therapy [6]. Next, we will list individual lncRNAs regarding their involvements in multiple aspects of GBMs. In the past few years, researchers largely focused on a few well-documented, abundantly expressed lncRNAs that have explicit indications for their participation in development and/or in tumorigenesis, with some of which showing diagnostic and prognostic potentials (Table 1).

3.1. HOTAIR (HOX Transcript Antisense Intergenic RNA). LncRNA *HOTAIR*, located at chromosome 12q13, was originally implicated in epigenetic silencing of genes at the *HOXD* locus and inhibits initiation of transcription by recruiting PRC2 (polycomb repressive complex 2) [59, 60]. Aberrant *HOTAIR* expression was closely related to cancer metastasis and was defined as a negative prognostic factor for patients with malignant tumor [61, 62]. *HOTAIR* expression in GBMs is significantly higher than that in normal brain tissues and low-grade gliomas and correlated with poor prognosis and glioma molecular subtype. Moreover, *HOTAIR* was an independent prognostic factor in GBM patients. A gene set enrichment analysis (GESA) revealed that *HOTAIR* expression primarily associated with genes involved in cell cycle progression. Consistently, *HOTAIR* maintains proliferation and tumorigenic potential of GBM cells [63]. *HOTAIR* regulates cell cycle progression of GBM cells probably through EZH2, the core component of the PRC2 (polycomb repressive complex 2) [64], and the Wnt/β-catenin pathway [65]. Reports also suggest that *HOTAIR* might behave as a competing endogenous RNA (ceRNAs) to regulate the levels of prooncogenic transcripts by buffering miRNAs [66, 67]. The mechanisms underlying elevated *HOTAIR* expression in glioma remain to be investigated,

Table 1: A list of deregulated lncRNAs in glioma with diagnostic and prognostic perspectives.

Category	LncRNA	Biological function/phenotypes	Molecular mechanisms/targets	Survival correlation	Others	References
Overrepresented in gliomas	*HOTAIR*	Maintains proliferation and tumorigenic potential of GBM cells	Associating with PRC2; regulating Wnt/β-catenin signaling; ceRNAs (competing endogenous RNAs) for miR-326 and miR-148b-3p	Yes	Preferentially expressed in classical and mesenchymal glioma	[64–67]
	CRNDE	Promotes glioma cell growth and migration in vitro and tumorigenesis in a xenograft mouse mode	ceRNAs for miR-136-5p, miR-186, and miR-384	Yes		[55, 71–73]
	NEAT1	Promotes glioma pathogenesis	ceRNAs for miR-449b-5p to upregulate *c-Met*; associating with EZH2	Yes		[78–80]
	XIST	Confers glioma cells' oncogenic and chemoresistant behaviors; *XIST* knockdown increased permeability of the brain-tumor barrier (BTB) and inhibited glioma angiogenesis	ceRNAs for miR-152, miR-429, miR-29c, and miR-137	/		[14, 83–86]
	H19	Enhances invasion, angiogenesis, stemness and tumorigenicity of GBM cells; depletion of *H19* inhibited glioma-provoked proliferation, migration, and tube formation of glioma endothelial cells (GECs)	ceRNA (miR-29a); negatively regulating *RB1* expression	Yes		[12, 89, 92, 93]
	TUG1	Maintains stemness and tumorigenic properties of GBM stem-like cells (GSCs); modulates blood-tumor barrier; and enhances glioma-induced angiogenesis	Associating with PRC2 and YY1; ceRNAs for miR-26a, miR-144, miR-299, and miR-145	/	Intravenous administration of ASOs against *TUG1* induces GSC differentiation and suppresses tumor growth intracranially	[46, 125–127]
	SOX2OT	Maintains proliferation, migration, invasion, and tumorigenesis of GSCs	ceRNAs for miR-122 and miR-194-5p	/		[99]
	UCA1	Promote glioma cell proliferation, invasion, and migration; modulates glioblastoma-associated stromal cell-mediated glycolysis and invasion of glioma cells	ceRNAs for miR-182 and miR-122	Yes		[165–168]

Table 1: Continued.

Category	LncRNA	Biological function/phenotypes	Molecular mechanisms/targets	Survival correlation	Others	References
Downregulated in gliomas	*MEG3*	Impairs *in vitro* glioma cell proliferation	MDM2-p53; ceRNAs for miR-19a and miR-93	/		[105–109]
	MALAT1	Tumor-suppressive function in glioma	Inactivating the ERK/MAPK signaling; enhancing the expression of tumor-suppressor *FBXW7*	Yes	*MALAT1* could be either a positive or a negative regulator in glioma tumorigenesis depending on cellular contexts	[115–119]
	ROR	Tumor-suppressive function in glioma	*ROR*'s expression is negatively correlated with the level of *KLF4*, a stem cell gene	/		[169]
	TUSC7	Suppresses cellular proliferation and invasion of glioma cells, accelerates cellular apoptosis, and inhibits TMZ resistance	ceRNAs for miR-23b and miR-10a	Yes		[170–172]
	CASC2	*CACS2* overexpression sensitizes glioma cells to TMZ	ceRNAs for miR-181a and miR-193-5p to upregulate *PTEN* and *mTOR* expression	Yes		[134–136]

and a study in breast cancer hints at both transcription and posttranscriptional regulations [47]. In summary, these findings reveal that *HOTAIR* may enhance the development of GBM through multiple regulatory signals, and its clinical value awaits further study.

3.2. CRNDE (Colorectal Neoplasia Differentially Expressed). LncRNA *CRNDE* is initially identified to be overrepresented in >90% of colorectal adenomas and adenocarcinomas [68, 69]. Later, *CRNDE* was also found to be highly expressed in brain cancers including GBM and astrocytomas [55, 69, 70]. Applying a microarray-mining approach, Zhang et al. reported that *CRNDE* was upregulated by 32-fold up in glioma tissues than that in nontumor brain tissues [58]. *CRNDE* overexpression promotes glioma cell growth and migration *in vitro* and tumorigenesis in a xenograft mouse model. Mechanistic studies suggested that *CRNDE* expression could be regulated by the mTOR signaling and the histone acetylation status in the promoter region [71]. *CRNDE* is enriched in the stem-like population of GBM cells and promotes tumor cell proliferation and migration and by sponging down miR-186 to derepressing the expression of *XIAP* (X-linked inhibitor of apoptosis) and *PAK7* [p21 protein-(Cdc42/Rac-) activated kinase 7], two prooncogenic molecules [72]. Similarly, *CRNDE* behaves ceRNAs for miR-384 to maintain the expression of *PIWIL4* (piwi-like RNA-mediated gene silencing 4), which promotes gliomagenesis probably by activating the STAT3 signaling [73]. Moreover, high *CRNDE* expression correlates with tumor progression and poor survival for glioma patients [55].

3.3. NEAT1 (Nuclear-Enriched Abundant Transcript 1). LncRNA *NEAT1* is crucial for the formation of paraspeckles, nuclear domains implicated in mRNA nuclear retention, and splicing [74–76]. *NEAT1* is upregulated in human GBM tissues [6] and glioma cell lines like U251 and U87 [77]. *NEAT1* expression was higher in glioma tissues than adjacent noncancerous tissues. Higher *NEAT1* expression correlated with a larger tumor size, higher WHO grade, recurrence rate, and unfavorable overall survival, supporting *NEAT1* as a potential prognostic predictor of glioma patients [78]. *NEAT1* has been implicated in gliomagenesis by promoting cell proliferation, invasion, and migration. Zhen et al. demonstrated that *NEAT1* could upregulate the expression of *c-Met* oncogene through buffering miR-449b-5p, a negative regulator of *c-Met* [79]. The latest study by Chen et al. showed *NEAT1* could be upregulated by the oncogenic EGFR pathway. Elevated *NEAT1* promotes GBM tumorigenesis by acting as a scaffold molecule to recruit the histone modification enzyme EZH2 to silence target-specific genes including *AXIN2*, *ICAT*, and *GSK3B*, thus leading to the activation of the canonical Wnt/β-catenin signaling. This study highlights the epigenetic role of lncRNAs in controlling the expression of tumorigenic components in GBM cells [80].

3.4. XIST (X-Inactive Specific Transcript). LncRNA *XIST* is the major effector of X inactivation in mammals to balance gene expression between the sexes, and the *XIST* RNA is exclusively transcribed from the inactive X chromosome [16, 17]. *XIST* loss in female mice leads to a highly aggressive myeloproliferative neoplasm and myelodysplastic syndrome (mixed MPN/MDS) [81]. *XIST* has been found to be up- or downregulated in a variety of human cancers [82]. Yao et al. found *XIST* expression was upregulated in glioma tissues and GSCs and knockdown of *XIST* suppresses GSC proliferation, migration, invasion, and tumorigenic potential by upregulating miR-152 [14]. A few other studies also indicated *XIST* confers glioma cell oncogenic and chemoresistant behaviors by serving as ceRNAs to suppress actions of microRNAs [83–85]. One recent study pointed out a novel role of *XIST* in regulating a glioma microenvironment. *XIST* was found to be overrepresented in glioma endothelial cells (GECs), and *XIST* knockdown increased permeability of the brain-tumor barrier (BTB) and inhibited glioma angiogenesis, which may have beneficial effects on GBM treatment. Mechanistically, *XIST* regulates the expression of the transcription factor Forkhead Box C1 (FOXC1) and Zonula occludens 2 (ZO-2), two molecules essential for maintaining the BTB integrity, by dampening miR-137 [86].

3.5. H19. LncRNA *H19*, transcribed only from the maternally inherited (imprinted) allele, is involved in the postnatal development and tumorigenesis [87, 88]. *H19* is upregulated in glioma tissues and was negatively associated with patient survival time [89]. *H19* overexpression enhances invasion, angiogenesis, stemness, and tumorigenicity of GBM cells, whereas *H19* depletion has opposite effects [90, 91]. *H19* might exert its prooncogenic function via the embedding miR-675, which could target the expression of tumor suppressor gene *RB1* [12, 89, 92]. Similar to *XIST*, *H19* was also found to be highly expressed in glioma-associated endothelial cells (GECs). Depletion of *H19* inhibited glioma-provoked GEC proliferation, migration, and tube formation. Knockdown of *H19* upregulates miR-29a, resulting in decreased expression of VASH2, an angiogenic factor [93].

3.6. SOX2OT (SOX2 Overlapping Transcript). The genomic region that transcribes lncRNA *SOX2OT* contains the *SOX2* gene, one of the major pluripotency regulators, in its intronic region [94]. Similar to *SOX2*, *SOX2OT* is highly expressed in embryonic stem cells and neural precursor cells but becomes downregulated upon differentiation. Elevated *SOX2OT* expression and the concomitant *SOX2* expression were also noticed in some carcinomas with an epithelial origin, including lung, breast, and esophageal cancer [95]. The transcriptional regulation of *SOX2* by *SOX2OT* has been highlighted in development and tumorigenic scenarios [96–98]. A recent study found *SOX2OT* is essential for proliferation, migration, invasion, and tumorigenesis of GBM stem-like cells (GSCs). The results also indicated *SOX2OT* might act as ceRNAs to maintain the expression of *SOX3* by buffering miR-122 and miR-194-5p. Moreover, in GSCs, SOX3 functions as an oncogene and transactivates the expression of *SOX2OT* and *TDGF-1*, thus forming a positive feedback loop [99].

3.7. MEG3 (Maternally Expressed Gene 3). LncRNA *MEG3*, also known as *Gtl2 (gene trap locus 2)* in mice, is transcribed from the imprinted maternal allele, with multiple isoforms generated by alternative splicing [100, 101]. *MEG3* expression is prevalent in human normal tissues, while it becomes diminished in most human tumors, and overexpression of *MEG3* inhibits the growth of human cancer cells [102–104]. DNA methylation at the promoter or the intergenic differentially methylated region of *MEG3* mediates silencing of the *MEG3* gene in tumors. Ectopic *MEG3* expression significantly elevates the level of tumor suppressor protein p53 in human cancer cell lines. The increased p53 level upon *MEG3* overexpression is partly due to the downregulation of MDM2, an E3 ubiquitin ligase that targets p53 for degradation [105]. *MEG3* is significantly downregulated in GBMs and behaves as a tumor suppressor in GBM cells in a p53-dependent manner [106, 107]. A recent study also suggested *MEG3* might act as competing endogenous RNAs (ceRNAs) of miR-19a and miR-93 to inhibit GBM cell growth [108, 109]. Li et al. provided evidence that hypermethylation at the *MEG3* promoter mediated by DNMT1 controls the expression of *MEG3* and subsequent p53 activity in glioma cells. Moreover, treating glioma cells with the DNA methylation inhibitor 5-AzadC inhibited the growth and promoted apoptosis of glioma cells, and its potential application in GBM animal models remains to be investigated [110].

3.8. MALAT1 (Metastasis-Associated Lung Adenocarcinoma Transcript 1). LncRNA *MALAT1*, also known as *NEAT2 (noncoding nuclear-enriched abundant transcript 2)*, was initially demonstrated to be positively associated with metastasis and shorter survival in non-small cell carcinoma (NSCLC) patients, specifically in the early stages of lung adenocarcinoma [111]. Analogous to *NEAT1*, *MALAT1/NEAT2* is majorly enriched in the paraspeckle, a nuclear structure essential for RNA storage and splicing [112]. But its role in alternative splicing might be species-specific [113]. In most solid tumors, *MALAT1* is highly expressed and is associated with poorer clinical parameters [114]. Interestingly, studies in glioma have inconsistent results regarding roles of *MALAT1*. *MALAT1* expression was reported to be lower in glioma tissues than that in noncancerous brain tissues, and its higher expression correlates with better patient survival, suggesting *MALAT1* may serve as an independent prognostic factor and act as a tumor suppressor in glioma [115]. Accordingly, the proliferation and invasion ability of glioma cells was significantly enhanced by *MALAT1* knockdown in glioma xenograft models, whereas *MALAT1* overexpression had opposite effects [115]. *MALAT1* might exert its tumor suppressor role by inactivation of the prosurvival ERK/MAPK signaling and/or enhance the expression of FBXW7, an antiproliferation protein [116]. However, a few other studies indicated *MALAT1* has prooncogenic (tumor-promoting) roles and knockdown *MALAT1* might confer beneficiary effects on glioma treatment [117–119]. Thus, *MALAT1* could be either a positive or a negative regulator in glioma tumorigenesis depending on cellular contexts.

3.9. TUG1 (Taurine-Upregulated Gene 1). *TUG1* was originally identified as a lncRNA required for the normal development of photoreceptors in the mouse retina [120]. As a chromatin-associated lncRNA, *TUG1* can regulate gene expression by interacting with the polycomb repressive complex 2 (PRC2) [121]. *TUG1* is extensively related to human malignancies, reported having either tumor promoting or tumor suppressing functions in different types of cancers [122–124]. *TUG1* is expressed at significantly higher levels in GBM tissues than in normal brain tissues [125]. In GSCs, *TUG1* expression is transcriptionally induced by the Notch signaling, a well-known oncogenic pathway. *TUG1* maintains stemness and tumorigenic properties of GSCs by two parallel mechanisms: in the cytosol, *TUG1* sponges miR-145 to maintain the expression of stemness-associated genes including *SOX2* and *MYC*; in the nucleus, TUG1 is able to associate with the PRC2 and YY1 transcription factor to suppress differentiation [46]. This study underscores the importance of the Notch-lncRNA axis in regulating self-renewal of GSCs and proposes a rationale for targeting *TUG1* as a potent therapeutic approach to treat GBMs by eradicating GSCs. *TUG1* could maintain the expression of tumor suppressor PTEN by sponging off its negative regulator miR-26a, but the functional implication is not clear [125]. *TUG1* was also found to be highly expressed in glioma endothelial cells (GECs). Serving as ceRNAs for miR-144 in GECs and miR-299 in GSCs, *TUG1* is able to modulate blood-tumor barrier and enhance glioma-induced angiogenesis, respectively [126, 127]. Therefore, depleting *TUG1* in glioma cells might facilitate a microenvironment detrimental for tumor growth while beneficial for drug delivery.

3.10. CASC2 (Cancer Susceptibility Candidate 2). LncRNA *CASC2* is first identified to be transcribed from an allelic loss region at chromosome 10q26 in human endometrial cancer [128, 129]. Later on, *CASC2* was unveiled to be a tumor suppressor gene in endometrial, colorectal, lung, and renal cancers and gliomas, probably behaving as ceRNAs by buffering miR-21 and miR-18a, two microRNAs with oncogenic effects [9, 130–133]. The status of *CASC2*'s low expression is positively correlated with advanced tumor grades, shorter survival time, and poorer TMZ response in glioma patients [134]. *CACS2* overexpression could sensitize glioma cells to temozolomide (TMZ) cytotoxicity by upregulating PTEN protein and downregulating p-AKT protein through regulating miR-181a or by inhibiting autophagy via sponging miR-193a-5p to increase mTOR expression [135, 136].

4. Potentials of lncRNAs in Diagnostic or Prognostic Applications

Like many other malignant tumors, the clinical diagnosis of glioma traditionally relies on symptoms, imaging findings from CT/MRI scans, and histological properties of resected tumor tissues. Recent advancement in high-throughput technology enables both clinicians and researchers to acquire genome, epigenome, transcriptome, and proteome data of tumor bulk or individual tumor cells [137–139]. Incorporation of these data generates molecular signatures of tumors much more comprehensively than traditional understanding. For instance, transcriptome signatures could classify GBMs into molecularly distinct subgroups including proneural and mesenchymal GBMs, which correspond to clinical and histological properties [140, 141]. More importantly, the current WHO classification of tumors of the central nervous system is now defined by both histology and molecular features, the latter including IDH1/2 mutation, 1p/19q co-deletion, and histone H3 K27-mutant [142]. The inclusion of these molecular signatures is due to their explicit indications for prognosis and/or targeted therapy. Furthermore, sampling and molecular description of glioma tissues at multiloci or overtime for individual tumors could allow the understanding of the heterogeneity features of GBMs, as well as their evolving path molecularly [143–145]. The accumulation of this knowledge would eventually lead to design patient-tailored glioma therapeutics.

As mentioned earlier, high-throughput transcriptome analyses of glioma tissues/cells identified numerous highly and/or differentially expressed lncRNAs, including *MALAT1*, *HOXA11-AS*, and *CRNDE*, which correlate with histological and molecular subclassification, and/or show prognostic values [15, 55]. Zhang et al. carried out lncRNA profiling from 213 GBMs using data from The Cancer Genome Atlas (TCGA) database and identified six lncRNAs including *KIAA0495*, *MIAT*, *GAS5*, *PART1*, *PAR5*, and *MGC21881*, whose weighted expressions are closely associated with the overall survival of GBM patients. Further analysis demonstrated that the six-lncRNA signature was an independent risk factor for the prognosis of GBM patients [146]. Similarly, using consensus clustering of 1970 lncRNAs from the Rembrandt dataset, a study classified gliomas into three molecular subtypes called LncR1, LncR2, and LncR3. Moreover, LncR1 subtype was associated with the poorest overall survival rate, while the LncR3 subtype correlated with the best prognosis [147]. These are interesting attempts showing potential application of lncRNAs in diagnostic and prognostic purposes for gliomas (Table 1). The drop of cost for molecular diagnosis based on high throughput techniques would greatly accelerate profiling of lncRNAs in glioma patients. These efforts will unveil practical lncRNA signatures for glioma, including their genomic, epigenetic, and expression features, which should be comparable with or even better than current molecular signatures. Next-generation diagnostics based on the CRISPR-Cas technique, including DETECTR (DNA endonuclease-targeted CRISPR trans reporter) [148] and SHERLOCK (specific high-sensitivity enzymatic reporter unlocking) [149, 150], will facilitate easier tumor detection and categorization molecularly, for example, using nucleic acids extracted from the cerebrospinal fluid (CSF). These emerging diagnostic tools will have to be robustly compared to standard diagnostics to ensure sensitivity and specificity.

5. Potentials of lncRNAs in Targeted Therapy against Glioma

Dismal outcomes for GBM patients are mostly due to high heterogeneity and aggressiveness of glioma cells, immune-privileged brain environment, lack of effective treatment, and poor BBB penetration for most drugs [151]. Although aforementioned lncRNAs are suggested to regulate these aspects, most evidence regarding their roles in tumorigenesis is collected using *in vitro* cultured GBM cells and xenografted animal models. Thus, unbiased whole-genome screening followed by the genetic manipulation of lncRNAs in glioma animal models would identify lncRNAs with explicit oncogenic or tumor suppressing roles, which could pave the path for translational application [152, 153]. Moreover, efficient delivering of small interfering RNAs (siRNAs) and antisense oligonucleotides (ASOs) that target lncRNAs in glioma mass requires advances in engineering and material science [154, 155]. For instance, Katsushima et al. revealed that intravenous injection of ASOs targeting *TUG1* in combination with a drug delivery system induced glioma stem cell differentiation and repressed tumor growth efficiently [46]. Alternatively, the CRISPR/Cas9-mediated editing of genome, epigenome, or RNA would be another promising approach to tackle glioma-expressed lncRNAs [156]. Notably, CRISPR/Cas9-based transcriptional activation or suppression can enhance or inhibit lncRNA expression epigenetically without modifying genomes, which could be advantageous to traditional gene-therapy approaches [157–159]. In addition, several epigenetic modulators that regulate oncogenic lncRNAs have recently emerged as novel therapeutic targets for GBM patients [110]. For example, Pastori et al. demonstrated that the inhibition of bromodomain protein BRD4 could alleviate the expression of oncogenic lncRNA *HOTAIR* in GBM patients, exerting an antiproliferation effect by inducing cell cycle arrest in GBM cells [160, 161].

6. Conclusion

A large variety of lncRNAs has been identified to be associated with deregulated gene expression and imbalanced biological processes in GBMs. LncRNAs were involved in nearly all facets of GBM malignancy, including cell proliferation, stemness, angiogenesis, migration, invasion, tumor immune responses, relapse, and drug resistance. However, it remains to be determined if certain deregulated lncRNAs are core causal factors in tumorigenesis and progression of GBMs. Further, using state-of-the-art biochemical and molecular approaches, we are able to precisely delineate how lncRNAs control molecular machinery and cellular functions. This knowledge is imperative for devising targeted therapeutics. In addition, the complex glioma milieu composed of microvessels, immune cells, extracellular vesicles, cytokines, and neural transmitters is indispensable for GBM propagation and invasion [42, 162]. Particularly, the latest progress using immune checkpoint inhibitors brings hopes for previously intractable tumors including GBMs [163, 164]. Hence, future studies need to identify lncRNAs that have essential roles in regulating the fate and behavior of microenvironment components in GBMs. In summary, the current understanding of GBM lncRNAs is only the tip of an iceberg, and continuing efforts will make possible developing novel RNA-based strategy to treat such a malignant tumor and bring new hopes for patients with GBM.

Conflicts of Interest

The authors declare that the research was conducted in the absence of any commercial or financial relationships that could be construed as a potential conflict of interest.

Authors' Contributions

Tao Zeng and Lei Li contributed equally to this work.

Acknowledgments

This work was supported by grants from the National Natural Science Foundation of China (no. 31671418 and no. 31471361) and Fundamental Research Funds for the Central Universities (2042017kf0205 and 2042017kf0242) to Yan Zhou.

References

[1] R. Stupp, W. P. Mason, M. J. van den Bent et al., "Radiotherapy plus concomitant and adjuvant Temozolomide for glioblastoma," *The New England Journal of Medicine*, vol. 352, no. 10, pp. 987–996, 2005.

[2] R. Stupp, S. Taillibert, A. Kanner et al., "Effect of tumor-treating fields plus maintenance Temozolomide vs maintenance Temozolomide alone on survival in patients with glioblastoma: a randomized clinical trial," *JAMA*, vol. 318, no. 23, pp. 2306–2316, 2017.

[3] Q. T. Ostrom, L. Bauchet, F. G. Davis et al., "The epidemiology of glioma in adults: a "state of the science" review," *Neuro-Oncology*, vol. 16, no. 7, pp. 896–913, 2014.

[4] R. Stupp, M. E. Hegi, M. R. Gilbert, and A. Chakravarti, "Chemoradiotherapy in malignant glioma: standard of care and future directions," *Journal of Clinical Oncology*, vol. 25, no. 26, pp. 4127–4136, 2007.

[5] P. D. Delgado-Lopez and E. M. Corrales-Garcia, "Survival in glioblastoma: a review on the impact of treatment modalities," *Clinical and Translational Oncology*, vol. 18, no. 11, pp. 1062–1071, 2016.

[6] Y. Zhang, N. Cruickshanks, M. Pahuski et al., "Noncoding RNAs in glioblastoma," in *Glioblastoma*, S. Vleeschouwer, Ed., pp. 95–130, Codon Publications, Brisbane, QLD, Australia, 2017.

[7] R. Abounader and J. Laterra, "Scatter factor/hepatocyte growth factor in brain tumor growth and angiogenesis," *Neuro-Oncology*, vol. 7, no. 4, pp. 436–451, 2005.

[8] F. B. Furnari, T. F. Cloughesy, W. K. Cavenee, and P. S. Mischel, "Heterogeneity of epidermal growth factor receptor signalling networks in glioblastoma," *Nature Reviews Cancer*, vol. 15, no. 5, pp. 302–310, 2015.

[9] Y. Yan, Z. Xu, Z. Li, L. Sun, and Z. Gong, "An insight into the increasing role of LncRNAs in the pathogenesis of gliomas," *Frontiers in Molecular Neuroscience*, vol. 10, p. 53, 2017.

[10] K. Ludwig and H. I. Kornblum, "Molecular markers in glioma," *Journal of Neuro-Oncology*, vol. 134, no. 3, pp. 505–512, 2017.

[11] W. Chen, X. K. Xu, J. L. Li et al., "MALAT1 is a prognostic factor in glioblastoma multiforme and induces chemoresistance to temozolomide through suppressing miR-203 and promoting thymidylate synthase expression," *Oncotarget*, vol. 8, no. 14, pp. 22783–22799, 2017.

[12] Y. Shi, Y. Wang, W. Luan et al., "Long non-coding RNA H19 promotes glioma cell invasion by deriving miR-675," *PLoS One*, vol. 9, no. 1, article e86295, 2014.

[13] C. Li, B. Lei, S. Huang et al., "H19 derived microRNA-675 regulates cell proliferation and migration through CDK6 in glioma," *American Journal of Translational Research*, vol. 7, no. 10, pp. 1747–1764, 2015.

[14] Y. Yao, J. Ma, Y. Xue et al., "Knockdown of long non-coding RNA XIST exerts tumor-suppressive functions in human glioblastoma stem cells by up-regulating miR-152," *Cancer Letters*, vol. 359, no. 1, pp. 75–86, 2015.

[15] Z. Peng, C. Liu, and M. Wu, "New insights into long noncoding RNAs and their roles in glioma," *Molecular Cancer*, vol. 17, no. 1, p. 61, 2018.

[16] G. F. Kay, G. D. Penny, D. Patel, A. Ashworth, N. Brockdorff, and S. Rastan, "Expression of *Xist* during mouse development suggests a role in the initiation of X chromosome inactivation," *Cell*, vol. 72, no. 2, pp. 171–182, 1993.

[17] G. D. Penny, G. F. Kay, S. A. Sheardown, S. Rastan, and N. Brockdorff, "Requirement for *Xist* in X chromosome inactivation," *Nature*, vol. 379, no. 6561, pp. 131–137, 1996.

[18] A. Matsumoto, A. Pasut, M. Matsumoto et al., "mTORC1 and muscle regeneration are regulated by the LINC00961-encoded SPAR polypeptide," *Nature*, vol. 541, no. 7636, pp. 228–232, 2017.

[19] D. M. Anderson, K. M. Anderson, C. L. Chang et al., "A micropeptide encoded by a putative long noncoding RNA regulates muscle performance," *Cell*, vol. 160, no. 4, pp. 595–606, 2015.

[20] S. Fang, L. L. Zhang, J. C. Guo et al., "NONCODEV5: a comprehensive annotation database for long non-coding RNAs," *Nucleic Acids Research*, vol. 46, no. D1, pp. D308–D314, 2018.

[21] L. Ma, V. B. Bajic, and Z. Zhang, "On the classification of long non-coding RNAs," *RNA Biology*, vol. 10, no. 6, pp. 925–933, 2013.

[22] M. Schlackow, T. Nojima, T. Gomes, A. Dhir, M. Carmo-Fonseca, and N. J. Proudfoot, "Distinctive patterns of transcription and RNA processing for human lincRNAs," *Molecular Cell*, vol. 65, no. 1, pp. 25–38, 2017.

[23] T. Derrien, R. Johnson, G. Bussotti et al., "The GENCODE v7 catalog of human long noncoding RNAs: analysis of their gene structure, evolution, and expression," *Genome Research*, vol. 22, no. 9, pp. 1775–1789, 2012.

[24] T. R. Mercer, I. A. Qureshi, S. Gokhan et al., "Long noncoding RNAs in neuronal-glial fate specification and oligodendrocyte lineage maturation," *BMC Neuroscience*, vol. 11, no. 1, p. 14, 2010.

[25] T. G. Belgard, A. C. Marques, P. L. Oliver et al., "A transcriptomic atlas of mouse neocortical layers," *Neuron*, vol. 71, no. 4, pp. 605–616, 2011.

[26] J. Aprea, S. Prenninger, M. Dori et al., "Transcriptome sequencing during mouse brain development identifies long non-coding RNAs functionally involved in neurogenic commitment," *The EMBO Journal*, vol. 32, no. 24, pp. 3145–3160, 2013.

[27] B. J. Molyneaux, L. A. Goff, A. C. Brettler et al., "DeCoN: genome-wide analysis of in vivo transcriptional dynamics during pyramidal neuron fate selection in neocortex," *Neuron*, vol. 85, no. 2, pp. 275–288, 2015.

[28] L. Lipovich, F. Dachet, J. Cai et al., "Activity-dependent human brain coding/noncoding gene regulatory networks," *Genetics*, vol. 192, no. 3, pp. 1133–1148, 2012.

[29] G. Barry, J. A. Briggs, D. P. Vanichkina et al., "The long noncoding RNA Gomafu is acutely regulated in response to neuronal activation and involved in schizophrenia-associated alternative splicing," *Molecular Psychiatry*, vol. 19, no. 4, pp. 486–494, 2014.

[30] T. R. Mercer, M. E. Dinger, and J. S. Mattick, "Long noncoding RNAs: insights into functions," *Nature Reviews Genetics*, vol. 10, no. 3, pp. 155–159, 2009.

[31] P. O. Angrand, C. Vennin, X. le Bourhis, and E. Adriaenssens, "The role of long non-coding RNAs in genome formatting and expression," *Frontiers in Genetics*, vol. 6, p. 165, 2015.

[32] I. Martianov, A. Ramadass, A. Serra Barros, N. Chow, and A. Akoulitchev, "Repression of the human dihydrofolate reductase gene by a non-coding interfering transcript," *Nature*, vol. 445, no. 7128, pp. 666–670, 2007.

[33] Y. Fang and M. J. Fullwood, "Roles, functions, and mechanisms of long non-coding RNAs in Cancer," *Genomics, Proteomics & Bioinformatics*, vol. 14, no. 1, pp. 42–54, 2016.

[34] J. M. Engreitz, J. E. Haines, E. M. Perez et al., "Local regulation of gene expression by lncRNA promoters, transcription and splicing," *Nature*, vol. 539, no. 7629, pp. 452–455, 2016.

[35] X. D. Fu, "Non-coding RNA: a new frontier in regulatory biology," *National Science Review*, vol. 1, no. 2, pp. 190–204, 2014.

[36] E. Hacisuleyman, L. A. Goff, C. Trapnell et al., "Topological organization of multichromosomal regions by the long intergenic noncoding RNA firre," *Nature Structural & Molecular Biology*, vol. 21, no. 2, pp. 198–206, 2014.

[37] A. Lin, C. Li, Z. Xing et al., "The *LINK-A* lncRNA activates normoxic HIF1α signalling in triple-negative breast cancer," *Nature Cell Biology*, vol. 18, no. 2, pp. 213–224, 2016.

[38] K. Katsushima and Y. Kondo, "Non-coding RNAs as epigenetic regulator of glioma stem-like cell differentiation," *Frontiers in Genetics*, vol. 5, p. 14, 2014.

[39] P.-M. Chua, H.-I. Ma, L.-H. Chen et al., "Deregulated microRNAs identified in isolated glioblastoma stem cells: an overview," *Cell Transplantation*, vol. 22, no. 4, pp. 741–753, 2013.

[40] M. Ousset, F. Bouquet, F. Fallone et al., "Loss of ATM positively regulates the expression of hypoxia inducible factor 1 (HIF-1) through oxidative stress: role in the physiopathology of the disease," *Cell Cycle*, vol. 9, no. 14, pp. 2886–2894, 2010.

[41] S. Kreth, N. Thon, and F. W. Kreth, "Epigenetics in human gliomas," *Cancer Letters*, vol. 342, no. 2, pp. 185–192, 2014.

[42] A. Bronisz, Y. Wang, M. O. Nowicki et al., "Extracellular vesicles modulate the glioblastoma microenvironment via a tumor suppression signaling network directed by miR-1," *Cancer Research*, vol. 74, no. 3, pp. 738–750, 2014.

[43] M. Cesana, D. Cacchiarelli, I. Legnini et al., "A long noncoding RNA controls muscle differentiation by functioning as a competing endogenous RNA," *Cell*, vol. 147, no. 4, p. 947, 2011.

[44] L. Salmena, L. Poliseno, Y. Tay, L. Kats, and P. P. Pandolfi, "A *ceRNA* hypothesis: the Rosetta stone of a hidden RNA language?," *Cell*, vol. 146, no. 3, pp. 353–358, 2011.

[45] Y. Tay, J. Rinn, and P. P. Pandolfi, "The multilayered complexity of ceRNA crosstalk and competition," *Nature*, vol. 505, no. 7483, pp. 344–352, 2014.

[46] K. Katsushima, A. Natsume, F. Ohka et al., "Targeting the notch-regulated non-coding RNA *TUG1* for glioma treatment," *Nature Communications*, vol. 7, article 13616, 2016.

[47] E. Pawlowska, J. Szczepanska, and J. Blasiak, "The long noncoding RNA HOTAIR in breast cancer: does autophagy play a role?," *International Journal of Molecular Sciences*, vol. 18, no. 11, p. 2317, 2017.

[48] A. Wang, J. Wang, Y. Liu, and Y. Zhou, "Mechanisms of long non-coding RNAs in the assembly and plasticity of neural circuitry," *Frontiers in Neural Circuits*, vol. 11, p. 76, 2017.

[49] K. C. Wang, Y. W. Yang, B. Liu et al., "A long noncoding RNA maintains active chromatin to coordinate homeotic gene expression," *Nature*, vol. 472, no. 7341, pp. 120–124, 2011.

[50] A. D. Ramos, A. Diaz, A. Nellore et al., "Integration of genome-wide approaches identifies lncRNAs of adult neural stem cells and their progeny in vivo," *Cell Stem Cell*, vol. 12, no. 5, pp. 616–628, 2013.

[51] L. Li and H. Y. Chang, "Physiological roles of long noncoding RNAs: insight from knockout mice," *Trends in Cell Biology*, vol. 24, no. 10, pp. 594–602, 2014.

[52] P. Grote, L. Wittler, D. Hendrix et al., "The tissue-specific lncRNA *Fendrr* is an essential regulator of heart and body wall development in the mouse," *Developmental Cell*, vol. 24, no. 2, pp. 206–214, 2013.

[53] E. Salta and B. De Strooper, "Non-coding RNAs with essential roles in neurodegenerative disorders," *The Lancet Neurology*, vol. 11, no. 2, pp. 189–200, 2012.

[54] Q. Wang, J. Zhang, Y. Liu et al., "A novel cell cycle-associated lncRNA, HOXA11-AS, is transcribed from the 5-prime end of the HOXA transcript and is a biomarker of progression in glioma," *Cancer Letters*, vol. 373, no. 2, pp. 251–259, 2016.

[55] S. Y. Jing, Y. Y. Lu, J. K. Yang, W. Y. Deng, Q. Zhou, and B. H. Jiao, "Expression of long non-coding RNA CRNDE in glioma and Its correlation with tumor progression and patient survival," *European Review for Medical and Pharmacological Sciences*, vol. 20, no. 19, pp. 3992–3996, 2016.

[56] L. Han, K. Zhang, Z. Shi et al., "LncRNA profile of glioblastoma reveals the potential role of lncRNAs in contributing to glioblastoma pathogenesis," *International Journal of Oncology*, vol. 40, no. 6, pp. 2004–2012, 2012.

[57] Y. Yan, L. Zhang, Y. Jiang et al., "LncRNA and mRNA interaction study based on transcriptome profiles reveals potential core genes in the pathogenesis of human glioblastoma multiforme," *Journal of Cancer Research and Clinical Oncology*, vol. 141, no. 5, pp. 827–838, 2015.

[58] X. Zhang, S. Sun, J. K. S. Pu et al., "Long non-coding RNA expression profiles predict clinical phenotypes in glioma," *Neurobiology of Disease*, vol. 48, no. 1, pp. 1–8, 2012.

[59] J. L. Rinn, M. Kertesz, J. K. Wang et al., "Functional demarcation of active and silent chromatin domains in human *HOX* loci by noncoding RNAs," *Cell*, vol. 129, no. 7, pp. 1311–1323, 2007.

[60] M. C. Tsai, O. Manor, Y. Wan et al., "Long noncoding RNA as modular scaffold of histone modification complexes," *Science*, vol. 329, no. 5992, pp. 689–693, 2010.

[61] B. Cai, X. Q. Song, J. P. Cai, and S. Zhang, "HOTAIR: a cancer-related long non-coding RNA," *Neoplasma*, vol. 61, no. 04, pp. 379–391, 2014.

[62] Y. Wu, L. Zhang, Y. Wang et al., "Long noncoding RNA HOTAIR involvement in cancer," *Tumour Biology*, vol. 35, no. 10, pp. 9531–9538, 2014.

[63] J.-X. Zhang, L. Han, Z. S. Bao et al., "HOTAIR, a cell cycle-associated long noncoding RNA and a strong predictor of survival, is preferentially expressed in classical and mesenchymal glioma," *Neuro-Oncology*, vol. 15, no. 12, pp. 1595–1603, 2013.

[64] K. Zhang, X. Sun, X. Zhou et al., "Long non-coding RNA HOTAIR promotes glioblastoma cell cycle progression in an EZH2 dependent manner," *Oncotarget*, vol. 6, no. 1, pp. 537–546, 2015.

[65] X. Zhou, Y. Ren, J. Zhang et al., "HOTAIR is a therapeutic target in glioblastoma," *Oncotarget*, vol. 6, no. 10, pp. 8353–8365, 2015.

[66] J. Ke, Y. L. Yao, J. Zheng et al., "Knockdown of long non-coding RNA HOTAIR inhibits malignant biological behaviors of human glioma cells via modulation of miR-326," *Oncotarget*, vol. 6, no. 26, pp. 21934–21949, 2015.

[67] L. Sa, Y. Li, L. Zhao et al., "The role of HOTAIR/miR-148b-3p/USF1 on regulating the permeability of BTB," *Frontiers in Molecular Neuroscience*, vol. 10, p. 194, 2017.

[68] L. D. Graham, S. K. Pedersen, G. S. Brown et al., "*Colorectal neoplasia differentially* expressed (CRNDE), a novel gene with elevated expression in colorectal adenomas and adenocarcinomas," *Genes & Cancer*, vol. 2, no. 8, pp. 829–840, 2011.

[69] B. C. Ellis, P. L. Molloy, and L. D. Graham, "CRNDE: a long non-coding RNA involved in cancer, neurobiology, and development," *Frontiers in Genetics*, vol. 3, p. 270, 2012.

[70] K. Kiang, X.-Q. Zhang, and G. Leung, "Long non-coding RNAs: the key players in glioma pathogenesis," *Cancer*, vol. 7, no. 3, pp. 1406–1424, 2015.

[71] Y. Wang, Y. Wang, J. Li, Y. Zhang, H. Yin, and B. Han, "CRNDE, a long-noncoding RNA, promotes glioma cell growth and invasion through mTOR signaling," *Cancer Letters*, vol. 367, no. 2, pp. 122–128, 2015.

[72] J. Zheng, X. D. Li, P. Wang et al., "CRNDE affects the malignant biological characteristics of human glioma stem cells by negatively regulating miR-186," *Oncotarget*, vol. 6, no. 28, pp. 25339–25355, 2015.

[73] J. Zheng, X. Liu, P. Wang et al., "CRNDE promotes malignant progression of glioma by attenuating miR-384/PIWIL4/STAT3 Axis," *Molecular Therapy*, vol. 24, no. 7, pp. 1199–1215, 2016.

[74] J. N. Hutchinson, A. W. Ensminger, C. M. Clemson, C. R. Lynch, J. B. Lawrence, and A. Chess, "A screen for nuclear transcripts identifies two linked noncoding RNAs associated with SC35 splicing domains," *BMC Genomics*, vol. 8, no. 1, p. 39, 2007.

[75] C. M. Clemson, J. N. Hutchinson, S. A. Sara et al., "An architectural role for a nuclear noncoding RNA: *NEAT1* RNA is essential for the structure of paraspeckles," *Molecular Cell*, vol. 33, no. 6, pp. 717–726, 2009.

[76] B. Yu and G. Shan, "Functions of long noncoding RNAs in the nucleus," *Nucleus*, vol. 7, no. 2, pp. 155–166, 2016.

[77] Q. Liu, S. Sun, W. Yu et al., "Altered expression of long noncoding RNAs during genotoxic stress-induced cell death in human glioma cells," *Journal of Neuro-Oncology*, vol. 122, no. 2, pp. 283–292, 2015.

[78] C. He, B. Jiang, J. Ma, and Q. Li, "Aberrant NEAT1 expression is associated with clinical outcome in high grade glioma patients," *APMIS*, vol. 124, no. 3, pp. 169–174, 2016.

[79] L. Zhen, L. Yun-hui, D. Hong-yu, M. Jun, and Y. Yi-long, "Long noncoding RNA NEAT1 promotes glioma pathogenesis by regulating miR-449b-5p/c-Met axis," *Tumour Biology*, vol. 37, no. 1, pp. 673–683, 2016.

[80] Q. Chen, J. Cai, Q. Wang et al., "Long noncoding RNA *NEAT1*, regulated by the EGFR pathway, contributes to glioblastoma progression through the WNT/β-catenin pathway by scaffolding EZH2," *Clinical Cancer Research*, vol. 24, no. 3, pp. 684–695, 2018.

[81] E. Yildirim, J. E. Kirby, D. E. Brown et al., "Xist RNA is a potent suppressor of hematologic cancer in mice," *Cell*, vol. 152, no. 4, pp. 727–742, 2013.

[82] S. M. Weakley, H. Wang, Q. Yao, and C. Chen, "Expression and function of a large non-coding RNA gene XIST in human cancer," *World Journal of Surgery*, vol. 35, no. 8, pp. 1751–1756, 2011.

[83] Z. Cheng, Z. Li, K. Ma et al., "Long non-coding RNA XIST promotes glioma tumorigenicity and angiogenesis by acting as a molecular sponge of miR-429," *Journal of Cancer*, vol. 8, no. 19, pp. 4106–4116, 2017.

[84] P. Du, H. Zhao, R. Peng et al., "LncRNA-XIST interacts with miR-29c to modulate the chemoresistance of glioma cell to TMZ through DNA mismatch repair pathway," *Bioscience Reports*, vol. 37, no. 5, 2017.

[85] Z. Wang, J. Yuan, L. Li, Y. Yang, X. Xu, and Y. Wang, "Long non-coding RNA XIST exerts oncogenic functions in human glioma by targeting miR-137," *American Journal of Translational Research*, vol. 9, no. 4, pp. 1845–1855, 2017.

[86] H. Yu, Y. Xue, P. Wang et al., "Knockdown of long non-coding RNA XIST increases blood–tumor barrier permeability and inhibits glioma angiogenesis by targeting miR-137," *Oncogene*, vol. 6, no. 3, article e303, 2017.

[87] Y. Hao, T. Crenshaw, T. Moulton, E. Newcomb, and B. Tycko, "Tumour-suppressor activity of H19 RNA," *Nature*, vol. 365, no. 6448, pp. 764–767, 1993.

[88] A. Gabory, M. A. Ripoche, T. Yoshimizu, and L. Dandolo, "The *H19* gene: regulation and function of a non-coding RNA," *Cytogenetic and Genome Research*, vol. 113, no. 1–4, pp. 188–193, 2006.

[89] T. Zhang, Y. R. Wang, F. Zeng, H. Y. Cao, H. D. Zhou, and Y. J. Wang, "LncRNA H19 is overexpressed in glioma tissue, is negatively associated with patient survival, and promotes tumor growth through its derivative miR-675," *European Review for Medical and Pharmacological Sciences*, vol. 20, no. 23, pp. 4891–4897, 2016.

[90] X. Jiang, Y. Yan, M. Hu et al., "Increased level of H19 long noncoding RNA promotes invasion, angiogenesis, and stemness of glioblastoma cells," *Journal of Neurosurgery*, vol. 2016, no. 1, pp. 129–136, 2016.

[91] W. Li, P. Jiang, X. Sun, S. Xu, X. Ma, and R. Zhan, "Suppressing H19 modulates tumorigenicity and stemness in U251 and U87MG glioma cells," *Cellular and Molecular Neurobiology*, vol. 36, no. 8, pp. 1219–1227, 2016.

[92] Y. Zheng, X. Lu, L. Xu, Z. Chen, Q. Li, and J. Yuan, "MicroRNA-675 promotes glioma cell proliferation and motility by negatively regulating retinoblastoma 1," *Human Pathology*, vol. 69, pp. 63–71, 2017.

[93] P. Jia, H. Cai, X. Liu et al., "Long non-coding RNA H19 regulates glioma angiogenesis and the biological behavior of glioma-associated endothelial cells by inhibiting microRNA-29a," *Cancer Letters*, vol. 381, no. 2, pp. 359–369, 2016.

[94] J. Fantes, N. K. Ragge, S. A. Lynch et al., "Mutations in SOX2 cause anophthalmia," *Nature Genetics*, vol. 33, no. 4, pp. 461–463, 2003.

[95] A. Shahryari, M. S. Jazi, N. M. Samaei, and S. J. Mowla, "Long non-coding RNA SOX2OT: expression signature, splicing patterns, and emerging roles in pluripotency and tumorigenesis," *Frontiers in Genetics*, vol. 6, p. 196, 2015.

[96] P. P. Amaral, C. Neyt, S. J. Wilkins et al., "Complex architecture and regulated expression of the *sox2ot* locus during vertebrate development," *RNA*, vol. 15, no. 11, pp. 2013–2027, 2009.

[97] M. E. Askarian-Amiri, V. Seyfoddin, C. E. Smart et al., "Emerging role of long non-coding RNA *SOX2OT* in *SOX2* regulation in breast cancer," *PLoS One*, vol. 9, no. 7, article e102140, 2014.

[98] A. Shahryari, M. R. Rafiee, Y. Fouani et al., "Two novel splice variants of SOX2OT, SOX2OT-S1, and SOX2OT-S2 are coupregulated with SOX2 and OCT4 in esophageal squamous cell carcinoma," *Stem Cells*, vol. 32, no. 1, pp. 126–134, 2014.

[99] R. Su, S. Cao, J. Ma et al., "Knockdown of SOX2OT inhibits the malignant biological behaviors of glioblastoma stem cells via up-regulating the expression of miR-194-5p and miR-122," *Molecular Cancer*, vol. 16, no. 1, p. 171, 2017.

[100] N. Miyoshi, H. Wagatsuma, S. Wakana et al., "Identification of an imprinted gene, *MEG3/Gtl2* and Its human homologue *MEG3*, first mapped on mouse distal chromosome 12 and human chromosome 14q," *Genes to Cells*, vol. 5, no. 3, pp. 211–220, 2000.

[101] X. Zhang, K. Rice, Y. Wang et al., "Maternally expressed gene 3 (MEG3) noncoding ribonucleic acid: isoform structure, expression, and functions," *Endocrinology*, vol. 151, no. 3, pp. 939–947, 2010.

[102] X. Zhang, Y. Zhou, K. R. Mehta et al., "A pituitary-derived MEG3 isoform functions as a growth suppressor in tumor cells," *The Journal of Clinical Endocrinology & Metabolism*, vol. 88, no. 11, pp. 5119–5126, 2003.

[103] X. Zhang, R. Gejman, A. Mahta et al., "*Maternally expressed gene 3*, an imprinted noncoding RNA gene, is associated with meningioma pathogenesis and progression," *Cancer Research*, vol. 70, no. 6, pp. 2350–2358, 2010.

[104] X. Cui, X. Jing, C. Long, J. Tian, and J. Zhu, "Long noncoding RNA MEG3, a potential novel biomarker to predict the clinical outcome of cancer patients: a meta-analysis," *Oncotarget*, vol. 8, no. 12, pp. 19049–19056, 2017.

[105] Y. Zhou, Y. Zhong, Y. Wang et al., "Activation of p53 by MEG3 non-coding RNA," *Journal of Biological Chemistry*, vol. 282, no. 34, pp. 24731–24742, 2007.

[106] P. Wang, Z. Ren, and P. Sun, "Overexpression of the long non-coding RNA MEG3 impairs in vitro glioma cell proliferation," *Journal of Cellular Biochemistry*, vol. 113, no. 6, pp. 1868–1874, 2012.

[107] Y. Zhou, X. Zhang, and A. Klibanski, "*MEG3* noncoding RNA: a tumor suppressor," *Journal of Molecular Endocrinology*, vol. 48, no. 3, pp. R45–R53, 2012.

[108] N. Qin, G. F. Tong, L. W. Sun, and X. L. Xu, "Long non-coding RNA MEG3 suppresses glioma cell proliferation, migration, and invasion by acting as a competing endogenous RNA of miR-19a," *Oncology Research Featuring Preclinical and Clinical Cancer Therapeutics*, vol. 25, no. 9, pp. 1471–1478, 2017.

[109] L. Zhang, X. Liang, and Y. Li, "Long non-coding RNA MEG3 inhibits cell growth of gliomas by targeting miR-93 and inactivating PI3K/AKT pathway," *Oncology Reports*, vol. 38, no. 4, pp. 2408–2416, 2017.

[110] J. Li, E.-B. Bian, X.-J. He et al., "Epigenetic repression of long non-coding RNA MEG3 mediated by DNMT1 represses the p53 pathway in gliomas," *International Journal of Oncology*, vol. 48, no. 2, pp. 723–733, 2016.

[111] P. Ji, S. Diederichs, W. Wang et al., "MALAT-1, a novel noncoding RNA, and thymosin beta 4 predict metastasis and survival in early-stage non-small cell lung cancer," *Oncogene*, vol. 22, no. 39, pp. 8031–8041, 2003.

[112] V. Tripathi, J. D. Ellis, Z. Shen et al., "The nuclear-retained noncoding RNA MALAT1 regulates alternative splicing by modulating SR splicing factor phosphorylation," *Molecular Cell*, vol. 39, no. 6, pp. 925–938, 2010.

[113] S. Nakagawa, J. Y. Ip, G. Shioi et al., "Malat1 is not an essential component of nuclear speckles in mice," *RNA*, vol. 18, no. 8, pp. 1487–1499, 2012.

[114] T. Gutschner, M. Hammerle, and S. Diederichs, "*MALAT1* — a paradigm for long noncoding RNA function in cancer," *Journal of Molecular Medicine*, vol. 91, no. 7, pp. 791–801, 2013.

[115] Y. Han, Z. Wu, T. Wu et al., "Tumor-suppressive function of long noncoding RNA MALAT1 in glioma cells by downregulation of MMP2 and inactivation of ERK/MAPK signaling," *Cell Death & Disease*, vol. 7, no. 3, article e2123, 2016.

[116] S. Cao, Y. Wang, J. Li, M. Lv, H. Niu, and Y. Tian, "Tumor-suppressive function of long noncoding RNA MALAT1 in glioma cells by suppressing miR-155 expression and activating FBXW7 function," *American Journal of Cancer Research*, vol. 6, no. 11, pp. 2561–2574, 2016.

[117] H. Li, X. Yuan, D. Yan et al., "Long non-coding RNA MALAT1 decreases the sensitivity of resistant glioblastoma cell lines to Temozolomide," *Cellular Physiology and Biochemistry*, vol. 42, no. 3, pp. 1192–1201, 2017.

[118] Z. Fu, W. Luo, J. Wang et al., "Malat1 activates autophagy and promotes cell proliferation by sponging miR-101 and upregulating STMN1, RAB5A and ATG4D expression in glioma," *Biochemical and Biophysical Research Communications*, vol. 492, no. 3, pp. 480–486, 2017.

[119] Z. Li, C. Xu, B. Ding, M. Gao, X. Wei, and N. Ji, "Long non-coding RNA MALAT1 promotes proliferation and suppresses apoptosis of glioma cells through derepressing rap1B by sponging miR-101," *Journal of Neuro-Oncology*, vol. 134, no. 1, pp. 19–28, 2017.

[120] T. L. Young, T. Matsuda, and C. L. Cepko, "The noncoding RNA taurine upregulated gene 1 is required for differentiation of the murine retina," *Current Biology*, vol. 15, no. 6, pp. 501–512, 2005.

[121] L. Yang, C. Lin, W. Liu et al., "ncRNA- and Pc2 methylation-dependent gene relocation between nuclear structures mediates gene activation programs," *Cell*, vol. 147, no. 4, pp. 773–788, 2011.

[122] E.-b. Zhang, D.-d. Yin, M. Sun et al., "P53-regulated long non-coding RNA TUG1 affects cell proliferation in human non-small cell lung cancer, partly through epigenetically

regulating HOXB7 expression," *Cell Death & Disease*, vol. 5, no. 5, article e1243, 2014.

[123] J. Sun, C. Ding, Z. Yang et al., "The long non-coding RNA TUG1 indicates a poor prognosis for colorectal cancer and promotes metastasis by affecting epithelial-mesenchymal transition," *Journal of Translational Medicine*, vol. 14, no. 1, p. 42, 2016.

[124] Z. Li, J. Shen, M. T. V. Chan, and W. K. K. Wu, "TUG1: a pivotal oncogenic long non-coding RNA of human cancers," *Cell Proliferation*, vol. 49, no. 4, pp. 471–475, 2016.

[125] J. Li, G. An, M. Zhang, and Q. Ma, "Long non-coding RNA TUG1 acts as a miR-26a sponge in human glioma cells," *Biochemical and Biophysical Research Communications*, vol. 477, no. 4, pp. 743–748, 2016.

[126] H. Cai, Y. Xue, P. Wang et al., "The long noncoding RNA TUG1 regulates blood-tumor barrier permeability by targeting miR-144," *Oncotarget*, vol. 6, no. 23, pp. 19759–19779, 2015.

[127] H. Cai, X. Liu, J. Zheng et al., "Long non-coding RNA taurine upregulated 1 enhances tumor-induced angiogenesis through inhibiting microRNA-299 in human glioblastoma," *Oncogene*, vol. 36, no. 3, pp. 318–331, 2017.

[128] G. Palmieri, P. Paliogiannis, M. C. Sini et al., "Long non-coding RNA CASC2 in human cancer," *Critical Reviews in Oncology/Hematology*, vol. 111, pp. 31–38, 2017.

[129] P. Baldinu, A. Cossu, A. Manca et al., "Identification of a novel candidate gene, *CASC2*, in a region of common allelic loss at chromosome 10q26 in human endometrial cancer," *Human Mutation*, vol. 23, no. 4, pp. 318–326, 2004.

[130] P. Wang, Y. H. Liu, Y. L. Yao et al., "Long non-coding RNA CASC2 suppresses malignancy in human gliomas by miR-21," *Cellular Signalling*, vol. 27, no. 2, pp. 275–282, 2015.

[131] G. Huang, X. Wu, S. Li, X. Xu, H. Zhu, and X. Chen, "The long noncoding RNA CASC2 functions as a competing endogenous RNA by sponging miR-18a in colorectal cancer," *Scientific Reports*, vol. 6, no. 1, article 26524, 2016.

[132] W. Zhang, W. He, J. Gao et al., "RETRACTED: the long noncoding RNA CASC2 inhibits tumorigenesis through modulating the expression of PTEN by targeting miR-18a-5p in esophageal carcinoma," *Experimental Cell Research*, vol. 361, no. 1, pp. 30–38, 2017.

[133] M. Harmalkar, S. Upraity, S. Kazi, and N. V. Shirsat, "Tamoxifen-induced cell death of malignant glioma cells is brought about by oxidative-stress-mediated alterations in the expression of BCL2 family members and is enhanced on miR-21 inhibition," *Journal of Molecular Neuroscience*, vol. 57, no. 2, pp. 197–202, 2015.

[134] R. Wang, Y. Li, G. Zhu et al., "Long noncoding RNA CASC2 predicts the prognosis of glioma patients and functions as a suppressor for gliomas by suppressing Wnt/β-catenin signaling pathway," *Neuropsychiatric Disease and Treatment*, vol. -Volume 13, pp. 1805–1813, 2017.

[135] Y. Liao, L. Shen, H. Zhao et al., "LncRNA CASC2 interacts with miR-181a to modulate glioma growth and resistance to TMZ through PTEN pathway," *Journal of Cellular Biochemistry*, vol. 118, no. 7, pp. 1889–1899, 2017.

[136] C. Jiang, F. Shen, J. du et al., "Upregulation of CASC2 sensitized glioma to temozolomide cytotoxicity through autophagy inhibition by sponging miR-193a-5p and regulating mTOR expression," *Biomedicine & Pharmacotherapy*, vol. 97, pp. 844–850, 2018.

[137] D. Capper, D. T. W. Jones, M. Sill et al., "DNA methylation-based classification of central nervous system tumours," *Nature*, vol. 555, no. 7697, pp. 469–474, 2018.

[138] R. Shai, T. Shi, T. J. Kremen et al., "Gene expression profiling identifies molecular subtypes of gliomas," *Oncogene*, vol. 22, no. 31, pp. 4918–4923, 2003.

[139] S. P. Niclou, F. Fack, and U. Rajcevic, "Glioma proteomics: status and perspectives," *Journal of Proteomics*, vol. 73, no. 10, pp. 1823–1838, 2010.

[140] H. S. Phillips, S. Kharbanda, R. Chen et al., "Molecular subclasses of high-grade glioma predict prognosis, delineate a pattern of disease progression, and resemble stages in neurogenesis," *Cancer Cell*, vol. 9, no. 3, pp. 157–173, 2006.

[141] R. G. Verhaak, K. A. Hoadley, E. Purdom et al., "Integrated genomic analysis identifies clinically relevant subtypes of glioblastoma characterized by abnormalities in *PDGFRA*, *IDH1*, *EGFR*, and *NF1*," *Cancer Cell*, vol. 17, no. 1, pp. 98–110, 2010.

[142] D. N. Louis, A. Perry, G. Reifenberger et al., "The 2016 World Health Organization classification of tumors of the central nervous system: a summary," *Acta Neuropathologica*, vol. 131, no. 6, pp. 803–820, 2016.

[143] A. Lai, "Evidence for sequenced molecular evolution of *IDH1* mutant glioblastoma from a distinct cell of origin," *Journal of Clinical Oncology*, vol. 29, no. 34, pp. 4482–4490, 2011.

[144] A. Sottoriva, I. Spiteri, S. G. M. Piccirillo et al., "Intratumor heterogeneity in human glioblastoma reflects cancer evolutionary dynamics," *Proceedings of the National Academy of Sciences of the United States of America*, vol. 110, no. 10, pp. 4009–4014, 2013.

[145] C. Hunter, R. Smith, D. P. Cahill et al., "A hypermutation phenotype and somatic *MSH6* mutations in recurrent human malignant gliomas after alkylator chemotherapy," *Cancer Research*, vol. 66, no. 8, pp. 3987–3991, 2006.

[146] X. Q. Zhang, S. Sun, K. F. Lam et al., "A long non-coding RNA signature in glioblastoma multiforme predicts survival," *Neurobiology of Disease*, vol. 58, pp. 123–131, 2013.

[147] R. Li, J. Qian, Y. Y. Wang, J. X. Zhang, and Y. P. You, "Long noncoding RNA profiles reveal three molecular subtypes in glioma," *CNS Neuroscience & Therapeutics*, vol. 20, no. 4, pp. 339–343, 2014.

[148] J. S. Chen, E. Ma, L. B. Harrington et al., "CRISPR-Cas12a target binding unleashes indiscriminate single-stranded DNase activity," *Science*, vol. 360, no. 6387, pp. 436–439, 2018.

[149] J. S. Gootenberg, O. O. Abudayyeh, J. W. Lee et al., "Nucleic acid detection with CRISPR-Cas13a/C2c2," *Science*, vol. 356, no. 6336, pp. 438–442, 2017.

[150] J. S. Gootenberg, O. O. Abudayyeh, M. J. Kellner, J. Joung, J. J. Collins, and F. Zhang, "Multiplexed and portable nucleic acid detection platform with Cas13, Cas12a, and Csm6," *Science*, vol. 360, no. 6387, pp. 439–444, 2018.

[151] M. Lacroix, "A multivariate analysis of 416 patients with glioblastoma multiforme: prognosis, extent of resection, and survival," *Journal of Neurosurgery*, vol. 95, no. 2, pp. 190–198, 2001.

[152] S. Zhu, W. Li, J. Liu et al., "Genome-scale deletion screening of human long non-coding RNAs using a paired-guide RNA CRISPR–Cas9 library," *Nature Biotechnology*, vol. 34, no. 12, pp. 1279–1286, 2016.

[153] R. D. Chow, "AAV-mediated direct in vivo CRISPR screen identifies functional suppressors in glioblastoma," *Nature Neuroscience*, vol. 20, no. 10, pp. 1329–1341, 2017.

[154] J. Y. Park, J. E. Lee, J. B. Park, H. Yoo, S. H. Lee, and J. H. Kim, "Roles of long non-coding RNAs on tumorigenesis and glioma development," *Brain Tumor Research and Treatment*, vol. 2, no. 1, pp. 1–6, 2014.

[155] M. Matsui and D. R. Corey, "Non-coding RNAs as drug targets," *Nature Reviews Drug Discovery*, vol. 16, no. 3, pp. 167–179, 2017.

[156] H.-X. Wang, M. Li, C. M. Lee et al., "CRISPR/Cas9-based genome editing for disease modeling and therapy: challenges and opportunities for nonviral delivery," *Chemical Reviews*, vol. 117, no. 15, pp. 9874–9906, 2017.

[157] I. B. Hilton, A. M. D'Ippolito, C. M. Vockley et al., "Epigenome editing by a CRISPR-Cas9-based acetyltransferase activates genes from promoters and enhancers," *Nature Biotechnology*, vol. 33, no. 5, pp. 510–517, 2015.

[158] L. A. Gilbert, M. H. Larson, L. Morsut et al., "CRISPR-mediated modular RNA-guided regulation of transcription in eukaryotes," *Cell*, vol. 154, no. 2, pp. 442–451, 2013.

[159] L. A. Gilbert, M. A. Horlbeck, B. Adamson et al., "Genome-scale CRISPR-mediated control of gene repression and activation," *Cell*, vol. 159, no. 3, pp. 647–661, 2014.

[160] C. Pastori, M. Daniel, C. Penas et al., "BET bromodomain proteins are required for glioblastoma cell proliferation," *Epigenetics*, vol. 9, no. 4, pp. 611–620, 2014.

[161] C. Pastori, P. Kapranov, C. Penas et al., "The Bromodomain protein BRD4 controls HOTAIR, a long noncoding RNA essential for glioblastoma proliferation," *Proceedings of the National Academy of Sciences of the United States of America*, vol. 112, no. 27, pp. 8326–8331, 2015.

[162] D. F. Quail and J. A. Joyce, "The microenvironmental landscape of brain tumors," *Cancer Cell*, vol. 31, no. 3, pp. 326–341, 2017.

[163] S. T. Beug, C. E. Beauregard, C. Healy et al., "Smac mimetics synergize with immune checkpoint inhibitors to promote tumour immunity against glioblastoma," *Nature Communications*, vol. 8, 2017.

[164] A. T. Yeo and A. Charest, "Immune checkpoint blockade biology in mouse models of glioblastoma," *Journal of Cellular Biochemistry*, vol. 118, no. 9, pp. 2516–2527, 2017.

[165] Z. He, Y. Wang, G. Huang, Q. Wang, D. Zhao, and L. Chen, "The lncRNA UCA1 interacts with miR-182 to modulate glioma proliferation and migration by targeting iASPP," *Archives of Biochemistry and Biophysics*, vol. 623-624, pp. 1–8, 2017.

[166] Z. He, C. You, and D. Zhao, "Long non-coding RNA UCA1/miR-182/PFKFB2 axis modulates glioblastoma-associated stromal cells-mediated glycolysis and invasion of glioma cells," *Biochemical and Biophysical Research Communications*, vol. 500, no. 3, pp. 569–576, 2018.

[167] Y. Sun, J. G. Jin, W. Y. Mi et al., "Long noncoding RNA UCA1 targets miR-122 to promote proliferation, migration, and invasion of glioma cells," *Oncology Research Featuring Preclinical and Clinical Cancer Therapeutics*, vol. 26, no. 1, pp. 103–110, 2018.

[168] W. Zhao, C. Sun, and Z. Cui, "A long noncoding RNA UCA1 promotes proliferation and predicts poor prognosis in glioma," *Clinical and Translational Oncology*, vol. 19, no. 6, pp. 735–741, 2017.

[169] S. Feng, J. Yao, Y. Chen et al., "Expression and functional role of reprogramming-related long noncoding RNA (lincRNA-ROR) in glioma," *Journal of Molecular Neuroscience*, vol. 56, no. 3, pp. 623–630, 2015.

[170] C. Shang, Y. Guo, Y. Hong, and Y. X. Xue, "Long non-coding RNA TUSC7, a target of miR-23b, plays tumor-suppressing roles in human gliomas," *Frontiers in Cellular Neuroscience*, vol. 10, 2016.

[171] C. Shang, W. Tang, C. Pan, X. Hu, and Y. Hong, "Long non-coding RNA TUSC7 inhibits temozolomide resistance by targeting miR-10a in glioblastoma," *Cancer Chemotherapy and Pharmacology*, vol. 81, no. 4, pp. 671–678, 2018.

[172] X. L. Ma, W. D. Zhu, L. X. Tian et al., "Long non-coding RNA TUSC7 expression is independently predictive of outcome in glioma," *European Review for Medical and Pharmacological Sciences*, vol. 21, no. 16, pp. 3605–3610, 2017.

15

Expression Analysis of Nitrogen Metabolism-Related Genes Reveals Differences in Adaptation to Low-Nitrogen Stress between Two Different Barley Cultivars at Seedling Stage

Zhiwei Chen,[1,2] Chenghong Liu,[1,2] Yifei Wang,[1,2] Ting He,[1,2] Runhong Gao,[1,2] Hongwei Xu,[1,2] Guimei Guo,[1,2] Yingbo Li,[1,2] Longhua Zhou,[1,2] Ruiju Lu,[1,2] and Jianhua Huang[1,2]

[1]*Biotechnology Research Institute of Shanghai Academy of Agricultural Sciences, 2901 Beidi Road, Minhang District, Shanghai 201106, China*
[2]*Shanghai Key Laboratory of Agricultural Genetics and Breeding, 2901 Beidi Road, Minhang District, Shanghai 201106, China*

Correspondence should be addressed to Ruiju Lu; luruiju62@163.com and Jianhua Huang; swl@saas.sh.cn

Academic Editor: Gunvant B. Patil

The excess use of nitrogen fertilizers causes many problems, including higher costs of crop production, lower nitrogen use efficiency, and environmental damage. Crop breeding for low-nitrogen tolerance, especially molecular breeding, has become the major route to solving these issues. Therefore, in crops such as barley (*Hordeum vulgare* L.), it is crucial to understand the mechanisms of low-nitrogen tolerance at the molecule level. In the present study, two barley cultivars, BI-04 (tolerant to low nitrogen) and BI-45 (sensitive to low nitrogen), were used for gene expression analysis under low-nitrogen stress, including 10 genes related to primary nitrogen metabolism. The results showed that the expressions of *HvNIA2* (nitrite reductase), *HvGS2* (chloroplastic glutamine synthetase), and *HvGLU2* (ferredoxin-dependent glutamate synthase) were only induced in shoots of BI-04 under low-nitrogen stress, *HvGLU2* was also only induced in roots of BI-04, and *HvGS2* showed a rapid response to low-nitrogen stress in the roots of BI-04. The expression of *HvASN1* (asparagine synthetase) was reduced in both cultivars, but it showed a lower reduction in the shoots of BI-04. In addition, gene expression and regulation differences in the shoots and roots were also compared between the barley cultivars. Taken together, the results indicated that the four above-mentioned genes might play important roles in low-nitrogen tolerance in barley.

1. Introduction

Nitrogen, one of the essential elements for crop growth and development, is a primary driver of crop production. Thus, many new crop varieties with high yields, dependent on high-nitrogen fertilizer input, were introduced into crop production in the 20th century according to the preferences of farmers and breeders [1]. However, the increased use of nitrogen fertilizer caused a number of problems, such as high input costs for crop production, a decrease in nitrogen use efficiency, nitrogen fertilizer loss, and environmental pollution [2]. As a result, there is now a consensus among plant and environmental scientists that it is important to balance the benefits of nitrogen application, mainly increased yield, against its disadvantages, and to minimize negative impacts by decreasing nitrogen fertilizer input and environmental pollution, while maintaining yields. Therefore, increasing nitrogen use efficiency or developing crops with the ability to tolerate low-nitrogen are important targets for future crop breeding [2]. Achieving these targets will require a comprehensive understanding of nitrogen metabolism under low-nitrogen condition, particularly the expression of genes involved in the adaptation to, or tolerance of, low-nitrogen stress.

Nitrogen physiology is complicated, comprising processes such as acquisition, assimilation, transportation, remobilization, and the metabolism of nitrogen-containing compounds [2–4]. The most commonly used external chemical nitrogen sources are nitrate and ammonium. Nitrate is taken up by low- and/or high-affinity nitrate transport systems (NRT1 and NRT2), while ammonium is taken up by ammonium transporters (AMT). The high-affinity nitrate transporters play important roles under nitrogen starvation or low-nitrogen stress [5]. The absorbed nitrate is firstly reduced to nitrite by nitrate reductase (NR) and then reduced to ammonium by nitrite reductase (NiR). Ammonium is assimilated into amino acids in a process that is catalyzed mainly by the GS-GOGAT (glutamine synthetase and glutamate synthase) pathway. Another important enzyme is asparagine synthetase (AS), which catalyzes the transfer of glutamine to asparagine. Asparagine is a major molecular nitrogen for nitrogen transportation in many plant species because it has a relatively high nitrogen : carbon ratio [6], and the translocation of nitrogen within plants is also very important for plant growth and seed development [7].

Barley (*Hordeum vulgare* L.) is a model plant for cereal research, as well as being an important crop, and many of its nitrogen metabolism-related genes have been cloned [8–18]. There have been reports comparing nitrogen metabolism-related gene expression in different plant genotypes with different responses to low-nitrogen stress [19–21]. Although there was a recent study concerning transcriptome analysis under low-nitrogen stress in the roots of two different barley genotypes [22], the different definition of low-nitrogen tolerance and only using root tissue for the transcriptome analysis might not be enough to fully understand their responses to low-nitrogen stress.

In the present study, we first determined a suitable degree of low-nitrogen stress to obtain significant differences in the phenotypes between two barley cultivars with different responses to low nitrogen. We compared the expression patterns of genes related to nitrogen metabolism in the shoots and roots of the two barley cultivars. The aim was to acquire information concerning the differential regulation of nitrogen metabolism-related genes between low-nitrogen-tolerant and low-nitrogen-sensitive barley cultivars under low-nitrogen treatment and to reveal their roles in adaptation to low-nitrogen stress.

2. Materials and Methods

2.1. Plant Growth and Low-Nitrogen Treatments. Barley BI-04 is a relatively low-nitrogen-tolerant cultivar, while barley BI-45 is a relatively low-nitrogen-sensitive cultivar [23, 24]. Seeds of the two cultivars were sterilized by immersion in 1% NaClO and germinated in an incubator at 25°C for one week. Seedlings were cultured in nutrient solution comprising 1.43 mM NH_4NO_3, 0.32 mM $NaH_2PO_4 \cdot 2H_2O$, 0.51 mM K_2SO_4, 1.00 mM $CaCl_2$, 3.36 mM MgSO4, 9.47 μM $MnSO_4 \cdot H_2O$, 0.08 μM $Na_2MoO_4 \cdot H_2O$, 19.42 μM H_3BO_3, 0.15 μM ZnSO4·7H_2O, 0.16 μM $CuSO_4 \cdot 5H_2O$, and 61.24 μM iron citrate (mainly according to [25]). Seedlings were transferred into nutrient solution with the NH_4NO_3 concentration reduced to 19.21 mg·L^{-1} (0.24 mM) at the fourth leaf stage. The pH was maintained at 6.2 ± 0.3. Plants in hydroponic growth boxes were kept in an artificial incubator with a 16/8 h (light/dark) cycle at 20°C ± 2°C and 70% relative humidity. Shoots and roots were harvested separately at 0, 1, 24, and 48 h after low-nitrogen treatment, frozen in liquid nitrogen, and kept at −80°C. There were three biological replicates for each sample. For biomass investigation, BI-04 and BI-45 plants were constantly cultured for another one week from the fourth leaf stage; one group of plants was grown with a normal nitrogen supply (1.43 mM NH_4NO_3), and the other group was grown under low-nitrogen stress (0.24 mM NH_4NO_3). The plants were then harvested, and shoots and roots were collected separately. There were 20 biological replicates of each variety under each treatment.

For biomass measurements, all shoots and roots of BI-04 and BI-45 were incubated at 105°C for 30 min and dried at 80°C for about 2 days until their weight remained constant weight, as determined using an electronic analytical balance.

2.2. RNA Extraction and cDNA Synthesis. Total RNA from shoot and root samples was isolated by using TRIzol (Invitrogen, Carlsbad, CA, USA) and treated with RQ1 RNase-Free DNase (Promega, Madison, WI, USA) to degrade any contaminating genomic DNA. cDNA was synthesized using SuperScript III Reverse Transcriptase (Invitrogen, USA) and checked for purity by using polymerase chain reaction (PCR) amplification with primers CATCAAGCTCAAGG ACGACA and GCCTTGTCCTTGTCAGTGAA, which anneal to sites flanking an intron within the *HvGAPDH* gene. The presence of contaminating genomic DNA would lead to amplification of a 229 bp product in addition to the 150 bp product amplified from the cDNA [26].

2.3. Quantitative Real-Time PCR (qPCR). Primers were designed using primer 3 (http://primer3.wi.mit.edu/) (Table 1). PCR reactions were performed in 96-well plates on a 7500 Real-Time PCR System (Applied Biosystems, Foster, CA, USA) using SYBR Select Master Mix (Applied Biosystems), according to the manufacturer's instructions. The reactions for sets of three biological replicate samples per time point were separated across three plates, thus forming statistical blocks for subsequent data analysis. Reactions contained 10 μL 2x mix, 0.6 μL of each primer (1 μM), and 100 ng of cDNA template in a final volume of 20 μL. The same thermal profile was used for all PCR reactions: 50°C for 2 min, 95°C for 2 min, and 45 cycles of 95°C for 15 s and 60°C for 1 min. Data collection was carried out during the 60°C step. Dissociation/melting curves were constructed after cycle 45.

2.4. Statistical Analysis. Biomass comparisons and gene expression comparisons between the shoots and the roots were analyzed statistically using a *t*-test in Excel 2007 software.

The efficiency of the PCR was estimated using the LinReg PCR program [27]. The cycle threshold (Ct) value was obtained using 7500 software v2.0.5 (Applied Biosystems), and the Ct and efficiency values were then used to calculate the relative quantity (RQ) and the normalized relative quantity (NRQ) of a target gene's expression with respect

Table 1: Primers for qRT-PCR.

Gene name	Accession number		Prime sequences (5′ to 3′)	Amplicon (bp)	Origin
HvNRT2.2	U34290.1	Forward	TCCTTCTTCACCTGCTTCGT	80	This study
		Reverse	TTGGCGAGGTTTAGGTTGTC		
HvNRT2.3	AF091115.1	Forward	ATGGCGTATTGCCTACTTCG	90	
		Reverse	TTCCCATCAGGGAGATCTTG		
HvNRT3.1	AY253448.1	Forward	GAACGTGAAGGTGAGCCTCT	96	
		Reverse	TGGCAGGTCTTGTCCTTCTT		
HvNRT3.3	AY253450.1	Forward	AAGGACGCCGACTACAAGAA	131	
		Reverse	TGCTGGGTGATCTTGAACTG		
HvNIA2	X57845.1	Forward	TGGCAAGAAGATCACACGAG	120	
		Reverse	CAGAAGCACCAGCACCAGTA		
HvNiR1	S78730.1	Forward	CTCACCGGGGTGTACAAGAA	114	
		Reverse	CTCCTCGTCCTCCTCCCTCT		
HvGS1_1	X69087.1	Forward	GTTCAGGGAGGGAAACAACA	112	
		Reverse	ATCGGGGTTGCTAAGGATCT		
HvGS2	X53580.1	Forward	ATAGCCGCATATGGTGAAGG	106	
		Reverse	GAATAGAGCAGCCACGGTTC		
HvGLU2	S58774.1	Forward	ACCAATGAGGTTGCTTGGAC	85	
		Reverse	TATTGTGGCTTCCCTTGACC		
HvASN1	AF307145.1	Forward	AAGGAGGGAGGCTTCAAGAG	146	
		Reverse	AGAACACCGAATGGAACGTC		
HvActin	AY145451.1	Forward	TGAGGCGCAGTCCAAGAGA	81	Chen et al. [26]
		Reverse	TCCATGTCATCCCAGTTGCTTA		
HvGAPDH	X60343.1	Forward	ACAGTTCACGGCCATTGGA	102	
		Reverse	AGGGTTCCTGACGCCAAAG		

to two reference genes, *HvActin* and *HvGAPDH*. The NRQ was calculated using the following formula:

$$\mathrm{NRQ} = \frac{E_{\mathrm{target}}^{-\mathrm{Ct,target}}}{\sqrt{E_{\mathrm{HvActin}}^{-\mathrm{Ct,HvActin}} \cdot E_{\mathrm{HvGAPDH}}^{-\mathrm{Ct,HvGAPDH}}}}. \quad (1)$$

Statistical analysis of the NRQ data was also according to Chen et al. [26].

3. Results

3.1. Effects of Low-Nitrogen Treatment on Plant Growth and Biomass in the Two Barley Cultivars. BI-04 was considered a low-nitrogen-tolerant barley cultivar, while BI-45 is a low-nitrogen-sensitive [23, 24]. In this study, the growth of BI-04 and BI-45 seemed to be suppressed, accompanied by chlorosis, under low-nitrogen stress, and the restriction was more serious in BI-45 (Figures 1(a) and 1(b)). Comparing the biomass, there was no significant difference in shoot dry weight of BI-04 between normal nitrogen supply and low-nitrogen stress, while there was a significant difference in BI-45 ($P < 0.05$), and there were no significant differences in root dry weight of the two barley cultivars between normal nitrogen supply and low-nitrogen stress (Figures 1(c) and 1(d)). The results indicated that the responses to low-nitrogen stress were different between BI-04 and BI-45 and that BI-04 was more tolerant to low-nitrogen stress than BI-45. The results also suggested that the responses to low-nitrogen stress were different between shoots and roots and the restriction of barley growth caused by low-nitrogen stress first happened in the shoots.

3.2. Identification of Genes Involved in Nitrogen Metabolism in Barley. The aim of this study was to analyze the expression levels of genes involved in nitrogen metabolism under low-nitrogen stress at the seedling stage in two barley cultivars, including genes encoding NRT2, NR, NiR, GS, GOGAT, and AS. All gene sequences were downloaded directly from the NCBI database, and the accession numbers are given below in parentheses.

The genes that were chosen for analysis included two high-affinity nitrate transporter genes, *HvNRT2.2* (gb|U34290.1) and *HvNRT2.3* (gb|AF091115.1) ([17]; Vidmar et al. [18]). Expression of these genes was studied only in the roots of the barley cultivars. Two other nitrate transporter genes, *HvNRT3.1* (gb|AY253448.1) and *HvNRT3.3* (gb|AY253450.1), which function with NRT2 as a two-component high-affinity nitrate uptake system [16], were also selected and assessed in the roots and shoots of the barley cultivars. NRT3 is much smaller than NRT2 and has fewer transmembrane domains [16].

One nitrate reductase gene, *HvNIA2* (gb| X57845.1), which encodes an NADH-specific nitrate reductase [13],

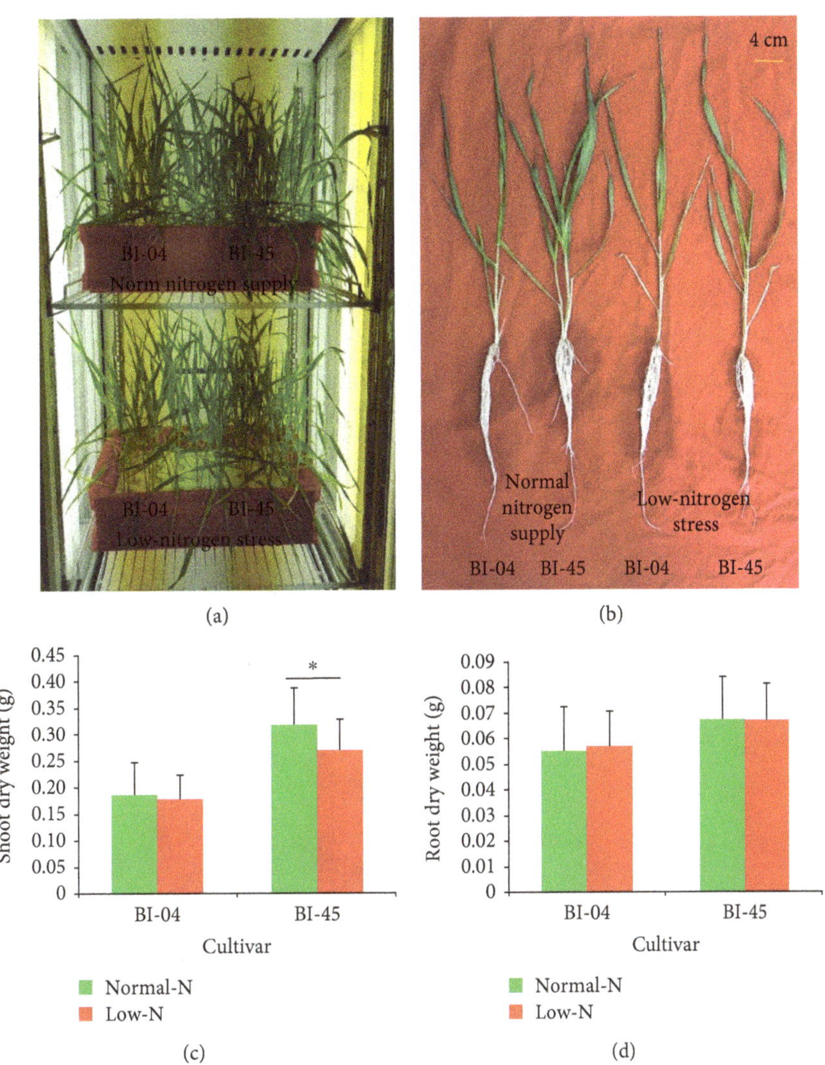

(a) (b) (c) (d)

Figure 1: Plant growth and performance of BI-04 (low-N tolerant) and BI-45 (low-N sensitive) under normal nitrogen supply and low-nitrogen stress. (a) Plant growth and treatment in an artificial climate incubator. (b) Plant performances under different nitrogen conditions. (c) Shoot dry weight (mean and SD, $n = 20$) under different nitrogen conditions. (d) Root dry weight (mean and SD, $n = 20$) under different nitrogen conditions. Significance levels of differences between normal nitrogen supply and low-nitrogen stress were estimated according to the two-tailed t-test method ($^*P < 0.05$).

was studied in both shoots and roots in the barley cultivars, as was a putative nitrite reductase-related gene, *HvNiR1* (gb|S78730.1) [9]. Barley contains another nitrate reductase gene, *HvNIA1* (gb| X60173.1), which encodes a NAD(P)H-bispecific nitrate reductase; however, this gene is normally expressed at very low levels, especially when *HvNIA2* is expressed ([11]; Sue et al. [15]).

Two glutamine synthetase genes, *HvGS1_1* (gb| X69087.1) which encodes cytoplasmic glutamine synthetase [10, 28] and *HvGS2* (gb| X53580.1) which encodes chloroplastic glutamine synthetase [14, 28], were included in the study and analyzed in the shoots and roots in the barley cultivars. One glutamate synthase gene, *HvGLU2* (gb|S58774.1), which encodes ferredoxin-dependent glutamate synthase [8], was also studied in the shoots and roots. There are two main types of GOGAT in higher plants, Fd-GOGAT and NADH-GOGAT, and the Fd-GOGAT activity is dominant in plants [29].

One asparagine synthetase gene, *HvASN1* (gb|AF307145.1), was studied in the shoots and roots of the barley cultivars [12, 28]. Two genes that encode asparagine synthetase were studied; however, the expression of *HvASN2* (gb|AY193714.1) was found to be very low and unstable, especially in roots, which is consistent with the report of Moller et al. [12], so it was not used for further expression analysis.

3.3. Expression Analyses in Shoots of Two Barley Cultivars. The expression levels of *HvNRT3.1*, *HvNRT3.3*, *HvNIA2*, *HvNiR1*, *HvGS1_1*, *HvGS2*, *HvGLU2*, and *HvASN1* were assessed in the shoots of the two barley cultivars under low-nitrogen stress (Figure 2). Analysis of variance (ANOVA) showed that the expression of *HvGS2* in the shoots was significantly different between the barley cultivars ($P < 0.05$), the expression level of *HvGS1_1* and *HvASN1* showed significant differences among different time points ($P < 0.05$), and there was a significant interaction in the *HvGS1_1*

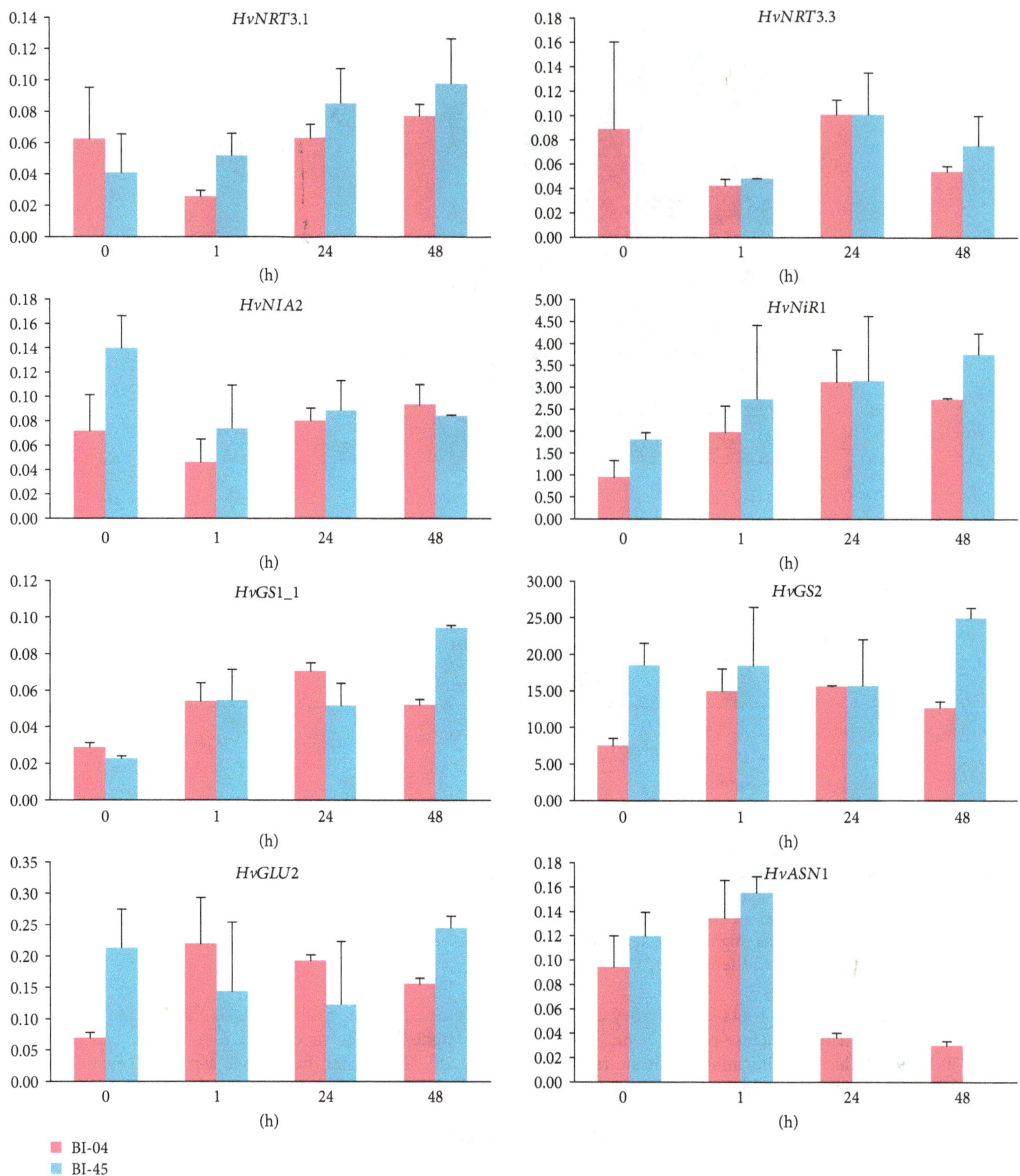

FIGURE 2: Differential expression of genes related to nitrogen metabolism in shoots of the two barley cultivars. Shoots were sampled at 0 h, 1 h, 24 h, and 48 h after low-nitrogen stress. Expression is represented as the normalized relative quantity (NRQ) of a target gene's expression with respect to the two reference genes: *HvActin* and *HvGAPDH*. Means and standard errors are shown from the analysis of three biological replicates.

between barley cultivars and time points ($P < 0.05$) (see Supplementary Table S1).

In the multiple comparison analysis of gene expression, *HvNRT3.1*, *HvNRT3.3*, and *HvNIA2* showed no significant changes in response to low-nitrogen stress in the shoots of BI-04, while *HvNRT3.1*, *HvNIA2*, *HvNiR1*, *HvGS2*, and *HvGLU2* showed no significant changes in BI-45 (see Supplementary Table S5). These results indicated that gene regulation was more sensitive in BI-04 than in BI-45 in the shoots under low-nitrogen stress. For BI-04, *HvGS1_1*,

HvGS2, and *HvGLU2* were significantly induced after 1 h of low-nitrogen treatment ($P < 0.05$), and *HvNiR1* was significantly induced after 24 h of low-nitrogen treatment ($P < 0.05$), while *HvASN1* was significantly reduced after 48 h of low-nitrogen treatment ($P < 0.05$). In BI-45, *HvNRT3.3* and *HvGS1_1* were significantly induced at 1 h after low-nitrogen treatment ($P < 0.05$), and *HvNRT3.3* was hardly detectable under normal nitrogen supply, while *HvASN1* was reduced and became undetectable from 24 h after low-nitrogen treatment.

Comparing the two barley cultivars, *HvNiR1*, *HvGS2*, and *HvGLU2* were only induced in BI-04. This suggested that these three genes might play important roles in the low-nitrogen tolerance of BI-04. In addition, *HvASN1* was reduced in both barley cultivars; however, the expression of *HvASN1* in BI-45 almost disappeared from 24 h after low-nitrogen treatment. The results suggested that the lower expression of *HvASN1* in BI-04 from 24 h after low-nitrogen stress might have a positive effect on low-nitrogen tolerance.

3.4. Expression Analyses in the Roots of Two Barley Cultivars. The expression level of ten genes, comprising *HvNRT2.2*, *HvNRT2.3*, *HvNRT3.1*, *HvNRT3.3*, *HvNIA2*, *HvNiR1*, *HvGS1_1*, *HvGS2*, *HvGLU2*, and *HvASN1*, was analyzed in the roots of the two barley cultivars (Figure 3). ANOVA showed that the expression levels of *HvNRT2.2*, *HvNRT3.1*, *HvNRT3.3*, *HvNiR1*, *HvGS1_1*, *HvGS2*, and *HvGLU2* were significantly different in roots between the barley cultivars ($P < 0.05$). The expression levels of all genes except *HvASN1* showed significant differences in the roots at different time points ($P < 0.05$), and the expression of *HvNRT3.3* had a significant interaction between barley cultivars and time points ($P < 0.05$) (see Supplementary Table S2).

Multiple comparison analyses of gene expression showed that the expression levels of *HvNRT3.3* and *HvASN1* showed no significant changes in response to the low-nitrogen stress in BI-04 and the expression levels of *HvGLU2* and *HvASN1* showed no significant changes in BI-45 (see Supplementary Table S5). These results indicated that the gene regulation was more sensitive in the roots than in the shoots under low-nitrogen stress, especially in BI-45. For BI-04, *HvGS2* and *HvGLU2* were significantly induced at 1 h after low-nitrogen treatment ($P < 0.05$), and *HvNRT2.2*, *HvNRT3.1*, *HvNIA2*, *HvNiR1*, and *HvGS1_1* were significantly induced at 24 h after low-nitrogen treatment ($P < 0.05$), while *HvNRT2.3* expression was reduced at 1 h after low-nitrogen treatment and then induced at 24 h after low-nitrogen treatment. While in BI-45, *HvNRT3.3* and *HvNIA2* were significantly induced at 1 h after low-nitrogen treatment ($P < 0.05$), and *HvNRT2.2*, *HvNRT2.3*, *HvNRT3.1*, *HvNiR1*, *HvGS1_1*, and *HvGS2* were significantly induced at 24 h after low-nitrogen treatment ($P < 0.05$).

Comparing the two barley cultivars, *HvNRT2.2*, *HvNRT3.1*, *HvNiR1*, and *HvGS1_1* showed similar inductions, *HvNIA2* and *HvGS2* were induced in both cultivars but at different time points, while *HvNIA2* showed a rapid response in BI-45, and *HvGS2* showed a rapid response in BI-04. However, *HvNRT2.3*, *HvNRT3.3*, and *HvGLU2* showed different responses to low-nitrogen stress: *HvNRT2.3* was reduced at 1 h after low-nitrogen treatment and then induced from 24 h after low-nitrogen treatment in BI-04 while it was induced from 24 h after low-nitrogen treatment in BI-45; *HvNRT3.3* was only upregulated in BI-45, while *HvGLU2* was only induced in BI-04. These results suggested that there were different responses to low-nitrogen stress in terms of gene expression between BI-04 and BI-45, although there were no significant differences in root dry weight of each barley cultivar after low-nitrogen stress, and these different gene expressions might also contribute different effects of the two barley cultivars on low-nitrogen tolerance.

3.5. Different Gene Expression Patterns between Shoots and Roots of Two Barley Cultivars. The expression levels of *HvNRT3.1*, *HvNRT3.3*, *HvNIA2*, *HvNiR1*, *HvGS1_1*, *HvGS2*, *HvGLU2*, and *HvASN1* were compared between the shoots and the roots of the two barley cultivars. ANOVA showed that the expression levels of all genes had significant differences between the shoots and the roots of BI-04, all genes except *HvNRT3.3* and *HvASN1* had significant differences in expression among different time points, and only *HvGS2* had significant interactions between tissues and time points ($P < 0.05$) (see Supplementary Table S3). Meanwhile, all genes except *HvNIA2* and *HvGLU2* had significant differences in expression between shoots and roots in BI-45 ($P < 0.05$); all genes except *HvGLU2* and *HvASN1* had significant differences in expression among different time points ($P < 0.05$); and *HvNIA2* and *HvGS2* had significant interactions between tissues and time points ($P < 0.05$) (see Supplementary Table S4).

Comparing the gene regulation between shoots and roots of BI-04, *HvNRT3.1* and *HvNIA2* were only induced in the roots, and *HvGS1_1* and *HvGLU2* were induced in both the shoots and roots, but at different time points, while *HvASN1* expression was only reduced in shoots (see Supplementary Table S5). The gene expression levels of *HvNRT3.1*, *HvNRT3.3*, *HvNiR1*, *HvGS2*, and *HvGLU2* were different between the shoots and the roots at all time points, while *HvASN1* was different between shoots and roots only at 1 h after low-nitrogen treatment (see Supplementary Table S6).

For BI-45, *HvNRT3.1*, *HvNIA2*, *HvNiR1*, and *HvGS2* were induced in the roots, but no changes in the shoots, and *HvGS1_1* was induced in both the shoots and the roots at different time points of low-nitrogen treatment, while *HvASN1* expression was reduced such that it almost disappeared (only in the shoots) (see Supplementary Table S5). The gene expression levels of *HvNRT3.1*, *HvNRT3.3*, and *HvGS2* were different between shoots and roots at all time points, *HvNiR1* and *HvGLU2* were different at 0 h and 48 h after low-nitrogen treatment, the expression of *HvASN1* was different at 24 h and 48 h after low-nitrogen treatment, and *HvNIA2* and *HvGS1_1* were different only at 0 h after low-nitrogen treatment (see Supplementary Table S6).

These results indicated that gene expression, both in terms of regulation and expression levels, was very different between shoots and roots. Therefore, it was necessary to investigate gene expressions in shoots and roots separately.

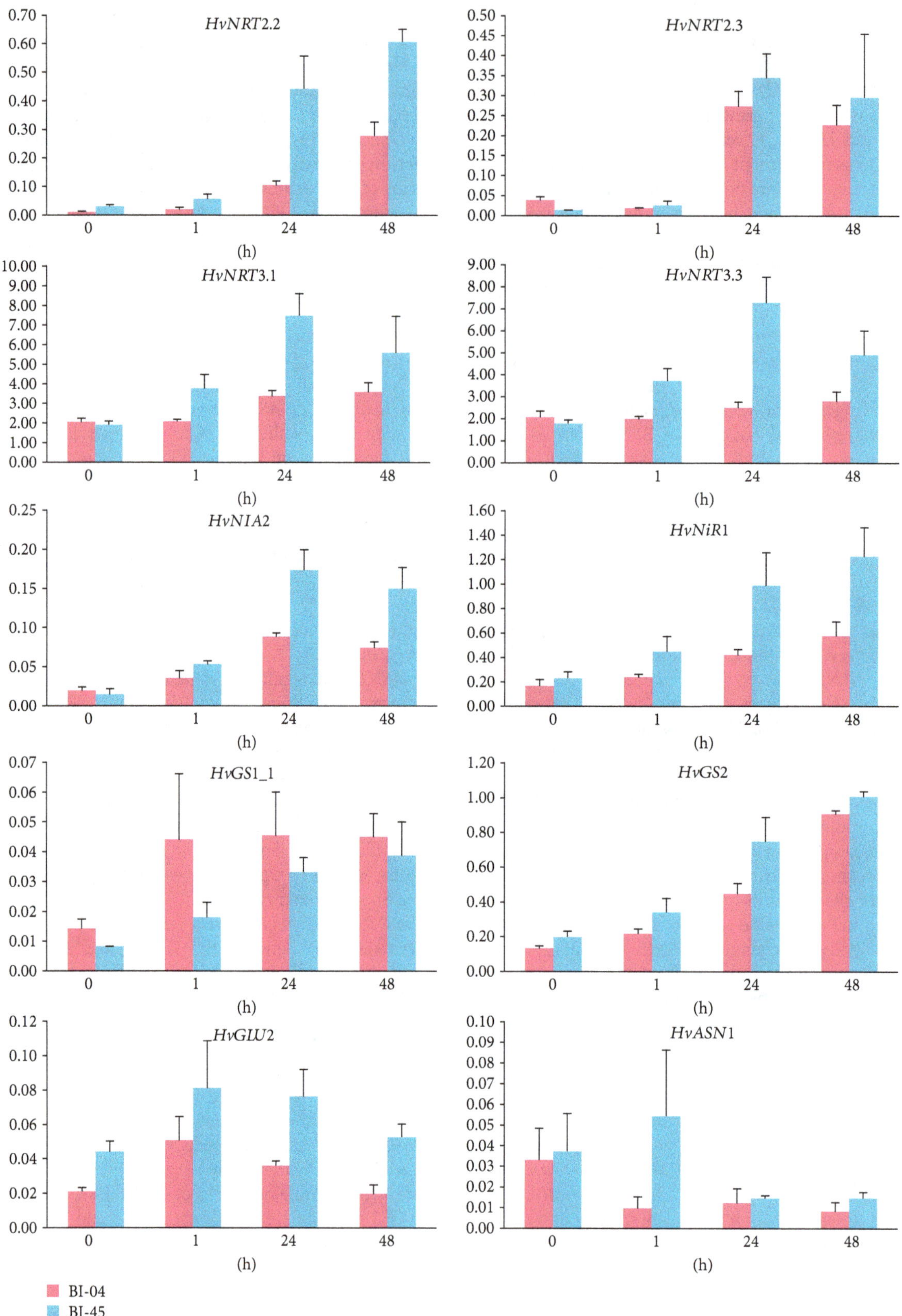

FIGURE 3: Differential expression of genes related to nitrogen metabolism in roots of the two barley cultivars. Roots were sampled at 0 h, 1 h, 24 h, and 48 h after low-nitrogen stress. Expression is represented as the normalized relative quantity (NRQ) of a target gene's expression with respect to the two reference genes: *HvActin* and *HvGAPDH*. Means and standard errors are shown from the analysis of three biological replicates.

4. Discussion

To identify the molecular mechanisms that are adopted by low-nitrogen-tolerant barley cultivars to adapt to low-nitrogen stress, we compared the differences in the expression levels of nitrogen metabolism-related genes between low-nitrogen-tolerant and low-nitrogen-sensitive barley cultivars. Kant et al. [19] compared gene expression levels in *Thellungiella halophila* with tolerance to low-nitrogen stress and *Arabidopsis* with sensitivity to low-nitrogen stress and suggested that *NR2*, *GS1*, *GS2*, *NRT2.1*, *NRT3.1*, and *NRT1.1* might be important in the adaptation to low-nitrogen stress in *Thellungiella*. In crops, Shi et al. [21] used two different rice cultivars to investigate their differences under low-nitrogen conditions and showed that *OsAMT1;1* and *OsNRT2;1* might play important roles in nitrogen acquisition. In trees, Luo et al. [20] also made a comparison of two contrasting *Populus* species and found that the strong responsiveness to limitation N supply by genes related to nitrogen metabolisms might be a good solution for acclimation to low-nitrogen stress in poplar.

In the present study, we compared the expression of genes related to nitrogen metabolism between a low-nitrogen-tolerant barley cultivar BI-04 and a low-nitrogen-sensitive barley cultivar BI-45 under low-nitrogen stress and found that *HvNiR1*, *HvGS2*, and *HvGLU2* were induced in shoots of BI-04, while *HvASN1* was reduced in both cultivars, and disappeared only in BI-45. In roots, we found that *HvGLU2* was only induced in BI-04, and *HvGS2* was induced from 1 h after low-nitrogen treatment in BI-04 while it was induced from 24 h after low-nitrogen treatment in BI-45. These results showed that *HvNiR1*, *HvGS2*, *HvGLU2*, and *HvASN1* might play important roles in low-nitrogen tolerance in BI-04, especially *HvGS2* and *HvGLU2* because of their induction in both shoots and roots of BI-04, and their stronger responses to low-nitrogen stress in the shoots of BI-04 than in BI-45, which might also be an important reason for BI-04's better adaptation to low-nitrogen stress.

The GS/GOGAT pathway is very important for primary nitrogen assimilation. This process changes inorganic nitrogen into organic nitrogen, which can then be directly absorbed by plants, and the inductions of *HvGS2* and *HvGLU2* in both shoots and roots might be one of the most important mechanisms underlying for the low-nitrogen tolerance of BI-04. A comparison of *Thellungiella halophila* with *Arabidopsis thaliana* showed that the former, as a low-nitrogen-tolerant species, had sustained the expression of *GS2* under low-nitrogen stress, while it was reduced in *Arabidopsis*, which grew poorly under N-limiting condition [19]. Furthermore, *GS2* was expressed in many tissues, including roots and leaves, while it was dominated in the leaves [30], and this phenomenon was also observed in our study. Additionally, Feraud et al. [31] showed that Fd-GOGAT was the most important enzyme in assimilation of photorespiratory and primary ammonium, especially in leaves. In our present study, the induction of *HvGS2* only in the shoots of BI-04 and the induction of *HvGLU2* in the shoots and roots of BI-04 might validate their predicted effects in the adaptation to low-nitrogen stress in barley.

Asparagine synthetase gene expression is dependent on nitrogen available and was reduced when nitrogen was limited [32, 33], and we also found that *HvASN1* was repressed only in shoots, indicating that the primary effects of nitrogen deficiency might appear initially in the shoots. Asparagine, which is synthesized by asparagine synthetase, is a key amino acid used to transport and store nitrogen in plants [7]. Overexpression of *ASN1* in *Arabidopsis* increased its tolerance to nitrogen-limiting stress [34]. Here, the lower repression of *HvASN1* in the shoots of BI-04 compared with that in BI-45 and the rapid induction of *HvASN1* in the roots of BI-04 might lead to better adaptation to low-nitrogen stress.

In addition, transgenic *Arabidopsis* with the spinach nitrite reductase gene showed an improvement in NO_2 assimilation in shoots [35]. Therefore, the induction of *HvNiR1* in the shoots of BI-04 might have some effects on incorporating NO_2 in the atmosphere to relieve low-nitrogen stress.

5. Conclusion

In this study, two barley cultivars with different adaptations to low-nitrogen stress were used to investigate the molecular mechanism of barley's response to low-nitrogen tolerance. Our results showed that the increased expression levels of *HvNiR1*, *HvGS2*, and *HvGLU2*, the less decreased expression of *HvASN1* in shoots under low-nitrogen stress, and the increased expression of *HvGLU2* and the rapid response of *HvGS2* in roots under low-nitrogen stress, could benefit adaptation to low-nitrogen stress in barley. The expressions of these genes will be preferentially detected to identify low-nitrogen-tolerant barley germplasms in the future. We also provided two important barley cultivars for exploring the in-depth molecular mechanism of low-nitrogen tolerance: one cultivar could maintain its biomass under early nitrogen deficiency, while the other could not. Furthermore, we also emphasized the importance of detecting gene expression in different barley tissues to completely reveal the mechanism of adaptation to low-nitrogen stress in barley.

Abbreviations

NRT1:	Low-affinity nitrate transport
NRT2:	High-affinity nitrate transport
AMT:	Ammonium transporters
NR:	Nitrate reductase
NiR:	Nitrite reductase
GS:	Glutamine synthetase
GOGAT:	Glutamate synthase
AS:	Asparagine synthetase
NRQ:	Normalized relative quantity.

Conflicts of Interest

The authors declare that they have no conflict of interest.

Authors' Contributions

Zhiwei Chen and Chenghong Liu contributed equally to this work.

Acknowledgments

This study was supported by the foundations of the Young Talent Development Plan of Shanghai Agriculture Committee of China [Grant no. 2015(1-28)], the Natural Science Foundation of Shanghai (17ZR1425300), Shanghai Seed Industry Development Foundation of China [Grant no. 2016(1-1)], and China Agriculture Research System (CARS-05-01A-02). The authors thank Professor Nigel Halford, Rothamsted Research, UK, and Professor Dayong Zhang, Nanjing Agricultural University of China, for revising this manuscript. The authors also thank Elixigen Company for providing the service of proofreading.

Supplementary Materials

Table S1: two-way ANOVA of Ct value of each gene expression at different time points after low-nitrogen stress in shoots of the two barley cultivars. Table S2: two-way ANOVA of Ct value of each gene expression at different time points after low-nitrogen stress in roots of the two barley cultivars. Table S3: two-way ANOVA of Ct value of each gene expression at different time points after low-nitrogen stress between shoots and roots of BI-04. Table S4: two-way ANOVA of Ct value of each gene expression at different time points after low-nitrogen stress between shoots and roots of BI-45. Table S5: Ct value of gene expression at different time points after low-nitrogen stress and the significant difference analysis in shoots and roots of the two barley cultivars. Table S6: Ct value of gene expression at different time points and significant difference analysis between shoots and roots of the two barley cultivars. *(Supplementary Materials)*

References

[1] D. N. Duvick, "The contribution of breeding to yield advances in maize (*Zea mays* L.)," *Advances in Agronomy*, vol. 86, pp. 83–145, 2005.

[2] G. Xu, X. Fan, and A. J. Miller, "Plant nitrogen assimilation and use efficiency," *Annual Review of Plant Biology*, vol. 63, no. 1, pp. 153–182, 2012.

[3] A. D. Glass, D. T. Britto, B. N. Kaiser et al., "The regulation of nitrate and ammonium transport systems in plants," *Journal of Experimental Botany*, vol. 53, no. 370, pp. 855–864, 2002.

[4] L. E. Jackson, M. Burger, and T. R. Cavagnaro, "Roots, nitrogen transformations, and ecosystem services," *Annual Review of Plant Biology*, vol. 59, no. 1, pp. 341–363, 2008.

[5] Z. Tang, X. Fan, Q. Li et al., "Knockdown of a rice stelar nitrate transporter alters long-distance translocation but not root influx," *Plant Physiology*, vol. 160, no. 4, pp. 2052–2063, 2012.

[6] P. J. Lea, L. Sodek, M. A. J. Parry, P. R. Shewry, and N. G. Halford, "Asparagine in plants," *The Annals of Applied Biology*, vol. 150, no. 1, pp. 1–26, 2007.

[7] R. A. Azevedo, M. Lancien, and P. J. Lea, "The aspartic acid metabolic pathway, an exciting and essential pathway in plants," *Amino Acids*, vol. 30, no. 2, pp. 143–162, 2006.

[8] C. Avila, A. J. Márquez, P. Pajuelo, M. E. Cannell, R. M. Wallsgrove, and B. G. Forde, "Cloning and sequence analysis of a cDNA for barley ferredoxin-dependent glutamate synthase and molecular analysis of photorespiratory mutants deficient in the enzyme," *Planta*, vol. 189, no. 4, pp. 475–483, 1993.

[9] E. Duncanson, A. F. Gilkes, D. W. Kirk, A. Sherman, and J. L. Wray, "*nir1*, a conditional-lethal mutation in barley causing a defect in nitrite reduction," *Molecular and General Genetics MGG*, vol. 236, no. 2-3, pp. 275–282, 1993.

[10] C. Marigo, F. Zito, and G. Casadoro, "Isolation and characterization of a cDNA coding for cytoplasmic glutamine synthetase of barley," *Hereditas*, vol. 118, no. 3, pp. 281–284, 1993.

[11] J. Miyazaki, M. Juricek, K. Angelis, K. M. Schnorr, A. Kleinhofs, and R. L. Warner, "Characterization and sequence of a novel nitrate reductase from barley," *Molecular and General Genetics MGG*, vol. 228, no. 3, pp. 329–334, 1991.

[12] M. G. Moller, C. Taylor, S. K. Rasmussen, and P. B. Holm, "Molecular cloning and characterisation of two genes encoding asparagine synthetase in barley (*Hordeum vulgare* L.)," *Biochimica et Biophysica Acta (BBA) - Gene Structure and Expression*, vol. 1628, no. 2, pp. 123–132, 2003.

[13] K. M. Schnorr, M. Juricek, C. X. Huang, D. Culley, and A. Kleinhofs, "Analysis of barley nitrate reductase cDNA and genomic clones," *Molecular and General Genetics MGG*, vol. 227, no. 3, pp. 411–416, 1991.

[14] P. Strøman, S. Baima, and G. Casadoro, "A cDNA sequence coding for glutamine synthetase in *Ordeum vulgare* L," *Plant Molecular Biology*, vol. 15, no. 1, pp. 161–163, 1990.

[15] K. Sueyoshi, A. Kleinhofs, and R. L. Warner, "Expression of NADH-specific and NAD(P)H-bispecific nitrate reductase genes in response to nitrate in barley," *Plant Physiology*, vol. 107, no. 4, pp. 1303–1311, 1995.

[16] Y. Tong, J. J. Zhou, Z. Li, and A. J. Miller, "A two-component high-affinity nitrate uptake system in barley," *The Plant Journal*, vol. 41, no. 3, pp. 442–450, 2005.

[17] L. J. Trueman, A. Richardson, and B. G. Forde, "Molecular cloning of higher plant homologues of the high-affinity nitrate transporters of *Chlamydomonas reinhardtii* and *Aspergillus nidulans*," *Gene*, vol. 175, no. 1-2, pp. 223–231, 1996.

[18] J. J. Vidmar, D. Zhuo, M. Y. Siddiqi, and A. D. M. Glass, "Isolation and characterization of *HvNRT2.3* and *HvNRT2.4*, cDNAs encoding high-affinity nitrate transporters from roots of barley," *Plant Physiology*, vol. 122, no. 3, pp. 783–792, 2000.

[19] S. Kant, Y. M. Bi, E. Weretilnyk, S. Barak, and S. J. Rothstein, "The Arabidopsis halophytic relative *Thellungiella halophila* tolerates nitrogen-limiting conditions by maintaining growth, nitrogen uptake, and assimilation," *Plant Physiology*, vol. 147, no. 3, pp. 1168–1180, 2008.

[20] J. Luo, H. Li, T. Liu, A. Polle, C. Peng, and Z. B. Luo, "Nitrogen metabolism of two contrasting poplar species during acclimation to limiting nitrogen availability," *Journal of Experimental Botany*, vol. 64, no. 14, pp. 4207–4224, 2013.

[21] W. M. Shi, W. F. Xu, S. M. Li, X. Q. Zhao, and G. Q. Dong, "Responses of two rice cultivars differing in seedling-stage nitrogen use efficiency to growth under low-nitrogen conditions," *Plant and Soil*, vol. 326, no. 1-2, pp. 291–302, 2010.

[22] X. Quan, J. Zeng, L. Ye et al., "Transcriptome profiling analysis for two Tibetan wild barley genotypes in responses to low nitrogen," *BMC Plant Biology*, vol. 16, no. 1, p. 30, 2016.

[23] Z. Chen, L. Zou, R. Lu et al., "Study on the relationship between the traits for low-nitrogen tolerance of different barley genotypes at seedling stage and grain yield," *Journal of Triticeae Crops*, vol. 30, no. 1, pp. 158–162, 2010.

[24] H. Xu, C. Liu, R. Lu et al., "The difference in responses to nitrogen deprivation and re-supply at seedling stage between two barley genotypes differing nitrogen use efficiency," *Plant Growth Regulation*, vol. 79, no. 1, pp. 119–126, 2016.

[25] S. Yoshida, D. A. Forno, J. H. Cock, and K. A. Gomez, *Laboratory Manual for Physiological Studies of Rice*, International Rice Research Institute, Los Baños, Philippines, 1976.

[26] Z. Chen, J. Huang, N. Muttucumaru, S. J. Powers, and N. G. Halford, "Expression analysis of abscisic acid (ABA) and metabolic signalling factors in developing endosperm and embryo of barley," *Journal of Cereal Science*, vol. 58, no. 2, pp. 255–262, 2013.

[27] C. Ramakers, J. M. Ruijter, R. H. L. Deprez, and A. F. M. Moorman, "Assumption-free analysis of quantitative real-time polymerase chain reaction (PCR) data," *Neuroscience Letters*, vol. 339, no. 1, pp. 62–66, 2003.

[28] L. Avila-Ospina, A. Marmagne, J. Talbotec, K. Krupinska, and C. Masclaux-Daubresse, "The identification of new cytosolic glutamine synthetase and asparagine synthetase genes in barley (Hordeum vulgare L.), and their expression during leaf senescence," *Journal of Experimental Botany*, vol. 66, no. 7, pp. 2013–2026, 2015.

[29] A. Suzuki and D. B. Knaff, "Glutamate synthase: structural, mechanistic and regulatory properties, and role in the amino acid metabolism," *Photosynthesis Research*, vol. 83, no. 2, pp. 191–217, 2005.

[30] Y. W. Deng, Y. D. Zhang, Y. Chen, S. Wang, D. M. Tang, and D. F. Huang, "Isolation and characterization of a *GS2* gene in melon (*Cucumis melo* L.) and its expression patterns under the fertilization of different forms of N," *Molecular Biotechnology*, vol. 44, no. 1, pp. 51–60, 2010.

[31] M. Feraud, C. Masclaux-Daubresse, S. Ferrario-Méry et al., "Expression of a ferredoxin-dependent glutamate synthase gene in mesophyll and vascular cells and functions of the enzyme in ammonium assimilation in *Nicotiana tabacum* (L.)," *Planta*, vol. 222, no. 4, pp. 667–677, 2005.

[32] F. Antunes, M. Aguilar, M. Pineda, and L. Sodek, "Nitrogen stress and the expression of asparagine synthetase in roots and nodules of soybean (*Glycine max*)," *Physiologia Plantarum*, vol. 133, no. 4, pp. 736–743, 2008.

[33] M. B. Herrera-Rodríguez, J. M. Maldonado, and R. Pérez-Vicente, "Light and metabolic regulation of *HAS1*, *HAS1.1* and *HAS2*, three asparagine synthetase genes in *Helianthus annuus*," *Plant Physiology and Biochemistry*, vol. 42, no. 6, pp. 511–518, 2004.

[34] H. M. Lam, P. Wong, H. K. Chan et al., "Overexpression of the *ASN1* gene enhances nitrogen status in seeds of Arabidopsis," *Plant Physiology*, vol. 132, no. 2, pp. 926–935, 2003.

[35] M. Takahashi, Y. Sasaki, S. Ida, and H. Morikawa, "Nitrite reductase gene enrichment improves assimilation of NO_2 in Arabidopsis," *Plant Physiology*, vol. 126, no. 2, pp. 731–741, 2001.

16

Regulation of Long Noncoding RNAs Responsive to Phytoplasma Infection in *Paulownia tomentosa*

Guoqiang Fan, [1,2] Yabing Cao, [1] and Zhe Wang [1]

[1] *Institute of Paulownia, Henan Agricultural University, Zhengzhou, Henan 450002, China*
[2] *College of Forestry, Henan Agricultural University, Zhengzhou, Henan 450002, China*

Correspondence should be addressed to Guoqiang Fan; guoqiangfan64@163.com

Academic Editor: Ferenc Olasz

Paulownia witches' broom caused by phytoplasma infection affects the production of Paulownia trees worldwide. Emerging evidence showed that long noncoding RNAs (lncRNA) play a protagonist role in regulating the expression of genes in plants. So far, the identification of lncRNAs has been limited to a few model plant species, and their roles in mediating responses to *Paulownia tomentosa* that free of phytoplasma infection are yet to be characterized. Here, whole-genome identification of lncRNAs, based on strand-specific RNA sequencing, from four *Paulownia tomentosa* samples, was performed and identified 3689 lncRNAs. These lncRNAs showed low conservation among plant species and some of them were miRNA precursors. Further analysis revealed that the 112 identified lncRNAs were related to phytoplasma infection. We predicted the target genes of these phytoplasma-responsive lncRNAs, and our analysis showed that 51 of the predicted target genes were alternatively spliced. Moreover, we found the expression of the lncRNAs plays vital roles in regulating the genes involved in the reactive oxygen species induced hypersensitive response and effector-triggered immunity in phytoplasma-infected Paulownia. This study indicated that diverse sets of lncRNAs were responsive to Paulownia witches' broom, and the results will provide a starting point to understand the functions and regulatory mechanisms of Paulownia lncRNAs in the future.

1. Introduction

Plant witches' broom is an epidemic disease, which is caused by specialized obligate bacteria (i.e., phytoplasma) and spread by insect vectors [1]. To date, over 1000 plant species have been found to be affected by phytoplasma worldwide, including Paulownia [2], mulberry [3], Chinese jujube [4], grape [5], and lime [6]. Phytoplasma-infected plants often undergo a series of physiological and biochemical changes, that induce drastic malformations, such as short internodes, dwarfism, proliferation of axillary buds (witches' broom), yellowing of leaves, flower sterility, and even dieback of plants [7], which have had a devastating effect on agriculture, forestry, and horticultural crop production.

Paulownia is a fast-growing deciduous hardwood tree species native to China, which plays a leading role in improving the ecological environment [8]. However, phytoplasma, which belongs to the aster yellows group "*Candidatus* Phytoplasma asteris" (16SrI-D), invasion in Paulownia often leads to slow growth and even death of trees, which results in enormous economic losses [9], and the genome sequencing of this phytoplasma has not yet been completed. Since the disease was first reported by Doi in 1967, lots of research have been carried out on the diagnosis [10], preservation [11], distribution and concentration changes with seasonal variation [12], and the molecular mechanisms of Paulownia witches' broom (PaWB) infection; other studies on phytoplasma genome and virulence factors have also been carried out [13, 14]. However, to date, no clear mechanisms behind its molecular regulation has been found, mostly due to the complexity of PaWB phytoplasma itself and the limitations of current technical methods. With the rapid development of high-throughput "omics" technologies, transcriptome, microRNA (miRNA), proteome, and metabolome data have become available and have been used to analyze variations in Paulownia after phytoplasma infection; as a result, a series of genes, miRNAs, proteins, and metabolites that are potentially related to the occurrence of PaWB have been identified [2, 15–23].

Despite this, the molecular mechanism of PaWB is still poorly understood. Because, though large numbers of genes, miRNAs, proteins, and metabolites have been reported, the correlation among them is low. Recent studies have shown that changes in the expression levels of long noncoding RNAs (lncRNAs) were closely related to plant growth and development and to biotic and abiotic stress responses in numerous organisms [24].

lncRNAs are defined as having more than 200 nucleotides and little protein-coding potential [25]. Usually, lncRNAs have time- or tissue-specific expression patterns and execute their functions in four main ways, that is, as signals, decoys, guides, or scaffolds [26]. To date, several plant species, including *Arabidopsis thaliana* [27], *Zea mays* [28], *Triticum aestivum* [29], *Oryza sativa* [30], *Populus tomentosa* [31], *Capsicum annuum* [32], *Selaginella moellendorffii* [33], and *Brassica napus* [34], have conducted to understand roles and mechanism of lncRNAs. In *Arabidopsis*, two lncRNAs, *COOLAIR* and *COLDAIR* can regulate flowering time through promoter interference and histone modification, respectively [35], and lncRNAs induced by phosphate starvation response 1 (PHR1) were also identified as the target gene of miRNA399 [36]. Furthermore, lncRNAs involved in responses to biotic stresses have also been found in plants. For instance, In *T. aestivum* infected with the fungus *Blumeria graminis* f. sp. *tritici*, two lncRNAs, *TaS1* and *TaS2*, were found to play important roles in the response to pathogen infection [37]. Similarly, lncRNAs as target mimics for miRNAs in tomatoes infected with tomato yellow leaf curl virus have also been identified [38]. All these results support the idea that lncRNAs may play significant roles in the regulation of plant growth and development, differentiation, and stress responses. However, the roles of lncRNAs in phytoplasma-infected woody plants are still unknown.

In the present study, we used high-throughput strand-specific RNA sequencing (RNA-seq) to elucidate the expression profiles of lncRNAs in healthy (PT), phytoplasma-infected (PTI), healthy 60 mg·L^{-1} MMS-treated (PT-MMS), and phytoplasma-infected 60 mg·L^{-1} MMS-treated (PTI-MMS) *Paulownia tomentosa* cuttings. We identified 112 lncRNAs that were differentially expressed in response to phytoplasma infection. This study will increase our understanding of phytoplasma-responsive lncRNAs and lay a solid foundation for further clarifying the functions of these phytoplasma-responsive lncRNAs.

2. Materials and Methods

2.1. Plant Material and Methyl Methanesulfonate (MMS) Treatment. All of the tissue-cultured cuttings used in this study were obtained from the Institute of Paulownia, Henan Agricultural University, Zhengzhou, Henan Province, China. Shoot tips from healthy *P. tomentosa* cuttings (PT) and PaWB-infected cuttings (PTI) were cultured for 30 days before being clipped. The uniform shoot tips from the PTs and PTIs were transferred into 1/2 MS culture medium containing 25 mg·L^{-1} sucrose and 8 mg·L^{-1} agar (Sangon, Shanghai, China) with 0 or 60 mg·L^{-1} MMS, respectively. PTs without MMS were used as the control. At least 120 samples, including 30 PTs, 30 PTIs, 30 PT-MMS, and 30 PTI-MMS, were prepared. The cultivation procedure was performed as described in Fan et al. [17]. Shoot tips about 1.5 cm in length were collected from these 30 cuttings and were mixed to form one biological replicate, then immediately frozen and ground in liquid nitrogen and stored at −80°C for RNA and DNA extraction.

2.2. PaWB Phytoplasma Detection. Total DNA was isolated from the PT, PTI, PT-MMS, and PTI-MMS using the cetyl trimethyl ammonium bromide (CTAB) (Beijing Chemical Co., Beijing, China) method as described by Zhang et al. [39], respectively. PaWB phytoplasma was detected by nested PCR. PCR amplification and agarose gel electrophoresis were performed as described by Fan et al. [40].

2.3. Strand-Specific RNA Library Construction and RNA Sequencing. Total RNAs were extracted from the PT, PTI, PT-MMS, and PTI-MMS using TRIzol reagent (Invitrogen, Carlsbad, CA, USA) according to the manufacturer's instruction, respectively. Total RNA quality and quantity were assessed using a NanoDrop ND-1000 spectrophotometer (Thermo Scientific, Waltham, USA) and Agilent Bioanalyzer 2100 (Agilent Technologies, Palo Alto, CA, USA) according to the manufacturer's instructions. RNA samples with an OD260/280 nm ratio of 2.0 to 2.1 were used for the later analysis. A Ribo-Zero™ Magnetic kit was used to remove the rRNA. The four RNA samples (PT, PTI, PT-MMS, and PTI-MMS) were used to construct strand-specific RNA-seq libraries according to the TruSeq RNA Sample Preparation Guide. The libraries were sequenced on an Illumina HiSeq™ 2000 platform at Beijing Genomics Institute (Shenzhen, China).

2.4. Transcript Assembling and lncRNA Prediction. High-quality pair-end RNA-seq reads were obtained by removing low-quality reads with more than 50% of the bases with a $Q \leq 10$, discarding reads with adaptor sequences, and reads with more than 10% "N" bases. To remove rRNA reads, the high-quality reads were aligned to sequences in the SILVA ribosomal RNA (rRNA) gene database (http://www.arb-silva.de/) using SOAP2, with a maximum of 5 mismatches allowed for each alignment to each read. All the remaining clean reads were assembled using Trinity. Because the complete *P. tomentosa* genome was not available at the time of this study, the clean reads were mapped to the *P. fortunei* reference genome (http://paulownia.genomics.cn) using TopHat2 with no more than 2 mismatches. The mapped reads were used to assemble the transcripts in each sample using Cufflinks 2.0 program [41]. Low-quality assemblies with transcript length < 200 bp were discarded. Novel transcripts were discovered by filtering out the known transcripts that mapped to the annotated reference gene sequence. Finally, the coding potential calculator was used to screen for putative lncRNAs (with coding potential calculator scores < 0) among the novel transcripts. Based on their location in the reference genome, these lncRNAs were categorized into five classes: intergenic (lncRNA located in intergenic region), intronic (lncRNA derived wholly from intron), sense, antisense (lncRNA overlapping one or more exons of another transcript on the same or opposite strand), and bidirectional (the expression of lncRNA and a neighboring coding transcript on the opposite strand).

2.5. Identification of Conserved lncRNAs. We set two different criteria to discover conserved lncRNAs in the lncRNA database CANTATAdb (http://yeti.amu.edu.pl/CANTATA/): (i) all the lncRNA sequences identified in this study were aligned against the lncRNA sequences in CANTATAdb using BLASTN with a cutoff E value $< 1E-5$ and (ii) lncRNA sequence identity to lncRNA sequences in other plant genomes was >20% [42]. In addition, lncRNAs that may act as the miRNAs precursors were predicted by aligning the lncRNA sequences to the miRNA sequences of Paulownia using BLASTN. PsRNATarget (http://plantgrn.noble.org/psRNATarget/) was used to predicted lncRNAs as targets of miRNAs with E value ≤ 3. Target mimics were predicted according to the rules described by Wu et al. [43].

2.6. Prediction of the Potential Target Genes of PaWB Responsive lncRNAs. In this study, two independent algorithms, cis- or trans-acting, were used to predict potential targets of the PaWB responsive lncRNAs in the PT, PTI, PT-MMS, and PTI-MMS cuttings according to their regulatory mechanism. The first algorithm predicted potential target genes of cisacting lncRNAs that were physically located within 10 kb upstream or 20 kb downstream of lncRNAs using a genome browser. The second algorithm predicted potential target genes of trans-acting lncRNAs based on the lncRNA-mRNA (target) sequence complementary and predicted lncRNA-mRNA duplex energy. First, BLASTN searches were performed to detect potential target mRNA sequences complementary to the lncRNA sequences with identity > 95% and E value $< 1E-5$. Then we used the RNAplex software to calculate the complementary energy between the lncRNAs and their potential transregulated target genes with RNAplex $- E^{-30}$.

2.7. Comparison of lncRNA Expression among the Different Samples. Several pairwise comparisons were carried out among the four samples to search for candidate lncRNAs related to PaWB formation (Figure S1). (i) Differentially expressed lncRNAs selected from the PTI-MMS versus PTI comparison may be related to the influence of the methylating agent MMS and PaWB; (ii) differentially expressed lncRNAs in PT-MMS versus PT comparison are likely to be related to the influence of MMS; (iii) differences between comparisons 1 and 2 may exclude differentially expressed lncRNAs related to the influence of MMS; (iv) differentially expressed lncRNAs in the PTI versus PT comparison are likely to be involved in PaWB; (v) and the common lncRNAs between comparisons 3 and 4 may be directly related to PaWB.

2.8. Screening of Differentially Expressed lncRNAs Related to PaWB Disease and Functional Prediction. The calculation of lncRNA expression levels from the four different samples was normalized to the FPKM value. The false discovery rate (FDR), a commonly used statistical method in multiple tests, was used to determine the threshold of the P value [44]. Differentially expressed lncRNAs were judged with an absolute value of log2 ratio > 1 or < -1 and a threshold FDR < 0.001. To better understand the biological processes regulated by these differentially expressed lncRNAs, the sequences of the potential target genes of these lncRNAs were aligned to the Nr, Swiss-Prot, KEGG, and COG databases using BLASTX (E value $< 1.0E-5$) and searched against the InterPro database using the InterProScan software package.

2.9. Quantitative Real-Time Polymerase Chain Reaction (qRT-PCR) Analysis. Total RNA (1 μg) obtained from leaves of four samples were reverse transcribed into cDNA for validating the expression of lncRNAs, miRNAs predicted to target lncRNAs, and potential target genes of lncRNAs by real-time quantitative PCR (qRT-PCR). qRT-PCR analyses were performed on a StepOne Plus real-time PCR system (Life Technologies, Burlington, ON, Canada) using FASTSYBR green mix from Kappa Biosystem (D Mark, Toronto, ON, Canada). All the primers were designed using Primer Express 5.0 (Applied Biosystems). The primer sequences used in the qRT-PCR analyses are provided in Table S1. All the amplifications were carried out in triplicate, with the standard reaction program (94°C for 3 min, followed by 40 cycles of 94°C for 10 s, and 58°C for 30 s, finally, 72°C for 30 s). The specificity of the amplified fragments are checked using the generated melting curve. The generated real-time data were analyzed using the Opticon Monitor Analysis Software 3.1 tool and standardized to the levels of 18S rRNA (lncRNA and there corresponding target genes) and U6 (miRNA) using the $2^{-\Delta\Delta Ct}$ method [45].

3. Results

3.1. Detection of Phytoplasma in P. tomentosa Cuttings Showing Symptoms of Witches' Broom. The phytoplasma-infected *P. tomentosa* (PTI) cuttings showed drastic malformations, including short internodes, proliferation of axillary buds, and yellowing leaves. When treated with 60 mg·L^{-1} MMS, the PTI-MMS cuttings regained a healthy morphology, while the healthy cuttings treated with 60 mg·L^{-1} MMS (PT-MMS) showed no obvious changes (Figure S2). Fragment of the 16S rDNA sequence from the PaWB phytoplasma genome was detected in the PTI samples but not in the PT, PT-MMS, or PTI-MMS samples (Figure S3). These results suggest that phytoplasmas have disappeared after the 60 mg·L^{-1} MMS treatment.

3.2. Genome-Wide Identification and Characterization of lncRNAs in Phytoplasma-Infected P. tomentosa. To systematically identify *P. tomentosa* lncRNAs responsive to phytoplasma infection, strand-specific RNA-seq was performed for RNA samples from healthy and phytoplasma-infected *P. tomentosa* leaves under 0 or 60 mg·L^{-1} MMS treatment. A total of 208,686,782 pair-end raw reads were obtained from the four libraries. After trimming, 204,943,862 clean reads were obtained. From these reads, 32,283 transcripts were assembled using Cufflink. Among these transcripts, 28,593 were completely aligned against the reference genome of

Paulownia fortunei, and the remaining transcripts that are not aligned were considered as novel transcripts. With FPKM > 0.5 as the cut-off and using the lncRNAs prediction standard, 3689 lncRNAs were identified in the four libraries (Table S2), these included 45 bidirectional lncRNA, 563 antisense lncRNAs, 3012 intergenic lncRNAs, and 69 sense lncRNAs. Thus, the intergenic lncRNAs made up 81.6% of the total *P. tomentosa* lncRNAs, which is consistent with the previous study [34]. The distribution of the lncRNAs in the genome is vital for the genetic manipulation required to adapt to the stress of the phytoplasma infection. By mapping these putative lncRNA sequences to the reference genome sequence, we found that they evenly distributed in each chromosome (Figure 1(a)). The size distribution of the potential lncRNAs identified in this study ranged from 200 to 18,769 bases, with most lncRNAs (68.4%) ranging from 200 to 1000 bases (Figure 1(b)). The average and median lengths of the lncRNAs were 1052 bp and 680 bp, respectively, which is longer than the *Arabidopsis* lncRNA transcripts (median length of 285 bp) but shorter than the lncRNA transcripts of rice (median length of 852 bp) and *Populus tomentosa* (median length of 736 bp) [42]. The average and median lengths of the protein-coding mRNA of Paulownia were 1528 bp and 1245 bp, respectively, which are longer than those of noncoding transcripts. Characterization of the genomic structure of these lncRNAs revealed that 2815 of them had only one exon, 512 had two exons, 219 had three exons, and the remaining lncRNAs had more than three exons (Figure 1(c)).

The conservation of lncRNA is considered to be lower than that of protein-coding genes, so if lncRNAs perform evolutionarily conserved functions, they may be conserved among different species. Thus, to detect conserved lncRNAs, all the lncRNA sequences were searched against the genomes of 10 representative plants (*A. thaliana, O. sativa, Glycine max, Selaginella, Chlamydomonas, Physcomitrella, Amborella, Solanum tuberosum, V. vinifera,* and *Z. mays*) using BLAST. The results showed that only a small number of the *P. tomentosa* lncRNAs were conserved across these 10 species (Table 1 and Table S3). The highest number matched to the known lncRNAs in *V. vinifera*, likely because, among these 10 species, Paulownia was most closely related to *V. vinifera* in evolution. To annotate the predicted lncRNAs from an evolutionary point of view, we used INFERNAL to classify them into different noncoding RNA families. Based on their consensus secondary structures, we identified 451 unique sequences belonging to 170 conserved lncRNA families (Table S4); among them, MIR families accounted for most of the conserved lncRNA families, followed by MIR families.

3.3. lncRNAs Function as Precursors or Target Mimics of miRNAs. lncRNAs may influence transcriptional, posttranscriptional, and epigenetic gene regulation through miRNAs [46]. We identified lncRNAs that could act as a precursor of known miRNAs in Paulownia. Six lncRNAs were predicted as precursors of 10 known miRNAs belonging to three miRNA families (Table 2). Thus, we speculated that these lncRNAs might function as miRNAs in response to phytoplasma infection. lncRNAs can also regulate gene expression and numerous biological processes by acting as miRNA targets or target mimics in plants [47]. To explore the possibility of lncRNAs as a target of miRNA, all the lncRNAs were aligned against the *P. tomentosa* miRNA sequences using PsRNATarget. Interestingly, 239 out of the 3689 lncRNAs were predicted to be targeted by 228 miRNAs of *Paulownia tomentosa*, including pt-miR156a/b-5p/c-3p/d/e/f-3p/q, pt-miR160a-3p/b/c, and pt-miR167a/b (Table S5). Moreover, in order to investigate the relationship between miRNAs and their target lncRNAs, qRT-PCR was used to measure their expression of four miRNA-lncRNA pairs. As shown in Figure 2(a), a negative relationship between miRNAs and their target lncRNAs was observed, suggesting that miRNAs may lead to the degradation of their corresponding target lncRNAs.

lncRNAs that potentially function as target mimics of miRNAs were predicted according to Wu et al. [43]. We identified 23 lncRNAs that may act as target mimics and may be bound by 33 miRNAs (26 known miRNAs and 7 novel miRNAs) to form 38 miRNA-lncRNA duplexes (Table S6). Among these miRNA target mimics, TCONS_00021785 was identified as the target mimic of the pt-miR319 family. Notably, three known miRNAs (pt-miR6173e-5p, pt-miR156m, and pt-miR156g-5p) and one novel miRNA (pt-mir30-5p) were predicted to be target mimic of two lncRNAs, respectively. That is to say, functions of these miRNAs may be inhibited. To validate it, the expression level of TCONS_00021785 and the potential target genes of pt-miR319a-3p were examined using qRT-PCR. As shown in Figure 2(b), the expression level of TCONS_00021785 was increased and the expression level of target gene PAU012728.1 was also increased in PTI, suggesting that TCONS_00021785 may increase the expression level of PAU012728.1 by interacting with pt-miR319a-3p.

3.4. Identification of Phytoplasma-Responsive lncRNAs. Emerging evidence has demonstrated that lncRNAs are involved in the regulation of the stress regulation response [29, 30]; thus, we analyzed the differentially expressed lncRNAs among the four samples. Differentially expressed lncRNAs were defined as log2 ratio > 1 or <−1, with FDR < 0.001. Accordingly, 728 differentially expressed lncRNAs (190 upregulated and 538 downregulated) were identified in PTI versus PT, 126 differentially expressed lncRNAs (72 upregulated and 54 downregulated) were identified in PTI-MMS versus PTI, and 211 differentially expressed lncRNAs (62 upregulated and 149 downregulated) were identified in PT-MMS versus PT. According to the comparison scheme of PaWB-related lncRNAs described in Materials and Methods, 112 lncRNAs were considered as related to phytoplasma infection (Figure 3, Table S7). The 112 phytoplasma-responsive lncRNAs comprised 32 antisense lncRNAs, 1 sense lncRNAs, and 79 intergenic lncRNAs. To confirm their phytoplasma-responsive expression, we selected 17 lncRNAs and validated their expression patterns by qRT-PCR (Table S8, Figure 4). As shown in Figure 4, the qRT-PCR results demonstrated that, except TCONS_00013163, the qRT-PCR results were consistent

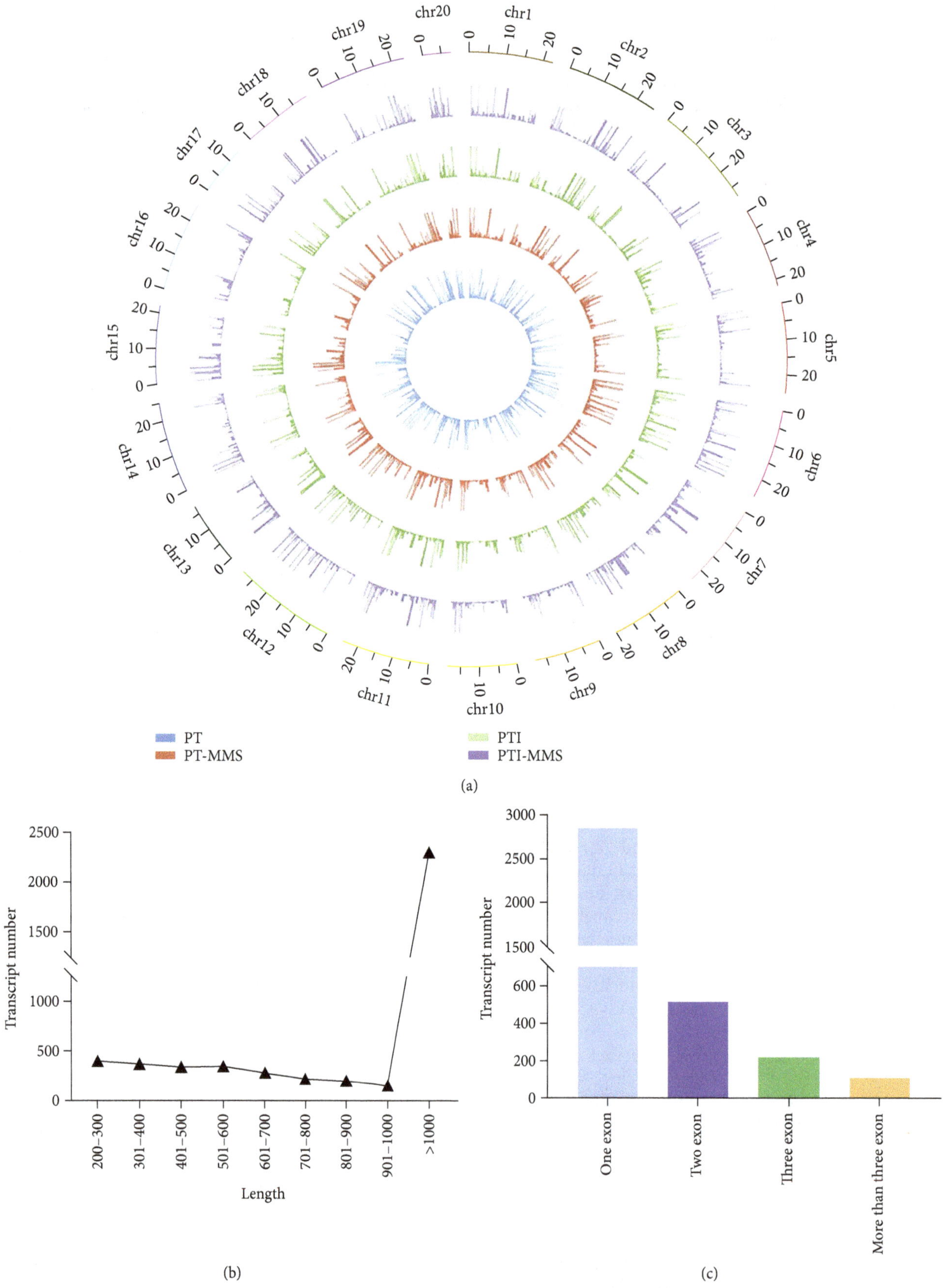

FIGURE 1: Features of lncRNAs in *P. tomentosa*. (a) Distribution of lncRNAs along each chromosome. (b) Length distribution of 3689 *P. tomentosa* lncRNAs. (c) Number of exons per transcript for all lncRNA transcripts.

Table 1: Summary of the conserved lncRNAs.

	Total number	Length	Identity	Coverage ≥ 10 number	Coverage ≥ 20 number
Selaginella	3	31–38	97–100	0	0
Potato	117	27–644	78–100	31	12
Vitis	117	27–335	80–100	38	16
Zea	17	26–282	82–100	2	1
Chlamydomonas	19	26–65	92–100	1	0
Physcomitrella	27	27–73	95–100	2	0
Glycine	36	29–178	80–100	10	4
Amborella	38	27–311	80–100	5	4
Oryza	59	26–215	79–100	7	2
Arabidopsis	36	25–161	82–100	4	1

Table 2: lncRNAs acted as precursors of known miRNAs in Paulownia.

lncRNA ID	miRNA ID	Pre-miRNA ID	Pre-miRNA length	Identity	Alignment length	*E* value
TCONS_00000284	pt-miR171d-5p	ssl-MIR171a-p5	90	100	90	2.00*E* − 46
TCONS_00000284	pt-miR171d-3p	mes-MIR171b	90	93	90	2.00*E* − 46
TCONS_00017319	pt-miR160a-5p	ptc-MIR160c	93	95	93	1.00*E* − 36
TCONS_00017319	pt-miR160a-3p	ptc-MIR160c	93	95	93	1.00*E* − 36
TCONS_00017319	pt-miR160c-5p	stu-MIR160a	84	88	84	4.00*E* − 43
TCONS_00017319	pt-miR160c-3p	stu-MIR160a	84	88	84	4.00*E* − 43
TCONS_00019806	pt-miR156e	ptc-MIR156i	100	92	96	2.00*E* − 50
TCONS_00019806	pt-miR156k	gma-MIR156g	142	90	124	4.00*E* − 67
TCONS_00019828	pt-miR156k	gma-MIR156g	142	91	142	1.00*E* − 77
TCONS_00019829	pt-miR156e	ptc-MIR156i	100	92	100	2.00*E* − 52
TCONS_00019829	pt-miR156k	gma-MIR156g	142	100	142	2.00*E* − 77
TCONS_00034513	pt-miR156e	ptc-MIR156i	100	97	91	3.00*E* − 42
TCONS_00034513	pt-miR156k	gma-MIR156g	142	97	92	7.00*E* − 43
TCONS_00034513	pt-miR156q	stu-MIR156c	149	100	149	1.00*E* − 81

with those from the RNA-seq data, despite some differences in expression levels.

To explore the function of the lncRNAs, we identified and analyzed their target genes. The computational analysis identified 157 potential target genes for 89 lncRNAs. Among them, 86 potential cis-regulated target genes and 71 potential transregulated target genes were predicted for 63 and 72 phytoplasma-responsive lncRNAs, respectively (Table S9). We found that one lncRNA could have more than one target gene, and one target gene could be targeted by one or more lncRNAs. Among these lncRNAs, 33 had one target gene, while two lncRNAs had as many as seven target genes (TCONS_00004908 and TCONS_00004911). Further analysis showed that of the 157 potential target genes, 27 were differentially expressed between the PT and PTI libraries ($P < 0.05$), 14 upregulated and 13 downregulated (Table 3). Moreover, we selected 18 genes and validated their expression by qRT-PCR. The expression patterns of the genes identified by qRT-PCR were consistent with those identified by RNA-seq (Figure 5 and Table S8). Besides, by comparing the expression trends of eight differentially expressed lncRNAs and their target genes, we found that among these differentially expressed lncRNAs-miRNAs pairs, three lncRNA-RNA had a positive correlation (TCONS_00034613/PAU030933.1, TCONS_00002625/PAU002322.1, and TCONS_00019890/PAU000284.1), one lncRNA-mRNA pair (TCONS_00026765/PAU030243.1) had an opposite expression pattern, and the last four lncRNA-RNA pairs (TCONS_00031692/LCONS_00023050, TCONS_00021207/PAU018908.1, TCONS_00007939/LCONS_00023050, and TCONS_00004908/LCONS_00004917) showed the mixed correlation (Figure 6). The same results had also reported in *Populus* [42]. This result suggested that lncRNAs may have various functions in regulating gene expression, and identification and analysis of the relationship between the expression patterns of the phytoplasma-responsive lncRNAs and their potential target genes may help in understanding the functions of these lncRNAs.

To confirm the functional annotations of these 27 target genes, BLAST was used to align their nucleotide sequences to the genes in other plants. The function of 24 of the target

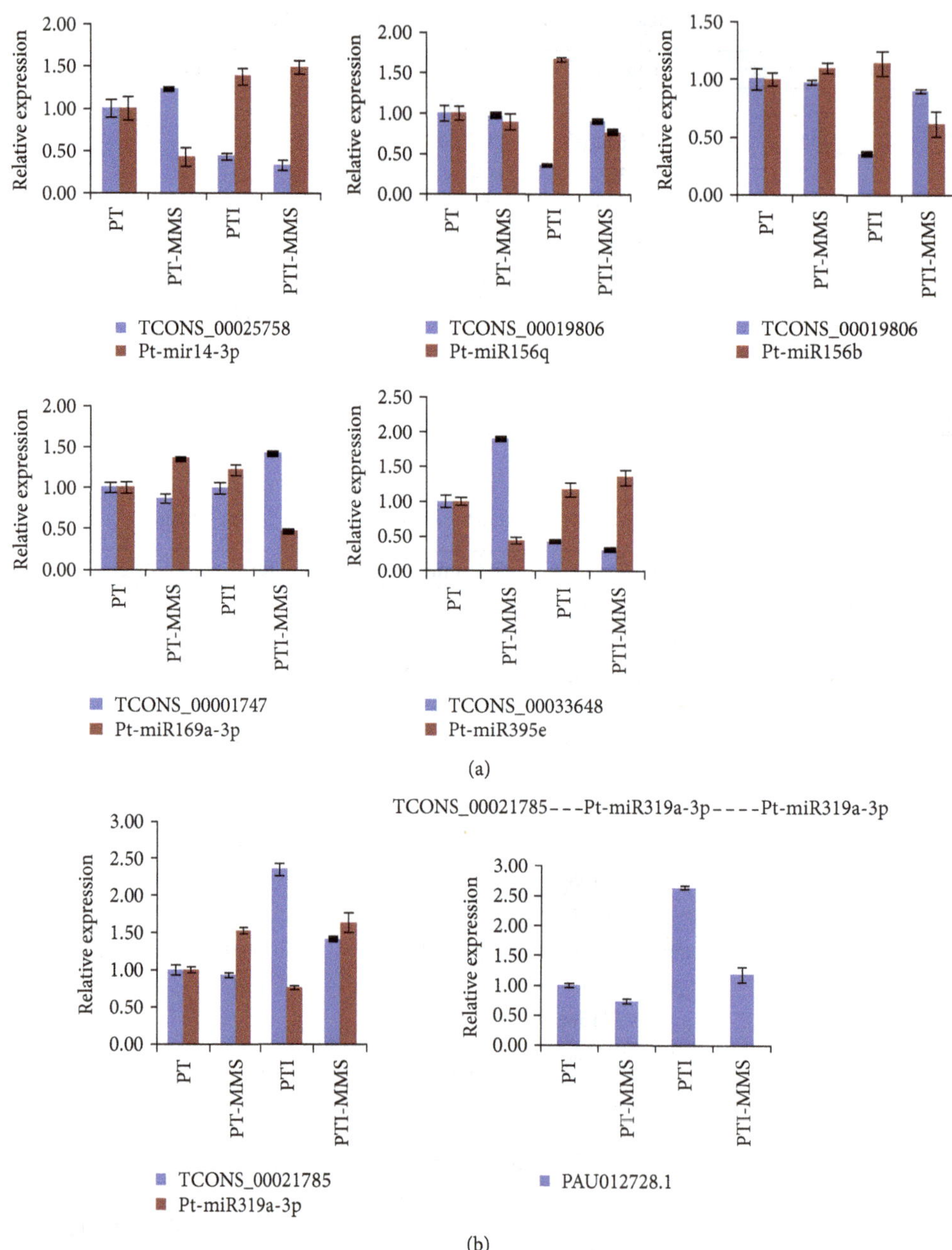

Figure 2: Expression analysis of lncRNAs as potential targets or target mimics of miRNAs. (a) Quantitative RT-PCR analysis of miRNAs and their potential target lncRNAs. (b) Quantitative RT-PCR analysis of TCONS_00021785 and one potential target gene of pt-miR319a-3p (PAU012728.1).

genes was confirmed in other plants (Table 3). Among them, eight genes were involved in stress resistance, namely, genes encoding glucan endo-1,3-beta-glucosidase 11 (LCONS_000 34335 and PAU030933.1), the acetyltransferase NATA1 (LCONS_00013095), zinc finger CCCH domain-containing protein 9 (LCONS_00004917 and LCONS_00022081), disease-resistance protein (PAU018908.1), protein SRC2 (PAU030243.1), and cytochrome P450 (LCONS_00023050). Seven genes were involved in growth, namely, genes encoding xyloglucan endo-transglycosylase/hydrolase (LCONS_00 022082, LCONS_00004912, and PAU019848.1), abscisic acid 8′-hydroxylase (PAU005580.1), zeaxanthin epoxidase (PAU011878.1), MADS-box transcription factor 27 (PAU003690.1), and protein bem46 (PAU021151.1). Four genes were involved in metabolism, namely, genes encoding ribonuclease 3-like protein 1 (LCONS_00019384), ribonuclease H protein (LCONS_00004913), histone-lysine N-methyltransferase (LCONS_00030149), and serine carboxypeptidase II (PAU022614.1). Two genes were involved in transport, namely, genes encoding ATP-binding cassette (PAU011882.1) and calcium-transporting ATPase 12 (PAU023543.1), and two genes were involved in photosynthesis, namely, genes encoding chlorophyll a-b binding protein (PAU002322.1) and photosystem II

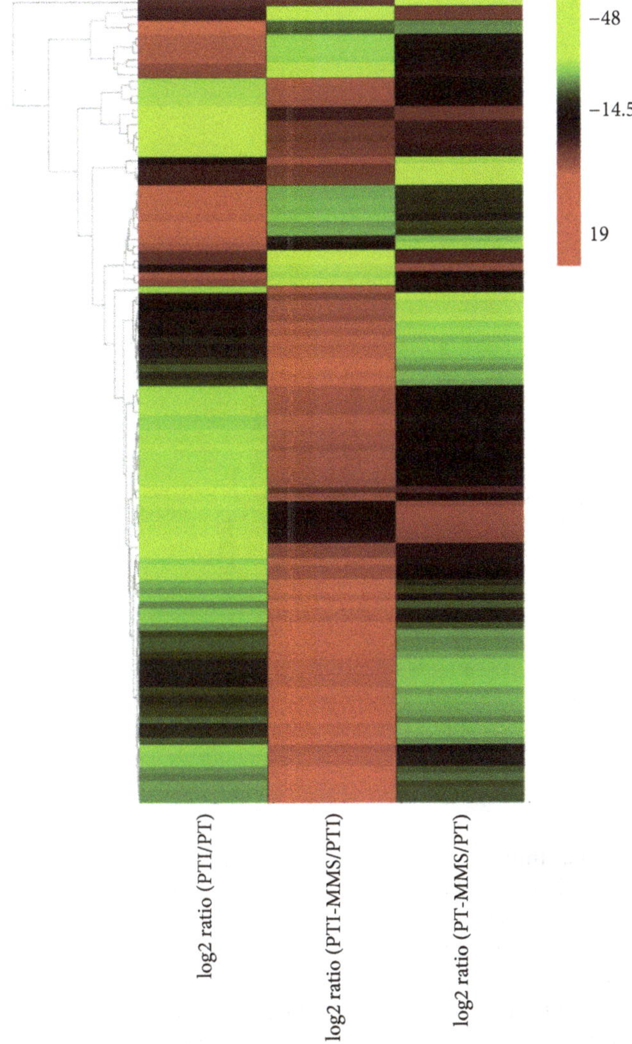

FIGURE 3: The heatmap of 112 PaWB-related lncRNAs in different comparison.

10 kDa polypeptide (PAU000284.1). Three of the PaWB-related target genes were annotated as an unknown function, and their functions are still to be verified.

3.5. Alternative Splicing Events. Alternative splicing is the key contributor to increasing the diversity of transcripts and proteins encoded in genomes. Many studies have shown that biotic and abiotic stresses can both influence splicing events and that alternative splicing is central for photosynthesis, defense responses, and the circadian clock of plants [66]. Alternative splicing of mRNAs is one of the most reported bioprocesses involving lncRNA; therefore, we calculated the numbers of alternative splicing events based on the Paulownia RNA-seq data and identified four types of alternative splicing: (i) exon skipping; (ii) intron retention; (iii) alternative 5′ splice site; and (iv) alternative 3′ splice site. Among these splicing types, the main patterns of alternative splicing were intron retention, which is consistent with the results of studies in other species [67]. Remarkably, the number of variable splicing in PTI is lower than that of PT, and the frequency of the occurrence of each splicing event is higher than that of PTI (Figure 7). These results indicated that complex variable splicing events had happened and potential differentially expressed proteins had emerged in Paulownia cuttings after phytoplasma infection, which may be due to the defense response of Paulownia triggered by phytoplasma infection. In addition, we found that among the 157 target genes of the PaWB-related lncRNAs, 51 genes were alternatively spliced, resulting in 315 transcripts (Table S10). Further analysis found that these splice variants mainly involved in photosynthesis and carbon metabolism. Genes involved in photosynthesis are known to play significant roles in phytoplasma-infected plants. Previous studies have demonstrated that phytoplasma infection might affect photosynthesis and resulted in yellow leaves. Together, these results suggested that alternative splicing might represent an additional level of gene regulation in response to phytoplasma infection. However, further comprehensive studies of the roles of alternative splicing events in phytoplasma-infected Paulownia are needed.

4. Discussion

Understanding the mechanism of gene regulation will provide a molecular basis for PaWB research in Paulownia and contribute to breeding Paulownia that are better adapted to stress conditions. Over the past decade, with the rapid development of sequencing technologies, RNA-seq has allowed the detection of novel types of noncoding transcripts, which has revealed the complexity of eukaryotic genome expression. To date, a large number of lncRNAs have been identified in different species [27–33]. However, phytoplasma invasion activates a set of physiological, biochemical, and molecular responses in host plants, but genome-wide identification and characterization of lncRNAs involved in these responses are poorly studied in Paulownia. In this study, strand-specific RNA-seq was performed to systematically identify and analyze lncRNAs dynamically regulated by phytoplasma infection. Under the strict screen criteria that we used, 3689 high-confidence lncRNAs were identified, of which 112 phytoplasma-responsive lncRNAs comprised 32 antisense lncRNAs, 1 sense lncRNAs, and 79 intergenic lncRNAs. This number of lncRNAs is far less than the numbers of lncRNAs identified in *Arabidopsis* or rice, likely because of the rigorous filtration criteria we used in this study. The structure analysis showed that the 3689 lncRNAs have a median length of 680 bp and usually contain only 1 exon. Our analysis generated a relatively robust list of potential lncRNAs for Paulownia that will be useful for functional genomics research.

4.1. Overall Insights into the Conservation of lncRNAs in P. tomentosa. Paulownia lncRNAs present low sequence conservation compared to the protein-coding genes, which is consistent with other studies [33, 34, 42]. In our study, most lncRNAs contain one exon and have more specific expression profiles than protein-coding genes. In addition, we found that only limited lncRNAs showed homologues with lncRNAs in other plant species. All these results suggested

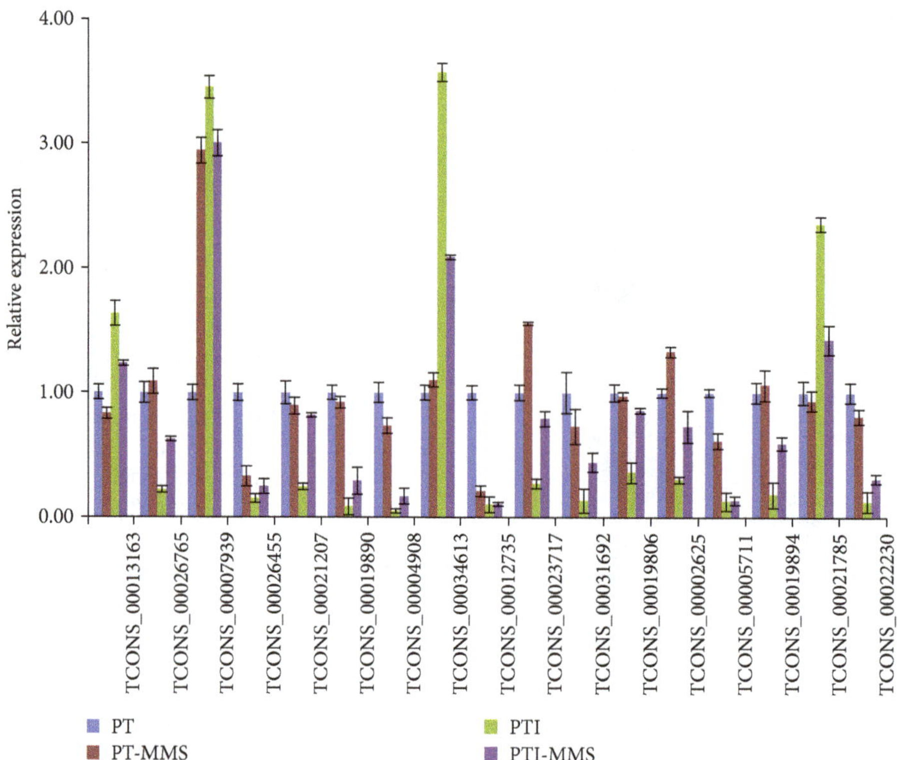

FIGURE 4: Changes in the relative expression levels of lncRNAs in *P. tomentosa*. Potential target genes of phytoplasma-responsive lncRNAs.

that lncRNAs identified in this study were not conserved. However, to date, thousands of conserved lncRNAs have been found, possibly owing to the more ancient origins of these lncRNAs giving their functions more time to be stabilized. In fact, the reasons for the limited conservation of lncRNAs are not surprising. First, unlike mRNAs, lncRNAs are not constrained by codon usage and do not have a single long open reading frame to prevent frame-shift mutations. They usually possess short conserved motifs that are not easily identifiable by BLAST and are constrained by structure or sequence-specific interactions [68]. Second, lncRNAs may have undergone recent and rapid adaptive selection. Moreover, some lncRNAs are associated with miRNAs, which can generate from short pairing fragments of lncRNAs that are less constrained in other parts of the transcripts.

4.2. Potential Function Roles for Phytoplasma-Responsive lncRNAs. lncRNAs can be targeted either to a nonsense-mediated mRNA decay pathway or to play direct functional roles as transcription regulators. Recent studies indicated that lncRNAs can also act as potential targets of miRNA [43]. In our study, we identified 239 lncRNAs as putative targets of 228 miRNAs. Among them, the pt-miR156q and pt-miR156b were upregulated in phytoplasma-infected cuttings (PTI) and downregulated in MMS-treated phytoplasma-infected cuttings (PTI-MMS). A previous study has reported that miR156 plays a vital role in plant growth and development, for example, in rice overexpressing miR156 showed dramatic morphological changes, including markedly increased number of axillary buds and dwarfism [69]. Similarly, in switchgrass, overexpression of miR156 reduced the apical dominance, delayed the flowering time, caused dwarfism and increased total leaf numbers [70]. Simultaneously, in our study, the lncRNA TCONS_00019806 was downregulated, suggesting that the pt-miR156q might trigger the degradation of lncRNAs TCONS_00019806 and lead to the dwarfism symptom in the phytoplasma-infected cuttings. In addition, miR395 was regulated under sulfate-limited conditions [71], and miR164, miR166, and miR482 have been found to play significant roles in plant microorganism interaction [72, 73]. All these results demonstrated that phytoplasma-responsive lncRNAs may participate in the response to stress.

Target mimicry, a newly identified regulatory mechanism of miRNAs, was first studied in plants and is used to block the interplay between miRNAs and their putative target genes by producing a false target transcript that cannot be cleaved [68]. The effectiveness of lncRNAs that function as putative target mimics for miRNAs has been confirmed in many plants [43]. In this study, we globally analyzed the regulatory network of miRNAs. By bioinformatics analysis, we identified 23 lncRNAs that can act as potential target mimics of 33 miRNAs in Paulownia. Among them, one phytoplasma-responsive lncRNA (TCONS_00021785) was predicted to be the target mimic of the pau-miR319 family. Expression analysis showed that one target gene of the pt-miR319a (PAU012728.1) was upregulated when the expression level of TCONS_00021785 increased after phytoplasma infection, indicating that TCONS_00021785 may regulate the expression of PAU012728.1 (encodes TCP transcription factor) by competing pt-miR319a. Transcription factors control a significant

TABLE 3: List of identified and characterized PaWB related to proteins in other species.

mRNA ID	PTI/PT	Nr annotation	Function classification	Reference	Species
LCONS_00019384**	−1.72	Ribonuclease 3-like protein	Metabolism	Kiyota et al. [48]	*Arabidopsis*
LCONS_00004913**	−1.49	Ribonuclease H protein	Metabolism	Cazenave et al. [49]	Wheat
LCONS_00030149**	−1.34	Histone-lysine N-methyltransferase	Metabolism	Pavankumar et al. [50]	*Arabidopsis*
PAU022614.1*	−1.57	Serine carboxypeptidase II-3	Metabolism	Bullock et al. [51]	Wheat
LCONS_00018738**	1.09	Proline-rich receptor-like protein kinase PERK1	Cell signal transduction	Silva et al. [52]	*Brassica napus*
PAU002322.1*	−1.42	Chlorophyll a-b binding protein of LHCII type 1	Photosynthesis	Chen et al. [53]	*Arabidopsis*
PAU000284.1*	−2.56	Photosystem II 10 kDa polypeptide	Photosynthesis	Allahverdiyeva et al. [54]	*Arabidopsis*
PAU011882.1*	−1.49	ATP-binding cassette, sub-family G (WBC)	Transport	Klein et al. [55]	*Secale cereale*
PAU023543.1*	1.28	Calcium-transporting ATPase	Transport	Boursiac et al. [56]	*Arabidopsis*
LCONS_00034335**	2.54	Glucan endo-1,3-beta-glucosidase 11	Stress resistance	Rol et al. [57]	*Sylvestris*; tobacco
PAU030933.1*	2.60	Glucan endo-1,3-beta-glucosidase 11	Stress resistance	Rol et al. [57]	*Sylvestris*; tobacco
LCONS_00013095**	2.26	Acetyltransferase NATA1-like	Stress resistance	Lou et al. [58]	*Arabidopsis*
LCONS_00004917**	1.37	Zinc finger CCCH domain-containing protein 9	Stress resistance	Maldonado-Bonilla et al. [59]	*Arabidopsis*
LCONS_00022081**	14.72	Zinc finger CCCH domain-containing protein 9	Stress resistance	Maldonado-Bonilla et al. [59]	*Arabidopsis*
PAU018908.1*	2.57	Disease-resistance protein	Stress resistance	Fan et al. [15]	*Paulownia*
PAU030243.1*	1.23	Protein SRC2	Stress resistance	Kim et al. [60]	Tobacco; pepper
LCONS_00023050**	1.83	Cytochrome P450 71A4	Stress resistance	Liu et al. [2]	*Paulownia*
LCONS_00022082**	14.33	Xyloglucan endo-transglycosylase/ hydrolase	Growth	Nishikubo et al. [61]	Poplar
LCONS_00004912**	−2.28	Xyloglucan endotransglucosylase/ hydrolase	Growth	Nishikubo et al. [61]	Poplar
PAU019848.1*	−2.14	Xyloglucan endotransglucosylase/ hydrolase	Growth	Nishikubo et al. [61]	Poplar
PAU005580.1*	−2.10	Abscisic acid 8′-hydroxylase 4	Growth	Saito et al. [62]	*Arabidopsis*
PAU003690.1*	5.03	MADS-box transcription factor	Growth	Martel et al. [63]	Tomato
PAU021151.1*	−1.12	Protein bem46-like	Growth	Ramírez et al. [64]	*Pleurotus ostreatus*
PAU011878.1*	−1.09	Zeaxanthin epoxidase	Growth	Audran et al. [65]	*Arabidopsis*
PAU005061.1*	2.03	Uncharacterized protein			
LCONS_00004914**	−1.83	Uncharacterized protein			
LCONS_00019769**	5.34	Hypothetical protein			

*Represents the known mRNA. **Represents the novel mRNA predicted in this study.

proportion of the defense response by regulating the defense gene. A previous study showed that the expression level of TCPs could regulate by the miR319 [74]. In addition, TCPs can directly determine the expression levels of LOX2, and mutation of TCP binding sites in the *LOX2* promoter strongly reduced its activity [74]. *LOX2* is the key enzyme in jasmonic acid biosynthesis, suggesting that the TCP transcription factor directly controls genes in the JA biosynthesis pathway. This result revealed the potential role of TCONS_00021785 in these processes. However, this hypothesis needs to be further validated, but our results suggest that crosstalk between miRNAs, mRNA, and phytoplasma-responsive lncRNAs may affect many different biological processes and provide useful information for further research into the function of lncRNAs in phytoplasma-infected *P. tomentosa*.

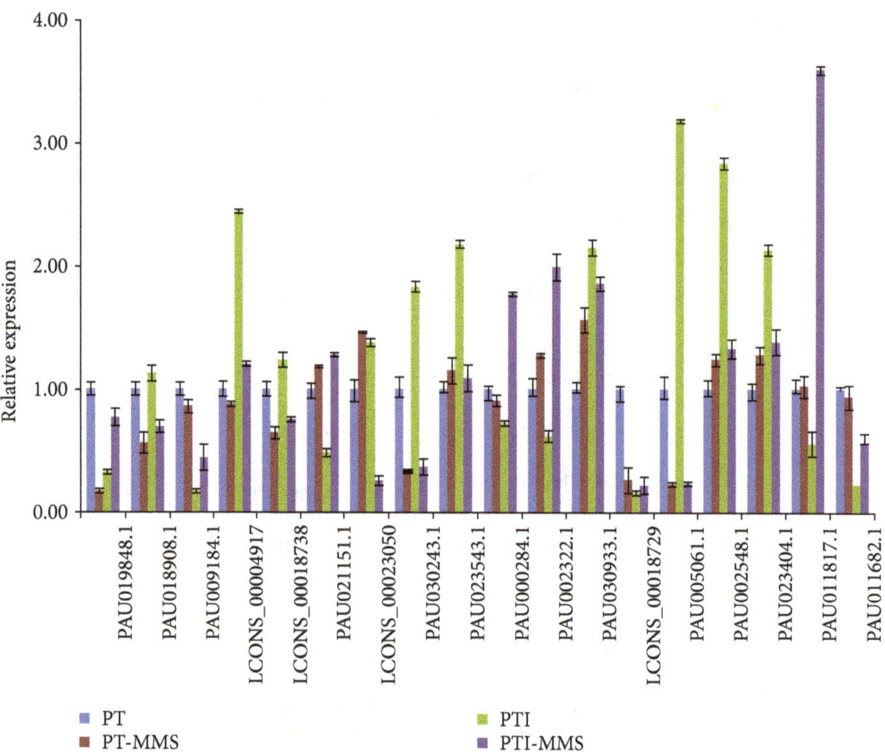

FIGURE 5: Changes in the relative expression levels of target genes in *P. tomentosa*.

4.3. Splice Variants in P. tomentosa May Be Related to PaWB. Generation of splice variants is a common mechanism to increase transcriptome plasticity and proteome diversity in eukaryotes. To date, there were few studies that have investigated alternative splicing in response to stress, and no alternative splicing events for genes in phytoplasma-infected Paulownia have been reported yet. In this study, we identified the genes with splice variants that are involved in photosynthesis, plant hormone signal transduction, and carbon metabolism, including genes encoding photosystem II 10 kDa polypeptide (PAU000284.1), phosphoglycerate kinase (PAU000324.1), chlorophyll a-b binding protein (PAU002322.1), abscisic acid 8′-hydroxylase (PAU005580.1), and auxin influx carrier (PAU000910.1). It has been demonstrated that in phytoplasma-infected plants, callose deposition is a common phenomenon and is associated with the accumulation of carbohydrates [75], which can accumulate free hexoses and further repress the synthesis of chlorophyll a-b binding proteins [54]. Chlorophyll a-b binding proteins capture solar energy for the primary light reactions of photosynthesis [76]. The decreased abundance of chlorophyll a-b binding proteins may influence the light-harvesting rate and induce the transfer of electrons [53]. In a previous study, a decreased chlorophyll a-b binding protein has been observed in plants infected with phytoplasma [22, 77]. Besides, photosystem II 10 kDa polypeptide (PsbR) is the main subunit of the oxygen-evolving complex of eukaryotic PSII, which participates in the water-splitting reaction and PSII electron transport [54]. Allahverdiyeva et al. [78] found that lacking of PsbR cloud leads to decrease rates of oxygen evolution and quinone reoxidation. Furthermore, mutation of PsbR in *Arabidopsis* leads to a decreased content of PsbP and PsbQ proteins [79], and plants with lacking PsbP awill be characterized with extensive defects of the thylakoid membrane [80], which is the main place for the transformation of light to the active chemical energy. All these results suggested the significant role of PsbR in PSII system. Moreover, a previous study has demonstrated that phytoplasma infection could lead to a decrease content of PsbR [81], and similar result has also been found in transcriptome and proteome research of phytoplasma-infected Paulownia plant [22, 23]. In this study, the expression of the alternative genes, which encodes photosystem II 10 kDa polypeptide (PAU000284.1) and chlorophyll a-b binding protein (PAU002322.1), was also downregulated (Figure 6), while the lncRNAs TCONS_00002625 and TCONS_00019890 were predicted to regulate the expression levels of the genes that encode chlorophyll a-b binding proteins and photosystem II 10 kDa polypeptide, respectively, implying that after phytoplasma infection, lncRNAs may influence the photosynthesis electron transfer chain in Paulownia.

4.4. Phytoplasma Infection Triggers the Immune Responses of Paulownia. Phytoplasma is a plant pathogen that induces drastic malformations, such as short internodes, dwarfism, proliferation of axillary buds, and yellowing leaves. The molecular basis for the pathogenicity of this disease is still poorly understood. Phytoplasma infection altered the expression of genes and proteins in Paulownia [15–23]; however, these observations were descriptive and an in-depth analysis of phytoplasma Paulownia interactions is lacking. Plants

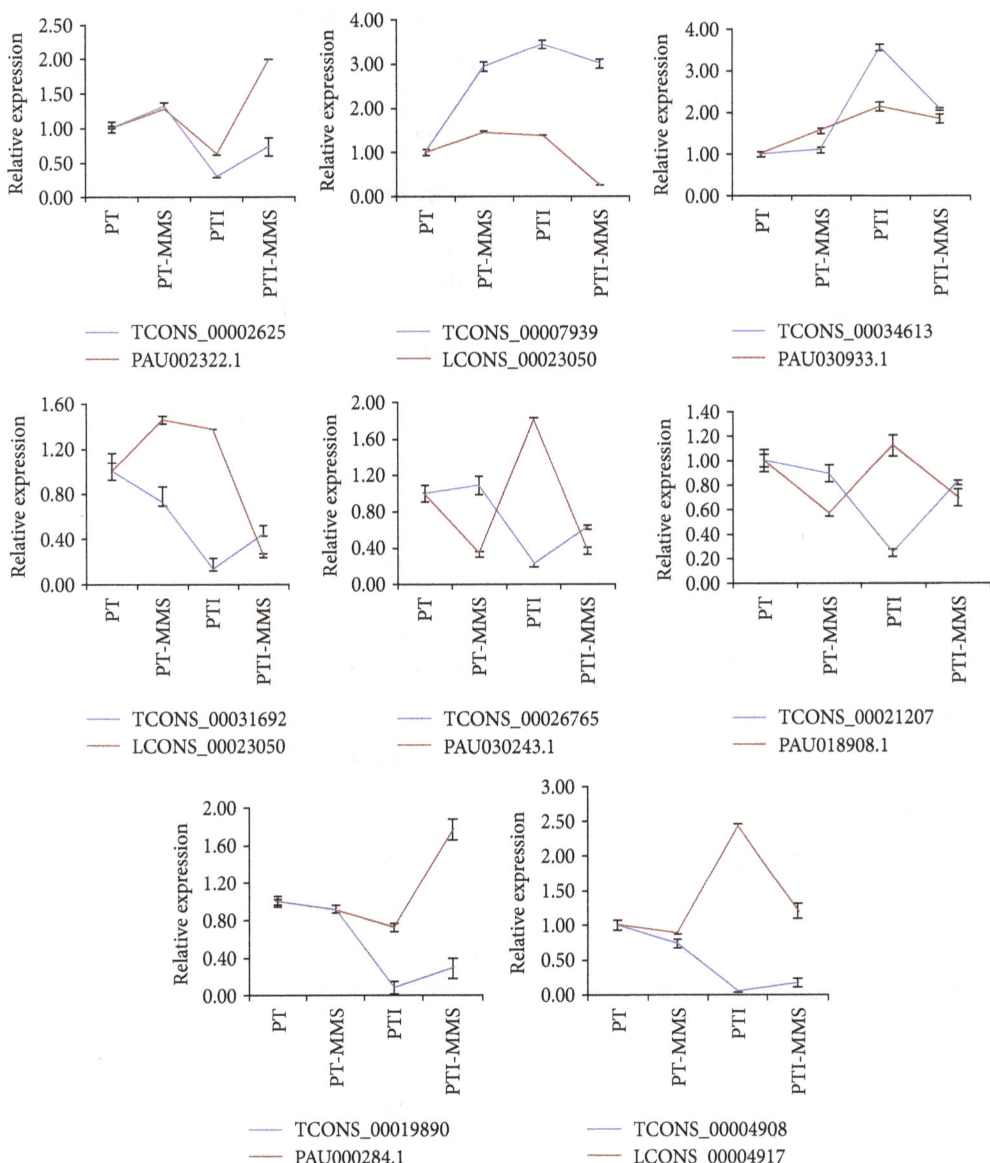

FIGURE 6: Relative expression of the target genes of eight *P. tomentosa* lncRNAs.

infected by phytoplasmas can produce potent strategies to defend themselves against invasion [82, 83]. First, plants perceive the presence of phytoplasma using pathogen-associated molecular patterns (PAMPs) that trigger immunity. In this stage, plants produce a large amount of reactive oxygen species (ROS) and antitoxin, which triggers hypersensitive response. At the same time, phytoplasmas secrete effectors through the Sec secretion system, such as SAP and TENGU, to interfere with the host PAMP-triggered immunity defense signaling transduction and successfully enhance colonization and facilitate their multiplication of themselves in the host plant cells. Plants also utilize cellular receptor proteins to recognize effectors and activate the effector-triggered immunity response, which activates MAPK cascades and induces the disease resistance protein. In the present study, the plant immune responses to phytoplasma infection were activated for we detected genes that encoded enzymes in the signaling pathways (e.g., proline-rich receptor-like protein kinase PERK1) as well as prominent marker genes involved in the associated activities (e.g., disease-resistance protein, glucan endo-1,3-beta-glucosidase, and protein SRC2). In necrotizing viruses-infected tobacco, the glucan endo-1,3-beta-glucosidase increased distinctly [57]. Similarly, in pathogen *Xanthomonas axonopodis* pv. glycines 8 ra-infected peppers, the gene encode for SRC2 was observed highly expressed [60]. Notably, PERK1, disease-resistance protein, glucan endo-1, 3-beta-glucosidase, and protein SRC2 were upregulated in the phytoplasma-infected cuttings (Figure 6). A previous study in Paulownia has also showed that the expression level of gene coding for disease-resistance protein is upregulated after phytoplasma infection [15], while, in this study, lncRNAs TCONS_00021207, TCONS_00034613, and TCONS_00026765 were predicted to regulate the expression levels of genes encoding disease-resistance

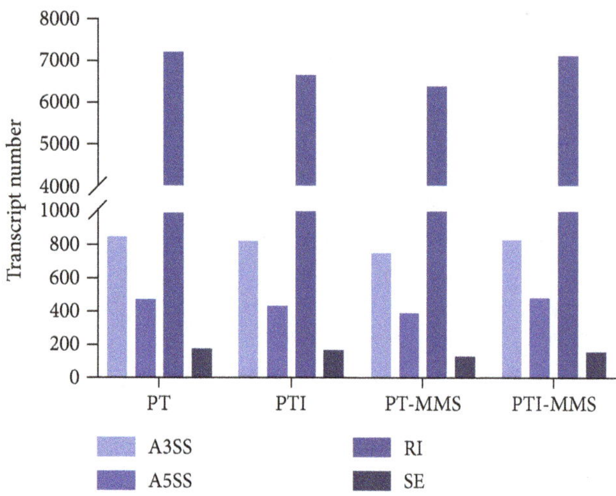

FIGURE 7: Alternative splicing events in Paulownia. A3SS: alternative 3′ splice site AS events; A5SS: alternative 5′ splice site AS events; RI: intron retention; ES: exon skipping.

protein, glucan endo-1, 3-beta-glucosidase, and protein SRC2, respectively. That is to say that these three lncRNAs are likely to play significant roles in phytoplasma-infected cuttings. In addition, the gene encoding cytochrome P450 was also elevated in the PTI cuttings (Figure 6). Cytochrome P450 can serve as an antioxidant to clean up the excess of ROS to reduce the cell damage. Interestingly, in a previous study, the cytochrome P450 level found to be elevated in phytoplasma-infected Paulownia plants [2]. TCONS_00031692 were downregulated and TCONS_00007939 were upregulated in the phytoplasma-infected cuttings, and they both regulated the expression level of gene LCONS_00023050, which encodes cytochrome P450 (Figure 6). Thus, it is clear that in phytoplasma-infected cuttings, lncRNAs might play vital roles in the ROS-induced hypersensitive response and the effector-triggered immunity.

In summary, by using computational analysis, for the first time, we identified 3693 putative Paulownia lncRNAs. These lncRNAs were not conserved among plant species, and some of them were miRNA precursors. Further, we identified 51 target genes of PaWB-related lncRNAs were alternatively spliced, resulting in 315 transcripts. Paulownia lncRNAs perform their function in various ways and their expressions play vital roles in ROS-induced hypersensitive response and the effector-triggered immunity in phytoplasma-infected Paulownia, suggesting the important roles of the lncRNAs in the regulation of biotic stresses. Our analysis also indicated the expression of some lncRNAs could be regulated by miRNAs, but this needs further investigation. The identification and expression analysis of the Paulownia lnRNAs will provide a starting point to understand their functions and regulatory mechanisms in the future.

Conflicts of Interest

The authors declare that there are no competing interests.

Authors' Contributions

Guoqiang Fan conceived and designed the experiments. Yabing Chao analyzed the data and wrote the paper. Zhe Weng performed the experiments.

Acknowledgments

The authors thank Margaret Biswas, PhD, from Liwen Bianji, Edanz Group China (http://www.liwenbianji.cn/ac), for editing the English text of a draft of this manuscript. This study was funded by the Key Science and Technology Program of Henan Province of China (152107000097) and by the Distinguished Talents Foundation of Henan Province of China (174200510001).

Supplementary Materials

Supplementary 1. Figure S1: comparison schemes of the four samples.

Supplementary 2. Figure S2: changes of morphology in Paulownia witches' broom cuttings. PT: sample of healthy *P. tomentosa*; PTI: sample of phytoplasma-infected *P. tomentosa*; PT-MMS: sample of 60 mg·L MMS-treated *P. tomentosa*; PTI-MMS: sample of 60 mg·L MMS-treated phytoplasma-infected *P. tomentosa.*

Supplementary 3. Figure S3: detection of phytoplasma 16S rRNA in the four samples. M: marker; 1: phytoplasma-infected *P. tomentosa* cuttings; 2: healthy cuttings of *P. tomentosa*; 3: 60 mg·L MMS-treated healthy *P. tomentosa* cuttings; 4: 60 mg·L MMS-treated phytoplasma-infected *P. tomentosa* cuttings; 5: ddH_2O.

Supplementary 4. Table S1: primers used for real-time quantitative PCR.

Supplementary 5. Table S2: the number of lncRNAs in each sample.

Supplementary 6. Table S3: conservation of *Paulownia tomentosa* lncRNAs.

Supplementary 7. Table S4: classification of predicted lncRNAs into different noncoding RNA families.

Supplementary 8. Table S5: lncRNAs predicted as a target of miRNAs.

Supplementary 9. Table S6: lncRNAs predicted as target mimicry of miRNAs.

Supplementary 10. Table S7: lncRNAs related to PaWB.

Supplementary 11. Table S8: the FPKM obtained by RNA-seq and the expression obtained by PCR of lncRNA and mRNA.

Supplementary 12. Table S9: the potential target genes of phytoplasma-responsive lncRNAs.

Supplementary 13. Table S10: the splice events of target genes between the healthy and phytoplasma-infected Paulownia.

References

[1] I. M. Lee, R. E. Davis, and D. E. Gundersen-Rindal, "Phytoplasma: phytopathogenic mollicutes," *Annual Review of Microbiology*, vol. 54, no. 1, pp. 221–255, 2000.

[2] R. Liu, Y. Dong, G. Fan et al., "Discovery of genes related to witches broom disease in *Paulownia tomentosa* × *Paulownia fortunei* by a *de novo* assembled transcriptome," *Plos One*, vol. 8, no. 11, article e80238, 2013.

[3] Y. P. Gai, X. J. Han, Y. Q. Li et al., "Metabolomic analysis reveals the potential metabolites and pathogenesis involved in mulberry yellow dwarf disease," *Plant, Cell & Environment*, vol. 37, no. 6, pp. 1474–1490, 2014.

[4] Z. G. Liu, Y. Wang, J. Xiao, J. Zhao, and M. J. Liu, "Identification of genes associated with phytoplasma resistance through suppressive subtraction hybridization in Chinese jujube," *Physiological and Molecular Plant Pathology*, vol. 86, pp. 43–48, 2014.

[5] P. Margaria, A. Ferrandino, P. Caciagli, O. Kedrina, A. Schubert, and S. Palmano, "Metabolic and transcript analysis of the flavonoid pathway in diseased and recovered Nebbiolo and Barbera grapevines (*Vitis vinifera* L.) following infection by Flavescence dorée phytoplasma," *Plant Cell & Environment*, vol. 37, no. 9, pp. 2183–2200, 2014.

[6] M. Mardi, F. L. Karimi, J. Gharechahi, and G. H. Salekdeh, "In-depth transcriptome sequencing of Mexican lime trees infected with *Candidatus* Phytoplasma aurantifolia," *PLoS One*, vol. 10, no. 7, article e0130425, 2015.

[7] S. Namba, "Molecular biological studies on phytoplasmas," *Journal of General Plant Pathology*, vol. 68, no. 3, pp. 257–259, 2002.

[8] S. Ates, Y. Ni, M. Akgul, and A. Tozluoglu, "Characterization and evaluation of Paulownia elongata as a raw material for paper production," *African Journal of Biotechnology*, vol. 7, no. 22, pp. 4153–4158, 2008.

[9] P. Weintraub and L. Beanland, "Insect vectors of phytoplasmas," *Annual Review of Entomology*, vol. 51, no. 1, pp. 91–111, 2006.

[10] Y. Doi, M. Teranaka, and Yora Ket al., "Mycoplasma- or PLT group-like microorganisms found in the phloem elements of plants infected with mulberry dwarf, potato witches' broom, aster yellows, or Paulownia witches' broom," *Japanese Journal of Phytopathology*, vol. 33, no. 4, pp. 259–266, 1967.

[11] I. M. Lee, R. W. Hammond, R. E. Davis, and D. E. Gundersen, "Universal amplification and analysis of pathogen 16S rDNA for classification and identification of mycoplasmalike organisms," *Phytopathology*, vol. 83, no. 8, pp. 834–842, 1993.

[12] K. Wang and C. Hiruki, "The molecular stability of genomic DNA of phytoplasma in the witches'-broom-affected paulownia tissues after microwave heat treatment," *Journal of Microbiological Methods*, vol. 33, no. 3, pp. 263–268, 1998.

[13] H. Nakamura, S. Ohgake, N. Sahashi, N. Yoshikawa, T. Kubono, and T. Takahashi, "Seasonal variation of paulownia witches'-broom phytoplasma in paulownia trees and distribution of the disease in the Tohoku district of Japan," *Journal of Forest Research*, vol. 3, no. 1, pp. 39–42, 1998.

[14] N. Minato, M. Himeno, A. Hoshi et al., "The phytoplasmal virulence factor TENGU causes plant sterility by downregulating of the jasmonic acid and auxin pathways," *Scientific Reports*, vol. 4, no. 1, article 7399, 2014.

[15] G. Fan, Y. Dong, M. Deng, Z. Zhao, S. Niu, and E. Xu, "Plant-pathogen interaction, circadian rhythm, and hormone-related gene expression provide indicators of phytoplasma infection in *Paulownia fortunei*," *International Journal of Molecular Sciences*, vol. 15, no. 12, pp. 23141–23162, 2014.

[16] G. Fan, S. Niu, Z. Zhao et al., "Identification of microRNAs and their targets in *Paulownia fortunei* plants free from phytoplasma pathogen after methyl methane sulfonate treatment," *Biochimie*, vol. 127, pp. 271–280, 2016.

[17] G. Fan, X. Cao, S. Niu, M. Deng, Z. Zhao, and Y. Dong, "Transcriptome, microRNA, and degradome analyses of the gene expression of Paulownia with phytoplasma," *BMC Genomics*, vol. 16, article 896, 2015.

[18] G. Fan, X. Cao, Z. Zhao, and M. Deng, "Transcriptome analysis of the genes related to the morphological changes of *Paulownia tomentosa* plantlets infected with phytoplasma," *Acta Physiologiae Plantarum*, vol. 37, no. 10, pp. 1-12, 2015.

[19] G. Fan, Y. Cao, M. Deng et al., "Identification and dynamic expression profiling of microRNAs and target genes of *Paulownia tomentosa* in response to *Paulownia* witches' broom disease," *Acta Physiologiae Plantarum*, vol. 39, no. 1, article 28, 2017.

[20] X. Cao, G. Fan, Y. Dong et al., "Proteome profiling of Paulownia seedlings infected with phytoplasma," *Frontiers in Plant Science*, vol. 8, article 342, 2017.

[21] Y. Cao, X. Zhai, M. Deng, Z. Zhao, and G. Fan, "Relationship between metabolites variation and Paulownia witches' broom," *Scientia Silvae Sinicae*, vol. 53, no. 6, pp. 85–93, 2017.

[22] H. Q. Mou, J. Lu, S. F. Zhu et al., "Transcriptomic analysis of *Paulownia* infected by Paulownia witches'-broom *phytoplasma*," *PLoS One*, vol. 8, no. 10, article e77217, 2013.

[23] Z. Wang, W. Liu, G. Fan et al., "Quantitative proteome-level analysis of Paulownia witches' broom disease with methyl methane sulfonate assistance reveals diverse metabolic changes during the infection and recovery processes," *PeerJ*, vol. 5, article e3495, 2017.

[24] C. P. Ponting, P. L. Oliver, and W. Reik, "Evolution and functions of long noncoding RNAs," *Cell*, vol. 136, no. 4, pp. 629–641, 2009.

[25] Y. Yang, L. Wen, and H. Zhu, "Unveiling the hidden function of long non-coding RNA by identifying its major partner-protein," *Cell & Bioscience*, vol. 5, no. 1, p. 59, 2015.

[26] K. C. Wang and H. Y. Chang, "Molecular mechanisms of long noncoding RNAs," *Molecular Cell*, vol. 43, no. 6, pp. 904–914, 2011.

[27] J. Liu, C. Jung, J. Xu et al., "Genome-wide analysis uncovers regulation of long intergenic noncoding RNAs in *Arabidopsis*," *Plant Cell*, vol. 24, no. 11, pp. 4333–4345, 2012.

[28] W. Zhang, Z. Han, Q. Guo et al., "Identification of maize long non-coding RNAs responsive to drought stress," *PLoS One*, vol. 9, no. 6, article e98958, 2014.

[29] M. Xin, Y. Wang, Y. Yao et al., "Identification and characterization of wheat long non-protein coding RNAs responsive to powdery mildew infection and heat stress by using microarray analysis and SBS sequencing," *BMC Plant Biology*, vol. 11, no. 1, article 61, 2011.

[30] Y. C. Zhang, J. Y. Liao, Z. Y. Li et al., "Genome-wide screening and functional analysis identify a large number of long noncoding RNAs involved in the sexual reproduction of rice," *Genome Biology*, vol. 15, no. 12, article 512, 2014.

[31] J. Chen, M. Quan, and D. Zhang, "Genome-wide identification of novel long non-coding RNAs in *Populus tomentosa* tension

wood, opposite wood and normal wood xylem by RNA-seq," *Planta*, vol. 241, no. 1, pp. 125–143, 2015.

[32] L. Ou, Z. Liu, Z. Zhang et al., "Noncoding and coding transcriptome analysis reveals the regulation roles of long non-coding RNAs in fruit development of hot pepper (*Capsicum annuum* L.)," *Plant Growth Regulation*, vol. 83, no. 1, pp. 141–156, 2017.

[33] Y. Zhu, L. Chen, C. Zhang, P. Hao, X. Jing, and X. Li, "Global transcriptome analysis reveals extensive gene remodeling, alternative splicing and differential transcription profiles in non-seed vascular plant *Selaginella moellendorffii*," *BMC Genomics*, vol. 18, article 1042, Supplement 1, 2017.

[34] R. K. Joshi, S. Megha, U. Basu, M. H. Rahman, and N. N. Kav, "Genome wide identification and functional prediction of long non-coding RNAs responsive to *Sclerotinia sclerotiorum* infection in *Brassica napus*," *PLoS One*, vol. 11, no. 7, article e0158784, 2016.

[35] S. Swiezewski, F. Liu, A. Magusin, and C. Dean, "Cold-induced silencing by long antisense transcripts of an *Arabidopsis* polycomb target," *Nature*, vol. 462, no. 7274, pp. 799–802, 2009.

[36] J. Yuan, Z. Ye, J. Dong et al., "Systematic characterization of novel lncRNAs responding to phosphate starvation in *Arabidopsis thaliana*," *BMC Genomics*, vol. 17, no. 1, article 655, 2016.

[37] S. H. Li, R. Dudler, R. Ji, M. L. Yong, Z. Y. Wang, and D. W. Hu, "Long non-coding RNAs in wheat are related to its susceptibility to powdery mildew," *Biologia Plantarum*, vol. 58, no. 2, pp. 296–304, 2014.

[38] J. Wang, W. Yu, Y. Yang et al., "Genome-wide analysis of tomato long non-coding RNAs and identification as endogenous target mimic for microRNA in response to TYLCV infection," *Scientific Reports*, vol. 5, no. 1, article 16946, 2015.

[39] Y. Zhang, X. Cao, X. Zhai, and G. Fan, "Study on DNA extraction of AFLP reaction system for Paulownia plants," *Journal of Henan Agricultural University*, vol. 43, no. 6, pp. 610–614, 2009.

[40] G. Fan, S. Zhang, X. Zhai, F. Liu, and Z. Dong, "Effects of antibiotics on the Paulownia witches' broom phytoplasmas and pathogenic protein related to witches' broom symptom," *Scientia Silvae Sinicae*, vol. 43, no. 3, pp. 138–142, 2007.

[41] C. Trapnell, A. Roberts, L. Goff et al., "Differential gene and transcript expression analysis of RNA-seq experiments with TopHat and Cufflinks," *Nature Protocols*, vol. 7, no. 3, pp. 562–578, 2012.

[42] J. Tian, Y. Song, Q. Du et al., "Population genomic analysis of gibberellin-responsive long non-coding RNAs in *Populus*," *Journal of Experimental Botany*, vol. 67, no. 8, pp. 2467–2482, 2016.

[43] H. Wu, Z. Wang, M. Wang, and X. J. Wang, "Widespread long noncoding RNAs as endogenous target mimics for micro-RNAs in plants," *Plant Physiology*, vol. 161, no. 4, pp. 1875–1884, 2013.

[44] Y. Benjamini and D. Yekutieli, "The control of the false discovery rate in multiple testing under dependency," *Annals of Statistics*, vol. 29, no. 4, pp. 1165–1188, 2001.

[45] T. D. Schmittgen and K. J. Livak, "Analyzing real-time PCR data by the comparative C_T method," *Nature Protocols*, vol. 3, no. 6, pp. 1101–1108, 2008.

[46] J. L. Rinn and H. Y. Chang, "Genome regulation by long non-coding RNAs," *Annual Review of Biochemistry*, vol. 81, no. 1, pp. 145–166, 2012.

[47] C. Fan, Z. Hao, J. Yan, and G. Li, "Genome-wide identification and functional analysis of lincRNAs acting as miRNA targets or decoys in maize," *BMC Genomics*, vol. 16, no. 1, article 793, 2015.

[48] E. Kiyota, R. Okada, N. Kondo, A. Hiraguri, H. Moriyama, and T. Fukuhara, "An Arabidopsis RNase III-like protein, AtRTL2, cleaves double-stranded RNA in vitro," *Journal of Plant Research*, vol. 124, no. 3, pp. 405–414, 2011.

[49] C. Cazenave, P. Frank, and W. Büsen, "Characterization of ribonuclease H activities present in two cell-free protein synthesizing systems, the wheat germ extract and the rabbit reticulocyte lysate," *Biochimie*, vol. 75, no. 1-2, pp. 113–122, 1993.

[50] P. Valencia-Morales Mdel, J. A. Camas-Reyes, J. L. Cabrera-Ponce, and R. Alvarez-Venegas, "The *Arabidopsis thaliana* SET-domain-containing protein ASHH1/SDG26 interacts with itself and with distinct histone lysine methyltransferases," *Journal of Plant Research*, vol. 125, no. 5, pp. 679–692, 2012.

[51] T. L. Bullock, K. Breddam, and J. S. Remington, "Peptide aldehyde complexes with wheat serine carboxypeptidase II: implications for the catalytic mechanism and substrate specificity," *Journal of Molecular Biology*, vol. 255, no. 5, pp. 714–725, 1996.

[52] N. F. Silva and D. R. Goring, "The proline-rich, extensin-like receptor kinase-1 (PERK1) gene is rapidly induced by wounding," *Plant Molecular Biology*, vol. 50, no. 4-5, pp. 667–685, 2002.

[53] Y. E. Chen, W. J. Liu, Y. Q. Su et al., "Different response of photosystem II to short and long-term drought stress in *Arabidopsis thaliana*," *Physiologia Plantarum*, vol. 158, no. 2, pp. 225–235, 2016.

[54] Y. Allahverdiyeva, M. Suorsa, F. Rossi et al., "Arabidopsis plants lacking PsbQ and PsbR subunits of the oxygen-evolving complex show altered PSII super-complex organization and short-term adaptive mechanisms," *The Plant Journal*, vol. 75, no. 4, pp. 671–684, 2013.

[55] M. Klein, E. Martinoia, G. Hoffmann-Thoma, and G. Weissenbock, "A membrane-potential dependent ABC-like transporter mediates the vacuolar uptake of rye flavone glucuronides: regulation of glucuronide uptake by glutathione and its conjugates," *The Plant Journal*, vol. 21, no. 3, pp. 289–304, 2000.

[56] Y. Boursiac, S. M. Lee, S. Romanowsky et al., "Disruption of the vacuolar calcium-ATPases in Arabidopsis results in the activation of a salicylic acid-dependent programmed cell death pathway," *Plant Physiology*, vol. 154, no. 3, pp. 1158–1171, 2010.

[57] B. S. Rol and F. Meins, "Physiological compensation in antisense transformants: specific induction of an "ersatz" glucan endo-1,3-β-glucosidase in plants infected with necrotizing viruses," *Proceedings of the National Academy of Sciences of the United States of America*, vol. 90, no. 19, pp. 8792–8796, 1993.

[58] Y. R. Lou, M. Bor, J. Yan, A. S. Preuss, and G. Jander, "Arabidopsis NATA1 acetylates putrescine and decreases defense-related hydrogen peroxide accumulation," *Plant Physiology*, vol. 171, no. 2, pp. 1443–1455, 2016.

[59] L. D. Maldonado-Bonilla, L. Eschen-Lippold, S. Gago-Zachert et al., "The Arabidopsis tandem zinc finger 9 protein binds RNA and mediates pathogen-associated molecular pattern-triggered immune responses," *Plant & Cell Physiology*, vol. 55, no. 2, pp. 412–425, 2014.

[60] Y. C. Kim, S. Y. Kim, D. Choi, C. M. Ryu, and J. M. Park, "Molecular characterization of a pepper C2 domain-containing SRC2 protein implicated in resistance against host and non-host pathogens and abiotic stresses," *Planta*, vol. 227, no. 5, pp. 1169–1179, 2008.

[61] N. Nishikubo, T. Awano, A. Banasiak et al., "Xyloglucan *endo*-transglycosylase (XET) functions in gelatinous layers of tension wood fibers in poplar–a glimpse into the mechanism of the balancing act of trees," *Plant & Cell Physiology*, vol. 48, no. 6, pp. 843–855, 2007.

[62] S. Saito, N. Hirai, C. Matsumoto et al., "Arabidopsis *CYP707A*s encode (+)-abscisic acid $8'$-hydroxylase, a key enzyme in the oxidative catabolism of abscisic acid," *Plant Physiology*, vol. 134, no. 4, pp. 1439–1449, 2004.

[63] C. Martel, J. Vrebalov, P. Tafelmeyer, and J. J. Giovannoni, "The tomato MADS-box transcription factor ripening inhibitor interacts with promoters involved in numerous ripening processes in a colorless ripening-dependent manner," *Plant Physiology*, vol. 157, no. 3, pp. 1568–1579, 2011.

[64] L. Ramírez, J. Oguiza, G. Pérez et al., "Genomics and transcriptomics characterization of genes expressed during postharvest at 4°C by the edible basidiomycete *Pleurotus ostreatus*," *International Microbiology*, vol. 14, no. 2, pp. 111–120, 2011.

[65] C. Audran, S. Liotenberg, M. Gonneau et al., "Localisation and expression of zeaxanthin epoxidase mRNA in Arabidopsis in response to drought stress and during seed development," *Functional Plant Biology*, vol. 28, no. 12, pp. 1161–1173, 2001.

[66] A. S. Reddy, Y. Marquez, M. Kalyna, and A. Barta, "Complexity of the alternative splicing landscape in plants," *The Plant Cell*, vol. 25, no. 10, pp. 3657–3683, 2013.

[67] J. Li, "Analysis of alternative splicing in whole transcriptome and its regulator SR protein under drought stress in seed of Zea mays," *Dissertation*, Zhengzhou University, Zhengzhou, China, 2013.

[68] J. M. Franco-Zorrilla, A. Valli, M. Todesco et al., "Target mimicry provides a new mechanism for regulation of microRNA activity," *Nature Genetics*, vol. 39, no. 8, pp. 1033–1037, 2007.

[69] K. Xie, C. Wu, and L. Xiong, "Genomic organization, differential expression, and interaction of SQUAMOSA promoter-binding-like transcription factors and microRNA156 in rice," *Plant Physiology*, vol. 142, no. 1, pp. 280–293, 2006.

[70] C. Fu, R. Sunkar, C. Zhou et al., "Overexpression of miR156 in switchgrass (*Panicum virgatum* L.) results in various morphological alterations and leads to improved biomass production," *Plant Biotechnology Journal*, vol. 10, no. 4, pp. 443–452, 2012.

[71] M. W. Jones-Rhoades and D. P. Bartel, "Computational identification of plant microRNAs and their targets, including a stress-induced miRNA," *Molecular Cell*, vol. 14, no. 6, pp. 787–799, 2004.

[72] E. Kaja, M. W. Szcześniak, P. J. Jensen, M. J. Axtell, T. McNellis, and I. Makałowska, "Identification of apple miRNAs and their potential role in fire blight resistance," *Tree Genetics & Genomes*, vol. 11, no. 1, article 812, 2015.

[73] S. Lu, Y. H. Sun, and V. L. Chiang, "Stress-responsive microRNAs in *Populus*," *The Plant Journal*, vol. 55, no. 1, article 131, 151 pages, 2008.

[74] C. Schommer, J. F. Palatnik, P. Aggarwal et al., "Control of jasmonate biosynthesis and senescence by miR319 targets," *PLoS Biology*, vol. 6, no. 9, article e230, 2008.

[75] N. M. Christensen, M. Nicolaisen, M. Hansen, and A. Schulz, "Distribution of phytoplasmas in infected plants as revealed by real-time PCR and bioimaging," *Molecular Plant-Microbe Interactions*, vol. 17, no. 11, pp. 1175–1184, 2004.

[76] A. Wehner, T. Grasses, and P. Jahns, "De-epoxidation of violaxanthin in the minor antenna proteins of photosystem II, LHCB4, LHCB5, and LHCB6," *Journal of Biological Chemistry*, vol. 281, no. 31, pp. 21924–21933, 2006.

[77] D. Rusjan, H. Halbwirth, K. Stich, M. Mikulič-Petkovšek, and R. Veberič, "Biochemical response of grapevine variety 'Chardonnay' (*Vitis vinifera* L.) to infection with grapevine yellows (Bois noir)," *European Journal of Plant Pathology*, vol. 134, no. 2, pp. 231–237, 2012.

[78] Y. Allahverdiyeva, F. Mamedov, M. Suorsa, S. Styring, I. Vass, and E. M. Aro, "Insights into the function of PsbR protein in *Arabidopsis thaliana*," *Biochimica et Biophysica Acta (BBA) - Bioenergetics*, vol. 1767, no. 6, article 677, 685 pages, 2007.

[79] K. Ido, K. Ifuku, Y. Yamamoto et al., "Knockdown of the PsbP protein does not prevent assembly of the dimeric PSII core complex but impairs accumulation of photosystem II supercomplexes in tobacco," *Biochimica et Biophysica Acta (BBA) - Bioenergetics*, vol. 1787, no. 7, pp. 873–881, 2009.

[80] X. Yi, S. R. Hargett, H. Liu, L. K. Frankel, and T. M. Bricker, "The PsbP protein is required for photosystem II complex assembly/stability and photoautotrophy in *Arabidopsis thaliana*," *Journal of Biological Chemistry*, vol. 282, no. 34, pp. 24833–24841, 2007.

[81] M. Bertamini, N. Nedunchezhian, F. Tomasi, and M. S. Grando, "Phytoplasma [stolbur-subgroup (bois noir-BN)] infection inhibits photosynthetic pigments, ribulose-1,5-bisphosphate carboxylase and photosynthetic activities in field grown grapevine (*Vitis vinifera* L. cv. Chardonnay) leaves," *Physiological and Molecular Plant Pathology*, vol. 61, no. 6, pp. 357–366, 2002.

[82] J. D. Jones and J. L. Dangl, "The plant immune system," *Nature*, vol. 444, no. 7117, pp. 323–329, 2006.

[83] V. Hegenauer, U. Fürst, B. Kaiser et al., "Detection of the plant parasite *Cuscuta reflexa* by a tomato cell surface receptor," *Science*, vol. 353, no. 6298, pp. 478–481, 2016.

17

Calibrating Transcriptional Activity Using Constitutive Synthetic Promoters in Mutants for Global Regulators in *Escherichia coli*

Ananda Sanches-Medeiros, Lummy Maria Oliveira Monteiro, and Rafael Silva-Rocha

Systems and Synthetic Biology Lab, FMRP - University of São Paulo, Ribeirão Preto, SP, Brazil

Correspondence should be addressed to Rafael Silva-Rocha; silvarochar@usp.br

Academic Editor: João Paulo Gomes

The engineering of synthetic circuits in cells relies on the use of well-characterized biological parts that would perform predicted functions under the situation considered, and many efforts have been taken to set biological standards that could define the basic features of these parts. However, since most synthetic biology projects usually require a particular cellular chassis and set of growth conditions, defining standards in the field is not a simple task as gene expression measurements could be affected severely by genetic background and culture conditions. In this study, we addressed promoter parameterization in bacteria in different genetic backgrounds and growth conditions. We found that a small set of constitutive promoters of different strengths controlling a short-lived GFP reporter placed in a low-copy number plasmid produces remarkably reproducible results that allow for the calibration of promoter activity over different genetic backgrounds and physiological conditions, thus providing a simple way to set standards of promoter activity in bacteria. Based on these results, we proposed the utilization of synthetic constitutive promoters as tools for calibration for the standardization of biological parts, in a way similar to the use of DNA and protein ladders in molecular biology as references for comparison with samples of interest.

1. Introduction

Understanding the logic underlying the genetics of a microorganism based on the dynamics of its promoters and transcription factors is essential for manipulation of other living systems. A way to study this logic is introducing synthetic circuits provided with a reporter gene into living cells and analyzing the results of the expression [1, 2]. However, the success of the implementation of complex circuits in living cells relies strongly on the correct production of the molecular components of the cells and is not limited to the influences of the promoters and transcription factors on gene expression. Several factors are responsible for controlling gene expression, including the rates of mRNA and protein production and their rates of degradation. However, synthesis of mRNA depends strongly on promoter strength, which determines how frequently the RNA polymerase (RNAP) is recruited to the promoter to initiate transcription [3]. On the other hand, the rate of protein production depends strongly on the strength of the ribosome binding site (RBS) in recruiting ribosomes for the translation of the target protein [4]. Additionally, the dilution or degradation of mRNA and proteins depends on the physiological state of the cell just as how their synthesis also relies on cell physiology with respect to the availability of nucleotides, amino acids, RNAP, and ribosomes [5]. In this way, changes in cell physiology and growth conditions can cause variability in gene expression in a manner that is independent of promoter regulation [5].

On account of these possible variations between the cells, several attempts have been made to establish biological standards for promoter activity, and the use of internal promoters as references has been proposed some years ago [6, 7]. More recently, the use of calibrated internal promoters has been proposed as an alternative for defining relative promoter activities during experimental measurements of transcription levels. In this method, an endogenous (or reference) promoter is placed in the same plasmid as the target promoter, each of them controlling the expression of a different fluorescent protein, and the intrinsic promoter activity is calculated as a ratio of the two outputs [8]. However, the expression of additional genes in the host bacterium can increase genetic load and influence gene expression as well. In this way, inserting a calibrated internal promoter would disturb cell functions [9]. Additionally, most methods have focused on the analysis of maximal promoter activity at fixed conditions or on linear expression range of promoter activity, limiting the utilization of standards on condition where cells are subjected to changing physiological regimens [8]. These requirements make the use of calibration methods for the analysis of regulated promoters extremely difficult.

In this study, we seek to analyze intrinsic promoter activity using a single reporter gene in different strains *of Escherichia coli* by using a simple and straightforward protocol. For the determination of intrinsic promoter activity, we used a low-copy number plasmid based on the p15a origin of replication (ori) and a short-lived GFP with LVA tag [10]. We analyzed four constitutive promoters available in the Registry of Standard Biological Parts and a wild-type *Plac* promoter as regulated system. As hosts, we used two strains of *E. coli* with mutant global regulatory proteins, *ihf* and *fis*, which are responsible for regulating the expression of hundreds of genes in this bacterium [11], obtained from the widely used Keio collection of *E. coli* mutants [12]. Additionally, glucose was used as the external source of variation, since all strains exhibited improved growth rates in its presence. Under the conditions of the analysis, we observed that the system we had used exhibited invariant promoter activities that were independent of the strains and growth conditions used, indicating it was able to demonstrate the intrinsic properties of the promoters analyzed. In addition, to prove that our calibrator works, we tested the natural promoter of *Pseudomonas putida Pm* promoter with different concentrations of 3-methylbenzoate (3MBz) [13] and calibrate it with our four constituent promoters in liquid and solid medium. In this way, it was possible to verify that the calibrator works and presents a potential application in synthetic biology. In this regard, we propose a simple, plasmid-based and single reporter method for promoter calibration that is compatible for use with regulated promoters and changes in growth conditions, which could be fundamental to the characterization of biological parts in synthetic biology.

2. Material and Methods

2.1. Bacterial Strains, Plasmids, and Growth Conditions. The bacterial strains, plasmids, and primers used in this study are listed in Table 1. *E. coli* DH5α was used for cloning the pMR1-*Pjx* (where *x* stands for 100, 106, 114, and 113) and pMR1-*Plac* vectors [14] by transformation using the heat-shock method, and *E. coli* DH10B was used for cloning pGLR2-*Pjx* vectors by electroporation and for GFP/Lux expression analysis [15]. For the calibration of promoter activity, *E. coli* BW25113 was used as wild-type strain, and *ihf* (Δ*ihf*) or *fis* (Δ*fis*) mutants (from the Keio collection) were used as mutant hosts with reduced growth rate. Plasmids pMR1 and pGLR2 were used as reporter systems for promoter analysis. Plasmid pMR1 has a low-copy p15a ori, a chloramphenicol-resistance marker, and two genes encoding fluorescent proteins oriented in opposite directions (*mCherry* and *gfplva*). Plasmid pGLR2 has a low-copy RK2 origin of replication, a kanamycin resistance marker, and two reporter genes oriented in the same direction, namely, the GFP gene followed by *luxCDABE.* Although the vector has GFP, in this work when pGLR2 was used, only the Lux was measured. The *E. coli* strains were grown at 37°C with aeration at an agitation rate of 180 rpm in LB medium (for overnight growth) or M9 minimal medium (containing 6.4 g/L $Na_2HPO_4 \bullet 7H_2O$, 1.5 g/L KH_2PO_4, 0.25 g/L NaCl, and 0.5 g/L NH_4Cl) supplemented with 2 mM $MgSO_4$, 0.1 mM $CaCl_2$, 0.1 mM casamino acids, and 1% glycerol as the sole carbon source (for growth during the analysis). When required, chloramphenicol (34 μg/mL), kanamycin (50 μg/mL), or glucose (0.4%) was added to the medium. In the minimal medium, the antibiotics were added at half of the previously mentioned concentrations.

2.2. Plasmid and Strain Construction. For the experiments, oligonucleotides were synthesized (*Exxtend*, Campinas, Brazil) based on the synthetic constitutive promoters from the iGEM BBa_J23104 set of promoters (http://parts.igem.org/Part:BBa_J23104), with an annealing site on pMR1 and restriction sites for *EcoRI* and *BamHI.* The promoters J23100, J23106, J23114, and J23113 (referred here as *Pj100*, *Pj106*, *Pj114*, and *Pj113*, resp.) were used (see Table 1). Once these fragments were amplified by PCR, they were digested by *EcoRI* and *BamHI* and inserted into the multiple cloning sites (MCS) of pMR1 and pGLR2, and thus, generating pMR1-*Pjx* and pGR2-*Pjx*. These plasmids were inserted into DH5α and DH10B strains, respectively, cloned, and sequenced. The plasmids, pMR1-*Pjx* were inserted into *E. coli* BW25113 and into *ihf* and *fis* mutants obtained from the Keio collection.

The *xylS*, *PxylS*, and *Pm* promoters were PCR amplified with Phusion High-Fidelity DNA polymerase (Thermo Fisher Scientific) using the primer pairs 5_xylS_EcoRI (5′-<u>GAA TTC</u> TCA AGC CAC TTC CTT TTT GCA TTG-3′) and 3_Pm_BamHI (5′-<u>GGA TCC</u> ATT ATT GTT TCT GTT GCA TAA AGC C-3′) and pSEVA438 vector (pBBR1 replication origin, Sm/Sp; Silva-Rocha and de Lorenzo [1]) as template. These primers introduced EcoRI and BamHI restriction sites (underlined) at the 5′ and 3′ ends, respectively. The PCR products were gel purified, digested with EcoRI/BamHI, and ligated to the pMR1 vector previously cut with the same restriction enzymes. The resulting plasmids were sequenced to check integrity and inserted to *E. coli* strains (*E. coli* BW25113and into *ihf* and *fis* mutants). The resulting plasmid was named pMR1-xylS-*Pm.*

2.3. GFP Fluorescence and Bioluminescence Assays and Data Processing. To analyze promoter activity, single colonies of recombinant strains containing pMR1-*Pjx* and *E. coli* DH10B containing pGLR2-*Pjx* were grown overnight in LB medium that was supplemented with chloramphenicol (34 μg/mL) or kanamycin (50 μg/mL) for plasmid selection at 37°C with aeration and agitated at 180 rpm. The strains

Table 1: Strains, plasmids and primers used in this study.

Strains, plasmids, and primers	Description	Reference
Strains		
E. coli DH5α	F^- *endA1 glnV44 thi-1 recA1 relA1 gyrA96 deoR nupG purB20* φ80d*lacZ*ΔM15 Δ(*lacZYA-argF*)U169, hsdR17($r_K^- m_K^+$), λ^-.	[12]
E. coli DH10B	*mcrA* Δ*mrr-hsdRMS-mcrBC)* φ *80lacZ*Δ*M15* Δ*lacX74 recA1 araD139* Δ (*ara-leu)7697 galU galK rpsL endA1 nupG* Δ*dcm.*	[12]
E. coli BW25113	*lacI*$^+$*rrnB*$_{T14}$ Δ*lacZ*$_{WJ16}$ *hsdR*514 Δ*araBAD*$_{AH33}$ Δ*rhaBAD*$_{LD78}$ *rph-1* Δ*(araB–D)567* Δ*(rhaD–B)568* Δ*lacZ4787(::rrnB-3) hsdR514 rph-1.*	[12]
E. coli JW1702	*E. coli BW25113* Δ*ihf* mutant	[12]
E. coli JW3229	*E. coli BW25113* Δ*fis* mutant	[12]
Plasmids		
pMR1	Cm^R, *ori p15a*. GFPlva promoter probe vector	[14]
pMR1-*Pj113*	pMR1 with *Pj113* cloned as EcoRI/BamHI fragment	This work
pMR1-*Pj114*	pMR1 with *Pj114* cloned as EcoRI/BamHI fragment	This work
pMR1-*Pj106*	pMR1 with *Pj106* cloned as EcoRI/BamHI fragment	This work
pMR1-*Pj100*	pMR1 with *Pj100* cloned as EcoRI/BamHI fragment	This work
pMR1-*Plac*	pMR1 with *Plac* promoter cloned as EcoRI/BamHI fragment	[28]
pGLR2	Km^R, *oriT, ori* RK2. SEVA-based vector with dual GFP-*lux* reporter	[15]
pGLR2-*Pj113*	pGLR2 with *Pj113* cloned as EcoRI/BamHI fragment	This work
pGLR2-*Pj114*	pGLR2 with *Pj114* cloned as EcoRI/BamHI fragment	This work
pGLR2-*Pj106*	pGLR2 with *Pj106* cloned as EcoRI/BamHI fragment	This work
pGLR2-*Pj100*	pGLR2 with *Pj100* cloned as EcoRI/BamHI fragment	This work
pMR1-*xylS-Pm*	pMR1 with *PxylS, xylS* and *Pm* cloned as EcoRI/BamHI fragment	This work
Primers		
Pj100-FW	GAATTCTTGACGGCTAGCTCAGTCCTAGG	This work
Pj100-RV	TACAGTGCTAGCAAGTGGATCCTTGCGATC	This work
Pj106-FW	GAATTCTTTACGGCTAGCTCAGTCCTAGGTA	This work
Pj106-RV	TAGTGCTAGCAAGTGGATCCTTGCGATC	This work
Pj114-FW	GAATTCTTTATGGCTAGCTCAGTCCTAGGT	This work
Pj114-RV	ACAATGCTAGCAAGTGGATCCTTGCGATC	This work
Pj113-FW	GAATTCCTGATGGCTAGCTCAGTCCTAGGG	This work
Pj113-RV	ATTATGCTAGCAAGTGGATCCTTGCGATC	This work
5_xylS_EcoRI	GAATTCTCAAGCCACTTCCTTTTTGCATTG	This work
3_Pm_BamHI	GGATCCATTATTGTTTCTGTTGCATAAAGCC	This work

grown overnight were washed with $MgSO_4$ (10 mM) buffer, resuspended in the same buffer, and diluted to a ratio of 1 : 20 with M9 minimal medium (containing 6.4 g/L $Na_2HPO_4 \bullet 7H_2O$, 1.5 g/L KH_2PO_4, 0.25 g/L NaCl, and 0.5 g/L NH_4Cl) supplemented with 2 mM $MgSO_4$, 0.1 mM $CaCl_2$, 0.1 mM casamino acids, chloramphenicol (17 μg/mL), and 1% glycerol as the sole carbon source. When required, glucose (0.4%) was supplemented to the medium. In total, 200 μL of the culture was placed in a 96-well plate and analyzed using a *Victor X3* plate reader (*PerkinElmer*) over several hours at 37°C. At 30-minute time intervals, the optical density at 600 nm (OD_{600nm}) and the fluorescence (excitation 485 nm and emission 535 nm) were measured for the strains containing pMR1-*Pjx*; the optical density at 600 nm (OD_{600nm}), the fluorescence (excitation 485 nm and emission 535 nm), and the bioluminescence were measured for DH10B containing pGLR2-*Pjx*. Promoter activities were expressed as fluorescence or bioluminescence normalized by the OD_{600nm} upon background normalization (fluorescence/OD_{600nm}). As a positive control for pMR1-*Pjx* analysis, wild type *lac* promoter (*Plac*), which is regulated by CRP, was used. Data analysis and representation was performed using *Microsoft Excel (2016)* and ad hoc *R* script. To prove that the pMR1-*Pjx* system works as a gene expression standard, the natural promoter of *P. putida Pm*, which is regulated by *xylS* when this regulator is induced by 3MBz, was used [13]. The pMR1-xylS-*Pm* construct contains the *xylS* promoter (*PxylS*), which in the presence of 3MBz leads to the expression of the XylS regulatory protein. The XylS regulator binds to 3MBz and activates the *Pm* promoter by inducing the expression of GFP. Data analysis and representation were performed using ad hoc *R* script.

3. Results and Discussion

3.1. Quantification of Constitutive Promoter Activity Using GFP and luxCDABE Reporters. For the analysis of promoter activities, we used two reporter plasmids based on a short-lived GFP variant placed into a narrow-host-range vector (pMR1 [14]) and a synthetic GFP-*luxCDABE* reporter system placed into a broad-host-range vector [15, 16] to measured Lux, as represented in Figure 1(a). In order to observe the effects of these differences (regarding use of different reporter systems) on the measurement of the activities of the promoters of interest, we analyzed the promoter activities of four BioBrick parts, namely Pj100, Pj106, Pj114, and Pj113, that contain mutations in the sequences at −35 or −10 and exhibit about 100%, 50%, 10%, and 1% activities, respectively (relative to *Pj100* activity). As shown in Figures 1(b) and 1(c), maximal promoter activity of the four synthetic promoters analyzed were virtually identical for both the short-lived GFPlva and the *luxCDABE* reporters, resulting in relative activity values that are closer to the expected value. When we analyzed the promoter activities during the growth period of *E. coli* using the GFPlva and Lux reporters, we observed that the differences were present throughout the growth period of the bacteria, with better differentiation of the intrinsic promoter activities achieved using the luminescent reporter system (Figures 1(d) and 1(e)). Although the luciferase reporter provided better differentiation, GFP reporter allows uses to perform single-cell experiments that cannot be made using light-emitting reporters. Since most synthetic biology works use the GFP reporter, and moreover, GFP provided sufficient resolution to analyze the promoters and also allowed for single-cell analysis (a possible calibrator approach), we focused on the section on the pMR1 reporter system containing the short-lived variant of the reporter protein.

3.2. Robust Calibration of Promoter Activities under Different Pleiotropic and Growth Conditions. During the experiments, wild type and mutant strains of *E. coli* were grown in minimal M9 medium with 1% glycerol and 1% glycerol plus 0.4% glucose in order that the cells were adapted to a richer physiological regimen. Figure 2(a) represents the critical steps in gene expression that were influenced by the bacterial hosts and growth conditions used. In this regard, the rate of mRNA synthesis (βm) was the main parameter controlled by a specific synthetic promoter, while the rate of protein synthesis (βp) was dependent on the strength of the RBS involved (which was the same in all constructs analyzed). Additionally, the rates of mRNA and protein degradation were dependent on the nature of the reporter sequence (i.e., due to differences in the sequence of reporter genes) and the growth rate of the bacteria (since fast-growing bacteria have higher dilution rates of mRNA and protein than slow-growing bacteria). In this method, the use of constitutive promoters with different strengths allowed for variations in βm and facilitated the analysis of its sensitivity to changes in the dilution or degradation rates of mRNA and proteins. As shown in Figures 2(b) and 2(c), both *ihf* and *fis* mutants exhibited reduced growth compared to the wild type for all constructs analyzed. However, in all cases, the addition of 0.4% glucose to the growth medium resulted in a stepwise improvement in the growth of the strains. In other words, in the presence of glucose, bacterial growth is faster, although there is no glucose effect on the final promoter activity. Our calibrator approach is an interesting way to avoid the genetic background differences, indicating that the calibrator can be used in several conditions to standardize promoter studies using different strains under a growth condition variety (glucose or 3MBz—performed below).

In order to observe the effects of the differences in strains and growth medium on the measurements of the activities of the promoters of interest, we analyzed four synthetic promoters that contain sequence differences at −35 or −10 (Figure 3(a)). As shown in Figure 3(b), the regulated *Plac* promoter (a natural promoter used as reference) exhibited strong activity in the three strains analyzed (wild type, Δ*ihf*, and Δ*fis*), and this activity was fully suppressed in the presence of glucose (due to the inactivation of CRP [17]). When we analyzed promoter activity of the four synthetic promoters, we observed that the promoter dynamics and steady state promoter activity were almost invariant in the different mutant strains and under the two growth conditions (Figure 3(c)), indicating that these promoters were not influenced by the drastic physiological variations regarding the different strains of *E. coli* (*E. coli* BW25113, and into *ihf* and *fis* mutants). It is noteworthy that the same was observed for the addition of glucose to the medium that did not compromise the promoter activity. Again showing that the internal calibrator can be used in different situations. When we performed a comparison of the observed promoter activities, *Pj106* exhibited an activity level very close to the expected value (45.7% observed versus 47% expected), whereas *Pj114* and *P113* exhibited promoter activities varying by~ 3% and 2.3%, respectively, from the value of the activity exhibited by *Pj100* (compared to the expected values, 10% and 1%, resp.). These expected values come from the previous analysis made by iGEM BBa_J23104 set of promoters (http://parts.igem.org/Part:BBa_J23104). These differences are due to the differences in the reporter, plasmids, and strains used for promoter characterization. Additionally, the *Plac* promoter exhibited activity of value about 30% under nonrepressive conditions when compared to *Pj100* reference promoter.

These results show that the use of the short-lived GFP reporter in combination with constitutive promoters is a simple way to calibrate promoter activity under user-specific experimental conditions, similar to the way that DNA and protein ladders are used in molecular biology techniques as references for the comparison of specific targets. In conclusion, our data shows how intrinsic promoter activity can be calibrated using single reporter genes and simple data processing without the need for using internal promoter references.

3.3. The Calibrator Can Be Applied to the Induction System xylS-Pm. In order to prove that the set of four promoters proposed in this work acts as an internal calibrator even when applied to an induced expression system, we analyzed

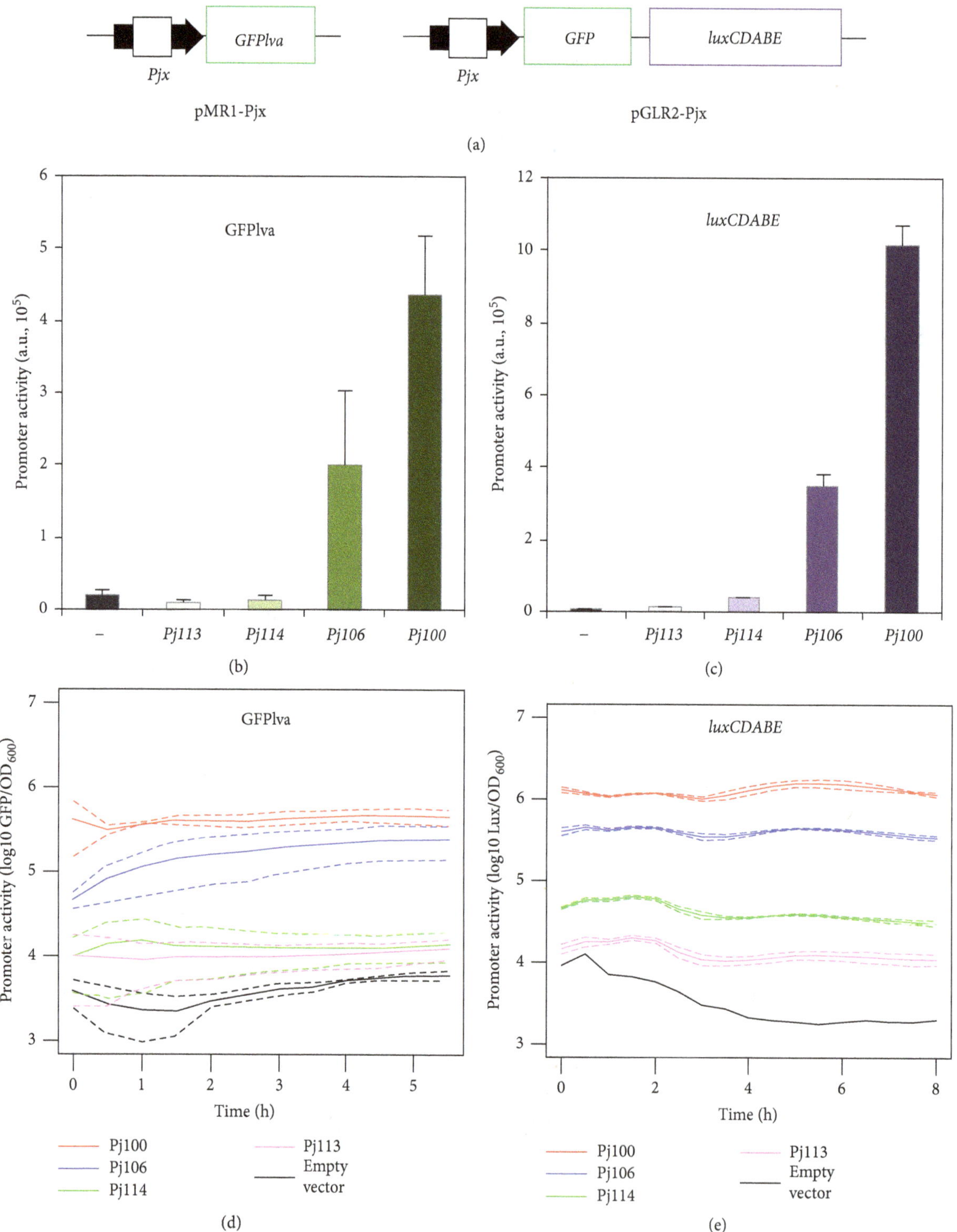

FIGURE 1: Construction and validation of the reporter systems. (a) Synthetic promoters were cloned into the plasmid pMR1, which contains a short-lived GFPlva variant, and pGLR2, a broad host range vector containing a GFP-*luxCDABE* reporter system. (b) Maximal promoter activity of the four promoters in pMR1 vector. (c) Maximal promoter activity analyzed by monitoring lux expression using pGLR2 constructions. (d) GFP expression profile along the growth curve from reporters cloned in pMR1 vector. (e) *lux* expression profile along the growth curve from reporters cloned in pGLR2 vector. The solid lines represent the average values calculated using data from three independent experiments while dashed lines represent standard deviation from the samples.

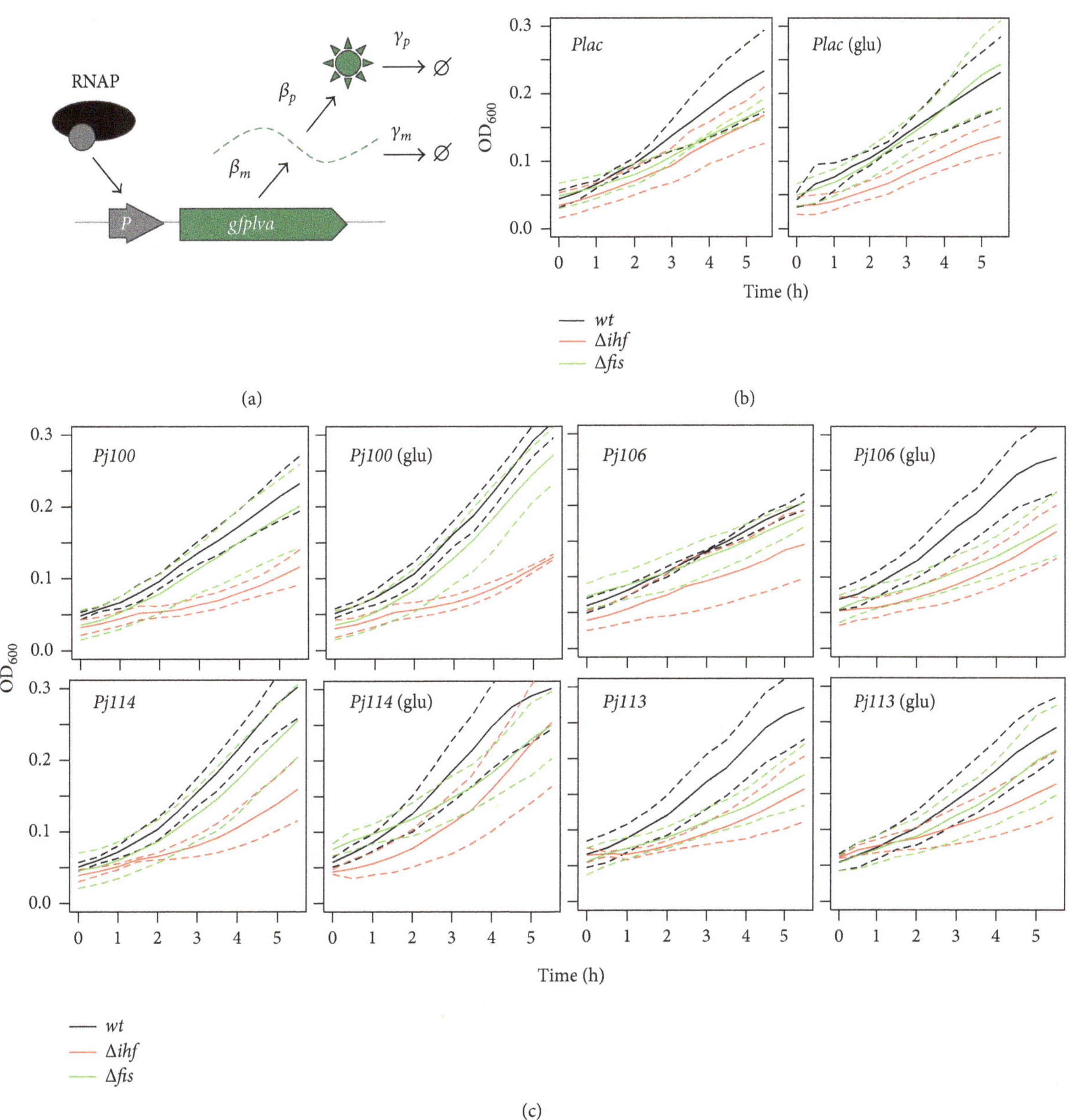

FIGURE 2: Quantification of growth variation in different *E. coli* strains under two physiological regimens. (a) Schematic representation of the main steps for gene expression in bacteria. The strength of the interaction between RNA polymerase (RNAP) and target promoter determines the rate of mRNA synthesis (β_m), while the RBS sequence determines the rate of protein translation (β_p). The rates of mRNA and protein dilution or degradation (γ_m and γ_p, *resp.*) depends on cell growth and physiological regimens of the cells. (b) Growth curve of *E. coli* strains harboring a *Plac::GFPlva* fusion in minimal medium with 1% glycerol (left) or 1%glycerol plus 0.4% glucose (right) as carbon source. (c) Growth curve of *E. coli* strains harboring different promoter fusions (*Pj100*, *Pj106*, *Pj114*, and *Pj113*) in minimal medium with 1% glycerol or 1%glycerol plus 0.4% glucose (labeled as glu) as carbon source. Solid lines represent average values calculated using data from three independent experiments for wild type (black), Δ*ihf* (red), and Δ*fis* (green) strains, while dashed lines represent the upper and lower limits of standard deviations.

the four synthetic promoters and pMR1-xylS-*Pm* system (Figure 4(a)) with increasing concentrations of 3MBz. In order to demonstrate that our calibrator is robust and even works in a blue light transilluminator, we analyzed colonies grown on petri dish contend medium LB plus 3MBz 1000 μM. As shown in Figure 4(b), pMR1-xylS-*Pm* displayed the same promoter activity as pMR1-P106. Next, we tested the system at increasing 3MBz inductor concentrations; they were carried out on the three *E. coli* strains previously used in this work. The data shown in Figure 4(d) are relative to 4.5 hours after the start of the induction. From Figure 4, it is possible to note that the increasing concentration of 3MBz did not promote differences in *Pjx* promoter activity (Figure 4(d)), neither during the 8 hours of the experiment (Figure 4(c)). On Figure 4(c), it is important to note that each color on the graph represents a different *Pjx* synthetic

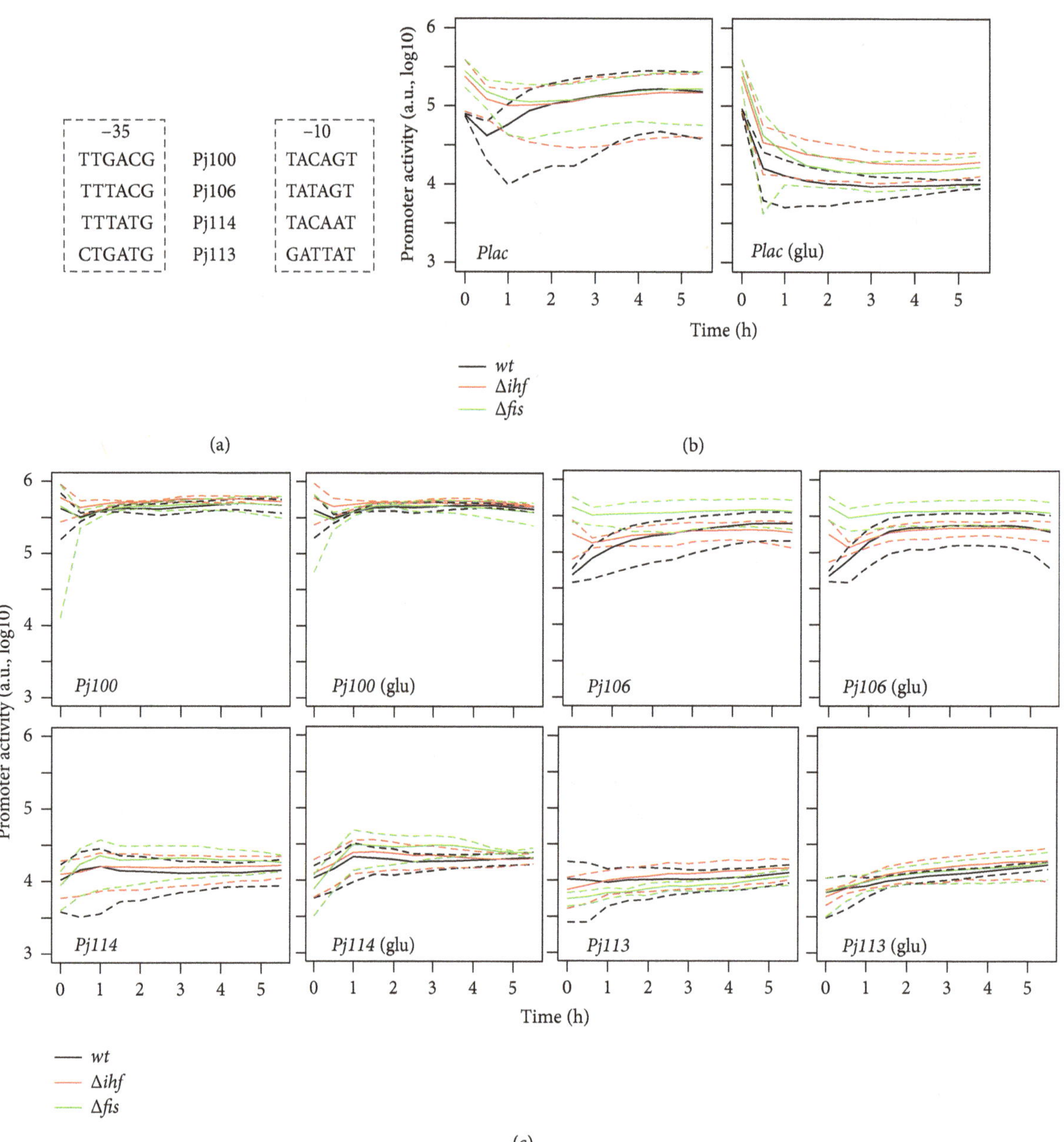

Figure 3: Quantification of promoter activities in different *E. coli* strains. (a) Representation of the sequences at −10 or −35 for the four constitutive promoters analyzed, using bold letters for bases conserved related to *Pj100* reference. (b) Promoter activity of *E. coli* strains harboring a *Plac::GFPlva* fusion in minimal medium with 1% glycerol (left) or 1%glycerol plus 0.4% glucose (right) as carbon source. (c) Promoter activity of *E. coli* strains harboring different promoter fusions (*Pj100*, *Pj106*, *Pj114*, and *Pj113*) in minimal media with 1% glycerol or 1%glycerol plus 0.4% glucose (labeled as glu) as carbon source. Solid lines represent the average values calculated using data from three independent experiments for wild type (black), Δ*ihf* (red), and Δ*fis* (green) strains, while dashed lines represent the upper and lower limits of standard deviations.

promoter and the set of lines belonging to each color group refers to different 3MBz concentrations. In this sense, it is possible to note that there are no differences between the set color lines. This result suggests that the 3MBz addition do no promote differences on promoter activity. On the other hand, for a 3MBz-induced system, the aromatic compound produced a change in promoter activity for a sigmoidal curve, proportional to the 3MBz increase concentration (Figure 4(d)).

A brief and simple conclusion can be made from Figure 4(d), regardless of the host strain used, in the range of 1 to 10 μM (0 to 1 on the x axis) concentration, the *Pm* promoter presents similar promoter activity to *Pj114* promoter. On the other hand, in the range of 10 to 100 μM (1 to 2 on the x-axis) concentration, *Pm* presents intermediate promoter activity to *Pj114* and *Pj106* promoters. Finally, in the range of 100 to 1000 μM (2 to 3.0 on the *x*-axis) concentrations, *Pm* presents promoter activity close

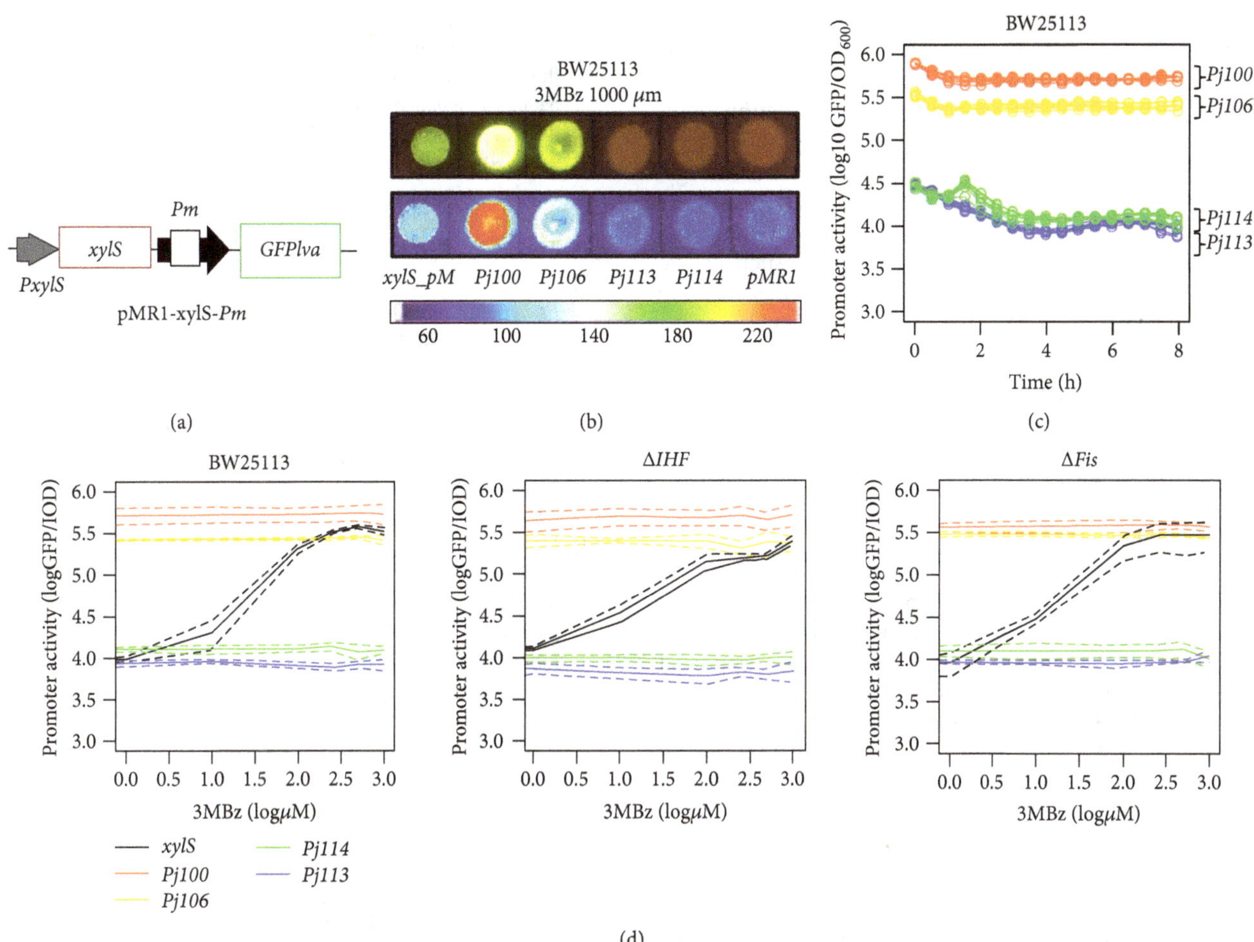

Figure 4: The calibrator can be applied to the induction system xylS-*Pm*. (a) xylS promoters (*PxylS*), xylS protein, and *Pm* promoter were cloned into the plasmid pMR1, which contains a short-lived GFP variant. (b) xylS-*Pm* calibration in LB solid medium with 1000 μM of 3MBz added. This calibration was performed in BW25113 *wt* strains. (c) *Pjx* promoter activity analyzed by 8 hours of experiment by monitoring GFPlva expression using pMR1 constructions. (d) GFP expression profile for 7 different 3MBz concentrations for *Pjx* and xylS-*Pm* in pMR1 vector, 4.5 hours after the induction. Solid lines represent the average values calculated using data from three independent experiments for wild type, Δ*ihf*, and Δ*fis* strains, while dashed lines represent the upper and lower limits of standard deviations.

to *Pj106* promoter activity. Additionally, we can safely confirm that the sigmoidal form for the xylS-*Pm* system is due to the 3MBz addition and not by environmental or host changes, since the calibration system does not change under these conditions.

4. Conclusions

The standardization of biological parts for the construction of complex circuits forms the basis of synthetic biology [18–21]. In this regard, failure in the implementation of constructed synthetic biological circuits may occur when poorly characterized parts are used, and several strategies have been proposed to mitigate this problem [6, 22–24]. In this report, we have highlighted that simple experimental techniques involving the use of a single fluorescent reporter and plasmids are sufficient to provide robust characterization of transcriptional elements without the necessity of using of dual markers and complicated mathematical treatments [8]. Additionally, plasmids provide an easy way of implementing synthetic circuits that accelerates design-build-test cycles in synthetic biology. Once a circuit has been effectively implemented and tested, the introduction of a single copy of the construct by using a chromosome insertion on a same region for all promoters is recommended in order to enhance the performance of the system as well as provide stable strains for final use because an insertion on different regions could modify the GFP expression [25–27]. In this sense, the use of this promoter on different chromosome regions could provide a way to standardize the variation of gene expression caused by variations on chromosome position and structure. At the same time, the use of low-copy number plasmids can provide similar results as can the use of single-copy set-ups under certain circumstances [1, 15]. In general, since each synthetic biology project has its own design and uses specific hosts and experimental conditions, the use of calibrators such as those described in this paper could provide a simple way to standardize the experimental conditions used. This would be similar to the use of molecular-weight size markers in molecular biology techniques as references for the comparison of samples of interest under varying experimental conditions. Although the calibration methods used in this study were implemented in strains of the Keio collection of *E. coli* mutants [12], we expect that the validations presented here will be adopted by other research groups studying synthetic biology as well as molecular microbiology. Additionally, the approaches used in this research study can be easily adopted using alternative plasmid standards for gram negative bacteria other than *E. coli* such as the vectors available at the SEVA database [16], thus creating a significant impact on research in microbiology.

Conflicts of Interest

The authors declare that there is no conflict of interest regarding the publication of this article.

Acknowledgments

This work was supported by the Young Investigator Award of Sao Paulo State Foundation (FAPESP, award number 2012/22921-8). Ananda Sanches-Medeiros was supported by a Scientific Initiation Scholarship (FAPESP, award number 2015/22386-3). Lummy Maria Oliveira Monteiro was supported by a FAPESP PhD fellowship (FAPESP, award number 2016/19179-9). The authors are thankful to the lab members for insightful discussions on this work.

References

[1] R. Silva-Rocha and V. de Lorenzo, "Chromosomal integration of transcriptional fusions," *Methods in Molecular Biology*, vol. 1149, pp. 479–489, 2014.

[2] A. Zaslaver, A. Bren, M. Ronen et al., "A comprehensive library of fluorescent transcriptional reporters for *Escherichia coli*," *Nature Methods*, vol. 3, no. 8, pp. 623–628, 2006.

[3] D. F. Browning and S. J. W. Busby, "Local and global regulation of transcription initiation in bacteria," *Nature Reviews Microbiology*, vol. 14, no. 10, pp. 638–650, 2016.

[4] L. Zelcbuch, N. Antonovsky, A. Bar-Even et al., "Spanning high-dimensional expression space using ribosome-binding site combinatorics," *Nucleic Acids Research*, vol. 41, no. 9, article e98, 2013.

[5] S. Klumpp and T. Hwa, "Bacterial growth: global effects on gene expression, growth feedback and proteome partition," *Current Opinion in Biotechnology*, vol. 28, pp. 96–102, 2014.

[6] J. R. Kelly, A. J. Rubin, J. H. Davis et al., "Measuring the activity of BioBrick promoters using an in vivo reference standard," *Journal of Biological Engineering*, vol. 3, no. 1, p. 4, 2009.

[7] B. Canton, A. Labno, and D. Endy, "Refinement and standardization of synthetic biological parts and devices," *Nature Biotechnology*, vol. 26, no. 7, pp. 787–793, 2008.

[8] T. J. Rudge, J. R. Brown, F. Federici et al., "Characterization of intrinsic properties of promoters," *ACS Synthetic Biology*, vol. 5, no. 1, pp. 89–98, 2016.

[9] M. Carbonell-Ballestero, E. Garcia-Ramallo, R. Montanez, C. Rodriguez-Caso, and J. Macia, "Dealing with the genetic load in bacterial synthetic biology circuits: convergences with the Ohm's law," *Nucleic Acids Research*, vol. 44, no. 1, pp. 496–507, 2016.

[10] J. B. Andersen, C. Sternberg, L. K. Poulsen, S. P. Bjorn, M. Givskov, and S. Molin, "New unstable variants of green fluorescent protein for studies of transient gene expression in bacteria," *Applied and Environmental Microbiology*, vol. 64, no. 6, pp. 2240–2246, 1998.

[11] A. Martinez-Antonio and J. Collado-Vides, "Identifying global regulators in transcriptional regulatory networks in bacteria," *Current Opinion in Microbiology*, vol. 6, no. 5, pp. 482–489, 2003.

[12] T. Baba, T. Ara, M. Hasegawa et al., "Construction of *Escherichia coli* K-12 in-frame, single-gene knockout mutants: the Keio collection," *Molecular Systems Biology*, vol. 2, 2006.

[13] R. Silva-Rocha and V. de Lorenzo, "Broadening the signal specificity of prokaryotic promoters by modifying *cis*-regulatory elements associated with a single transcription factor," *Molecular BioSystems*, vol. 8, no. 7, pp. 1950–1957, 2012.

[14] M. E. Guazzaroni and R. Silva-Rocha, "Expanding the logic of bacterial promoters using engineered overlapping operators for global regulators," *ACS Synthetic Biology*, vol. 3, no. 9, pp. 666–675, 2014.

[15] I. M. Benedetti, V. de Lorenzo, and R. Silva-Rocha, "Quantitative, non-disruptive monitoring of transcription in single cells with a broad-host range GFP-*luxCDABE* dual reporter system," *PLoS One*, vol. 7, no. 12, article e52000, 2012.

[16] R. Silva-Rocha, E. Martínez-García, B. Calles et al., "The Standard European Vector Architecture (SEVA): a coherent platform for the analysis and deployment of complex prokaryotic phenotypes," *Nucleic Acids Research*, vol. 41, no. D1, pp. D666–D675, 2013.

[17] A. Schmitz, "Cyclic AMP receptor protein interacts with lactose operator DNA," *Nucleic Acids Research*, vol. 9, no. 2, pp. 277–292, 1981.

[18] A. S. Khalil and J. J. Collins, "Synthetic biology: applications come of age," *Nature Reviews Genetics*, vol. 11, no. 5, pp. 367–379, 2010.

[19] R. P. Shetty, D. Endy, and T. F. Knight Jr, "Engineering BioBrick vectors from BioBrick parts," *Journal of Biological Engineering*, vol. 2, no. 1, p. 5, 2008.

[20] E. Andrianantoandro, S. Basu, D. K. Karig, and R. Weiss, "Synthetic biology: new engineering rules for an emerging discipline," *Molecular Systems Biology*, vol. 2, 2006.

[21] C. A. Voigt, "Genetic parts to program bacteria," *Current Opinion in Biotechnology*, vol. 17, no. 5, pp. 548–557, 2006.

[22] J. A. N. Brophy and C. A. Voigt, "Principles of genetic circuit design," *Nature Methods*, vol. 11, no. 5, pp. 508–520, 2014.

[23] V. Singh, "Recent advancements in synthetic biology: current status and challenges," *Gene*, vol. 535, no. 1, pp. 1–11, 2014.

[24] A. de Las Heras, C. A. Carreño, E. Martinez-Garcia, and V. de Lorenzo, "Engineering input/output nodes in prokaryotic regulatory circuits," *FEMS Microbiology Reviews*, vol. 34, no. 5, pp. 842–865, 2010.

[25] R. C. Brewster, F. M. Weinert, H. G. Garcia, D. Song, M. Rydenfelt, and R. Phillips, "The transcription factor titration effect dictates level of gene expression," *Cell*, vol. 156, no. 6, pp. 1312–1323, 2014.

[26] J. W. Lee, A. Gyorgy, D. E. Cameron et al., "Creating single-copy genetic circuits," *Molecular Cell*, vol. 63, no. 2, pp. 329–336, 2016.

18

Multitarget Effects of Danqi Pill on Global Gene Expression Changes in Myocardial Ischemia

Qiyan Wang,[1] Hui Meng,[2] Qian Zhang,[1] Tianjiao Shi,[1] Xuefeng Zhang,[1] Mingyan Shao,[1] Linghui Lu,[3] Jing Wang,[4] Wei Wang,[3] Chun Li,[2] and Yong Wang[1]

[1]*School of Life Sciences, Beijing University of Chinese Medicine, Beijing 100029, China*
[2]*Modern Research Center for Traditional Chinese Medicine, Beijing University of Chinese Medicine, Beijing 100029, China*
[3]*School of Chinese Medicine, Beijing University of Chinese Medicine, Beijing 100029, China*
[4]*Staidson (Beijing) Biopharmaceuticals Co., Ltd., Beijing 100176, China*

Correspondence should be addressed to Chun Li; lichun19850204@163.com and Yong Wang; doctor_wangyong@sina.com

Academic Editor: Martine A. Collart

Danqi pill (DQP) is a widely prescribed traditional Chinese medicine (TCM) in the treatment of cardiovascular diseases. The objective of this study is to systematically characterize altered gene expression pattern induced by myocardial ischemia (MI) in a rat model and to investigate the effects of DQP on global gene expression. Global mRNA expression was measured. Differentially expressed genes among the sham group, model group, and DQP group were analyzed. The gene ontology enrichment analysis and pathway analysis of differentially expressed genes were carried out. We quantified 10,813 genes. Compared with the sham group, expressions of 339 genes were upregulated and 177 genes were downregulated in the model group. The upregulated genes were enriched in extracellular matrix organization, response to wounding, and defense response pathways. Downregulated genes were enriched in fatty acid metabolism, pyruvate metabolism, PPAR signaling pathways, and so forth. This indicated that energy metabolic disorders occurred in rats with MI. In the DQP group, expressions of genes in the altered pathways were regulated back towards normal levels. DQP reversed expression of 313 of the 516 differentially expressed genes in the model group. This study provides insight into the multitarget mechanism of TCM in the treatment of complex diseases.

1. Introduction

Acute myocardial ischemia (MI) occurs when blood flow stops to a part of the heart causing damage to the myocardial tissue, and MI is one of the leading causes of death worldwide [1–3]. How to reduce MI-caused mortality is a major challenge to the entire medical community. Conventional managements of MI include intravenous thrombolysis, percutaneous coronary intervention, optimization of oxygenation, and pain control [4]. Danqi pill (DQP), composed of *Salvia miltiorrhiza* Bunge and *Panax notoginseng*, is a widely prescribed traditional Chinese medicine (TCM) in the treatment of a variety of cardiovascular diseases [5, 6]. DQP has been used as an alternative or complementary medicine in the prevention and treatment of MI in China. There are multiple components of DQP that have potential regulative effects on multiple targets in the treatment of cardiovascular diseases [7–10]. However, the overall regulative effect of DQP has not been explored yet. Investigations into the effects of DQP on global gene expressions will further our understanding on the mechanisms of TCM in treating complex diseases.

In recent years, transcriptome sequencing technologies have been developing very fast. Digital Gene Expression (DGE) technique is one of the RNA-sequencing methods for the analysis of differentially expressed genes in different samples [11]. DGE has more accuracy in quantification of gene expression levels compared with microarray technologies. It is able to provide quantitative readout of mRNA expression levels in samples [12]. Illumina's sequencing platform was applied in our present study.

Acute MI model was induced by ligation of left anterior descending (LAD) coronary artery in rats. The rats were divided into three groups: sham-operated group, model group, and DQP-treated group. Gene expressions in the three groups were measured by RNA-sequencing technology, and differentially expressed genes and signaling pathways among different groups were analyzed. This study will expand our knowledge on the multitarget mechanism of DQP in the treatment of ischemic heart disease.

2. Materials and Methods

2.1. Grouping of Animals. Sprague-Dawley (SD) male rats, weighing 220 ± 10 g, were randomly divided into three groups: sham-operated group, acute MI model group, and DQP-treated group. Each group contained 10 rats. The animals were purchased from Beijing Vital River Laboratory Animal Technology Co., Ltd. This study complied with the China Physiological Society's "Guiding Principles in the Care and Use of Animal" and got the approval of Animal Care Committee of Beijing University of Chinese Medicine.

MI model was induced in model and DQP group by direct LAD artery ligation as previously described [13, 14]. Briefly, SD rats were anaesthetized intraperitoneally with pentobarbital sodium (1%, 50 mg/kg). The left anterior descending coronary artery was ligated proximal to its main branching point with a 5-0 polypropylene suture. Sham-operated rats went through identical thoracotomy procedure but their coronary arteries were not ligated. After the operations, all of the rats were fed with a standard diet and were maintained on a 12 h light and dark cycle for 28 days. Animals in the DQP group were treated with concentrated DQP (purchased from Tongren Tang, Beijing, China) dissolved in pure water, with the daily dosage of 1.5 g/kg, for 28 consecutive days beginning from the day after the operation. At the end of the study, all rats were anaesthetized using pentobarbital sodium following an overnight fast. Rats were sacrificed and the left ventricle was carefully dissected to keep only the viable myocardium in the marginal zone of the infarct region in the DQP group and MI model group. The same region in the sham group was also dissected. Dissected heart tissues were frozen and stored in a freezer at −80°C before RNA extraction.

2.2. Echocardiographic Assessment of Heart Function. Echocardiography (Vevo 2100, Visual Sonics, Canada) was applied to assess the cardiac function-related parameters, including left ventricular ejection fraction (LVEF), left ventricular internal diameter at end-diastole (LVID;d) and at end-systole (LVID;s), and left ventricular fractional shortening (LVFS). LVEF and LVFS were calculated automatically by the software.

2.3. RNA Preparation. Heart tissues of the 3 rats in each group were homogenized in liquid nitrogen for RNA sequencing. Total RNA of the heart tissues were extracted using TRIzol Reagent® (Invitrogen, Carlsbad, CA), following the instruction of the manufacturer. Extracted RNA was treated with DNase to remove potential genomic DNA contamination. The quality of RNA was examined by Agilent 2100 Bioanalyzer (Agilent Technologies, Palo Alto, CA, USA).

2.4. cDNA Library Preparation and RNA Sequencing. Sequence tag was prepared using Illumina's Digital Gene Expression Tag Profiling Kit according to the manufacturer's protocol. Briefly, mRNA was purified from total RNA by binding mRNA to a magnetic oligo(dT) bead. Oligo(dT) was then used as a primer to synthesize cDNA. The bead-bound cDNA was subsequently digested with restriction enzyme NlaIII, which recognizes and cuts off the CATG sites. The fragments apart from the 3′ cDNA fragments connected to oligo(dT) beads are washed away and the Illumina adaptor 1 is ligated to the sticky 5′ end of the digested bead-bound cDNA fragments. Mme I, a type of endonuclease with separated recognition sites and digestion sites, was used to cut at 17 bp downstream of the CATG site, producing tags with adaptor 1. After removing 3′ fragments with magnetic beads precipitation, Illumina adaptor 2 is ligated to the 3′ ends of tags. A tag library with different adaptors of both ends was thus formed.

After PCR amplification of 12 cycles, fragments were purified by a 6% Novex TBE PAGE Gel electrophoresis. In the end, the purified cDNA tags were sequenced using Illumina HiSeq™ 2000 at BGI-Shenzhen.

2.5. Analysis of Sequencing Data and Tag Mapping. During the quality control steps, Agilent 2100 Bioanalyzer was used in quantification and qualification of the sample library. Low-quality tags and tags of copy number less than two were filtered to produce clean tags. The clean tags were classified according to their copy number in the library. Clean tags were aligned to the rat reference sequences and annotated. Clean tag numbers corresponding to each gene were counted in each sample.

2.6. Analysis of Differentially Expressed Genes. The number of clean tags in each sample was normalized to transcripts per million (TPM) and TPM values were used to calculate fold change and false discovery rate across different groups. To avoid possible noise signal from high-throughput sequencing, the genes with TPM less than 3 in 2 or more samples were excluded. Analysis of differentially expressed genes among different groups was performed in R (version 3.0.2) with edgeR Bioconductor package [15]. EdgeR uses empirical Bayes estimation and exact tests based on the negative binomial distribution. The resulting P values for all genes were corrected for multiple tests using a FDR adjustment. In this study, the fold change larger than 2 and FDR less than 0.01 were used to define the differentially expressed gene.

2.7. Gene Functional and Enrichment Analysis. The gene ontology (GO) enrichment analysis and Kyoto Encyclopedia of Genes and Genomes (KEGG) pathway analysis of all differentially expressed genes were carried out using the DAVID Functional Annotation Tool [16]. According to the transcription factor database (TFdb) (http://genome.gsc.riken.jp/TFdb/), transcription factors that had altered expressions induced by ischemia were selected. The target genes of these transcription factors were searched through databases, including Transcriptional Regulatory Element Database (http://rulai.cshl.edu/TRED), Human Transcriptional Regulation interaction database (http://www.lbbc.ibb.unesp.br/htri/index.jsp), and Transcription factor checkpoint (http://www.tfcheckpoint.org/index.php/search) [17]. Regulation networks were constructed using Cytoscape software [18].

2.8. Messenger RNA Expression by Quantitative Real-Time PCR. Real-time PCR was applied to validate differentially expressed genes in six samples in each group. First-strand cDNA was synthesized from total RNA with a RevertAid First Strand cDNA Synthesis Kit (Thermo Scientific, USA, lot number: K1622) according to the manufacturer's instruction. Quantitative real-time PCR assays were performed using C1000 Thermal Cycler PCR machine (Bio-Rad, USA). The reaction volume was 20 μl including 1 μl forward and reverse primer pairs, 2 μl cDNA, 10 μl FastStart Universal SYBR Green Master (Roche, Germany, lot number: 04913914001), and 7 μl RNase free water. The PCR procedures were as follows: 15 s at 95°C for denaturation and 1 min at 60°C for annealing and extension. Ct values were obtained after 40 cycles of reactions. Primer sequences of each gene were listed in Table 1. Ct values of targeted mRNA were normalized to the Ct values of GAPDH. Relative expressions of these genes were calculated by the $2^{-\Delta\Delta CT}$ method.

Table 1: Nucleotide sequences of primers used in real-time PCR.

Gene	Primers	Lengths (bp)	Temp (°C)
LPL	CGCTCCATCCATCTCTTC	159	49.8
	GGCTCTGACCTTGTTGAT		49.2
ACSL1	GCAGTTCATCGGCATCTT	109	50.6
	GGTTCCAAGCGTGTCATA		49.8
ACADM	ATTACGGAAGAGTTGGCATA	166	54.15
	GTTCTGTCACGCAGTAGG		55.12
ME1	AAGAACCTAGAAGCCATTGT	104	54.23
	GCAGCCATATCCTTGAGAA		54.62
TFB2M	AGAATGCGGATGGAGAGT	141	55.20
	CTGCTGACCAAGGAACTG		55.27
MMP2	AAGTCTGAAGAGTGTGAAGT	180	53.86
	GTGAAGGAGAAGGCTGATT		54.45
ALOX15	CTCAGGCTTGCTACTTCAT	158	54.57
	CTTCTCCATTGTTGCTTCC		53.97
BAX	GATGATTGCTGATGTGGATAC	86	50.50
	AGTTGAAGTTGCCGTCTG		50.40
GPX1	CAATCAGTTCGGACATCAG	133	53.52
	AGCCTTCTCACCATTCAC		54.14
MMP23	GATGGTCCTACAGGTGAAC	195	54.59
	CTGGTCTTGCTGTGAGTG		55.30
GAPDH	TTCAACGGCACAGTCAAG	116	50.70
	TACTCAGCACCAGCATCA		50.50

3. Results

3.1. Effects of DQP on Cardiac Function. Twenty-eight days after surgery, echocardiography showed that LVEF and LVFS of rats that underwent ligation in model group were downregulated significantly by 63.7% and 72.7%, compared with those of the sham group (Figure 1, $P < 0.01$), indicating that cardiac function of rats in the model group was impaired and a MI model was established. LVID;s and LVID;d increased by 204.9% and 44.3% in the model group compared with those in the sham group. After treatment with DQP, the LVEF and LVFS were upregulated by 56.3% and 72.1%, compared with those in the model group ($P < 0.05$). LVID;s and LVID;d also decreased by 23.3% and 2.4% after treatment with DQP, suggesting that DQP could improve cardiac functions in the MI model (Figure 1).

3.2. High-Throughput Sequencing Data and Differentially Expressed Genes. In total, we obtained four million raw tags and over 3.7 high-quality clean tags from each sample. 51.8%, 47.8%, and 52.5% tags were mapped to annotated rat genomes in the DQP-treated, sham-operated, and model rats, respectively. About 5% of the tags were unknown ones. Altogether, 10,813 genes were detected, and 9537, 8907, and 9344 genes were detected and quantified in the DQP-treated, sham-operated, and model rats, respectively (Supplement Table 1).

Redundancy and heterogeneity are characteristics of mRNA expression. The majority of mRNAs have low expression level, whereas a minority of mRNAs has high abundance of expression. In this study, among the distinct clean tags in the nine libraries, the majority of distinct clean tags (60.21%–63.31%) had 2–5 copies. Only 3.54%–4.47% clean tags had more than 100 copies. The distribution of distinct clean tags showed a similar tendency, indicating that gene expression patterns among the libraries were similar.

Differentially expressed genes were analyzed using the following criteria: fold change > 2 and FDR < 0.01. The analysis showed that compared with the sham-operated group, 339 genes were upregulated and 177 genes were downregulated significantly in the model group, demonstrating that gene expression pattern was altered under ischemic condition. Compared with the model group, 886 genes were upregulated and 1082 genes were downregulated in the DQP-treated group.

Genes were expressed at different abundances, and expressions of some genes were altered remarkably in ischemic heart tissues. The top 20 upregulated and 20 downregulated genes in regard to abundance in the model compared with the sham-operated group and their expressions in DQP group were shown in Figure 2. NPPA, which encodes

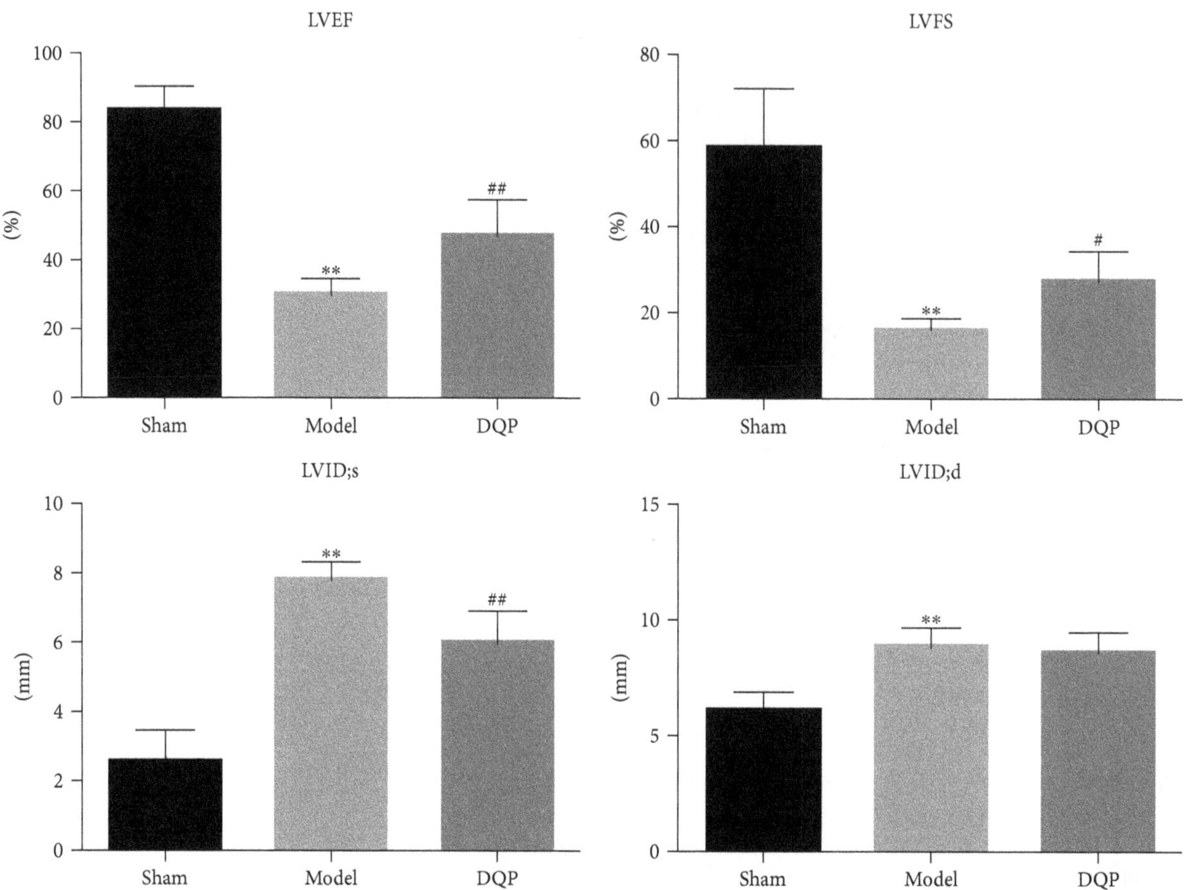

FIGURE 1: Indicators of heart functions in three groups of rats. LVEF, LVFS, LVID;s, and LVID;d in the model group were significantly changed compared with those in the sham group ($N = 10$ in each group, $P < 0.01$). In the DQP-treated group, LVEF and LVFS were significantly upregulated compared with the model group ($P < 0.05$). LVID;d and LVID;s in the DQP-treated group were also downregulated, though the difference of LVID;d was not statistically significant. $^{**}P < 0.01$ versus sham group; $^{\#}P < 0.05$ versus model group; $^{\#\#}P < 0.01$ versus model group.

atrial natriuretic factor precursor, was upregulated greatly in ischemic heart. Insulin-like growth factor binding protein 7 (IGFBP7) was also upregulated. Other genes that were upregulated in great abundance encode collagen (COL3A1, COL1A1), light chain of myosin (MYL7), metallopeptidase (MMP2), and so forth. The top 20 downregulated genes in the model group are involved in fatty acid metabolism (ACAA2, ACSL1, ACADM, and DCI) and glycolysis (ALDOA, LDHB, MDH2, and PDHA1). Most of the products of the downregulated genes are located in the mitochondria. Among the top 20 genes that were upregulated or downregulated in the model group, DQP was able to regulate the expression of 19 and 14 genes back towards normal levels, respectively (Figure 2).

3.3. Biological Pathways and Processes Affected in Ischemia Heart. To investigate the global gene expression changes in ischemic heart in the model group, we studied the biological pathways and processes affected in ischemic heart by KEGG pathway enrichment and GO enrichment analysis. The upregulated KEGG pathways and biological processes mainly include complement and coagulation cascades, ECM-receptor interaction, dilated cardiomyopathy, response to wounding, coagulation, immune response, extracellular matrix organization, focal adhesion, skeletal system development, and cell growth (Tables 2 and 3). The downregulated biological processes and pathways mainly involve in nutrient metabolism and energy supply, such as fatty acid metabolism, pyruvate metabolism, amino acid metabolism, carbohydrate catabolism, citrate cycle, and oxidation reduction (Tables 2 and 3). This analysis illustrated that the energy supply of the heart was seriously compromised and the myocardial remodeling took place to compensate for the short supply of energy in the ischemic process.

3.4. Genes and Signaling Pathways Regulated by DQP. Expressions of 516 genes were significantly affected by MI. In the DQP-treated group, expressions of 313 out of these 516 genes were reversed by DQP. These overlapped genes are considered as DQP-responsive genes. Of the 339 genes that were upregulated in the model group, 212 were significantly downregulated in the DQP treatment group. These genes were enriched in infection and immunity, cell growth, extracellular matrix deposition, and so forth (Table 4). 46% of proteins encoded by the downregulated genes were located in extracellular space. Among the 177 downregulated genes

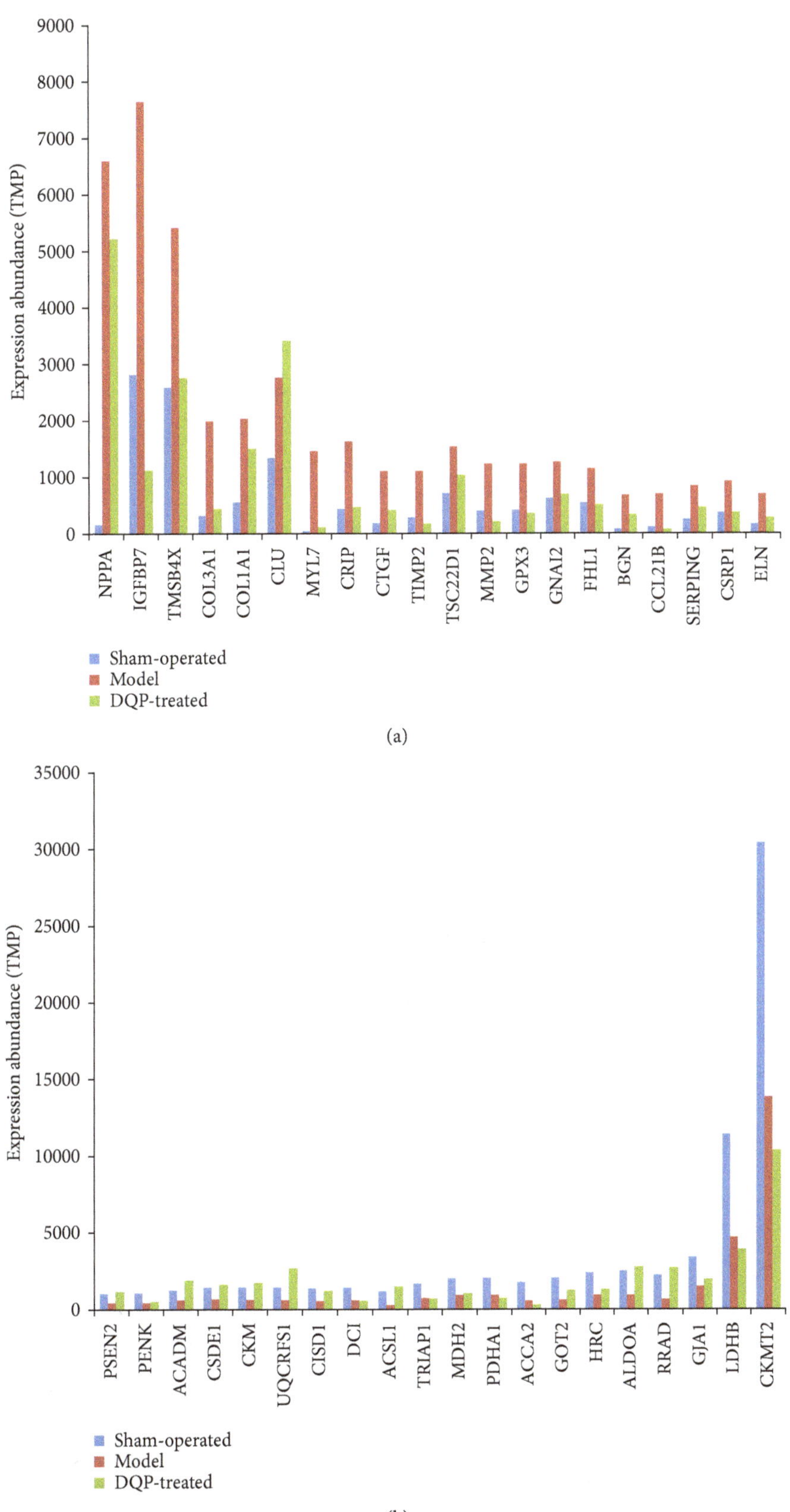

FIGURE 2: Expression abundance changes. (a) Top 20 genes upregulated in the model group compared with the sham-operated group and their expression abundances in the DQP-treated group. (b) Top 20 genes downregulated in the model group and their expression abundances in the DQP-treated group.

Table 2: Significantly enriched KEGG pathways among upregulated and downregulated genes in ischemic heart. The pathways are ranked according to the order of the increasing P values. Gene count referred to the number of genes among all the pathway member genes.

	KEGG pathway ID and name	Gene count	P value
Upregulated	rno04610: complement and coagulation cascades	11	$4.59E-07$
	rno04670: leukocyte transendothelial migration	9	0.001
	rno05322: systemic lupus erythematosus	8	0.001
	rno04510: focal adhesion	11	0.003
	rno04512: ECM-receptor interaction	7	0.004
	rno05200: pathways in cancer	14	0.005
	rno05414: dilated cardiomyopathy	7	0.006
Downregulated	rno00071: fatty acid metabolism	6	$7.55E-05$
	rno00280: valine, leucine, and isoleucine degradation	6	$1.18E-04$
	rno00620: pyruvate metabolism	5	$7.45E-04$
	rno03320: PPAR signaling pathway	6	$9.11E-04$
	rno00330: arginine and proline metabolism	4	0.018834
	rno00020: citrate cycle (TCA cycle)	3	0.04066
	rno05412: arrhythmogenic right ventricular cardiomyopathy	4	0.044529
	rno00640: propanoate metabolism	3	0.048358
	rno00071: fatty acid metabolism	6	$7.55E-05$
	rno00280: valine, leucine, and isoleucine degradation	6	$1.18E-04$

Table 3: Top ten significantly enriched GO biological processes among upregulated and downregulated genes in ischemic heart. The GO terms are ranked according to the order of the increasing P values. Gene count referred to the number of genes among all the GO member genes.

	Gene ontology ID and terms	Gene count	P value
Upregulated	GO:0009611~response to wounding	39	$4.65E-16$
	GO:0030198~extracellular matrix organization	16	$6.64E-11$
	GO:0001501~skeletal system development	24	$7.24E-10$
	GO:0060348~bone development	17	$1.11E-09$
	GO:0032535~regulation of cellular component size	22	$1.92E-09$
	GO:0001503~ossification	16	$2.21E-09$
	GO:0001558~regulation of cell growth	19	$2.34E-09$
	GO:0040008~regulation of growth	24	$7.57E-09$
	GO:0043062~extracellular structure organization	16	$4.00E-08$
	GO:0008361~regulation of cell size	18	$6.08E-08$
Downregulated	GO:0006091~generation of precursor metabolites and energy	12	$1.60E-05$
	GO:0044275~cellular carbohydrate catabolic process	7	$6.08E-05$
	GO:0006007~glucose catabolic process	6	$2.01E-04$
	GO:0055114~oxidation reduction	17	$2.14E-04$
	GO:0046365~monosaccharide catabolic process	6	$2.37E-04$
	GO:0019320~hexose catabolic process	6	$2.37E-04$
	GO:0046395~carboxylic acid catabolic process	7	$3.17E-04$
	GO:0016054~organic acid catabolic process	7	$3.17E-04$
	GO:0016052~carbohydrate catabolic process	7	$3.36E-04$
	GO:0043648~dicarboxylic acid metabolic process	5	$4.94E-04$

in the model group, DQP significantly reversed expressions of 101 genes. The enriched pathways and biological processes were involved in energy embolisms, such as carbon metabolism, fatty acid degradation, and pyruvate metabolism (Table 5). 25% of the proteins encoded by the downregulated genes were located in the mitochondrion. These results demonstrated that DQP can regulate multiple pathways altered under ischemic stimulus.

TABLE 4: Enriched downregulated pathways and biological processes by DQP treatment.

Term	%	*P* value	Genes
KEGG pathway			
Staphylococcus aureus infection	3.6	0.000015	C1QA, C1QB, C5AR1, LOC498276, CFH, RT1-DMA, C1QC
Phagosome	4.1	0.003	ACTB, RAB5C, NCF4, LOC498276, SCARB1, RT1-DMA, TUBA1C, SEC61A1
Complement and coagulation cascades	2.6	0.006	C1QA, C1QB, C5AR1, CFH, C1QC
African trypanosomiasis	2.1	0.006	VCAM1, F2RL1, LOC100134871, HBB
Leukocyte transendothelial migration	3.1	0.007	ACTB, VCAM1, MYL7, NCF4, CLDN5, MMP2
Tuberculosis	3.6	0.010	LSP1, RAB5C, LOC498276, TGFB3, FCER1G, RT1-DMA, LBP
Platelet activation	3.1	0.011	ACTB, GP1BB, LOC498276, COL3A1, FCER1G, COL5A2
ECM-receptor interaction	2.6	0.012	GP1BB, COL3A1, COL5A2, SPP1, FN1
Glutathione metabolism	2.1	0.020	GSTA3, GPX3, GSTT1, GPX7
Malaria	2.1	0.021	VCAM1, LOC100134871, TGFB3, HBB
GO_biological process			
Cell adhesion	6.7	0.000017	IGFBP7, COL16A1, VCAM1, WISP2, CTGF, GP1BB, FBLN5, VCAN, GPNMB, SPON1, FN1, AOC3, SPP1
Regulation of cell growth	3.6	0.000045	WISP2, CREB3, CTGF, IGFBP7, FBLN5, IGFBP6, IGFBP4
Aging	6.2	0.00042	VCAM1, C1QB, GSTA3, LITAF, CTGF, ELN, COL3A1, TGFB3, TIMP2, TIMP3, MMP2, AOC3
Collagen fibril organization	2.6	0.001	COL3A1, LOXL2, COL5A2, ANXA2, DPT
Positive regulation of fibroblast proliferation	3.1	0.001	FBLN1, TGIF1, AQP1, MYC, ANXA2, FN1
Integrin-mediated signaling pathway	3.1	0.002	FBLN1, CTGF, ADAMTS15, COL3A1, FCER1G, TYROBP
Elastic fiber assembly	1.5	0.003	FBLN5, ELN, MFAP4
Neutrophil chemotaxis	2.6	0.003	C5AR1, LGALS3, FCER1G, CCL19, SPP1
Extracellular matrix organization	3.1	0.003	FBLN1, LGALS3, FBLN5, ELN, CCDC80, FN1
Ossification	3.1	0.003	ALOX15, CTGF, MGP, COL5A2, SPP1, FN1

TABLE 5: Enriched upregulated pathways and biological processes by DQP treatment.

Term	%	*P* value	Genes
KEGG pathway			
Carbon metabolism	7.5	0.000027	ME1, ALDOA, GOT2, ME3, ACADM, ENO3, SUCLA2
Biosynthesis of antibiotics	6.5	0.004	ALDOA, GOT2, ACADM, ENO3, SUCLA2, HADHB
Metabolic pathways	15.1	0.008	NDUFA4, ALDOA, ME1, ME3, ACADM, CHKB, UQCRFS1, HADHB, GOT2, ACSL1, CKM, MCCC1, ENO3, SUCLA2
Fatty acid degradation	3.2	0.023	ACSL1, ACADM, HADHB
Fatty acid metabolism	3.2	0.030	ACSL1, ACADM, HADHB
Valine, leucine, and isoleucine degradation	3.2	0.031	ACADM, MCCC1, HADHB
GO_biological process			
Pyruvate metabolic process	3.2	0.001	ME1, ME3, PFKFB2
Response to hormone	5.4	0.001	ME1, BDNF, ACADM, ADRA1B, UQCRFS1
Response to drug	8.6	0.013	BDNF, ACSL1, ACADM, ADRA1B, ENO3, AQP7, UQCRFS1, SOD2
Glycolytic process	3.2	0.015	ALDOA, PFKFB2, ENO3
Negative regulation of fat cell differentiation	3.2	0.021	VEGFA, INSIG1, SOD2
Regulation of synaptic plasticity	3.2	0.021	BDNF, PSEN2, RAPGEF2
Muscle contraction	3.2	0.024	TRDN, MYOM2, TMOD4
Response to cold	3.2	0.026	ACADM, VEGFA, SOD2
Intraciliary transport involved in cilium morphogenesis	2.2	0.033	IFT81, PCM1
Oxidation-reduction process	8.6	0.036	ME1, NDUFA4, ME3, ACADM, L2HGDH, UQCRFS1, OXR1, HADHB

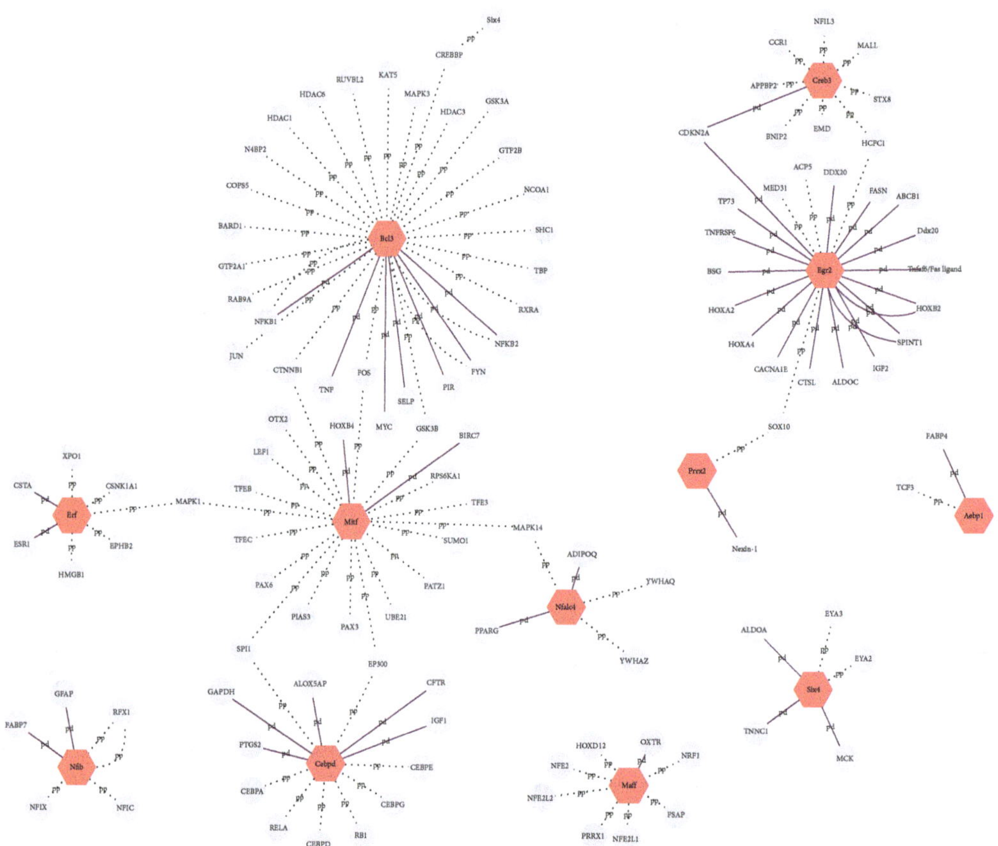

Figure 3: Regulative network by altered transcription factors. Transcription factors were represented as red octagons. Regulated proteins were represented as blue circles. Dot lines represented protein-protein interaction, and solid lines represented protein-gene interactions.

3.5. Transcriptional Regulations by DQP. Transcription factors play critical roles in regulating expressions of target genes. According to the transcription database, we searched transcription factors that had altered expressions induced by ischemic stimuli. Among the dysregulated 516 genes in the MI model group, 12 were transcription factors, including CEBPD, CREB3, BCL3, SIX4, NFATC4, MAFF, ERF, NFIB, MITF, AEBP1, EGR2, and PRRX2. The target genes of these transcription factors were also searched out through databases, and the regulative networks were constructed (Figure 3). Expressions of SIX4, NFIB, and MITF were downregulated in the model group compared with the sham-operated group and the expressions of the nine genes were downregulated. The deregulated transcription factors play roles in inflammation, lipid metabolism, mitochondrial metabolism, and cardiac fibrosis. DQP treatment regulated expressions of the 12 transcription factors towards normal levels.

3.6. Real-Time PCR Validation Analyses. To validate the RNA-sequencing results, we measured the expression patterns of 10 genes by quantitative real-time PCR analyses. Five genes involved in fatty acid metabolism pathways were chosen for validations. Real-time PCR results showed that expressions of all of these five genes were downregulated in the model group and upregulated by DQP treatment, consistent with RNA-sequencing results (Figure 4(a)). Expressions of another five genes involved in ventricular remodeling, cellular apoptosis, and inflammation were also validated. Real-time PCP results showed that expressions of these five genes were upregulated in the model group and downregulated by DQP treatment, which were also consistent with RNA-sequencing results (Figure 4(b)).

4. Discussion

MI is one of the cardiovascular conditions that threaten people's health, and TCM has been shown to be effective in attenuating symptoms of MI. The aim of this study was to systematically analyze the effects of DQP on altered global gene expression patterns induced by MI. An MI rat model was induced by coronary artery ligation, and Illumina's RNA-seq platform was applied in this study. The results showed that heart functions were impaired in the MI model rat and DQP treatment protected heart functions at 28 days after the operation. 10,813 genes were detected and quantified in the heart tissues of three groups of rats. Compared with the sham-operated group, 516 genes were differentially expressed in the MI model group. In the DQP-treated group, expressions of 313 of the altered 516 genes in MI model were regulated back towards normal levels. DQP treatment could regulate the altered signaling pathways in ischemic heart tissues and exert an overall regulative effect on multiple targets in signaling pathways.

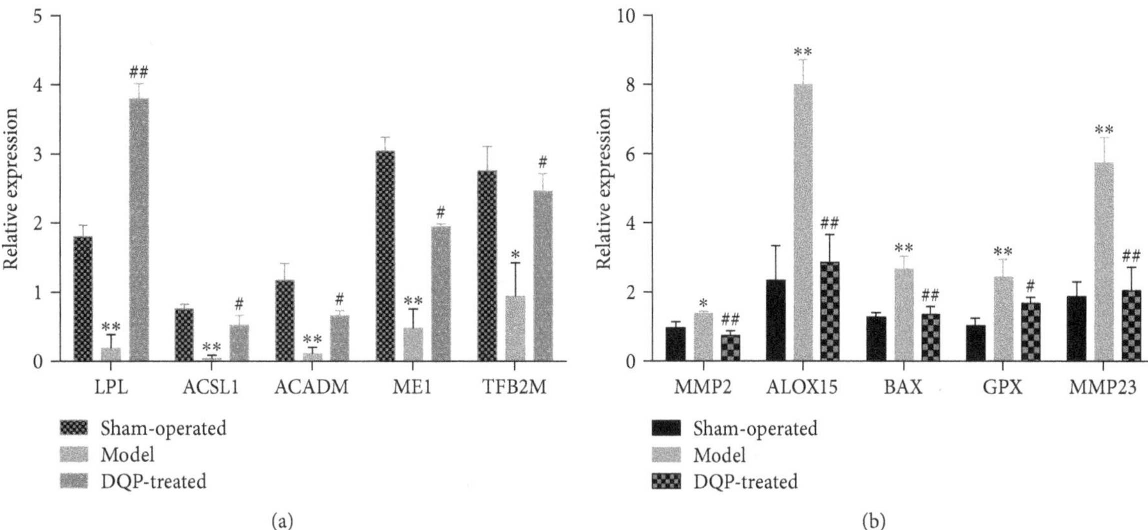

FIGURE 4: Validations of expressions of genes by real-time PCR. (a) Expressions of five genes involved in fatty acids were downregulated in the model group and upregulated by DQP treatment. (b) Expressions of five genes were increased in the MI model group and reduced by DQP treatment. Each group contained six samples. $^{*}P < 0.05$ versus sham group; $^{**}P < 0.01$ versus sham group; $^{\#}P < 0.05$ versus model group; $^{\#\#}P < 0.01$ versus model group.

DQP is composed of *Salvia miltiorrhiza* Bunge and *Panax notoginseng*. Our results showed that DQP had a remarkable cardioprotective effect, as demonstrated by improved LVEF and LVFS. Previous studies have demonstrated that DQP has cardioprotective effects, and the mechanism is related to its regulation on energy metabolism pathways [5, 19, 20]. DQP has also been shown to have anti-inflammatory effect in heart failure model of animals [6]. *Salvia miltiorrhiza* Bunge and *Panax notoginseng* are the major components of DQP, and their effects on cardiovascular diseases have been investigated by numerous studies [21–26]. The major potential effective components of *Salvia miltiorrhiza* Bunge and *Panax notoginseng* include salvianolic acids and *Panax notoginseng* saponins, and they may exert cardioprotective effects through multiple targets in a synergistic way [27, 28]. The effects of DQP on global gene expression patterns in ischemic heart tissues were investigated in this study.

Myocardial ischemic stimulus induced global gene expression changes in the border zone of the heart tissues. In response to a short supply of oxygen and nutrient, several signaling pathways were downregulated, including fatty acid metabolism, amino acid degradation, pyruvate metabolism, PPAR signaling pathway, and citrate cycle. The proteins encoded by these genes were enriched in mitochondria. The top genes with the greatest abundance reductions in the MI model group were enriched in fatty acid metabolism and glycolysis (Figure 2). For example, long-chain acyl-coenzyme A synthetase 1 (ACSL1) could interact with fatty acid transport proteins (FATP) and contribute to the efficient cellular uptake of long-chain fatty acids through vectorial acylation. Inhibition of ACSL1 activity in ischemia heart impairs fatty acid uptake [29]. Acyl-coenzyme A dehydrogenase for medium-chain fatty acids (ACADM) catalyzes the initial step of the mitochondrial fatty acid beta-oxidation pathway [30]. Reduced expressions of these genes suggested that energy metabolism is disrupted in ischemic heart tissues. Among the 177 downregulated genes, DQP treatment upregulated expressions of 101 of these genes, demonstrating that DQP could modulate energy metabolism under ischemic conditions. Our previous study showed that the major components of DQP, salvianolic acids, and *Panax notoginseng* saponins, had effects on energy metabolism [19].

In the MI model group, expressions of 339 genes were upregulated compared with the sham group. According to gene ontology terms, these genes were enriched in response to wounding, extracellular matrix organization, skeletal system development, bone development, regulation of cellular component size, and so forth. Expressions of gene encoding collagens (COL3A1, COL1A1) were greatly increased. Gene encoding natriuretic peptide A (NPPA) implicated in the control of extracellular fluid volume and electrolyte homeostasis was also greatly upregulated. Levels of metalloproteinases (MMP2, MMP23) were upregulated in the model group [31]. These results demonstrated that ventricular remodeling had occurred in response to MI. DQP treatment reversed expressions of these genes towards normal levels. Oxidative stress could produce reactive oxygen species (ROS). ROS could damage endothelial cells and heart tissues, contributing to myocardial fibrosis [32, 33]. Superoxide dismutase (SOD) is a major defense mechanism against ROS [32]. Mitochondrial SOD2 was repressed in ischemic rat models and was upregulated by DQP, suggesting that DQP has a protective antioxidative effect on the heart. Components of DQP, such as salvianolic acid A, salvianolic acid B, ginsenoside Rg1, and so on, have been shown to have antioxidative effects and attenuate hypertrophy [34–37]. Transcription factors involved in inflammation, lipid metabolism, mitochondrial metabolism, and cardiac fibrosis were also regulated back towards normal levels by DQP.

In conclusion, this study demonstrated that DQP could exert cardioprotective effects in the MI model by regulating global gene expression pattern. One limitation of this study is that the sample size was small. The implications of this study warrant further studies with larger sample size to validate the synergistic effects of Chinese medicine. Furthermore, this kind of study will provide insight into the effective components of Chinese medicine and their respective targets in treating complex diseases.

Disclosure

The manuscript was presented as an abstract in the 26th Great Wall International Congress of Cardiology.

Conflicts of Interest

The authors declare that there are no conflicts of interest regarding the publication of this manuscript.

Authors' Contributions

Qiyan Wang, Hui Meng, and Qian Zhang contribute equally to this paper.

Acknowledgments

This work was financially supported, in part, by the grants from the National Natural Science Foundation of China (nos. 81503379, 81673712, 81530100, and 81473456), Fok Ying Tung Education Foundation (no. 151044), Beijing Nova program (Z171100001117028), and the excellent young scientist foundation of BUCM (nos. 2016-JYB-XJ003, 2015-JYB-QNJSZX001, and 2015-JYB-XYQ001).

References

[1] S. M. Dai, S. Zhang, K. P. Chen, W. Hua, F. Z. Wang, and X. Chen, "Prognostic factors affecting the all-cause death and sudden cardiac death rates of post myocardial infarction patients with low left ventricular ejection fraction," *Chinese Medical Journal*, vol. 122, no. 7, pp. 802–806, 2009.

[2] L. C. Slobbe, O. A. Arah, A. de Bruin, and G. P. Westert, "Mortality in Dutch hospitals: trends in time, place and cause of death after admission for myocardial infarction and stroke. An observational study," *BMC Health Services Research*, vol. 8, no. 1, p. 52, 2008.

[3] H. Brønnum-Hansen, T. Jørgensen, M. Davidsen et al., "Survival and cause of death after myocardial infarction: the Danish MONICA study," *Journal of Clinical Epidemiology*, vol. 54, no. 12, pp. 1244–1250, 2001.

[4] P. T. O'Gara, F. G. Kushner, D. D. Ascheim et al., "2013 ACCF/AHA guideline for the management of ST-elevation myocardial infarction: a report of the American College of Cardiology Foundation/American Heart Association Task Force on Practice Guidelines," *Circulation*, vol. 127, no. 4, pp. e362–e425, 2013.

[5] H. Chang, Q. Wang, T. Shi et al., "Effect of DanQi pill on PPARα, lipid disorders and arachidonic acid pathway in rat model of coronary heart disease," *BMC Complementary and Alternative Medicine*, vol. 16, no. 1, p. 103, 2016.

[6] Y. Wang, C. Li, Z. Liu et al., "DanQi pill protects against heart failure through the arachidonic acid metabolism pathway by attenuating different cyclooxygenases and leukotrienes B4," *BMC Complementary and Alternative Medicine*, vol. 14, no. 1, p. 67, 2014.

[7] K. Takahashi, X. Ouyang, K. Komatsu et al., "Sodium tanshinone IIA sulfonate derived from Danshen (*Salvia miltiorrhiza*) attenuates hypertrophy induced by angiotensin II in cultured neonatal rat cardiac cells," *Biochemical Pharmacology*, vol. 64, no. 4, pp. 745–750, 2002.

[8] J. Y. Han, J. Y. Fan, Y. Horie et al., "Ameliorating effects of compounds derived from *Salvia miltiorrhiza* root extract on microcirculatory disturbance and target organ injury by ischemia and reperfusion," *Pharmacology & Therapeutics*, vol. 117, no. 2, pp. 280–295, 2008.

[9] Y. Y. Jiang, L. Wang, L. Zhang et al., "Characterization, antioxidant and antitumor activities of polysaccharides from *Salvia miltiorrhiza* Bunge," *International Journal of Biological Macromolecules*, vol. 70, pp. 92–99, 2014.

[10] Y. Zhang, X. Li, and Z. Wang, "Antioxidant activities of leaf extract of *Salvia miltiorrhiza* Bunge and related phenolic constituents," *Food and Chemical Toxicology*, vol. 48, no. 10, pp. 2656–2662, 2010.

[11] A. Mortazavi, B. A. Williams, K. McCue, L. Schaeffer, and B. Wold, "Mapping and quantifying mammalian transcriptomes by RNA-Seq," *Nature Methods*, vol. 5, no. 7, pp. 621–628, 2008.

[12] Z. Wang, M. Gerstein, and M. Snyder, "RNA-Seq: a revolutionary tool for transcriptomics," *Nature Reviews Genetics*, vol. 10, no. 1, pp. 57–63, 2009.

[13] Y. Wang, C. Li, Y. Ouyang et al., "Cardioprotective effects of Qishenyiqi mediated by angiotensin II type 1 receptor blockade and enhancing angiotensin-converting enzyme 2," *Evidence-based Complementary and Alternative Medicine*, vol. 2012, Article ID 978127, 9 pages, 2012.

[14] Y. Wang, Z. Liu, C. Li et al., "Drug target prediction based on the herbs components: the study on the multitargets pharmacological mechanism of qishenkeli acting on the coronary heart disease," *Evidence-based Complementary and Alternative Medicine*, vol. 2012, Article ID 698531, 10 pages, 2012.

[15] M. D. Robinson, D. J. McCarthy, and G. K. Smyth, "edgeR: a bioconductor package for differential expression analysis of digital gene expression data," *Bioinformatics*, vol. 26, no. 1, pp. 139-140, 2010.

[16] W. Huang da, B. T. Sherman, and R. A. Lempicki, "Bioinformatics enrichment tools: paths toward the comprehensive functional analysis of large gene lists," *Nucleic Acids Research*, vol. 37, no. 1, pp. 1–13, 2009.

[17] L. A. Bovolenta, M. L. Acencio, and N. Lemke, "HTRIdb: an open-access database for experimentally verified human transcriptional regulation interactions," *BMC Genomics*, vol. 13, no. 1, p. 405, 2012.

[18] M. E. Smoot, K. Ono, J. Ruscheinski, P. L. Wang, and T. Ideker, "Cytoscape 2.8: new features for data integration and network visualization," *Bioinformatics*, vol. 27, no. 3, pp. 431-432, 2011.

[19] Q. Wang, C. Li, Q. Zhang et al., "The effect of Chinese herbs and its effective components on coronary heart disease through PPARs-PGC1α pathway," *BMC Complementary and Alternative Medicine*, vol. 16, no. 1, p. 514, 2016.

[20] Y. Wang, C. Li, Q. Wang et al., "Danqi pill regulates lipid metabolism disorder induced by myocardial ischemia through FATP-CPTI pathway," *BMC Complementary and Alternative Medicine*, vol. 15, no. 1, p. 28, 2015.

[21] Y. Zhang, H. Wang, L. Cui et al., "Continuing treatment with *Salvia miltiorrhiza* injection attenuates myocardial fibrosis in chronic iron-overloaded mice," *PLoS One*, vol. 10, no. 4, article e0124061, 2015.

[22] X. Y. Ji, B. K. Tan, and Y. Z. Zhu, "Salvia miltiorrhiza and ischemic diseases," *Acta Pharmacologica Sinica*, vol. 21, no. 12, pp. 1089–1094, 2000.

[23] J. Sun, S. Hong Huang, B. K. H. Tan et al., "Effects of purified herbal extract of *Salvia miltiorrhiza* on ischemic rat myocardium after acute myocardial infarction," *Life Sciences*, vol. 76, no. 24, pp. 2849–2860, 2005.

[24] S. Y. Han, H. X. Li, X. Ma et al., "Evaluation of the anti-myocardial ischemia effect of individual and combined extracts of *Panax notoginseng* and *Carthamus tinctorius* in rats," *Journal of Ethnopharmacology*, vol. 145, no. 3, pp. 722–727, 2013.

[25] S. Chen, J. Liu, X. Liu et al., "Panax notoginseng saponins inhibit ischemia-induced apoptosis by activating PI3K/Akt pathway in cardiomyocytes," *Journal of Ethnopharmacology*, vol. 137, no. 1, pp. 263–270, 2011.

[26] J. W. Guo, L. M. Li, G. Q. Qiu et al., "Effects of Panax notoginseng saponins on ACE2 and TNF-alpha in rats with post-myocardial infarction-ventricular remodeling," *Zhong Yao Cai*, vol. 33, no. 1, pp. 89–92, 2010.

[27] J. Chen, F. Wang, F. S. C. Lee, X. Wang, and M. Xie, "Separation and identification of water-soluble salvianolic acids from *Salvia miltiorrhiza* Bunge by high-speed counter-current chromatography and ESI-MS analysis," *Talanta*, vol. 69, no. 1, pp. 172–179, 2006.

[28] J. B. Wan, C. M. Lai, S. P. Li, M. Y. Lee, L. Y. Kong, and Y. T. Wang, "Simultaneous determination of nine saponins from *Panax notoginseng* using HPLC and pressurized liquid extraction," *Journal of Pharmaceutical and Biomedical Analysis*, vol. 41, no. 1, pp. 274–279, 2006.

[29] M. R. Richards, J. D. Harp, D. S. Ory, and J. E. Schaffer, "Fatty acid transport protein 1 and long-chain acyl coenzyme a synthetase 1 interact in adipocytes," *Journal of Lipid Research*, vol. 47, no. 3, pp. 665–672, 2006.

[30] E. M. Maier, B. Liebl, W. Röschinger et al., "Population spectrum of *ACADM* genotypes correlated to biochemical phenotypes in newborn screening for medium-chain acyl-CoA dehydrogenase deficiency," *Human Mutation*, vol. 25, no. 5, pp. 443–452, 2005.

[31] G. L. Brower, S. P. Levick, and J. S. Janicki, "Inhibition of matrix metalloproteinase activity by ACE inhibitors prevents left ventricular remodeling in a rat model of heart failure," *American Journal of Physiology-Heart and Circulatory Physiology*, vol. 292, no. 6, pp. H3057–H3064, 2007.

[32] G. K. Asimakis, S. Lick, and C. Patterson, "Postischemic recovery of contractile function is impaired in $SOD2^{+/-}$ but not $SOD1^{+/-}$ mouse hearts," *Circulation*, vol. 105, no. 8, pp. 981–986, 2002.

[33] M. Ohashi, M. S. Runge, F. M. Faraci, and D. D. Heistad, "MnSOD deficiency increases endothelial dysfunction in ApoE-deficient mice," *Arteriosclerosis, Thrombosis, and Vascular Biology*, vol. 26, no. 10, pp. 2331–2336, 2006.

[34] B. Jiang, D. Li, Y. Deng et al., "Salvianolic acid A, a novel matrix metalloproteinase-9 inhibitor, prevents cardiac remodeling in spontaneously hypertensive rats," *PLoS One*, vol. 8, no. 3, article e59621, 2013.

[35] G. R. Zhao, H. M. Zhang, T. X. Ye et al., "Characterization of the radical scavenging and antioxidant activities of danshensu and salvianolic acid B," *Food and Chemical Toxicology*, vol. 46, no. 1, pp. 73–81, 2008.

[36] C. Y. Li, W. Deng, X. Q. Liao, J. Deng, Y. K. Zhang, and D. X. Wang, "The effects and mechanism of ginsenoside Rg1 on myocardial remodeling in an animal model of chronic thromboembolic pulmonary hypertension," *European Journal of Medical Research*, vol. 18, no. 1, p. 16, 2013.

[37] C. H. Lee and J. H. Kim, "A review on the medicinal potentials of ginseng and ginsenosides on cardiovascular diseases," *Journal of Ginseng Research*, vol. 38, no. 3, pp. 161–166, 2014.

Permissions

All chapters in this book were first published in IJG, by Hindawi Publishing Corporation; hereby published with permission under the Creative Commons Attribution License or equivalent. Every chapter published in this book has been scrutinized by our experts. Their significance has been extensively debated. The topics covered herein carry significant findings which will fuel the growth of the discipline. They may even be implemented as practical applications or may be referred to as a beginning point for another development.

The contributors of this book come from diverse backgrounds, making this book a truly international effort. This book will bring forth new frontiers with its revolutionizing research information and detailed analysis of the nascent developments around the world.

We would like to thank all the contributing authors for lending their expertise to make the book truly unique. They have played a crucial role in the development of this book. Without their invaluable contributions this book wouldn't have been possible. They have made vital efforts to compile up to date information on the varied aspects of this subject to make this book a valuable addition to the collection of many professionals and students.

This book was conceptualized with the vision of imparting up-to-date information and advanced data in this field. To ensure the same, a matchless editorial board was set up. Every individual on the board went through rigorous rounds of assessment to prove their worth. After which they invested a large part of their time researching and compiling the most relevant data for our readers.

The editorial board has been involved in producing this book since its inception. They have spent rigorous hours researching and exploring the diverse topics which have resulted in the successful publishing of this book. They have passed on their knowledge of decades through this book. To expedite this challenging task, the publisher supported the team at every step. A small team of assistant editors was also appointed to further simplify the editing procedure and attain best results for the readers.

Apart from the editorial board, the designing team has also invested a significant amount of their time in understanding the subject and creating the most relevant covers. They scrutinized every image to scout for the most suitable representation of the subject and create an appropriate cover for the book.

The publishing team has been an ardent support to the editorial, designing and production team. Their endless efforts to recruit the best for this project, has resulted in the accomplishment of this book. They are a veteran in the field of academics and their pool of knowledge is as vast as their experience in printing. Their expertise and guidance has proved useful at every step. Their uncompromising quality standards have made this book an exceptional effort. Their encouragement from time to time has been an inspiration for everyone.

The publisher and the editorial board hope that this book will prove to be a valuable piece of knowledge for researchers, students, practitioners and scholars across the globe.

List of Contributors

Ilaria Laudadio, Sara Formichetti, Giuseppe Macino, Claudia Carissimi and Valerio Fulci
Dipartimento di Biotecnologie Cellulari ed Ematologia, Sez Genetica Molecolare, Sapienza Università di Roma, Rome, Italy

Silvia Gioiosa
Istituto di Biomembrane e Bioenergetica (IBBE), CNR, Bari, Italy

Filippos Klironomos and Nikolaus Rajewsky
Laboratory for Systems Biology of Gene Regulatory Elements, Berlin Institute for Medical Systems Biology, Max-Delbrück Center for Molecular Medicine, Berlin, Germany

Chunyang Feng, Junxue Dong, Weiqin Chang, Manhua Cui and Tianmin Xu
The Second Hospital of Jilin University, Jilin, Changchun 130041, China

Marco Ragusa, Cristina Barbagallo, Duilia Brex, Angela Caponnetto, Matilde Cirnigliaro, Rosalia Battaglia, Davide Barbagallo, Cinzia Di Pietro and Michele Purrello
BioMolecular, Genome and Complex Systems BioMedicine Unit (BMGS Unit), Section of Biology and Genetics G Sichel, Department of BioMedical Sciences and Biotechnology, University of Catania, Catania, Italy

Marco Ragusa
IRCCS Associazione Oasi Maria S.S., Institute for Research on Mental Retardation and Brain Aging, Troina, Enna, Italy

Alexandru Burcea, Gina-Oana Popa, Iulia Elena Florescu (Gune), Andreea Dudu, Sergiu Emil Georgescu and Marieta Costache
Department of Biochemistry and Molecular Biology, Faculty of Biology, University of Bucharest, Bucharest 050095, Romania

Marilena Maereanu
S.C. Danube Research-Consulting S.R.L., Isaccea 825200, Romania

Qing Chen, Xunju Liu, Yueyang Hu, Bo Sun, Haoru Tang and Yan Wang
College of Horticulture, Sichuan Agricultural University, Chengdu, Sichuan 611130, China

Yaodong Hu
Science and Technology Management Division, Sichuan Agricultural University, Chengdu, Sichuan 611130, China

Xiaorong Wang
Institute of Pomology and Olericulture, Sichuan Agricultural University, Chengdu, Sichuan 611130, China

Jiaxing He, Weiqin Chang, Chunyang Feng, Manhua Cui and Tianmin Xu
The Second Hospital of Jilin University, Jilin, Changchun 130041, China

Ayalew Ligaba-Osena and Bertrand Hankoua
College of Agriculture and Related Sciences, Delaware State University, 1200 N DuPont Highway, Dover, DE 19901, USA

Kay DiMarco
22217 Earth and Engineering Sciences, Pennsylvania State University, University Park, PA 16802, USA

Tom L. Richard
Agricultural and Biological Engineering, Pennsylvania State University, 132 Land and Water Research Building, PA 16802, USA

Kyle J. Burghardt and Brittany N. Lines
Department of Pharmacy Practice, Wayne State University Eugene Applebaum College of Pharmacy and Health Sciences, 259 Mack Avenue, Suite 2190, Detroit, MI 48201, USA

Jacyln M. Goodrich
Department of Environmental Health Sciences, University of Michigan School of Public Health, 1415 Washington Heights Ann Arbor, MI 48109, USA

Vicki L. Ellingrod
Department of Clinical Social and Administrative Sciences, College of Pharmacy, University of Michigan, 428 Church Street, Ann Arbor, MI 48109, USA
Department of Psychiatry, School of Medicine, University of Michigan, 1301 Catherine, Ann Arbor, MI 48109, USA

Riya R. Kanherkar, Thomas Heinbockel and Antonei B. Csoka
Epigenetics Laboratory, Department of Anatomy, Howard University, 520 W St. NW, Washington, DC 20059, USA

Bruk Getachew and Yousef Tizabi
Department of Pharmacology, Howard University, 520 W St. NW, Washington, DC 20059, USA

Joseph Ben-Sheetrit
Tel-Aviv Brüll Community Mental Health Center, Clalit Health Services, 9 Hatzvi St., 6719709 Tel-Aviv, Israel

Sudhir Varma
HiThru Analytics LLC, 1001 Spring St. No. 219, Silver Spring, MD 20910, USA

Jing Han, Yueying Yuan and Yan Wu
Institute of Chinese Medicine, Beijing University of Chinese Medicine, Beijing 100029, China

Zhenglin Wang, Wei Xing, Tiantian Lv and Hongliang Wang
College of Basic Medicine, Key Laboratory of Ministry of Education (Syndromes and Formulas), Key Laboratory of Beijing (Syndromes and Formulas), Beijing University of Chinese Medicine, Beijing 100029, China

Yi Zhang
Modern Research Center for Traditional Chinese Medicine, Beijing University of Chinese Medicine, Beijing 100029, China

Yonggang Liu
College of Chinese Medicine, Beijing University of Chinese Medicine, Beijing 100029, China

Zhen Wei, Jingting Zhu, Zhi-Liang Lu, Rong Rong and Jia Meng
Department of Biological Sciences, Xi'an Jiaotong-Liverpool University, Suzhou, Jiangsu 215123, China

Zhen Wei
Integrative Genomics of Ageing Group, Institute of Ageing and Chronic Disease, University of Liverpool, L7 8TX Liverpool, UK

Subbarayalu Panneerdoss, Santosh Timilsina, Tabrez A. Mohammad and Manjeet K. Rao
Greehey Children's Cancer Research Institute, University of Texas Health Science Center at San Antonio, San Antonio, TX 78229, USA

Subbarayalu Panneerdoss, Santosh Timilsina and Manjeet K. Rao
Department of Cellular Structural Biology, University of Texas Health Science Center at San Antonio, San Antonio, TX 78229, USA

Jingting Zhu, Zhi-Liang Lu, Rong Rong and Jia Meng
Institute of Integrative Biology, University of Liverpool, L7 8TX Liverpool, UK

Yidong Chen and Yufei Huang
Department of Epidemiology and Biostatistics, University of Texas Health Science Center at San Antonio, San Antonio, TX 78229, USA

Yufei Huang
Department of Electrical and Computer Engineering, University of Texas at San Antonio, San Antonio, TX 78249, USA

Beata Narożna, Wojciech Langwiński, Zuzanna Stachowiak and Aleksandra Szczepankiewicz
Laboratory of Molecular and Cell Biology, Department of Pediatric Pulmonology, Allergy and Clinical Immunology, Poznan University of Medical Sciences, Poznan, Poland

Claire Jackson, Peter M. Lackie and John W. Holloway
Clinical and Experimental Sciences, Faculty of Medicine, University of Southampton, Southampton, UK

John W. Holloway
Human Development and Health, Faculty of Medicine, University of Southampton, Southampton, UK

Monika Dmitrzak-Węglarz
Department of Psychiatric Genetics, Poznan University of Medical Sciences, Poznan, Poland

Minjung Kim, Youngseok Yu, Ji-Hoi Moon and Jae-Hyung Lee
Department of Life and Nanopharmaceutical Sciences, Kyung Hee University, Seoul, Republic of Korea

Ji-Hoi Moon and InSong Koh
Department of Maxillofacial Biomedical Engineering, School of Dentistry, and Institute of Oral Biology, Kyung Hee University, Seoul, Republic of Korea

In Song Koh
Department of Physiology, College of Medicine, Hanyang University, Seoul, Republic of Korea
Department of Biomedical Informatics, Hanyang University, Seoul, Republic of Korea

Tao Zeng, Lei Li and Liang Gao
Department of Neurosurgery, Shanghai Tenth People's Hospital, Tongji University School of Medicine, No. 301 Middle Yanchang Road, Shanghai 200072, China

Yan Zhou
Medical Research Institute, College of Life Sciences, Wuhan University, Wuhan 430071, China

Zhiwei Chen, Chenghong Liu, Yifei Wang, Ting He, Runhong Gao, Hongwei Xu, Guimei Guo, Yingbo Li, Longhua Zhou, Ruiju Lu and Jianhua Huang
Biotechnology Research Institute of Shanghai Academy of Agricultural Sciences, 2901 Beidi Road, Minhang District, Shanghai 201106, China
Shanghai Key Laboratory of Agricultural Genetics and Breeding, 2901 Beidi Road, Minhang District, Shanghai 201106, China

Guoqiang Fan, Yabing Cao and Zhe Wang
Institute of Paulownia, Henan Agricultural University, Zhengzhou, Henan 450002, China

Guoqiang Fan
College of Forestry, Henan Agricultural University, Zhengzhou, Henan 450002, China

Ananda Sanches-Medeiros, Lummy Maria Oliveira Monteiro and Rafael Silva-Rocha
Systems and Synthetic Biology Lab, FMRP - University of São Paulo, Ribeirão Preto, SP, Brazil

Qiyan Wang, Qian Zhang, Tianjiao Shi, Xuefeng Zhang, Mingyan Shao and Yong Wang
School of Life Sciences, Beijing University of Chinese Medicine, Beijing 100029, China

Hui Meng and Chun Li
Modern Research Center for Traditional Chinese Medicine, Beijing University of Chinese Medicine, Beijing 100029, China

Wei Wang and Linghui Lu
School of Chinese Medicine, Beijing University of Chinese Medicine, Beijing 100029, China

Jing Wang
Staidson (Beijing) Biopharmaceuticals Co., Ltd., Beijing 100176, China

Index

www.ingramcontent.com/pod-product-compliance
Lightning Source LLC
Chambersburg PA
CBHW080253230326
41458CB00097B/4426

* 9 7 8 1 6 4 1 1 6 2 6 3 0 *